Springer Proceedings in Business and Economics

Springer Proceedings in Business and Economics brings the most current research presented at conferences and workshops to a global readership. The series features volumes (in electronic and print formats) of selected contributions from conferences in all areas of economics, business, management, and finance. In addition to an overall evaluation by the publisher of the topical interest, scientific quality, and timeliness of each volume, each contribution is refereed to standards comparable to those of leading journals, resulting in authoritative contributions to the respective fields. Springer's production and distribution infrastructure ensures rapid publication and wide circulation of the latest developments in the most compelling and promising areas of research today.

The editorial development of volumes may be managed using Springer Nature's innovative EquinOCS, a proven online conference proceedings submission, management and review system. This system is designed to ensure an efficient timeline for your publication, making Springer Proceedings in Business and Economics the premier series to publish your workshop or conference volume.

This book series is indexed in SCOPUS.

Damianos P. Sakas • Dimitrios K. Nasiopoulos
Yulia Taratuhina
Editors

Computational and Strategic Business Modelling

IC-BIM 2021, Athens, Greece

Springer

Editors
Damianos P. Sakas
Department of Agribusiness and Supply Chain Management
Agricultural University of Athens
Athens, Greece

Dimitrios K. Nasiopoulos
Department of Agribusiness and Supply Chain Management
Agricultural University of Athens
Athens, Greece

Yulia Taratuhina
Business and Management
National Research University Higher School of Economics
Moscow, Russia

ISSN 2198-7246 ISSN 2198-7254 (electronic)
Springer Proceedings in Business and Economics
ISBN 978-3-031-41370-4 ISBN 978-3-031-41371-1 (eBook)
https://doi.org/10.1007/978-3-031-41371-1

This Springer imprint is published by the registered company Springer Nature Switzerland AG
The registered company address is: Gewerbestrasse 11, 6330 Cham, Switzerland

Preface: 9th International Conference on Business Intelligence and Modelling (IC-BIM)

Aims and Scope of the Conference

The International Conference on Business Intelligence and Modelling took place in Athens, Greece. At the same time, an online collaborative platform was used, in order to conduct the online part of the conference. IC-BIM is an international interdisciplinary conference focusing on the theoretical approach of the contemporary issues evolved in Business Intelligence and Modelling and the integration of theory and practice.

More than 280 pre-registered authors submitted their work in the conference. IC-BIM finally accepted and hosted about 55 original research papers, after a double-blinded peer review process. During the conference, many workshops were held in order to advance and contribute to specific research areas in the field of Business Intelligence and Modelling.

The conference aims at creating a forum for further discussion for a Business Intelligence field incorporating a series of issues and/or related organizations that manage Business Intelligence in their everyday operations. Therefore, the call for papers was addressed to scholars and/or professionals of the fields of Business Intelligence, Business Informatics, Business Model Innovation, Simulation Modelling, Digital Marketing, Management Information Systems (MIS), Decision Support Systems (DSS), E-Business, Enterprise and Conceptual Modelling, etc. IC-BIM provides a common ground for the exchange of insights on business issues in the science of Business Intelligence, ensuring significant contributions to this field.

Grouping emerging technologies in the Business Intelligence field together in a close examination of practices, problems and trends, IC-BIM and its emphases on business presented the state of the art in the field. This annual event is addressed jointly to academics and practitioners and provides a forum for a number of perspectives, based on either theoretical analyses or empirical case studies that foster the dialogue and the exchange of ideas.

Topics

Business Intelligence, Business Informatics, Business Model Innovation, Digital Marketing, Management Information Systems (MIS), Decision Support Systems (DSS), Simulation Modelling, E-Business, E-Learning, Business Process Engineering, Enterprise Architecture, Enterprise and Conceptual Modelling, Modelling Methods, Collaborative (Enterprise) Modelling, Business Management, Enterprise and Business Transformation, Methodologies for Business Transformation, Service Innovation, Business Data Engineering, Knowledge Integration, Knowledge Management, Data Science, Business IT, Business Value, Business Analytics, Social Networks Analysis, Social Computing, Tools for Monitoring Risk, Human, Computer Interaction, E-Market, Mobile Computing, Service-Oriented Computing, Machine-to-Machine Communications (M2M), Smarter Planet, Information Technology, Computer Science, Business Processes in Cloud (BP-Cloud), Cloud Computing, Security and Compliance in Business Processes, Emerging Technologies, Business Process Modeling, Workflow Management, Model Driven Architecture (MDA), IS Modelling, Business Process Management, Modeling Process Architectures, Business Rules Management, Modelling Methods, Languages and Tools, Enterprise Modelling, Business Model and Service Innovation, Data and Knowledge Integration, Model-Driven Computing, Business Economics, Economics of Business Strategy, Global Business, Social Informatics, Business Informatics Security, Management Systems, Social media marketing, Business administration, Business Modelling for Logistics

Paper Peer Review

More than 280 original researches had been submitted for consideration in IC-BIM. All papers submitted to the Conference were reviewed using a double-blind peer-review process. The Conference Scientific Committee decided about the acceptance or not of the submitted papers, with the contribution of competent and expertised reviewers.

Thanks

We would like to thank all members that participated in any way in the ICBIM Conference and especially:

- The famous publication house Springer for their communication sponsorship
- The co-organizing universities and institutes for their support and development of a high-quality Conference scientific level and profile

- The members of the Scientific Committee that honored the Conference with their presence and provided a significant contribution to the review of papers as well as for their indications for the improvement of the conference
- All members of the Organizing Committee for their help, support and spirit participation before, during and after the conference
- The Session Organizers for their willing to organize sessions of high importance and for their editorial work, contributing in the development of valued services to the conference
- Ms Antonia Veltsista, editorial assistant

Athens, Greece — Damianos P. Sakas, Dimitrios K. Nasiopoulos

Moscow, Russia — Yulia Taratuhina

Conference Details

Chairs

Damianos P. Sakas, Agricultural University of Athens, Greece
Dimitrios K. Nasiopoulos, Agricultural University of Athens, Greece
Yulia Taratuhina, Russian Academy of Sciences, Russia

Scientific Committee

- **Gustaf Neumann**, Vienna University of Economics and Business Administration, Austria Program Transformation, Data Modelling, Information Systems Technology, E-Learning Applications
- **Konstadinos Kutsikos**, University of the Aegean, Greece Business Transformation, Innovation Management, Corporate Entrepreneurship, Business Incubation
- **Vaughan Michell**, University of Reading, UK Business Architecture, Business Technology Consulting, Knowledge Intensive Process Optimization
- **Yannakoudakis Emmanuel**, Athens University of Economics & Business, Greece Database Design & Management, Information Retrieval, Databases for Artificial Intelligence and Psychometric Testing
- **Joseph Tan**, McMaster University, Canada Strategic E-Business and E-Health Models, Individual or Group Decision Effectiveness
- **Roger Chiang**, University of Cincinnati, USA Business Intelligence, Data and Knowledge Management, Intelligent Systems, Database Integration, Data and Text Mining
- **Angelo Sifaleras**, University of Macedonia, Greece Mathematical Programming, Network Optimization, Combinatorial Optimization, Operations Research, Optimization Software

- **Achilleas Boukis**, University of Sussex, UK Innovation Management, Luxury Marketing, Marketing Research, Service Encounter, Services Marketing
- **Georgia Koloniari**, University of Macedonia, Greece Social Networks, Graph Databases, Distributed Systems
- **Laurent Moccozet**, University of Geneva, Switzerland 3D Modelling, Technology-Enhanced Learning, Digital Literacy, CMS Systems, Virtual Environments
- **Panagiotis Trivellas**, Agricultural University of Athens, Greece Strategic Leadership, Corporate Culture, HRM, Quality Assurance, TQM, and 6 Sigma, Quality Assurance in Higher Education, Strategic Management
- **Dimitrios Tsagdis**, University of Hull, UK International business, strategy and policy is interdisciplinary grounded in a holistic approach involving different levels of organizational complexity
- **Panagiotis Reklitis**, Agricultural University of Athens, Greece Business Administration, Human Resource Management, Organizational Behavior Management Innovations, Organizational Culture and Supply Chain Management
- **Yulia Taratukhina**, Russian Academy of Sciences, Russia Smart Education, Smart E-Learning Strategies, Hardware and Software Systems
- **Xiaodong Liu**, Edinburgh Napier University, UK Cloud Computing, Semantic Model, Sustainable and Personalizable E-Business Services
- **Zinaida Avdeeva**, National Research University, Russia Smart Education, Smart E-Learning Strategies, Hardware and Software Systems
- **Joan Montllor-Serrats**, Universitat Autònoma de Barcelona, Spain Economics, Finance Management, Business Systems Economics, Business Management
- **Giannis Tsoulfas**, Agricultural University of Athens, Greece Supply Chain – Logistics
- **Rima Tamosiuniene**, Mykolas Romeris University, Lithuania Decision Methods, Cases on Business Economics, Business Economics
- **Joao Alvaro Carvalho**, Universidade do Minho, Portugal Modelling Enterprises, Modelling IT Applications, Business Processes Models
- **Nikolajus Markevicius**, Mykolas Romeris University, Lithuania Business Economics, Business Policy and Strategy, Decision Methods
- **Linas Sadzius**, Mykolas Romeris University, Lithuania Corporate Strategy, Cases on International Business Strategy, Decision Techniques
- **Vicky Manthou**, University of Macedonia, Greece Management Information Systems, Decision Support Systems, Information Systems Strategy, ERP – Enterprise Resource-Integrated Systems
- **Rainer Alt**, University of Leipzig, Germany Business Application Systems, Customer Relationship Management, Supply Chain Management, Methods for Management of Interorganizational Processes
- **George Boda**, Corvinus University in Budapest, Hungary Business Economics, Business Policy and Strategy, Decision Methods

- **Witold Abramowicz**, The Poznan University of Economics, Poland Adaptive Services Grid, Enhanced Information Retrieval and Filtering for Analytical Systems, Filtering Web Services, Mobile Information Filtering, The Distance Learning Platform
- **Thomas Fotiadis**, Democritus University of Thrace, Greece Marketing, High-Tech Marketing, International Marketing, Strategic Marketing
- **Dimitris Folinas**, International Hellenic University, Greece Logistics Management, Supply Chain Management, Business Information Systems
- **Theodore H. Kaskalis**, University of Macedonia, Greece Computer Science, Information Science, Embedded Systems, Computer Architecture, Distributed Systems, Parallel and Distributed Computing
- **Gediminas Macys**, Mykolas Romeris University, Lithuania Global Strategy and Competitiveness, International Forwarding and Logistics, Cases on Business Economics
- **Nima Kordzadeh**, Idaho State University, USA Business Management, Healthcare information
- **Andrey Dmitriev**, National Research University, Russian Federation Business Informatics, Business Economics, Dynamic Systems
- **Elio Shijaku**, Universitat Autònoma de Barcelona, Spain Economics, Finance Management, Business Systems Economics, Business Management
- **Kanellos Toudas**, Agricultural University of Athens, Greece Accounting – Financial Analysis
- **Gediminas Davulis**, Mykolas Romeris University, Lithuania Economics, Finance Management, Business Systems Economics, Business Management
- **Matjaz B. Juric**, University of Ljubljana, Slovenia Service-Oriented Architectures – SOA, Business Process Modeling and Execution (BPEL, BPMN), Business to Business (B2B) and Business to Consumer (B2C) Integration, Patterns, particularly for integration, business processes and component models
- **Fotis Kitsios**, University of Macedonia, Greece Strategic Business Management, Innovation Management, Development of New Products and Services, Marketing Management – Market Research – Customer Satisfaction, Quantitative Methods – Data Analysis, Operations Research – Multicriteria Analysis
- **Saulius Gudas**, Vilnius University, Lithuania Business Informatics, Business Process (Enterprise) Modelling Methodologies and Techniques, Enterprise Software Systems Development
- **Gintautas Garšva**, Vilnius University, Lithuania Business Environment Factors, Information Society Technologies

Organizing Committee

Nikolaos Th. Giannakopoulos, Agricultural University of Athens, Greece
Ioannis Dimitrios Kamperos, Agricultural University of Athens, Greece
Dimitrios P. Reklitis, Agricultural University of Athens, Greece
Marina C. Terzi, Agricultural University of Athens, Greece
Nikos Kanellos, Agricultural University of Athens, Greece
Dimitrios A. Arvanitidis, Agricultural University of Athens, Greece
Dimitrios M. Mastrakoulis, Agricultural University of Athens, Greece

Invited Speakers

Konstadinos Kutsikos is a Professor of Technology and Innovation Management at the Business School of the University of the Aegean (UotA), Greece. He holds an MBA and a PhD in Computer Science from the University of Southern California, USA, and an MS in Computer Science from Stanford University, USA. Before joining UotA, Dr. Kutsikos had a 10-year international career balancing academic, industry and entrepreneurial activities, topped by a Managing Consultant position with PA Consulting Group (a Carlyle Group company; NASDAQ:CG) in London, UK. Based on this background, Dr. Kutsikos' current activities are focused on business transformation in the context of organizational innovation (entrepreneurship, corporate venturing) and service innovation (service systems, value analysis). Dr. Kutsikos is mentor at Stanford University, and active member of scientific, professional and civic groups – including the Academy of Management, the UK Institute of Directors and the Stanford Club.

Panagiotis Reklitis is a Professor of "Organization and Management" in the Department of Agribusiness and Supply Chain Management of the Agricultural University of Athens and a member of the Greek Open University SEP in the "Business and Organizations Management" module. He has held positions of responsibility in Public Organizations. He is a Doctor of NTUA in the field of "Analysis, Design and Development of Processes and Systems" on Innovation and Competitiveness. With years of teaching experience in Higher Education at both undergraduate and postgraduate levels, his main research interests are in the scientific field of business management, human resources management, organizational behavior management innovations, organizational culture and supply chain management. He has published articles in internationally renowned scientific journals and book series. He is a Fellow of the National Center for Public Administration and has taught both in the ECHR and the INEP for a number of years. He has been invited to participate in training programs in organizations and companies such as ELTA SA, the Hellenic Foreign Trade Organization (OPE), etc.

Yulia Taratuhina is a member of SIETAR EUROPA from 2018. Associate Professor, PhD, Senior Researcher in the Institute of Control of Sciences Russian Academy of Sciences. Her research interests focus on cross-cultural didactics, intercultural communication, business communication and emotional intelligence and educational coaching. She has more than 70 research papers including 25 articles in journals, articles in foreign peer-reviewed journals (indexed in Scopus WoS databases), chapters in books (Springer publishing house), 10 textbooks and teaching aids for higher and secondary educational institutions, 5 monographs, etc.

Panagiotis Trivelas is a Professor of "Organization and Management" in the Department of Agribusiness and Supply Chain Management of the Agricultural University of Athens, where he teaches in the field of Strategic Human Resources Management. He has taught at the Athens University of Economics and Business (OPA), the University of Piraeus, the Hellenic Open University (EAP) and the TEI Deputy (ex-Larissa) (undergraduate and postgraduate programs) in business administration and strategy. He has been an Award Leader in the Postgraduate Program (MSc) of the Department of Finance and Accounting of TEI, Thessaly (formerly Larissa), in partnership with Coventry University in Great Britain. Since 2003, he has been teaching in the Master's Degree Programs at TEI of Larissa in partnership with Staffordshire University, UK, entitled Master of Science in Management and Master of Business Administration – MBA). As a researcher or lead researcher, he has participated in a number of funded research programs on Organizational Behavior,

Management and Business Strategy on: Business Administration and Organization, Quality (Total Quality Management, Corporate Management, Systems Management). He has also been involved in the design, implementation and evaluation of a large number of educational programs.

Kanellos S. Toudas graduated from the Department of Business Administration of the University of Piraeus. He was awarded his postgraduate diploma in Business Administration from the same University. In 2006 he obtained his PhD in the field of Financial Management and Accounting from the Athens University of Economics and Business. Moreover, Dr Toudas furthered his research interests with a Postdoctoral study at the National and Kapodistrian University of Athens

He is working as an Assistant Professor in Accounting and Financial Analysis at the Agricultural University of Athens. He has also worked as an Adjunct Lecturer in the Athens University of Economics and Business, the University of Patras and in the National and Kapodistrian University of Athens and in the Hellenic Open University.

Dr Toudas has participated in numerous conferences, and he has been an ongoing author and co-author in many research articles published in reputable conferences proceedings and journals.

Contents

Part VIII Core Workshop

Part I
Digital Marketing and Human Behavior (Να μου απαντήσει ο Δαμιανός)

Damianos P. Sakas

Description: Digital Marketing science, using simulation modeling methods, provides a powerful methodology for research on complex human behaviors. This session aims to promote the understanding of Digital Marketing methodology and to develop an appreciation of its potential contributions to human behavior by describing the nature of Digital Marketing special problems, as well as some uses of computational methods in management research.

D. P. Sakas
Agricultural University of Athens, Athens, Greece

Decentralized Payment Networks Digital Marketing Campaigns' Evaluation Through Estimating Web Page Factors

Damianos P. Sakas and Nikolaos T. Giannakopoulos

1 Introduction

Marketing assists businesses in comprehending and explaining the benefit that consumers recognize and obtain from a product or service [1]. Various strategies used by a company may change depending on the market. A combination of digital marketing and technological tools has improved how businesses contact and connect with their consumers [2]. Nonetheless, the primary goals and problems of customer involvement maintain the very same.

Big data is described as a tremendous volume of knowledge containing massive concentration and diversity, whose mobility exceeds the capabilities of present technologies to handle appropriately [3]. Cryptocurrencies are a comparatively fresh form of payment that gives corporate Web pages a strategic advantage [4]. Cryptocurrencies emphasize the need for further practical research and inquiry into digital currencies, including new findings focusing on consumer aspirations to accept cryptocurrency [5] for a myriad of purposes including payment, among others [6].

The following is how the article is structured: The first component includes a literature analysis and overview of the critical subjects, such as DPNs' digital marketing advancement and web analytics role, as well as the resolution of study hypotheses. The second part is about using statistical tools to harvest relevant data for creating a diagnostic and dynamic model in the next step. Finally, reflections regarding the significance of Web page audits for DPN enterprises' digital marketing effectiveness are presented.

D. P. Sakas · N. T. Giannakopoulos (✉)
BICTEVAC LABORATORY: Business Information and Communication Technologies in Value Chains Laboratory, Department of Agribusiness and Supply Chain Management, School of Applied Economics and Social Sciences, Agricultural University of Athens, Athens, Greece
e-mail: n.giannakopoulos@aua.gr

D. P. Sakas et al. (eds.), *Computational and Strategic Business Modelling*,
Springer Proceedings in Business and Economics,
https://doi.org/10.1007/978-3-031-41371-1_1

2 Related Background

Digital marketing is exploiting modern social networking platforms to give businesses with creative, imaginative, cost-effective, and impactful ways to communicate with clients [7]. Clients, in fact, are constituting a vital component of the developing engagement discussion and increasing their sway over the marketing strategy [8].

Marketing executives must comprehend the potential of blockchain technologies like an interprocess networking that symbolizes the move from the Internet of content to the Internet of security and trust [9]. Blockchain may very well be considered as a revolutionary method of verifying assets utilized in transactions and may be included to a wide range of corporate operations and services [10].

A plethora of literature reviews has focused on blockchain technology and its applications in various markets and sectors of the economy. Most recently, studies have been aiming to analyze the role of big data in assessing cryptocurrency organizations' digital marketing efficacy. Toward this direction, there has been found that behavioral data of websites tend to impact the brand name and digital marketing results of airline firms [11], while examination of cryptocurrency trade organizations' website analytics could also benefit airlines' digital brand name value [12]. Moreover, behavioral data, customers' device preference, and website traffic sources have been estimated as important factors in enhancing cryptocurrency organizations' digital marketing campaigns [13].

Through the referred literature, it becomes obvious that no study has been made regarding DPNs' digital marketing campaigns' evaluation, based on their Web page factors. It can be stated that each factor's impact on DPNs' marketing campaigns is unique, provoking the need for more in-depth analysis. Moreover, efficient evaluation of DPNs' digital marketing campaigns should consider all Web page factors that affect each variable of the campaigns. Thus, the paper's research hypotheses are formed:

H1: "DPNs' visibility can be affected by specific webpage factors."

H2: "The more factors arise to DPN's websites, the more their campaigns' costs rise."

H3: "The number of keywords that lead to DPNs' websites is related to specific webpage factors."

3 Methodology

Development of the context of the paper's methodology is merely based on robust tools' utilization. The authors developed a diagnostic-exploratory model, as well as a dynamic model, by inserting coefficients and correlations of the regression analysis. So, the evaluation of DPNs' digital marketing campaigns can be performed, with the combination of the models to illustrate a representative situation. Starting with the deployment of the diagnostic model, the authors shaped a cognitive context connecting all of the analysis factors [14], using as inputs the coefficients of the regression and correlation analysis. Moving forward, the authors harvested the dynamic modeling process (DM), which contains the dynamic relationships among the analysis' chosen variables, in order to calculate the factor fluctuation throughout a 100-day projection time frame [15–18]. Throughout this direction, simulation findings can be assessed and shaped into an analytical framework for DPN organizations to follow, regarding:

- Giving the necessary information to DPNs' digital marketers could improve the efficacy of their marketing efforts and drive more prospective clients to their websites.
- Collaborate and acquire the necessary big data analytics for strategic decisions to quantify the impact of Web page factors on their digital marketing initiatives and develop management plans.

The authors gathered Web page analytic data from the top 5 dominant cryptocurrency organizations in the market [19] for the objectives of the report. Such statistics are based on everyday monitoring and collection of their website traffic and factors, utilizing decision support systems based on online platforms [20]. Originally mentioned at the start of the conceptual approach, the collected data is examined in the following parts.

4 Results

The results section provides the necessary tools for the next steps of the analysis. In order to obtain the results needed, the authors performed three linear regression models, using as dependent variables organic traffic, costs, and keywords. As independent variables, 12 individual Web page factors were used, like page load time, total checks, crawl depth, etc., summing up some of the most common factors observed in a website.

Doing so, the outcomes of Table 1 accrue, where DPNs organic traffic's regression appears to be insignificantly impacted in total, from the Web page factors, with p-value $= 0.116 > a = 0.05$ and an $R^2 = 0.671$. Only two independent variables were flagged as significant, raw internal links and incoming internal links with p-values of 0.043 and 0.036 accordingly, being below the 0.05 level of significance. So, the paper's first research hypothesis (H1) is verified, with specific Web page factors being able to affect the visibility of DPNs' websites.

In the second regression, organic costs are used as the dependent variable, which is also not generally verified with low $R^2 = 0.475$ and p-value $= 0.566 < a = 0.05$ level of significance. No other independent variable has a significant p-value, below 0.05 level of significance. This means that the second research hypothesis (H2) is rejected, showing that there is no connection between the rising numbers of Web page factors and DPNs' campaign costs.

Table 1 Centralized payment networks' regressions

Variables	Standardized coefficient	R^2	F	p-value
Dependent: organic traffic				
Constant	–			0.116
Raw internal links	5.610			0.043*
Crawl depth	1.884			0.149
HTTP status	0.364			0.948
Page load time	−0.216			0.610
Incoming internal links	−4.104			0.036*
Outcoming internal links	−2.522	0.671	2.037	0.591
Outcoming external links	2.623			0.569
JS CSS number	0.199			0.869
JS CSS size	−0.967			0.870
Website issues	−0.207			0.636
Failed checks	0.129			0.770
Total checks	−0.043			0.924
Dependent: organic costs				
Constant	–			0.566
Raw internal links	6.399			0.064
Crawl depth	2.410			0.144
HTTP status	2.474			0.725
Page load time	−0.508			0.348
Incoming internal links	−4.239			0.077
Outcoming internal links	−1.237	0.475	0.907	0.834
Outcoming external links	1.628			0.778
JS CSS number	1.659			0.288
JS CSS size	−1.014			0.891
Website issues	−0.051			0.926
Failed checks	−0.169			0.763
Total checks	0.262			0.648
Dependent: organic keywords				
Constant	–			0.001**
Raw internal links	8.263			0.000**
Crawl depth	2.905			0.002**
HTTP status	4.938			0.154
Page load time	−0.437			0.101
Incoming internal links	−6.009			0.000**
Outcoming internal links	−4.181	0.883	7.537	0.151
Outcoming external links	4.265			0.137
JS CSS number	0.878			0.237
JS CSS size	1.879			0.594
Website issues	0.176			0.502
Failed checks	−0.171			0.521
Total checks	0.319			0.251

* and ** indicate statistical significance at the 95% and 99% levels, respectively

Finally, linear regression of organic keywords is verified in total, with $R^2 = 0.883$ and p-value $= 0.001 < a = 0.01$ level of significance. The independent variables of raw internal links, crawl depth, and incoming internal links appear to have a significant impact on organic keywords, with p-values of 0.000, 0.002, and 0.000, below 0.01. As a result, the paper's third and final research hypothesis (H3) is verified, meaning that specific Web page factors affect the number of organic keywords that lead to a DPN's websites.

5 Diagnostic Model

As stated in the Methodology section, in order to deploy the diagnostic model, required for the research, fuzzy cognitive maps (FCM) online platform application from Mental Modeler [21] has been harvested, through FCM applications, the depiction and explanation of fuzzy contexts, and the vast segments of causative variables and their intercorrelations. The matrix below design allows for systematic causal dissemination, namely, forward and backward cabling, and the extension of data sources by connecting diverse components [14]. As seen in Fig. 1, there are substantial positive relationships (blue bold arrows) between organic traffic, keywords, and costs. Negative relationships are depicted by red arrows.

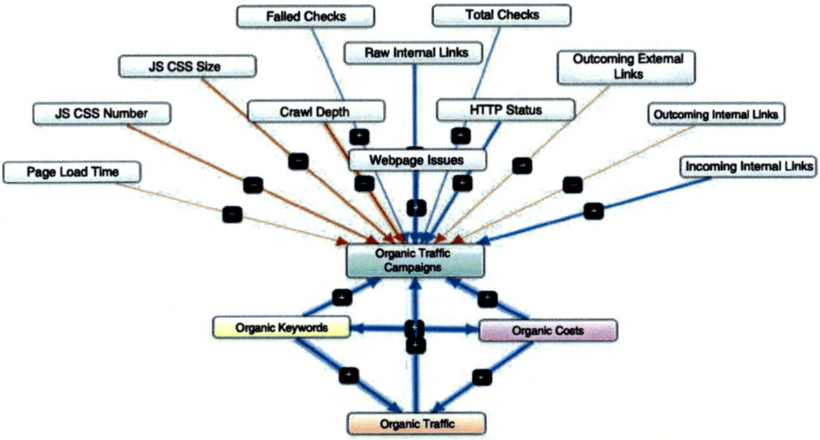

Fig. 1 Fuzzy cognitive mapping of DPNs website factors

6 Dynamic Model

This paper has capitalized on the dynamic model to deploy the simulation of DPN organizations' Web page factors' impact on their organic campaign metrics. The duration of the model is set to be 100 days. Through each day of the simulation, the model's Web page factors are taking specific values, arranged by following the normal distribution, based on the means and standard deviations collected from the sample. For each value they get, according to the regression results' significance, the dynamic variables of organic traffic, keywords, and costs are affected [15]. Figure 2 depicts the designed model for assessing each Web page factor's dynamic impact on DPNs' organic campaign metrics in the simulation period. The oscillations of Web page factors, organic traffic, keywords, and costs were created by running this model.

By running the dynamic model, specific outcomes emerged that tend to verify some of the research hypotheses, as shown in Fig. 3. For instance, it can be discerned from the variables' variation that raw internal links, crawl depth, and incoming internal links appear to fluctuate in accordance with organic traffic, keywords, and costs' variations in the specified simulation period. Therefore, the evaluation of DPNs' digital marketing campaigns, expressed by the organic traffic, keywords, and costs metrics, can be performed by examining the referred Web page factors, since their variation is in accordance.

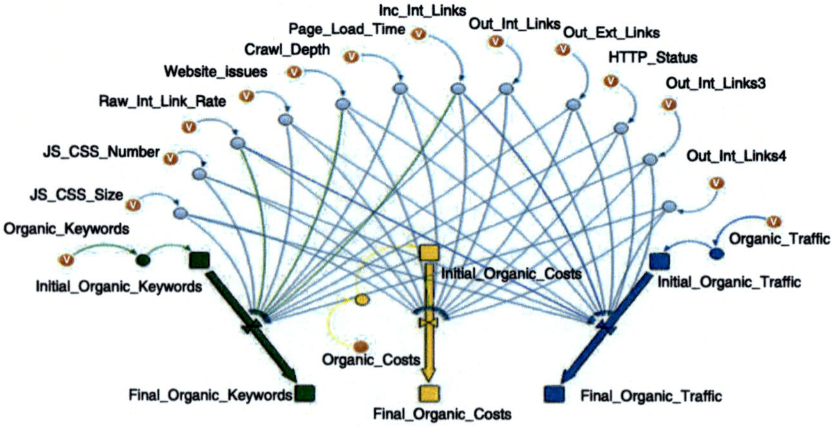

Fig. 2 Dynamic modeling of DPNs' website factors and organic campaigns

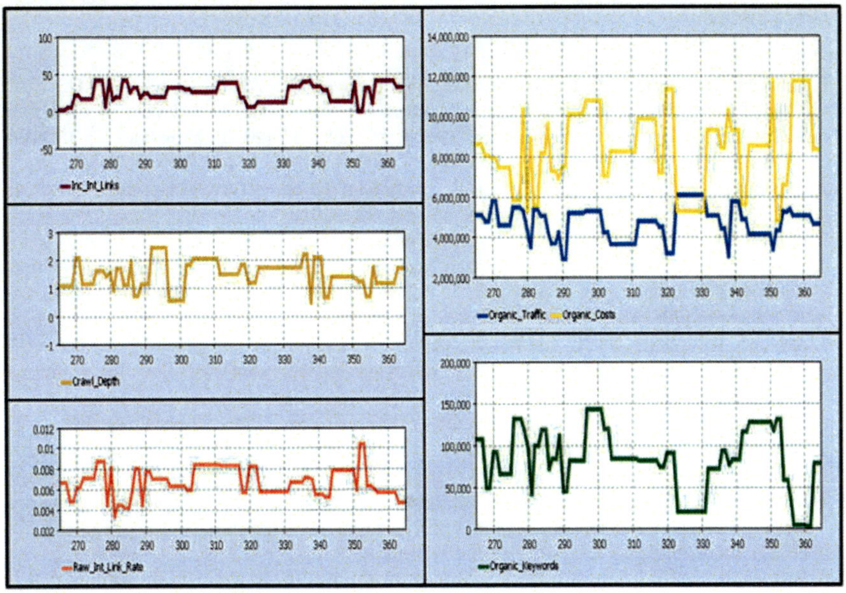

Fig. 3 Model's factors and organic campaign outcomes in 100 days

7 Conclusions

The purpose of the research has been the provision of Web page factors as a driver for evaluating DPN organizations' digital marketing campaigns. Three Web page factors surfaced as important in terms of impacting DPN organizations' digital marketing campaigns. These results were also verified in the simulation process, apart from the regression analysis. The only exception was the organic costs, which did not have significant regression outcomes, but through the simulation process, it had an identical variation to the examined Web page factors.

8 Practical Implications

The contribution of this paper discerns the importance of specific Web page factors in evaluating DPNs' digital marketing campaigns. Visitors and customers of DPN organizations enter their website ad interact with it, thus leading to enhanced organic campaign results. For DPN organizations to improve organic campaign results (increased traffic, keywords, and lower costs), Web page factors like crawl depth, raw internal links, and incoming internal links should be monitored and optimized.

References

1. Larivière, B., Joosten, H., Malthouse, E.C., Van Birgelen, M., Aksoy, P., Kunz, W.H., et al.: Value fusion: the blending of consumer and firm value in the distinct context of mobile technologies and social media. J. Serv. Manag. **24**, 268–293 (2013). https://doi.org/10.1108/09564231311326996
2. Cvitanovic, P.L.: New technologies in marketing as competitive advantage. In: 2018 ENTRENOVA Conference Proceedings, pp. 294–302 (2018)
3. Chaffey, D., Ellis-Chadwick, F.: Digital Marketing. Pearson (2020)
4. ElBahrawy, A., Alessandretti, L., Kandler, A., Pastor-Satorras, R., Baronchelli, A.: Evolutionary dynamics of the cryptocurrency market. R. Soc. Open Sci. **4**, 170623 (2017). https://doi.org/10.1098/rsos.170623
5. Al-Amri, R., Zakaria, N.H., Habbal, A., Hassan, S.: Cryptocurrency adoption: current stage, opportunities, and open challenges. Int. J. Adv. Comput. Res. **9**(44), 293–307 (2019). https://doi.org/10.19101/IJACR.PID43
6. Kanellos, N. by Sakas, D.P, Kamperos, I.D.G., Reklitis, D.P., Giannakopoulos, N.T., Nasiopoulos, D.K., Terzi, M.C.: The effectiveness of centralized payment network advertisements on digital branding during the COVID-19 crisis. Sustainability. **14**(6)
7. Melewar, T.C., Foroudi, P., Gupta, S., Kitchen, P.J., Foroudi, M.M.: Integrating identity, strategy and communications for trust, loyalty and commitment. Eur. J. Mark. **51**, 572–604 (2017). https://doi.org/10.1108/EJM-08-2015-0616
8. Berman, S.J., McClellan, B.E.: Ten strategies for survival in the attention economy. Strategy Leadersh. **30**, 28–33 (2002). https://doi.org/10.1108/10878570210697513
9. Zamani, E.D., Giaglis, G.M.: With a little help from the miners: distributed ledger technology and market disintermediation. Ind. Manag. Data Syst. **118**, 637–652 (2018). https://doi.org/10.1108/IMDS-05-2017-0231
10. Rejeb, A., Keogh, J.G., Treiblmaier, H.: Leveraging the internet of things and blockchain technology in supply chain management. Future Internet. **11**(161) (2019). https://doi.org/10.3390/fi11070161
11. Sakas, D.P., Giannakopoulos, N.T.: Big data contribution in desktop and mobile devices comparison regarding airlines' digital brand name effect. Big Data Cogn. Comput. **5**(4), 48 (2021). https://doi.org/10.3390/bdcc5040048
12. Sakas, D.P., Giannakopoulos, N.T., Reklitis, D.P., Dasaklis, T.K.: The effects of cryptocurrency trading websites on airlines' advertisement campaigns. J. Theor. Appl. Electron. Commer. Res. **16**(7), 3099–3119 (2021). https://doi.org/10.3390/jtaer16070169
13. Sakas, D.P., Giannakopoulos, N.T., Kanellos, N., Tryfonopoulos, C.: Digital marketing enhancement of cryptocurrency websites through customer innovative data process. PRO. **10**, 960 (2022). https://doi.org/10.3390/pr10050960
14. Mpelogianni, V., Groumpos, P.P.: Re-approaching fuzzy cognitive maps to increase the knowledge of a system. AI & Soc. **33**, 175–188 (2018). https://doi.org/10.1007/s00146-018-0813-0
15. Dehghani, M., Trojovský, P.: Hybrid leader based optimization: a new stochastic optimization algorithm for solving optimization applications. Sci. Rep. **12**, 5549 (2022). https://doi.org/10.1038/s41598-022-09514-0
16. Nasiopoulos, D.K.: The role of digital marketing in the development of a distribution and logistics network of information technology companies. Bus. Intell. Model. **6**, 267–276
17. Dimitrios, N.K., Sakas, D.P., Vlachos, D.S.: The contribution of dynamic simulation model of depiction of knowledge, in the leading process of high technology companies. Key Eng. Mater. **543**, 406–409

18. Nasiopoulos, D.K., Sakas, D.P., Masselos, K.: Open source web applications. How it spread through the Internet and their contribution to education. In: 4th Symposium on Business and Management and Dynamic Simulation Models Supporting Management Strategies
19. Komarraju, A.: Top 10 cryptocurrencies to buy & invest in today in May 2021 (2021). https://www.analyticsinsight.net/top-10-cryptocurrencies-to-buy-invest-in-today-in-may-2021/
20. Semrush: (2022, March 24). https://www.semrush.com/
21. MentalModeler: (2022, March 24). https://dev.mentalmodeler.com/

Modeling Decentralized Payment Services' Digital Marketing Performance Through Webpage Auditing Strategy

Damianos P. Sakas and Nikolaos T. Giannakopoulos

1 Introduction

Blockchain has shown potential as a decentralized digitized database system. Decentralized payment networks have given the ability to people for performing decentralized payments and transactions [1]. Decentralized payment channel networks [2], a newly suggested idea, include a particularly elevated alternative for blockchain-based payment solutions. Clients' need for rapid and tailored services is growing as a result of such advancements [3]. The emergence of the Web resulted in the growth of e-commerce, digital trading, and decentralized payment organizations, which are all now considered part of a broad client's understanding [4]. Due to the growing demand for decentralization, decentralized payment organizations face the need to further develop their websites, to attract more customers for their services and engage them more [5].

Furthermore, a rising number of businesses have implemented digital analytics to enhance their promotional efforts, and academics have acknowledged the possibilities of participatory multiple digital interaction and cooperation [6]. Advertising researchers, for instance, emphasize that growing analytic tools may improve the transfer of information among customers and suppliers more efficiently [7] and stimulate successful marketing operations and procedures. However, notwithstanding its undeniable usefulness and salience in B2B, existing research on the utilization of analytical tools by such companies is in an infant stage, only with a few studies addressing the advertising possibilities in corporate sectors [8].

D. P. Sakas · N. T. Giannakopoulos (✉)
BICTEVAC Laboratory: Business Information and Communication Technologies in Value Chains Laboratory, Department of Agribusiness and Supply Chain Management, School of Applied Economics and Social Sciences, Agricultural University of Athens, Athens, Greece
e-mail: n.giannakopoulos@aua.gr

D. P. Sakas et al. (eds.), *Computational and Strategic Business Modelling*, Springer Proceedings in Business and Economics,
https://doi.org/10.1007/978-3-031-41371-1_2

The rest of the paper presents the following topics: the first phase concentrates on providing a sufficient number of recent studies and research outcomes of digital marketing concepts in the payment networks sector and big data analytics. In the second phase, the authors deployed linear regression models, followed by a diagnostic and a hybrid modeling process, which utilized the regressions' outputs. At last, in the results section, valuable information regarding the optimization of decentralized payment companies' digital marketing performance is deducted.

2 Related Background

Electronic commerce customization is the method of changing the design and operation of a digital store, whether it is open to the general market. Each change a company introduces to its storefront involves software, layout, and other artistic labor [9]. Webpage analytics may provide engaging material on e-commerce systems through continually giving data, prediction models, and special discounts, depending on a customer's online activity and perhaps other personal characteristics.

User experience analyses have been undertaken in sectors like e-commerce, e-government, smartphone applications, and university portals [10]. Considering data indicating the usefulness of employing web analytics to improve interface and engagement ratio, businesses can profit by implementing such skills to achieve their site's objectives [4]. Practical research from many industries (e.g., aerospace) suggests that the considered utility of analytic data inside B2B enterprises is influenced by several characteristics, including brand, usefulness, and perceived hurdles. Interactions among businesses and consumers, and also client relations, are predicted to pique the focus of both researchers and scientists [11].

A variety of strategies have been deployed, regarding the enhancement of digital marketing in various sectors. These strategies have utilized websites' traffic analytic metrics and showed valuable results for the centralized [12] and decentralized [13] sectors, as well as for the air forwarding companies [14]. Just a few research have investigated whether analytic data adoption in service firms varies from digital marketing communication techniques adopted in the industrial arena [15].

After the elaboration of the current literature, a gap in the research framework has been spotted, regarding the effect of payment network websites' factors on their digital marketing performance. Each of the factors of payment network websites should be analyzed. Their impact on specific digital marketing metrics, such as global rank, unique visitors, and bounce rate, should be estimated. As a consequence, the paper's research hypotheses are listed below to cover the referred research gap.

H1: "Decentralized Payment services' internet ranking can be altered by their website's performance and visibility."

H2: "The number of new visitors to Decentralized Payment services websites is affected by their webpage factors."

H3: "The more factors a Decentralized Payment webpage has the more often visitors abandon the website."

3 Methodology

In this section of the paper, the authors present the procedure for establishing the methodological context of the research. For research purposes, the authors developed a hybrid model consisting of dynamic and agent-based modeling. To fuel these models, linear regression coefficients were used. Through this process, the presentation of decentralized payment companies' digital marketing performance can be achieved, as well as potential ways for optimizing it may accrue. In the next stage, after the regression analysis, the authors harvested the fuzzy cognitive mapping models to deploy a diagnostic framework as a basis for the analysis [16], for noticing variables' linkage and relationships. The development of the hybrid model was supported by the exertion of dynamic and agent-based models, to predict the fluctuation of key variables in 100-simulation days [17]. Throughout the presented framework various insights into decentralized payment companies' strategies can be provided:

- Digital marketers could obtain useful information for assessing the performance of a company's marketing and the impact of specific webpage factors on it.
- Insights regarding webpage factors' addressing can be discerned, thus leading to a more enhanced website visitor experience and less bounce rate.

As for the methodological context creation, the authors selected the top five leading decentralized payment companies worldwide [18] for extracting the required data. The referred data were observed daily and collected en masse from an online website information system [19] and concerned visitors' metrics and webpage factors. In the next stage, the data was analyzed, and important information was produced.

4 Results

The main results of the research have been produced by utilizing the linear regression model. As seen in Table 1, the authors performed three linear regression models with dependent variables the global rank, unique visitors, and bounce rate of decentralized payment companies' websites. Independent variables consist of several webpage factors like webpage factors, internal-external links, etc. Starting with the global rank regression, the model is not overall verified, since it has a p-value $= 0.110 > a = 0.05$ level of significance and a low $R^2 = 0.213$. None of the independent variables of webpage factors seem to significantly impact global rank with p-values below 0.05 level of significance. So, the paper's first research

Table 1 Decentralized payment companies' regressions

Variables	Standardized coefficient	R^2	F	*p*-value
Dependent: Global rank				
Constant	–			0.110
Unique visits	0.006			0.959
Bounce rate	−0.109			0.352
JS CSS number	−0.163			0.526
JS CSS size	−0.124			0.688
Webpage issues	0.141	0.213	1.603	0.298
Total checks	0.167			0.131
Incoming internal links	−0.245			0.367
Outcoming internal links	−0.076			0.633
Outcoming external links	−0.109			0.518
Page load time	−0.154			0.296
Crawl depth	−0.096			0.489
Dependent: unique visits				
Constant	–			0.590
JS CSS number	−0.416			0.121
JS CSS size	0.350			0.280
Webpage issues	0.030			0.831
Total checks	0.049	0.115	0.851	0.671
Incoming internal links	−0.400			0.158
Outcoming internal links	−0.017			0.918
Outcoming external links	−0.203			0.247
Page load time	−0.102			0.507
Crawl depth	0.072			0.623
Dependent: bounce rate				
Constant	–			0.189
JS CSS number	0.203			0.435
JS CSS size	−0.191			0.542
Webpage issues	−0.269			0.047*
Total checks	0.007			0.948
Incoming internal links	−0.587	0.177	1.407	0.030*
Outcoming internal links	−0.214			0.182
Outcoming external links	−0.327			0.051
Page load time	0.133			0.371
Crawl depth	−0.105			0.457

* and ** indicate statistical significance at the 95% and 99% levels, respectively

hypothesis (H1) is rejected, as the decentralized payment service's Internet ranking is not affected by the visibility and performance of its website.

Moving to unique visits regression, the low R^2 equal to 0.115, combined with regression's *p*-value of $0.590 > a = 0.05$, shows the overall non-verification of the model. Moreover, none of the webpage factors' variables have *p*-values below 0.05 significance level. Hence, the second research hypothesis (H2) of the research is also rejected, meaning that the number of new visitors to decentralized payment services websites is not impacted by their webpage factors.

Finally, the regression of decentralized payment services' bounce rate is also not verified overall, since it has p-value $= 0.189 > a = 0.05$ and due to low $R^2 = 0.117$. Furthermore, two variables appeared to impact significantly the bounce rate variables, the webpage issues and the incoming internal links with p-values $= 0.047$ and 0.030, respectively, being below the $a = 0.05$ level of significance. Thus, the paper's last research hypothesis (H3) has been verified, meaning that the more specific factors of a decentralized payment company's webpage increase, the more their visitors tend to abandon the website.

5 Diagnostic Model

Paper's deployed model for performing diagnostical analysis has been formed by harvesting MentalModeler's [20] onsite application for creating fuzzy cognitive maps (FCM). This mapping procedure can represent the possible interconnections of a research's variables and provide a coherent report. Due to their fuzzy cognitive illustration, the causative links among the elements are depicted by arrows with red and blue colors [17]. The red color is for negative correlation and blue for positive, which is discerned by the width of the arrow. In Fig. 1, the above model and its factors and information can be seen.

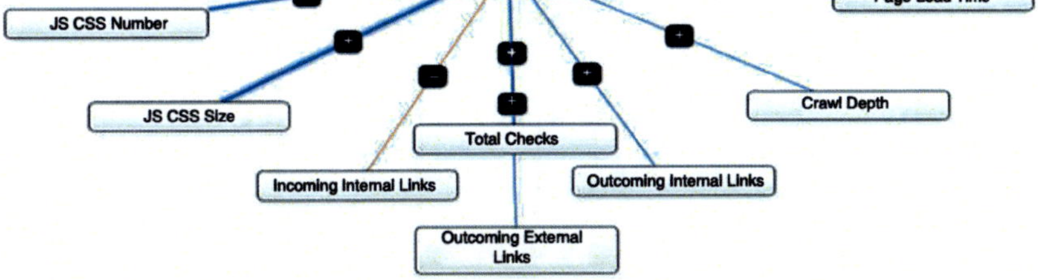

Fig. 1 Fuzzy cognitive mapping of decentralized payment companies' webpage auditing

6 Predictive Model

For this research, the hybrid modeling (HM) process, consisting of both agent-based (ABM) and dynamic models (DM), have been utilized. The produced model begins with agents reaching the potential visitors statechart, leading to unique decentralized payment websites' visitors' statechart, with bounce rate and global rank following the normal distribution of the sample. At the same time, with each visitor entering the website, webpage factors start to variate and affect the variables of bounce rate, unique visitors, and global rank [18]. Through these interactions and interconnections, the fluctuations of the model's metrics are produced (Fig. 2).

Deployment of the above model provides the following results. The regression analysis discerned the fact that only the bounce rate of decentralized payment companies' websites gets affected by webpage factors. So, the simulation results showed that the fluctuation of bounce rate, as unique visitors tend to rise in time, appears to affect the variation of the global rank. This means that since webpages factors pose a significant impact on decentralized payment companies' website bounce rate, their global rank variable, which indicates digital marketing performance, seems to variate accordingly with the bounce rate variable (Fig. 3).

7 Conclusions

The paper's main results' extraction is focused on distinguishing the specific webpage factors that affect decentralized payment companies' digital marketing performance. The bounce rate metric of decentralized payment companies' websites is highly connected with the number of webpage factors and the incoming internal

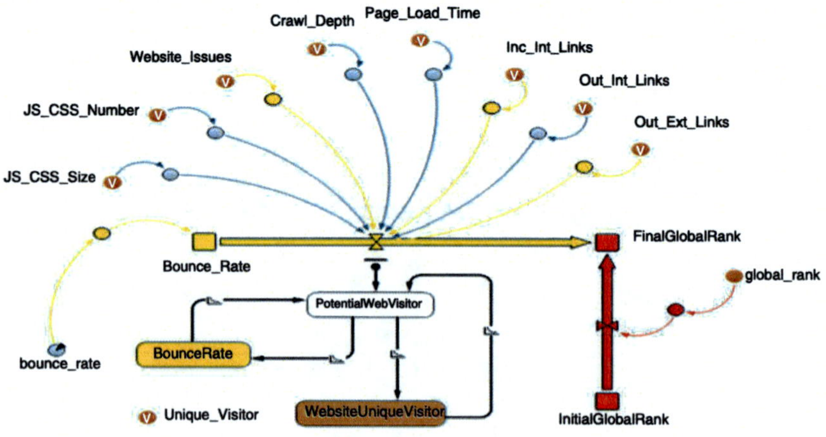

Fig. 2 Hybrid modeling of decentralized payment companies' webpage auditing and traffic

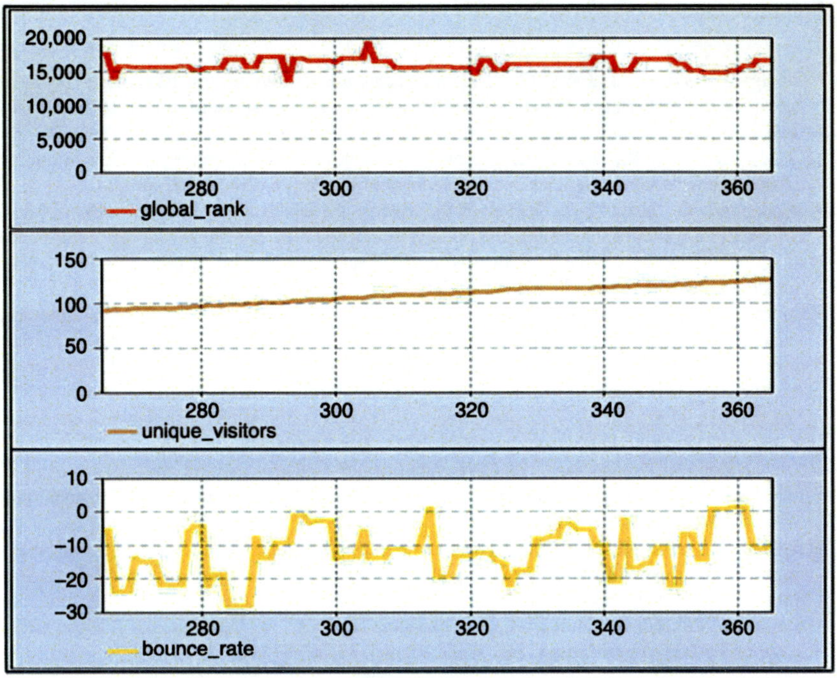

Fig. 3 Model's variables and traffic outcomes in 100 days

links. Therefore, via the 100 days simulation, decentralized payment companies' global rank appears to have a similar variation through time with the bounce rate variable, indicating a strong positive connection between them. At this point, there should be reminded that optimally, both global rank and bounce rate metrics are more enhanced the lesser their values get.

8 Practical Implications

Decentralized payment services and their digital marketing performance, as observed through this paper, are highly connected with specific factors of their websites. Worsening of specific webpage factors cause website visitors to abandon the website, thus affecting decentralized payment companies' digital marketing performance. This paper gives a velar view of the exact webpage factors, decentralized payment companies should optimize in order to improve their digital marketing performance. Other scientists with similar cases are studying similar problems, using modeling and simulation [21–23].

References

1. Zhang, X., Shi, S. & Qian, C.: Scalable decentralized routing for Blockchain payment networks. In: Third International Symposium on Foundations and Applications of Blockchain (FAB '20), May 1, 2020, Santa Cruz, California, USA (2020)
2. Gilad, Y., Hemo, R., Micali, S., Vlachos, G., Zeldovich, N.: Algorand: scaling byzantine agreements for cryptocurrencies. In: Proceedings of the 26th SOSP, pp. 51–68 (2017)
3. Manogaran, G., Hsu, C.-H., Rawal, B.S., Muthu, B., Mavromoustakis, C.X., Mastorakis, G.: ISOF: information scheduling and optimization framework for improving the performance of agriculture systems aided by industry 4.0. IEEE Internet Things J. **8**(5), 3120–3129 (2021). https://doi.org/10.1109/JIOT.2020.3045479
4. Fundingsland, E.L., Fike, J., Calvano, J., Beach, J., Deborah Lai, D., He, S.: Methodological guidelines for systematic assessments of health care websites using web analytics: tutorial. J. Med. Internet Res. **24**(4), 28291 (2022). https://doi.org/10.2196/28291
5. Kanellos, N. by Sakas, D.P, Kamperos, I.D.G., Reklitis, D.P., Giannakopoulos, N.T., Nasiopoulos, D.K., Terzi, M.C.: The effectiveness of centralized payment network advertisements on digital branding during the COVID-19 crisis. Sustainability. **14**(6)
6. Michaelidou, N., Siamagka, N.T., Christodoulides, G.: Usage, barriers and measurement of social media marketing: an exploratory investigation of small and medium B2B brands. Ind. Mark. Manag. **40**(7), 1153–1159 (2011)
7. Agnihotri, R., Kothandaraman, P., Kashyap, R., Singh, R.: Bringing 'social' into sales: the impact of salespeople's social media use on service behaviors and value creation. J. Pers. Sell. Sales Manag. **32**(3), 333–348 (2012)
8. Siamagka, N.T., Christodoulides, G., Michaelidou, N., Valvi, A.: Determinants of social media adoption by B2B organizations. Ind. Mark. Manag. **51**, 89–99 (2015)
9. Zhan, J., Dong, S., Hu, W.: IoE-supported smart logistics network communication with optimization and security. Sustain. Energy Technol. Assess. **52**, 102052 (2022). https://doi.org/10.1016/j.seta.2022.102052
10. Huang, Z., Benyoucef, M.: Usability and credibility of e-government websites. Gov. Inf. Q. **31**(4), 584–595 (2014). https://doi.org/10.1016/j.giq.2014.07.002
11. Swani, K., Milne, G.R., Brown, B.P., Assaf, A.G., Donthu, N.: What messages to post? Evaluating the popularity of social media communications in business versus consumer markets. Ind. Mark. Manag. **62**, 77–87 (2017)
12. Sakas, D.P., Kamperos, I.D.G., Reklitis, D.P., Giannakopoulos, N.T., Nasiopoulos, D.K., Terzi, M.C., Kanellos, N.: The effectiveness of centralized payment network advertisements on digital branding during the COVID-19 crisis. Sustainability. **14**, 3616 (2022). https://doi.org/10.3390/su14063616
13. Sakas, D.P., Giannakopoulos, N.T., Kanellos, N., Tryfonopoulos, C.: Digital marketing enhancement of cryptocurrency websites through customer innovative data process. PRO. **10**, 960 (2022). https://doi.org/10.3390/pr10050960
14. Sakas, D., Giannakopoulos, N.: Harvesting crowdsourcing platforms' traffic in favour of air forwarders' brand name and sustainability. Sustainability. **13**(15), 8222 (2021). https://doi.org/10.3390/su13158222
15. Buratti, N., Parola, F., Satta, G.: Insights on the adoption of social media marketing in B2B services. The TQM J. **30**(5), 490–529 (2018). https://doi.org/10.1108/TQM-11-2017-0136
16. Mpelogianni, V., Groumpos, P.P.: Re-approaching fuzzy cognitive maps to increase the knowledge of a system. AI & Soc. **33**, 175–188 (2018). https://doi.org/10.1007/s00146-018-0813-0
17. Uddin, M.N., Chi, H.-L., Wei, H.-H., Lee, M., Ni, M.: Influence of interior layouts on occupant energy-saving behaviour in buildings: an integrated approach using Agent-Based Modelling, System Dynamics and Building Information Modelling. Renew. Sust. Energ. Rev. **161**, 112382 (2022). https://doi.org/10.1016/j.rser.2022.112382

18. Komarraju, A: Top 10 cryptocurrencies to buy & invest in today in May 2021. https://www.analyticsinsight.net/top-10-cryptocurrencies-to-buy-invest-in-today-in-may-2021/
19. Semrush: (2022, March 16). https://www.semrush.com/
20. MentalModeler: (2022, March 16). https://dev.mentalmodeler.com/
21. Nasiopoulos, D.K.: The role of digital marketing in the development of a distribution and logistics network of information technology companies. Bus. Intell. Model. **6**, 267–276
22. Dimitrios, N.K., Sakas, D.P., Vlachos, D.S.: The contribution of dynamic simulation model of depiction of knowledge, in the leading process of high technology companies. Key Eng. Mater. **543**, 406–409
23. Nasiopoulos, D.K., Sakas, D.P., Masselos, K.: Open source web applications. How it spread through the Internet and their contribution to education. In: 4th Symposium on Business and Management and Dynamic Simulation Models Supporting Management Strategies

Airlines' Low-Demand Risk Management Implications of COVID-19 Crisis-Induced Immobility Utilizing Big Data

Damianos P. Sakas, Ioannis Dimitrios G. Kamperos, Marina C. Terzi, and Athanasios Kriemadis

1 Literature Review

1.1 COVID-19 Effect on Airlines' Websites' User Engagement Metrics

It has been 2 years since the World Health Organization (WHO) declared the COVID-19 pandemic. In response to the pandemic and in an attempt to eliminate the unprecedented impact on health, economic, social, and fiscal dimension [1], national authorities have forced travel restrictions, flights suspension, and entry bans around the globe [2]. This novel crisis has caused severe damage and great uncertainty to the aviation sector [3]. According to ICAO, the pandemic reduced the passengers' traffic by 50% from 2019 to 2020 and 40% from 2020 to 2021 [4], leading to operations' breakdown and huge revenue losses [5].

The challenges of health crises are not new to the airline industry, such as the H1N1 flu in 2009 and the Ebola virus disease in 2014. The aviation sector was already facing mounting challenges [6], but due to COVID-19, the wreaked financial devastation [7] and labor shortage [8], among others, are only a few of the new challenges that have been emerged from the crisis. The widespread lockdowns, grounded aircrafts, self-isolation, quarantine, and customers' outflow led to a

D. P. Sakas · I. D. G. Kamperos (✉) · M. C. Terzi
BICTEVAC LABORATORY Business Information and Communication Technologies in Value Chains Laboratory, Department of Agribusiness and Supply Chain Management, School of Applied Economics and Social Sciences, Agricultural University of Athens, Athina, Greece
e-mail: kamperosdigese@aua.gr

A. Kriemadis
Department of Management Science and Technology, School of Economics and Technology, University of Peloponnese, Tripolis, TK, Greece

D. P. Sakas et al. (eds.), *Computational and Strategic Business Modelling*,
Springer Proceedings in Business and Economics,
https://doi.org/10.1007/978-3-031-41371-1_3

domino effect that provoked massive shifts in users' online behavior. Nevertheless, the crisis has also offered opportunities for strategic change [9] and accelerated digital transformation across the industry. The reason behind this is to achieve business sustainability and resilience [10] and to redefine customer experience, based on technological innovations [11].

Since both air passenger traffic and air freight traffic are expected to more than double in the next two decades [4], businesses have to adopt a digital reorientation and focus on digital day-to-day activities. To this end, airline companies have to adapt to the "new normal," especially when digital branding empowerment and company performance come to matter. In this aspect, getting a 360-degree view of consumers' online behavior is a key factor that leads to innovations in brand management [12] and technical quality development opportunities for their digital presence [13].

1.2 Big Data and the Aviation Sector

Multiple studies have been devoted to the investigation of the COVID-19 crisis on economic dynamics [14], airfares [15], airline employment [16], and strategic responses [17], among others, demonstrating the devasting and unprecedented impact on the air transport ecosystem. The majority of the studies focus on traditional methods so as to investigate public viewpoints regarding COVID-19 and its impact [18]. Sternberg et al. [19] conducted their research on analyzing customer engagement of Turkish airlines, using big social data. Social media analysis is considered an obligation, since it paves the way for online users to express their feelings and opinions on major health issues [20]. In this direction, the current research focuses on customers' engagement metrics to evaluate online attitudes and purchasing behaviors, based on big data analysis of major airlines' websites, during and after the COVID-19 crisis.

Millions of flights operate every day, generating a massive amount of data. Unintentionally, online users leave their fingerprints due to their interaction with digital brands, a process that reflects their online purchasing behavior during and after quarantine periods [21]. The nature of the structure and unstructured data provides ample opportunities for marketers to empower airline companies' digital branding and optimize their marketing campaigns [22, 23]. Big data analysis (BDA) is a valuable tool able to decipher the crowd's sentiment toward specific brands. In addition, it has the potential to minimize customers' risk perception during and after a novel crisis [24] and build trust strategies [25] in a fast and efficient way, since it produces automated reports, which, if decoded properly, add great value to companies.

In response to the pandemic, the aviation industry has to invest in new technologies as an essential part of its recovery and resilience planning. Having as a pivotal aim digital brand awareness, airlines' marketers and strategists need to invest in behavioral KPIs analysis so as to acquire a deeper understanding of the factors affecting airlines' organic traffic and user engagement [24]. Although BDA has been the point of several research studies by academia [26, 27], the importance of behavioral KPIs has not been extensively researched, given the value of the indirect information on users' online behavior and experience [28]. Olaganathan [29] highlights the importance of BDA in making better strategic decisions and creating a competitive advantage. However, the research has a severe limitation: it uses BDA for global, regional, subregional, and country analyses. The author suggests future research on consumer online behavior based on BDA.

In the absence of sufficient research regarding customers' online behavior during the COVID-19 crisis, the present paper provides a comprehensive understanding of five major airlines' analysis, setting as a point of research customers' behavioral KPIs. The purpose of the study is to thoroughly investigate the effects of uncertainty on online users' purchasing behavior in light of the current pandemic. Specifically, with the use of BDA, the authors investigate the connection between behavioral KPIs and crisis escalation, during a novel crisis.

2 Materials and Methods

2.1 Problem Formulation and Research Hypotheses

In the digital era, data is valuable and profitable when used analytically, since it is able to enrich and personalize customer online experience and predict future online behaviors. Therefore, the major challenge is for airline companies to shift from a corrective to a predictive approach. Adopting such an approach will minimize risk perceptions and thus contribute to the advancement of the aviation industry at an economical, strategic, and operational level [24]. Through the analysis of Web analytics, the aviation sector could create huge potential in obtaining key behavioral intelligence to manage its digital brand effectively. By formulating a new cohesive strategy, adapted to the available behavioral KPIs, airline companies will ensure their brand thrives in this new era of digitalization. For the purpose of the current study, the authors have identified four behavioral KPIs: pageviews, time on site, reach percentage, and bounce rate. To track COVID-19 crisis escalation, authors also utilize the number of global daily new COVID-19 infections number [30].

Even if airline companies track customer engagement metrics daily, it can be struggling to identify which ones to use, since it relies on the companies' campaigns, channels, and goals. Behavioral KPIs can elucidate the give-and-take relationship between e-businesses and online users to precisely measure the success of marketing activities, the structure of the website, and the quality of the content. The first purpose of this study is to track users' engagement metrics that serve as an indication of the customers' behavior and interaction with the organization's domain. This process provides important insights into online users' behavior and gives important information to marketers regarding actions that could yield higher conversion rates, during a crisis.

In order to serve the research purposes of this study, two research questions were formulated:

Hypothesis 1 (H1): *User engagement KPIs will be significantly different during the COVID-19 period compared to the pre-COVID-19 period.*

COVID-19 period is characterized by health regulations, self-isolation, and travel restrictions. This crisis-induced demand decrease for airline travel is expected to significantly alter not only the total number of users visiting the airlines' Web domains but also the online behavior of the existing visitors as well.

Hypothesis 2 (H2): *During the COVID-19 pandemic period, crisis escalation expressed through global COVID-19 cases will significantly affect user engagement KPIs.*

COVID-19 new infection cases serve as the main indicator of crisis level of escalation. According to the psychometric paradigm [31], it is expected that people's risk perception for the pandemic hazard will significantly increase as infection cases rise. This result, combined with self-isolation and travel restriction regulation, is expected to highly affect the demand for airline travel services and consequently affect the user engagement behavior of the online customers.

2.2 Sample Selection and Data Collection

Our sample was developed by taking into account the 2019 top airlines by RPK (revenue-passengers-kilometers) and FTK (freight-tonne-kilometers) [32]. We chose the top five airlines to constitute a representative sample of the respected business area.

We chose 1-year periods for our research (Fig. 1). Because most countries had begun implementing border control laws and self-isolating practices by that date, we chose March 1, 2020, as the commencement date for our research. The first phase (March 1, 2019, to February 29, 2020) corresponds to the year before the COVID-19 epidemic (pre-COVID period). The first year of the COVID-19 pandemic (COVID period) is marked by the second phase (1 March 2020 to 28 February 2021).

To achieve more efficient data processing and statistical analysis, we collected every KPI value from the sample companies through Web analytics platform [33]

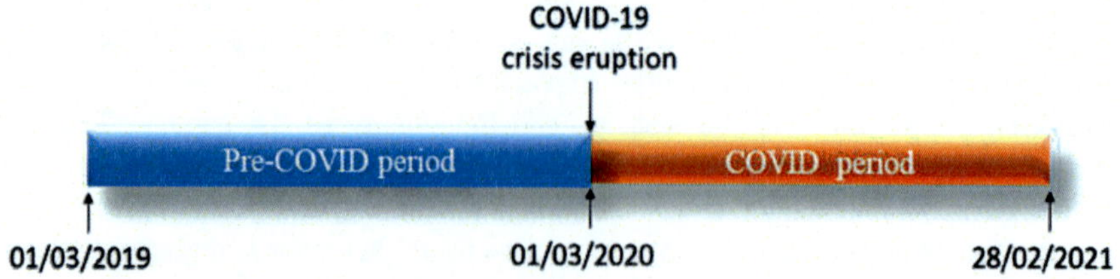

Fig. 1 Timeline presenting the periods studied in this article

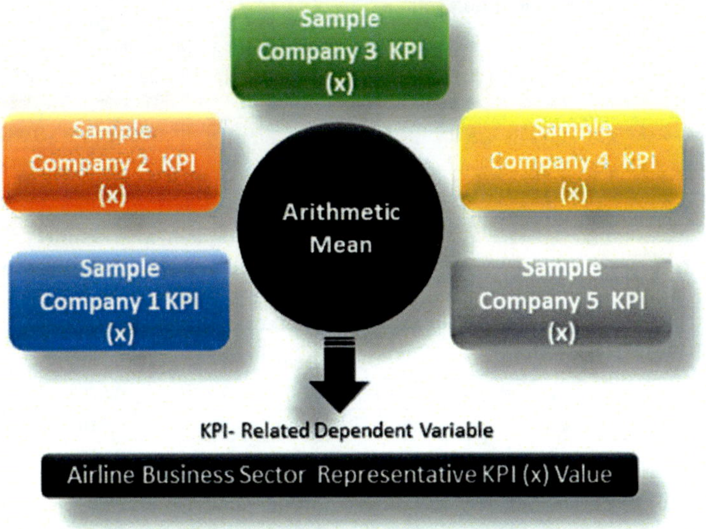

Fig. 2 Big data processing methodology used in the presented study

big data mining techniques and formed a representative KPI-related variable based on the average value. This process (Fig. 2) was applied to all studied KPIs.

For our study, we acquired data regarding the global daily confirmed cases of COVID-19 as recorded by the STATISTA database [30]. For better accuracy and compatibility with the dependent variables of the study, these data were processed into an average daily value for every month, forming the independent variable of the study (COVID cases).

3 Results

3.1 *Statistical Analysis*

Independent samples t-tests were applied between the two datasets, to establish statistically significant differences between the pre-COVID and the COVID period. Results indicated statistically significant decreases for three of the four studied variables (Table 1).

We conducted a Pearson correlation coefficient (PCC–r) statistical analysis on our datasets to elucidate any apparent linear relationships between pairs of variables over the COVID period. There were several statistically significant associations found, with some of them indicating a very strong linear correlation. (Table 2). PCC analysis indicated five statistically significant correlations between variables, three at the 0.01 level of significance and two at the 0.05 level (2-tailed).

Table 1 Independent samples t-tests results

Variables	Levene's test for equality of variances significance	t-Test for equality of means Significance (2-tailed)	Mean difference	Standard error difference
Reach %	0.7350	*0.000	0.0248	0.0026
Pageviews per user	0.1560	*0.004	0.3897	0.1184
Bounce rate	0.1810	0.1570	0.0143	0.0097
Time on site	0.8880	*0.000	69.6000	11.9967

The symbol "*" indicates a statistically significant result

Table 2 Pearson correlation coefficient (PCC–r) statistical analysis results

Variables			Pearson correlation	Significance (2-tailed)
COVID cases	and	Pageviews per user	0.751	0.010
COVID cases	and	Time on site	0.756	0.010
Reach %	and	Pageviews per user	0.668	0.050
Reach %	and	Time on site	0.686	0.050
Pageviews per user	and	Time on site	0.968	0.010

3.2 *Development of a Diagnostic Explanatory Model*

Following the statistical analysis, we created a fuzzy cognitive mapping (FCM) to visualize all of the cause-effect interactions between the COVID-19 and KPI-related factors investigated in this study (Fig. 3).

FCMs (fuzzy cognitive mappings) are fuzzy-graph structures that are used to illustrate causal reasoning. Because of their foggy character, they can have varying degrees of causality among various causal actors [34]. It's a "soft computing" technique for system modeling that incorporates both fuzzy logic and neural networks. Data regarding the impact of user engagement metrics on the global ranking of the website is adopted from previous research [35].

In order to further investigate the interaction between the studied variables, we developed three hypothetical scenarios regarding three different stages of the pandemic. For scenario 1 (Fig. 4), we assumed a level of 0.6 of maximal pandemic escalation (low cases scenario). For scenario 2 (Fig. 5), we introduced to our model a level of 0.8 of maximum pandemic escalation (medium cases scenario), and for scenario 3 (Fig. 6), we suggested a high-level crisis at a level 1.0 (high cases scenario).

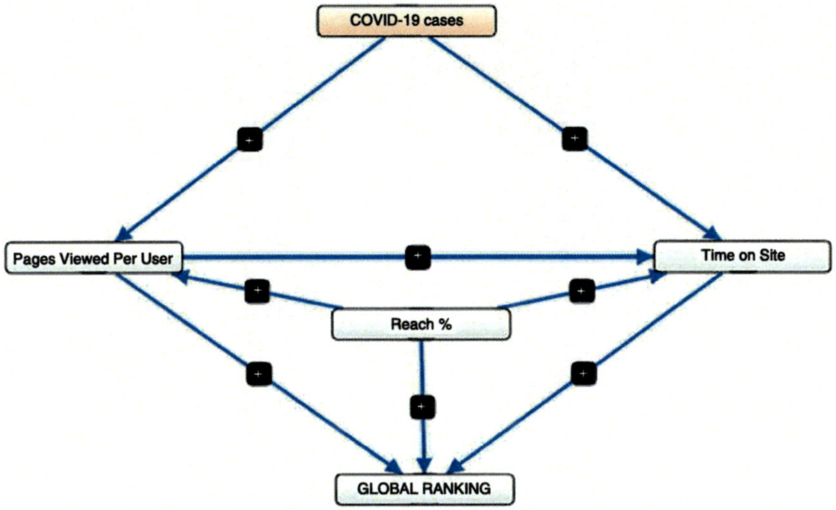

Fig. 3 Fuzzy cognitive mapping of correlated variables

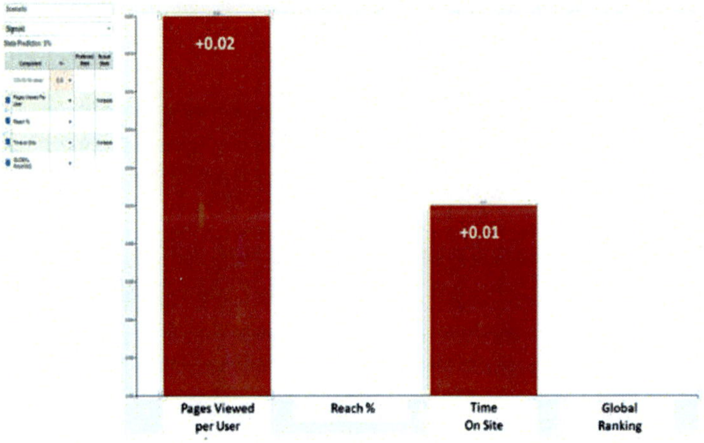

Fig. 4 Scenario 1 – low cases scenario

4 Discussion

The purpose of this study is to analyze the impact of a crisis with the characteristics of a pandemic on airlines' customer behavior through monitoring Web domain user engagement metrics. Based on the results of our research, customers have significantly decreased their engagement with the airlines' Web domains during the studied

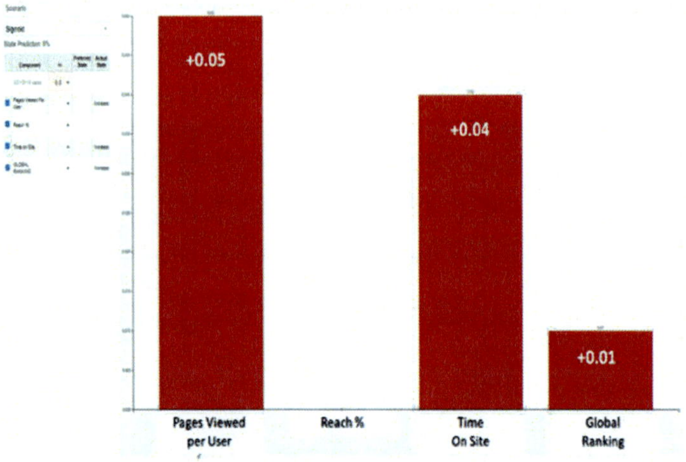

Fig. 5 Scenario 2 – medium cases scenario

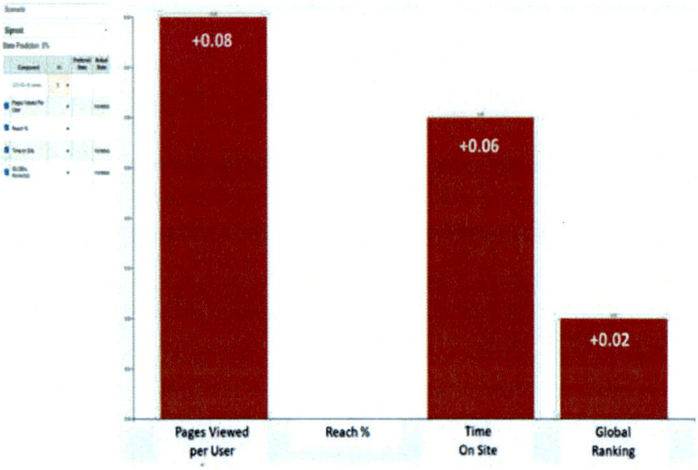

Fig. 6 Scenario 3– high cases scenario

COVID-19 period, confirming H1. These results are aligned with the findings reported by Abdullah et al. [12].

The escalation of the crisis, expressed through worldwide confirmed new COVID-19 infections, is suggested to highly correlate with the fluctuation of online users' engagement. Although users decrease dramatically their engagement after the start of the pandemic, they tend to increase it, for three of our four investigated variables, in substantial correlation with the escalation of the crisis, resulting in digital branding enhancement. This direct connection between crisis escalation and customer behavior confirms H2 of the study and supports findings reported by previous research [24].

Although results are also supported by the FCM scenarios outcomes, scenarios results suggest a possible low-level negative impact of crisis escalation on digital branding, proposing that further investigation is required.

5 Conclusion

COVID-19 pandemic caused financial devastation across the transportation supply chain, causing persistent weakness for the airlines. The crisis significantly affected the demand for airline services, resulting in direct increase of the financial risk through revenue losses and uncertainty for future sales. The transportation industry should adapt to the new economic parameters induced by the pandemic by developing low-demand risk management plans to control financial disasters for their organizations [39–41].

Airline companies are highly relied on Web-based platforms to reach customers and increase their revenues and profits [36]. Utilizing knowledge provided by high technology resources is a skill that needs to be acquired in a wide range of entrepreneurship [37, 38]. This challenge, combined with understanding and forecasting online customer behavior during a crisis can give the company a survival advantage.

The current study suggests that, during the escalation of crises, airlines should invest in marketing activities so as to optimize their digital strategy in order to maintain existing customers, to effectively generate new traffic, and to establish brand awareness. To this end, more research should be accomplished on how big data could contribute to the optimization of the conversion funnel to convert new online users into actual clients. Therefore, other scientists with similar cases are studying similar problems, using modeling and simulation.

References

1. OECD: The territorial impact of COVID-19: managing the crisis across levels of government. https://read.oecd-ilibrary.org/view/?ref=128_128287-5agkkojaaa&title=The-territorial-impact-of-covid-19-managing-the-crisis-across-levels-of-government (2020)
2. New COVID-19 Surges Keep Travel Restrictions in Place.: https://www.unwto.org/news/new-covid-19-surges-keep-travel-restrictions-in-place. Last accessed 2022/05/10
3. Dube, K., Nhamo, G., Chikodzi, D.: COVID-19 pandemic and prospects for recovery of the global aviation industry. J. Air Transp. Manag. **92**, 102022 (2021). https://doi.org/10.1016/j.jairtraman.2021.102022
4. ICAO: Economic impacts of COVID-19 on civil aviation. https://www.icao.int/sustainability/Pages/Economic-Impacts-of-COVID-19.aspx. Last accessed 2022/05/02
5. IATA - Losses reduce but challenges continue - Cumulative $201 billion losses for 2020–2022. https://www.iata.org/en/pressroom/2021-releases/2021-10-04-01/. Last accessed 2022/05/10

6. Stamolampros, P., Korfiatis, N.: Airline service quality and economic factors: an ARDL approach on US airlines. J. Air Transp. Manag. **77**, 24–31 (2019). https://doi.org/10.1016/j.jairtraman.2019.03.002
7. Gole, I., Dobrea, R.C., Gombos, C.C.: Aviation industry - challenges and uncertainties after the Covid-19 pandemic. SHS Web Conf. **92**, 01010 (2021). https://doi.org/10.1051/shsconf/20219201010
8. McLaughlin, H., Fearon, C.: Labor in the aviation industry. In: The Air Transportation Industry, pp. 55–76. Elsevier (2022). https://doi.org/10.1016/B978-0-323-91522-9.00013-0
9. Adwan, M.: COVID-19 on aviation: insights on challenges facing the airlines; managing uncertainty, and the potential role of leadership in dealing with a crisis. Bus.Educ.: Working Paper Series of the Centre for Management Learning and Business Education. **1**, 61–78 (2021)
10. Sun, X., Wandelt, S., Zhang, A.: Technological and educational challenges towards pandemic-resilient aviation. Transp. Policy. **114**, 104–115 (2021). https://doi.org/10.1016/j.tranpol.2021.09.010
11. Serrano, F., Kazda, A.: The future of airports post COVID-19. J. Air Transp. Manag. **89**, 101900 (2020). https://doi.org/10.1016/j.jairtraman.2020.101900
12. Abdullah, M., Dias, C., Muley, D., Shahin, M.: Exploring the impacts of COVID-19 on travel behavior and mode preferences. Transp. Res. Interdiscip. Perspect. **8**, 100255 (2020). https://doi.org/10.1016/j.trip.2020.100255
13. Sakas, D.P., Reklitis, D.P.: Predictive model for estimating the impact of technical issues on consumers interaction in agri-logistics websites. In: Bochtis, D.D., Pearson, S., Lampridi, M., Marinoudi, V., Pardalos, P.M. (eds.) Information and Communication Technologies for Agriculture—Theme IV: Actions, pp. 269–283. Springer International Publishing, Cham (2021). https://doi.org/10.1007/978-3-030-84156-0_14
14. Eichenbaum, M., Rebelo, S., Trabandt, M.: The Macroeconomics of Epidemics. National Bureau of Economic Research, Cambridge, MA (2020). https://doi.org/10.3386/w26882
15. Wozny, F.: The impact of COVID-19 on airfares—a machine learning counterfactual analysis. Econometrics. **10**, 8 (2022). https://doi.org/10.3390/econometrics10010008
16. Sobieralski, J.B.: COVID-19 and airline employment: insights from historical uncertainty shocks to the industry. Transp. Res. Interdiscipl. Perspect. **5**, 100123 (2020). https://doi.org/10.1016/j.trip.2020.100123
17. Amankwah-Amoah, J.: Note: Mayday, Mayday, Mayday! Responding to environmental shocks: insights on global airlines' responses to COVID-19. Transp. Res. Part E: Logist. Transp. Rev. **143**, 102098 (2020). https://doi.org/10.1016/j.tre.2020.102098
18. Alfalih, A.A.: Customer engagement design during the COVID 19 pandemic, mutual trust and intelligent automation: a conceptual perspective. J. Innov. Entrep. **11**, 32 (2022). https://doi.org/10.1186/s13731-022-00222-7
19. Sternberg, F., Hedegaard Pedersen, K., Ryelund, N.K., Mukkamala, R.R., Vatrapu, R.: Analysing customer engagement of Turkish airlines using big social data. In: 2018 IEEE International Congress on Big Data (BigData Congress), pp. 74–81. IEEE, San Francisco (2018). https://doi.org/10.1109/BigDataCongress.2018.00017
20. Yousefinaghani, S., Dara, R., Mubareka, S., Papadopoulos, A., Sharif, S.: An analysis of COVID-19 vaccine sentiments and opinions on Twitter. Int. J. Infect. Dis. **108**, 256–262 (2021). https://doi.org/10.1016/j.ijid.2021.05.059
21. Gu, S., Ślusarczyk, B., Hajizada, S., Kovalyova, I., Sakhbieva, A.: Impact of the COVID-19 pandemic on online consumer purchasing behavior. JTAER. **16**, 2263–2281 (2021). https://doi.org/10.3390/jtaer16060125
22. Sakas, D.P., Reklitis, D.P.: The impact of organic traffic of crowdsourcing platforms on airlines' website traffic and user engagement. Sustainability. **13**, 8850 (2021). https://doi.org/10.3390/su13168850
23. Drivas, I.C., Sakas, D.P., Giannakopoulos, G.A., Kyriaki-Manessi, D.: Optimization of paid search traffic effectiveness and users' engagement within websites. In: Sakas, D.P.,

Nasiopoulos, D.K., Taratuhina, Y. (eds.) Business Intelligence and Modelling, pp. 17–30. Springer International Publishing, Cham (2021). https://doi.org/10.1007/978-3-030-57065-1_2

24. Sakas, D.P., Kamperos, I.D.G., Reklitis, P.: Estimating risk perception effects on courier companies' online customer behavior during a crisis, using crowdsourced data. Sustainability. **13**, 12725 (2021). https://doi.org/10.3390/su132212725
25. Akpan, S.S., Etuk, S.G.: (PDF) Risk reduction and trust building strategy for consumers: an explanatory power of branding. https://www.researchgate.net/publication/283443406_Risk_Reduction_and_Trust_Building_Strategy_for_Consumers_An_Explanatory_Power_of_Branding. Last accessed 2022/05/10
26. Chen, S., Huang, Y., Huang, W.: Big data analytics on aviation social media: the case of China southern airlines on Sina Weibo. In: 2016 IEEE Second International Conference on Big Data Computing Service and Applications (BigDataService), pp. 152–155. IEEE, Oxford, UK (2016). https://doi.org/10.1109/BigDataService.2016.51
27. Borgi, T., Zoghlami, N., Abed, M.: Big data for transport and logistics: a review. In: 2017 International Conference on Advanced Systems and Electric Technologies (IC_ASET), pp. 44–49. IEEE, Hammamet (2017). https://doi.org/10.1109/ASET.2017.7983742
28. Sakas, D.P., Reklitis, D.P., Trivellas, P., Vassilakis, C., Terzi, M.C.: The effects of logistics websites' technical factors on the optimization of digital marketing strategies and corporate brand name. PRO. **10**, 892 (2022). https://doi.org/10.3390/pr10050892
29. Olaganathan, R.: Impact of COVID-19 on airline industry and strategic plan for its recovery with special reference to data analytics technology. Global J. Eng. Tech. Adv. **7**, 033–046 (2021). https://doi.org/10.30574/gjeta.2021.7.1.0050
30. COVID-19 new cases worldwide by day. https://www.statista.com/statistics/1103046/new-coronavirus-covid19-cases-number-worldwide-by-day/. Last accessed 2022/03/22
31. Fischhoff, B., Slovic, P., Lichtenstein, S., Read, S., Combs, B.: How safe is safe enough? A psychometric study of attitudes toward technological risks and benefits. Policy. Sci. **9**, 127–152 (1978). https://doi.org/10.1007/BF00143739
32. IATA: WATS-World Air Transport Statistics 2019. https://www.labeline.com/product/iata-world-air-transport-statistics-wats-2019-electronic-2-year-airline-data-download-code-iata9443-63/ (2019)
33. Semrush - Online Marketing Can Be Easy.: https://www.semrush.com/. Last accessed 2022/03/21
34. Kosko, B.: Fuzzy cognitive maps. Int. J. Man-Mach. Stud. **24**, 65–75 (1986). https://doi.org/10.1016/S0020-7373(86)80040-2
35. Sakas, D.P., Kamperos, I.D.G., Reklitis, D.P., Giannakopoulos, N.T., Nasiopoulos, D.K., Terzi, M.C., Kanellos, N.: The effectiveness of centralized payment network advertisements on digital branding during the COVID-19 crisis. Sustainability. **14**, 3616 (2022). https://doi.org/10.3390/su14063616
36. Sakas, D.P., Giannakopoulos, N.T.: Big data contribution in desktop and mobile devices comparison, regarding airlines' digital brand name effect. BDCC. **5**, 48 (2021). https://doi.org/10.3390/bdcc5040048
37. Terzi, M.C., Sakas, D.P., Seimenis, I.: The contribution of the scientific committee in the development of conferences. Procedia. Soc. Behav. Sci. **73**, 373–382 (2013). https://doi.org/10.1016/j.sbspro.2013.02.064
38. Kanellos, N.: A conceptual framework for analyzing knowledge-based entrepreneurship. undefined. (2011)
39. Nasiopoulos, D.K.: Logistics companies reengineering. Change of leadership and human resources planning through management information systems. Bus. Intell. Model. **6**, 207–217 (2021)
40. Nasiopoulos, D.K.: Web search applications in Logistics companies in the context of evolution of Management Information Systems. Bus. Intell. Model. **6**, 241–254 (2021)
41. Nasiopoulos, D.K.: Email and E-branding applications in logistics companies as a field of management information systems. Bus. Intell. Model. **6**, 255–265 (2021)

Reputational Crisis Risk Management: The Immediate Effect of a Plane Crash on Airlines' Digital Branding

Damianos P. Sakas, Marina C. Terzi, Ioannis Dimitrios G. Kamperos, and Athanasios Kriemadis

1 Literature Review

1.1 Reputational Crisis in Air Disasters

Business success is a result of imperative decisions, which are made under pressure, uncertainty, and unpredictable conditions [1]. Especially when air disasters occur, there is an urgent call for action, not only to mitigate the huge impact of the operational crisis [2] but also to deactivate the negative reputational alarm. In such cases, a widespread allocation of blame and responsibility takes place [3], raising fears regarding reputational concerns in air passengers' risk perceptions toward a brand.

Typically considered one of the safest forms of travel, the airline industry had an average fatality risk of 0.13 in 2020 [4]. Airline tragedies are rare, due to the high standards set by international authorities and the International Air Transport Association (IATA). In simple words, it takes 461 years for someone, traveling every day, to experience an accident with at least one fatality [4].

Numerous studies have been devoted to airlines' crises [2, 5, 6]. Medina-Munoz et al. [7] have concluded that the most important attribute in affecting customers' choice, behavior, and loyalty is safety and punctuality. Chang et al. [8] have developed a four-dimensional framework with 31 attributes for effective air crisis

D. P. Sakas · M. C. Terzi · I. D. G. Kamperos (✉)
Bictevac Laboratory Business Information and Communication Technologies in Value Chains Laboratory, Department of Agribusiness and Supply Chain Management, School of Applied Economics and Social Sciences, Agricultural University of Athens, Athina, Greece
e-mail: kamperosdigese@aua.gr

A. Kriemadis
Department of Management Science and Technology, School of Economics and Technology, University of Peloponnese, Tripoli, Greece

D. P. Sakas et al. (eds.), *Computational and Strategic Business Modelling*, Springer Proceedings in Business and Economics,
https://doi.org/10.1007/978-3-031-41371-1_4

management. Seo and Park [9] acknowledged the impact of an airline's reputation on brand preference. To this end, communication strategy ensures reputation even if an air disaster occurs [10, 11].

The reputational crisis is directly related to the halo effect in corporate ability [12]. Well-reputed firms tend to have higher probabilities of receiving positive post-crisis reviews [13]. However, when air fatal accidents occur, the safety of the airline's reputation should be the core of crisis management [8], since customers have higher expectations from well-established brands in terms of liability [14].

1.2 Reputational Crisis Effect on Online Customers' Behavior

When disasters occur, giving rise to static and predictable steps, without taking into consideration the ever-changing environment of e-commerce, is the first step to failure. Just in seconds after the crash, the information is going to make the rounds on internet news and social media sites with incredible speed [15]. In essence, it instantly sends shock waves across the globe, which means that reputational damage occurs in seconds.

Nowadays, there are plenty of sites that show flight tracks, origin and destination details, and aircraft type, among others, in real-time. When an emergency notification broadcasts, those sites will tweet an alert to millions of followers. As a consequence, the demand for air travel is likely to fall [16]. Nevertheless, because air crashes are mercifully rare, compared to the number of people flying each year, no air passengers' choice model has been developed [17].

However, the need for air travel will remain at high levels. Online users willing to travel, either on a business trip or for leisure, will predominantly have to overcome their risk perceptions, while at the same time will demand more information. This is a cognitive process to minimize their risk perceptions on air travel safety.

1.3 Reputational Crisis and Big Data

Crisis responsibility, crisis history, and prior relational reputation are the key factors that influence potential damage to existing reputations [18]. All three factors allow crisis managers to understand customers' perceptions and reactions to crisis. Online users tend to adopt supply chain reputation as an incentive that greatly influences their purchasing behavior and attitude [19].

Grundy and Moxon [2] have researched the effectiveness of British Airways' crisis management on brand protection. They have developed a quantitative brand metric that measures customers' perceptions on a daily basis. As support to their research, the present paper examines the impact of reputational crisis after airline disasters and the effects on online users' behavior, based on big data.

Big data offers great opportunities to crisis managers due to its efficiency in analyzing online users' daily attitudes and interactions with a brand [20]. Lately, Web analytics attract the attention of academia and have been the point of numerous research on the supply chain [21–24]. Web analytics approaches enable companies to leverage desktop and mobile usage metrics, such as bounce rate, time on site, traffic rank, unique visitors, and pages per visit, among others, to optimize their digital marketing strategy, acknowledge online trends, and improve digital branding [25].

2 Materials and Methods

2.1 Problem Formulation and Research Hypotheses

Lasting success is largely a delusion, if taking into consideration the competitiveness of the e-marketplace. Especially when crisis occur, there is no formula that can be easily applied as a plug-and-play solution that ensures reputation recovery. In fact, it is evident that airlines crashes will also affect the performance of non-crash airlines in large-scale disasters [26].

The data collected from the airlines' websites will give valuable feedback to:

1. Airlines companies to easily understand the level of reputational threat based on online users' behavior, after an air disaster.
2. Crisis managers to respond instantly and develop marketing strategies for the improvement of the reputational risk, in favor of brand optimization.

Two research questions were established to support the objective of his study:

Hypothesis 1 (H1) *User engagement KPIs and traffic rank will be significantly different between the "pre-crash" and the "immediate reputation crisis" period.*

An airline disaster, especially when the dimension of passengers' death is involved, is expected to significantly affect people's trust toward an airline's brand name and thus increase their risk perception, in the light of lacking safety factors [27]. The increase in risk perception is expected to disrupt online users' behavior within the supply chain [28]. Investigating this research question will help aviation organizations to clarify the immediate effect of a serious accident on users' online behavior following a plane crash and provides valuable information so as airlines' companies to immediate act based on the guidelines of their risk mitigation strategy.

Hypothesis 2 (H2) *During the "immediate reputation crisis" period, user engagement KPIs' and traffic rank fluctuations will be significantly affected by the time that has passed since the plane crash.*

Destabilizing customers' trust is expected to recede as time passes and the crisis is de-escalating. However, the time needed for an aviation-safety crisis to "fade out" is not well established by research. It is expected that the level of customers' trust is

Fig. 1 Timeline presenting the periods studied in this article

strongly related with airlines' branding prior to the air tragedy. This research question will help aviation corporations to gain a better understanding on customers' online behavior right after an airplane crash (immediate reputation crisis period). The acknowledgment of the extent and depth of customers' search will enable companies to better allocate resources on marketing activities in an attempt to eliminate their reputational threats and optimize their digital branding.

2.2 Sample Selection and Data Collection

Our sample was collected by taking into consideration five airlines that had one fatal plane disaster between 2019 and 2022. All selected airlines heavily relied on Web-based platforms for sales revenue and customers' interaction. Two 30-day periods were selected to serve the need of the current research (Fig. 1). Day "1" is defined as the date of the air crash. The period of 30 days after the crash are labeled as the "immediate reputation crisis" period, and the one of 30 days before the crash is defined as the "pre-crash" period.

Every KPI value of the sample was collected using a Web analytics platform [29] and big data mining techniques to accomplish more efficient data processing and statistical analysis and generate a representative KPI-related variable based on the average value. This method (Fig. 2) was used to analyze all the KPIs in question.

3 Results

3.1 Statistical Analysis

Independent samples t-tests were used to determine statistically significant changes between the "pre-crash" and the "immediate reputational crisis" periods. Only one of the KPIs variables (Traffic Rank) indicated a significant difference between the two datasets (Table 1). Statistically significant results are indicated with "*".

We used a Pearson correlation coefficient (PCC–r) statistical analysis on our datasets to see if there were any apparent linear correlations between the pairs of variables. Several statistically significant relationships were discovered, with a

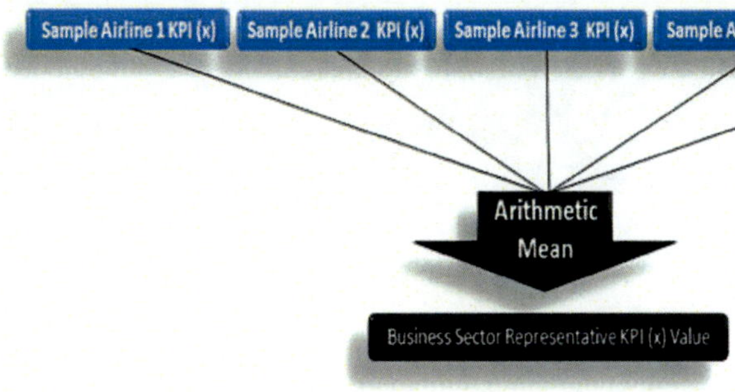

Fig. 2 Big data processing methodology used in the presented study

Table 1 Independent samples t-test results

Variables	Levene's test for equality of variances sig.	t-Test for equality of means		
		Significance (2-tailed)	Mean difference	Standard error difference
Bounce rate	0.1830	0.1160	−0.0172	0.0107
Pages viewed per user	0.2430	0.5190	0.19417	0.2994
Time on site	0.9120	0.2580	17.2500	15.1802
Traffic rank	0.0000	*0.0000	2455.5583	182.9207
Reach %	0.0700	0.7020	0.0002	0.0004

Table 2 Pearson correlation coefficient (PCC–r) statistical analysis results

Variables			Pearson correlation	Significance (2-tailed)
Reach %	and	traffic rank	0.496	0.01
Reach %	and	time on site	−0.599	0.01
Reach %	and	bounce rate	0.732	0.01
Reach %	and	days after crash	0.456	0.05
Traffic rank	and	days after crash	−0.972	0.01
Bounce rate	and	time on site	−0.582	0.05

number of them indicating a very strong linear link (Table 2). Six statistically significant correlations between the variables were discovered, using PCC analysis, four at the 0.01 level and two at the 0.05 level (2-tailed).

3.2 Development of a Diagnostic Explanatory Model

After the statistical analysis of our data, we add a level of analysis by developing a fuzzy cognitive mapping (FCM) to depict all the cause-effect interactions between the variables in this study (Fig. 3). Fuzzy cognitive mappings (FCMs) are fuzzy-

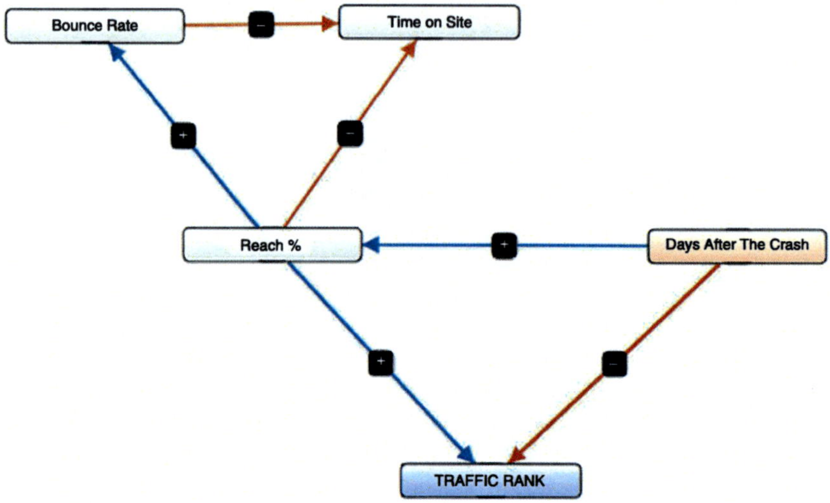

Fig. 3 Fuzzy cognitive mapping of correlated variables

graph structures that exhibit causal reasoning. FCM is considered a "soft computing" system modeling technique that combines fuzzy logic and neural networks, being in this way, the closest simulation of human thinking. FCMs can have varying degrees of causation among distinct causal players due to their vague nature [30].

In order to further investigate the interaction between the studied variables, we developed three hypothetical scenarios regarding three different stages of the crisis. For scenario 1 (Fig. 4), we assumed a level of 0.6 of passed time (18 day scenario). For scenario 2 (Fig. 5), we introduced to our model a level of 0.8 of passed time (24 day scenario), and for scenario 3 (Fig. 6), we simulated that all 30 days have passed (30 days scenario).

4 Discussion

This study aims at evaluating the immediate impact of a reputational crisis triggered by a fatal airline accident on customers' online behavior, through monitoring Web domain user engagement metrics, and the level of impact on airlines' digital branding. Results suggest that during the first 30 days following the crash, the overall users' online behavior resulted in a significantly improvement of airlines' digital Web domain ranking, partially confirming H1. However, their engagement with the Web page itself was not significantly altered compared to the period before the accident, contradicting H1 regarding the user engagement sub-hypothesis.

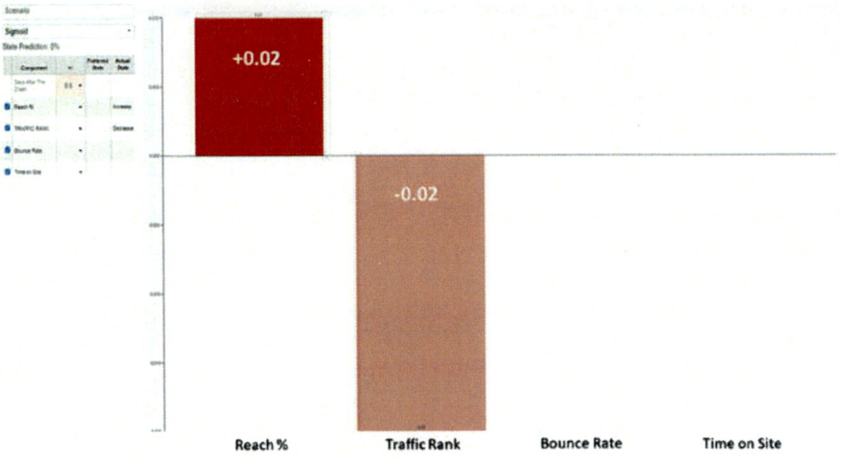

Fig. 4 Scenario 1 – 18 days after the crash

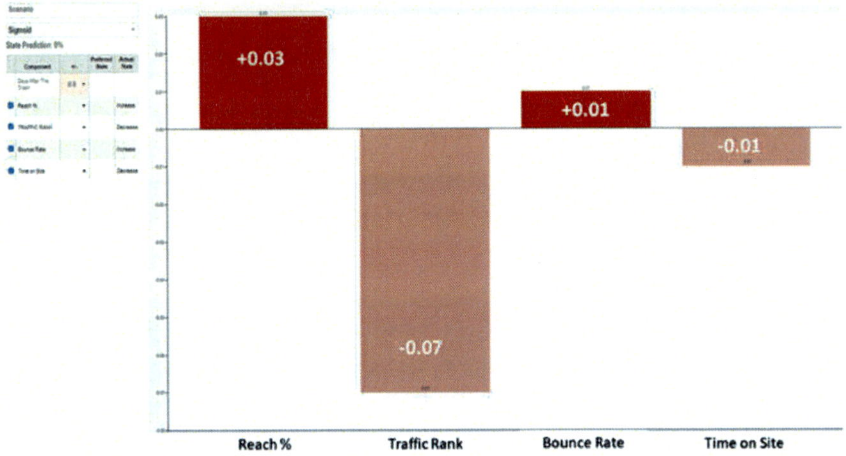

Fig. 5 Scenario 2 – 24 days after the crash

Although user engagement metrics were not substantially different, there is a tendency that the number of users who land on the airline's-under-crisis Web domain (reach percentage KPI) to increase as days from the accident passes by. This predisposition is strongly imprinted on the traffic rank. The reach-percentage KPI correlation (−0.972) further suggests that after an airplane crash, the increase in digital branding is related to the amount of the days following the crash, with an almost linear relationship [31]. These results confirm H2 in terms of traffic rank.

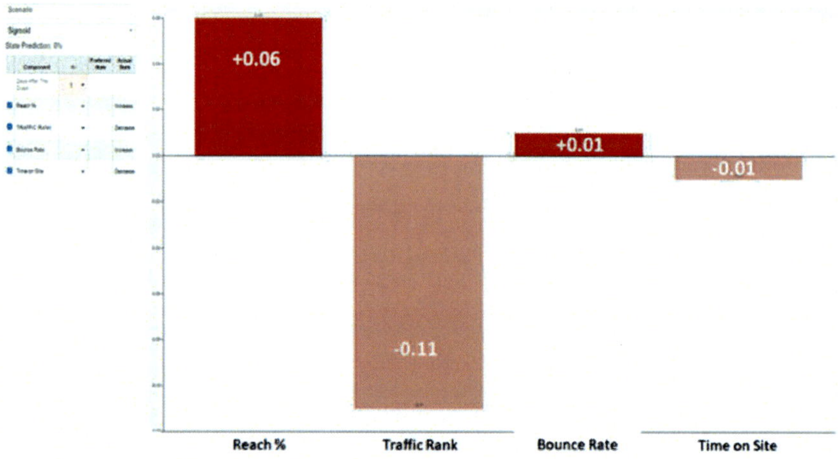

Fig. 6 Scenario 3 – 30 days after the crash

FCM scenarios results also support the findings of the study, clearly suggesting the close correlation between digital branding and time after the disaster strongly introducing that a disaster crisis provides the airline operator with a significant platform that could be very useful for reputational damage control strategies [32].

5 Conclusion

In the market place, corporate reputation is a reflection of a company's offering, communication and action [18]. Especially when an aviation disaster occurs, the company management should have procedures in place to help them respond to the crisis quickly and effectively. By doing so, airline managers can protect the company's reputation and thus avoid any negative consequences in terms of business performance and operations stability.

One important outcome of the current study is that right after an air crash (immediate reputation crisis period), digital ranking is significantly increased. Since the airline company is rapidly growing its online footprint, managers should release on the website a damage control development content, based on risk mitigation, in an attempt to reverse the existing negative reputational situation.

With any brand crisis, there is always an opportunity to effectively stage a remarkable turnaround. Another significant outcome of the current research is that as the days are passing by and the reputational threat increases, new visitors are landing on the airline-under-crisis websites. This is an excellent opportunity for manages to put into action the crisis response strategy so as to change online users' current beliefs and expectations, built trust relationships, and protect the company's brand against negative reactions.

Overall, the present paper clearly demonstrated that after an air tragedy, the company should immediately allocate resources on effective communication strategies, in favor to reputational recovery and risk mitigation strategies, since the level of online visibility is significantly increased.

References

1. Rosenzweig, P.: Misunderstanding the nature of company performance: the halo effect and other business delusions. Calif. Manag. Rev. **49**, 6–20 (2007). https://doi.org/10.2307/41166403
2. Grundy, M., Moxon, R.: The effectiveness of airline crisis management on brand protection: a case study of British Airways. J. Air Transp. Manag. **28**, 55–61 (2013). https://doi.org/10.1016/j.jairtraman.2012.12.011
3. Akyildirim, E., Corbet, S., O'Connell, J.F., Sensoy, A.: The influence of aviation disasters on engine manufacturers: an analysis of financial and reputational contagion risks. Int. Rev. Financ. Anal. **74**, 101630 (2021). https://doi.org/10.1016/j.irfa.2020.101630
4. IATA: IATA releases 2020 safety report, details airline safety performance. https://www.iata.org/en/pressroom/pr/2021-03-25-01/. Last accessed 11 May 2022
5. Hätty, H., Hollmeier, S.: Airline strategy in the 2001/2002 crisis—the Lufthansa example. J. Air Transp. Manag. **9**, 51–55 (2003). https://doi.org/10.1016/S0969-6997(02)00064-9
6. Suk, M., Kim, W.: COVID-19 and the airline industry: crisis management and resilience. Tour. Rev. **76**, 984–998 (2021). https://doi.org/10.1108/TR-07-2020-0348
7. Medina-Muñoz, D.R., Medina-Muñoz, R.D., Suárez-Cabrera, M.Á.: Determining important attributes for assessing the attractiveness of airlines. J. Air Transp. Manag. **70**, 45–56 (2018). https://doi.org/10.1016/j.jairtraman.2018.01.002
8. Chang, Y.-H., Yeh, C.-H., Wu, P.-S.: Evaluating airline crisis management performance: the cases of flights GE222 and GE235 crash accidents. J. Air Transp. Manag. **70**, 62–72 (2018). https://doi.org/10.1016/j.jairtraman.2018.04.017
9. Seo, E.-J., Park, J.-W.: A study on the impact of airline corporate reputation on brand loyalty. Int. Bus. Res. **10**, 59 (2016). https://doi.org/10.5539/ibr.v10n1p59
10. Siomkos, G.J.: Managing airline disasters. J. Air Transp. Manag. **6**, 101–108 (2000). https://doi.org/10.1016/S0969-6997(99)00030-7
11. Boamah, M.: Analysing crisis communication strategies of airline companies in United States: a case study of Southwest Airline 2016 power outage crisis. Stud. Media Commun. **7**, 7 (2019). https://doi.org/10.11114/smc.v7i1.4190
12. Sohn, Y.J., Lariscy, R.W.: Understanding reputational crisis: definition, properties, and consequences. J. Public Relat. Res. **26**, 23–43 (2014). https://doi.org/10.1080/1062726X.2013.795865
13. Kiambi, D.M., Shafer, A.: Corporate crisis communication: examining the interplay of reputation and crisis response strategies. Mass Commun. Soc. **19**, 127–148 (2016). https://doi.org/10.1080/15205436.2015.1066013
14. Helm, S., Tolsdorf, J.: How does corporate reputation affect customer loyalty in a corporate crisis?: Corporate reputation, customer loyalty and crises. J. Conting. Crisis Manag. **21**, 144–152 (2013). https://doi.org/10.1111/1468-5973.12020
15. Rodriguez-Toubes, D., Dominguez-Lopez, Y.: Social media in crisis communication: Germanwings flight 4U9525. In: Freire, F.C., Rúas Araújo, X., Martínez Fernández, V.A., García, X.L. (eds.) Media and Metamedia Management, pp. 243–249. Springer International Publishing, Cham (2017). https://doi.org/10.1007/978-3-319-46068-0_31
16. Liu, H., Zeng, J.: Airline passenger fatality and the demand for air travel. Appl. Econ. **39**, 1773–1781 (2007). https://doi.org/10.1080/00036840701371822

17. Molin, E., Blangé, J., Cats, O., Chorus, C.: Willingness to pay for safety improvements in passenger air travel. J. Air Transp. Manag. **62**, 165–175 (2017). https://doi.org/10.1016/j.jairtraman.2017.04.002
18. Coombs, W.T.: Protecting organization reputations during a crisis: the development and application of situational crisis communication theory. Corp. Reput. Rev. **10**, 163–176 (2007). https://doi.org/10.1057/palgrave.crr.1550049
19. Lemke, F., Petersen, H.L.: Teaching reputational risk management in the supply chain. Supply Chain Manag. **18**, 413–429 (2013). https://doi.org/10.1108/SCM-06-2012-0222
20. Sakas, D.P., Giannakopoulos, N.T., Reklitis, D.P., Dasaklis, T.K.: The effects of cryptocurrency trading websites on airlines' advertisement campaigns. J. Theor. Appl. Electron. Commer. Res. **16**, 3099–3119 (2021). https://doi.org/10.3390/jtaer16070169
21. Sakas, D.P., Reklitis, D.P.: Predictive model for estimating the impact of technical issues on consumers interaction in agri-logistics websites. In: Bochtis, D.D., Pearson, S., Lampridi, M., Marinoudi, V., Pardalos, P.M. (eds.) Information and Communication Technologies for Agriculture—Theme IV: Actions, pp. 269–283. Springer International Publishing, Cham (2021). https://doi.org/10.1007/978-3-030-84156-0_14
22. Sakas, D.P., Giannakopoulos, N.T.: Big data contribution in desktop and mobile devices comparison, regarding airlines' digital brand name effect. Big Data Cogn. Comput. **5**, 48 (2021). https://doi.org/10.3390/bdcc5040048
23. Gunasekaran, A., Papadopoulos, T., Dubey, R., Wamba, S.F., Childe, S.J., Hazen, B., Akter, S.: Big data and predictive analytics for supply chain and organizational performance. J. Bus. Res. **70**, 308–317 (2017). https://doi.org/10.1016/j.jbusres.2016.08.004
24. Nguyen, T., Zhou, L., Spiegler, V., Ieromonachou, P., Lin, Y.: Big data analytics in supply chain management: a state-of-the-art literature review. Comput. Oper. Res. **98**, 254–264 (2018). https://doi.org/10.1016/j.cor.2017.07.004
25. Sakas, D.P., Kamperos, I.D.G., Reklitis, D.P., Giannakopoulos, N.T., Nasiopoulos, D.K., Terzi, M.C., Kanellos, N.: The effectiveness of centralized payment network advertisements on digital branding during the COVID-19 crisis. Sustainability. **14**, 3616 (2022). https://doi.org/10.3390/su14063616
26. Ho, J.C., Qiu, M., Tang, X.: Do airlines always suffer from crashes? Econ. Lett. **118**, 113–117 (2013). https://doi.org/10.1016/j.econlet.2012.09.031
27. Fischhoff, B., Slovic, P., Lichtenstein, S., Read, S., Combs, B.: How safe is safe enough? A psychometric study of attitudes toward technological risks and benefits. Policy Sci. **9**, 127–152 (1978). https://doi.org/10.1007/BF00143739
28. Sakas, D.P., Kamperos, I.D.G., Reklitis, P.: Estimating risk perception effects on courier companies' online customer behavior during a crisis, using crowdsourced data. Sustainability. **13**, 12725 (2021). https://doi.org/10.3390/su132212725
29. Semrush: Online marketing can be easy. https://www.semrush.com/. Last accessed 21 Mar 2022
30. Kosko, B.: Fuzzy cognitive maps. Int. J. Man Mach. Stud. **24**, 65–75 (1986). https://doi.org/10.1016/S0020-7373(86)80040-2
31. Nasiopoulos, D.K.: Email and E-branding applications in Logistics companies as a field of management information systems. In: Business Intelligence and Modelling, vol. 6, pp. 255–265. Springer, Cham (2021)
32. Nasiopoulos, D.K.: Web search applications in Logistics companies in the context of evolution of management information systems. In: Business Intelligence and Modelling, vol. 6, pp. 241–254. Springer, Cham (2021)

Cruise Industry Crisis Risk Management and Recovery Strategies Utilizing Crowdsourcing Data

Damianos P. Sakas, Marina C. Terzi, Ioannis Dimitrios G. Kamperos, and Athanasios Kriemadis

1 Literature Review

1.1 Risk Perceptions Within the Cruise Industry

Cruise tourism was considered a growing segment of global tourism [1] with 30 million passengers in 2019 [2]. However, the pandemic has been catastrophic for the industry. Based on the CLIA (Cruise Lines International Association) 2020 report, the financial loss of the cruise sector accounted for $50 billion, far away from the remarkable amount of $150 billion that the industry was annually generating until then [3], leading to revenue shrinkage up to 91% compared to pre-pandemic levels [4].

In February 2020, an outbreak was reported in the British-registered luxury cruise ship "Princess Diamond", resulting in 619 out of 3700 infected passengers and 6 deaths [5]. In March 2020, the CDC (Center for Disease Control and Prevention) issued a no-sail order, preventing cruise ships from operating [6]. These are only two of the blows that the cruise sector has suffered, accelerating the negative publicity of being perfect disease hubs due to the confined environment onboard [7]. This negative sentiment still prevails in the cruise industry, as both people who have experienced a cruise and those who haven't are negative about using cruise services in the future [8].

D. P. Sakas · M. C. Terzi · I. D. G. Kamperos (✉)
BICTEVAC LABORATORY Business Information and Communication Technologies in Value Chains Laboratory, Department of Agribusiness and Supply Chain Management, School of Applied Economics and Social Sciences, Agricultural University of Athens, Athina, Greece
e-mail: kamperosdigese@aua.gr

A. Kriemadis
Department of Management Science and Technology, School of Economics and Technology, University of Peloponnese, Tripoli, Greece

D. P. Sakas et al. (eds.), *Computational and Strategic Business Modelling*, Springer Proceedings in Business and Economics,
https://doi.org/10.1007/978-3-031-41371-1_5

As tourism continues to face challenges regarding COVID-19, the cruise community has to remove their pandemic reputation as floating disease carriers to be seen as a safe, sustainable, and environmentally friendly business model. Since risk perception greatly influences customers' purchasing intention and retention [9], the cruise market's challenges for vitality, viability, and sustainability are of the essence in order to avoid total wipeouts. But as cruise companies look to a sustainable future, it is clear that much more work needs to be done on post-crisis recovery strategies.

1.2 Recovery Strategies in the Post-Crisis Era

Even if recovery strategies involve the dimension of risk, when applied properly, it can achieve operational stability and business revitalization after crises [10]. During challenging economic situations, service development and corporate social responsibility are among the five categories of investments to overcome obstacles and maintain a sustainable competitive advantage [11]. To this end, innovation is imperative not only to a company's vitality but to further ensure performance recovery, as an antidote to mitigate customers' risk perceptions [12].

Developing a turnaround strategy signals an upturn period, an organization's reshaping to push innovation and service excellence [13]. Turnaround strategies are concentrated on revenue so as to achieve a company's reposition in the long run, by introducing new pricing models or developing new products [14]. An interesting research, within the cruise sector, has been accomplished by Tianyu Pan et al. [15] regarding the perception of financial-affected and non-affected groups, highlighting the effectiveness of perceived crisis management on trust manipulation. As such, financial strategies, such as huge discounts, significantly affect customers' purchasing satisfaction, behavior, and intention for cruise services in the post-pandemic era [16]. Financial bonding strategies significantly influence customer loyalty and satisfaction during periods of uncertainty [17].

The cruise sector needs to invest in recovery strategies based not only on new service developments or financial activities but also on current trends, arising from the pandemic. As people are becoming environmentally conscious, travelers tend to choose companies with environmental commitments [18]. As the cruise industry takes shape in the post-pandemic era, the development of responsible environmental practices is a must-do, as an integrated part of recovery strategies [19]. However, the main challenge during or after the implementation of turnaround strategies is the lack of communication, mainly when feedback channels are the pain point [20].

1.3 Crowdsourcing Data Within the Cruise Sector

Given the devastating nature of the pandemic to the cruise sector, it is more than important to understand COVID-19 impact on customers' perceptions and feelings, regarding cruise services. Customers' behaviors tend to change following crises and disasters, and so do online users' behavior [21]. Holland et al. [8] identify a significant research gap regarding the extent to which the cruise community understands customers' risk perception and its effect on their intention to cruise. Franklyn-Green et al. [22] concluded that the sustainability of the cruise industry relies on the reexamination of customers' interactions. As consumers get more critical about the impact of the pandemic, tracking and analyzing users' online behavior has emerged as a key trend within the sector, as an incentive for the development of targeted recovery strategies. Utilizing business data empowers cruise companies to gain more information on users' online behavior and make smarter decisions regarding risk mitigation, performance recovery, and brand optimization.

If a cruise company aims at implementing a successful recovery strategy, it has to innovatively orchestrate processes and technology to create opportunities and meet market needs. Notably, integrating big data from several sources, such as the website or social media platforms, offers real-time process performance that enables companies to be proactive. Motivated by Muritala et al. study [23] on customers' online perceptions of cruises, based on Twitter, the present research investigates users' online behavior, based on Web analytics. Specifically, the authors evaluate customers' risk perceptions for cruises, shaped by crisis escalation and track their online attitudes, based on behavioral factors, in favor of business recovery and brand optimization. It is a novel topic that will elucidate the purchasing behavior of online users and provide a solution to the pain point of turnaround strategies. Cruise companies can overcome the roadblocks that stem from a lack of communication by monitoring users' behavior, based on big data. The identification and examination of crowdsourced key performance indicators (KPIs) are crucial for the efficacy of users' decision-making process [24], mitigation of customers' risk perception [25], digital marketing enhancement [26], and digital branding optimization [27].

2 Materials and Methods

2.1 Problem Formulation and Research Hypotheses

When data is used analytically to enhance and customize client online experiences and predict future online activities, it is valuable and profitable in the digital world. As an imminent result, the transportation and touristic industry have a significant challenge in transitioning from a corrective to a predictive approach. Adopting such a plan will help the cruise industry thrive economically, strategically, and operationally by reducing customers' risk perceptions [25].

The cruises industry could gain a considerable advantage in gaining vital knowledge regarding consumers' behavior to make effective managing decisions by reviewing Web analytics. Cruise operators will secure their organizations' viability and financial growth during the era of "free data" by implementing a novel approach that allows the utilization of the available crowdsourced key performance indicators

(KPIs) like page views per user, time on site, reach percentage, and bounce rate, which is the metrics investigated in this article, as well as the number of globally confirmed daily new COVID-19 infections [28].

Web metrics can help to measure the efficacy of marketing initiatives, website architecture, and content characteristics by underlining the customer – Web domain interaction mechanism. The ability for businesses to foresee client behavior during a crisis gives them a significant advantage in risk management strategy.

The main objective of this study is to clarify the fluctuations of consumers' online behavior during a pandemic crisis through crowdsourced Web metrics. This technique facilitates marketers with critical insights into Internet users' activity as well as information on steps that could lead to higher conversion rates and improved viability.

In order to promote the study's goals, two research questions were developed:

Hypothesis 1 (H1) *Digital branding of the cruise operators' Web domains as well as user engagement metrics of the "before-crisis" period will be significantly different from the "crisis" period.*

Research findings strongly suggest that customers have adapted their behavior to the new parameters induced by the COVID-19 pandemic [25]. Digital branding of the supply chain organizations responds differently, depending on the business sector and the related low-demand risk generated by the nature of the provided services. Knowledge of the way that these parameters behave during a crisis will facilitate cruise operators to follow effective risk management tactics beforehand and build strong brand recovery strategies during and after the crisis.

Hypothesis 2 (H2) *Fluctuation of global COVID-19 confirmed new infection cases will significantly affect cruises operators' visibility through traffic ranking during the "crisis" period.*

Crisis escalation is a very important parameter that affects customers' and organizations' risk perception regarding the crisis-related hazard. The most common metric used to estimate COVID-19 crisis escalation is daily confirmed cases, either globally or regionally. An increase in that metric is expected to affect users' engagement KPIs by triggering the dread-risk perception cognitive mechanism suggested by Fischhoff et al. [29]. The extent of application of the psychometric paradigm in the cruise line industry can help cruise operators forecast customers' behavior and plan efficient brand recovery strategies during a pandemic crisis.

2.2 Sample Selection and Data Collection

To investigate our hypotheses, we formed a typical sample of the global cruise industry by unifying the data of five companies representing 59.7% of the passengers served globally for 2018 [30]. After forming our sample, we extracted passively crowdsourced big data regarding the studied KPIs from a Web analytics platform [31].

The final dataset was created by estimating the monthly arithmetic mean for every parameter for two distinctive periods. The 12-month period before the eruption of the COVID-19 pandemic (March 1, 2019 to February 19, 2020) is labeled the "before-crisis" period and the 12-month period after the eruption of the COVID-19 pandemic (March 1, 2020 to February 28, 2021) labeled "crisis" period. Starting date of the crisis was determined based on the dates that most countries had started implementing self-isolating and traveling restrictions.

We utilized data from the STATISTA database [28] regarding the globally daily confirmed new infection cases of the COVID-19 virus.

For greater precision and consistency with the dependent variables, these data were processed into an average daily value for each month, establishing the study's independent variable (COVID cases).

3 Results

3.1 Statistical Analysis

Between the two datasets, independent samples t-tests were used to identify statistically significant differences between the COVID-free and COVID-crisis periods. Four of the five variables analyzed showed statistically significant variations (Table 1).

We conducted a Pearson correlation coefficient (PCC–r) statistical analysis on our datasets to elucidate any apparent linear relationships between pairs of variables over the COVID-crisis period. There were three statistically significant associations found (Table 2), one at the 0.01 level of significance and two at the 0.05 level (two-tailed).

Table 1 Results of the independent samples t-tests

		t-Test for equality of means		
Variables	Levene's test for equality of variances significance	Significance (2-tailed)	Mean difference	Standard error difference
Reach %	0.3930	*0.000	0.0034	0.0002
Pageviews per user	0.0060	*0.047	0.5208	0.2381
Bounce rate	0.2060	0.417	−0.0067	0.0080
Time on site	0.0620	*0.033	47.4709	20.9112
Traffic rank	0.0000	*0.0000	−20334.0784	3179.4714

The symbol "*" indicates a statistically significant result

Table 2 Results of the Pearson correlation coefficient (PCC–r) statistical analysis

Variables			Pearson correlation	Significance (2-tailed)
Page views per user	and	time on site	0.853	0.010
COVID cases	and	traffic rank	0.637	0.050
Reach %	and	traffic rank	−0.661	0.050

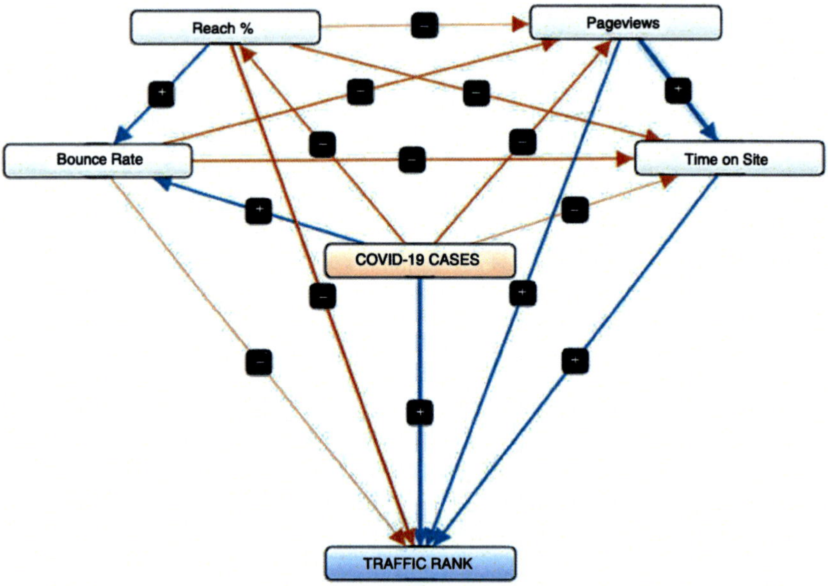

Fig. 1 Fuzzy cognitive mapping (FCM) of the variables investigated in the present research

3.2 Development of a Diagnostic Explanatory Model

For the next level of our analysis, we developed a fuzzy cognitive mapping (FCM) to visualize all of the cause-effect relationships between the COVID-19 and KPI-related factors investigated in this study (Fig. 1). FCMs (fuzzy cognitive mappings) are graph structures that are used to illustrate causal reasoning using connecting lines between variables, with different colors and widths representing different kinds and degrees of causal interaction. It incorporates in a "soft computing" technique, neural networks, and fuzzy logic in order to simulate human thinking.

An FCM scenario was created in order to provide an additional level of analysis to our data. We created three scenarios simulating three different stages of crisis escalation. For scenario 1 (Fig. 2), we introduced a level of 0.6 of maximal pandemic escalation. For scenario 2 (Fig. 3), we assumed a level of 0.8 of maximum pandemic escalation, and for scenario 3 (Fig. 4), we simulated a high-level crisis at a level 0.9.

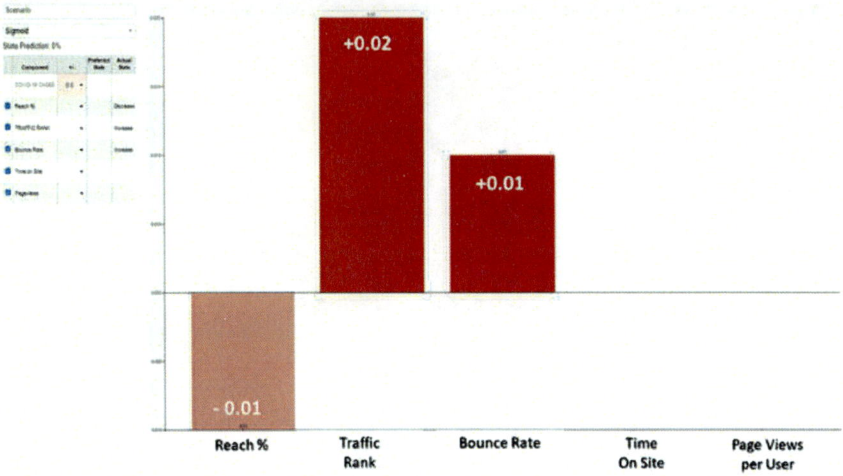

Fig. 2 FCM scenario 1 results

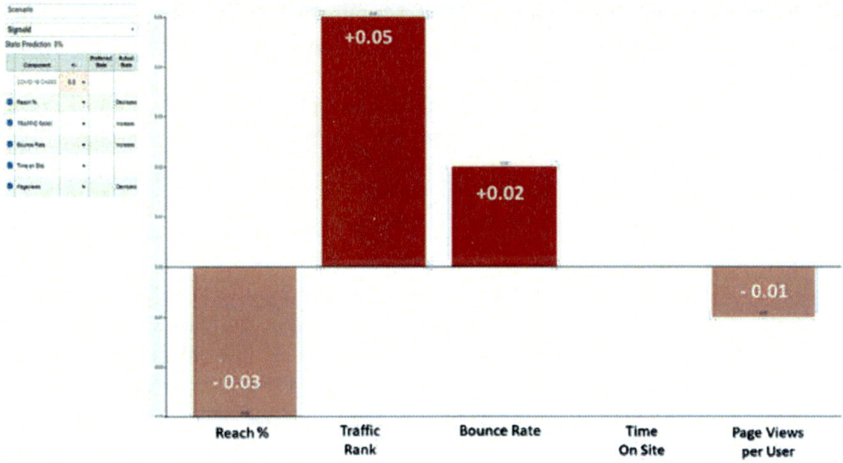

Fig. 3 FCM scenario 2 results

4 Discussion

This research aims to examine the influence of a pandemic crisis on cruise consumer online behavior by tracking Web domain user engagement key performance indicators. Results of our research clearly suggest that online user engagement significantly declined for almost all of our studied variables (except bounce rate). This fluctuation is reflected in traffic ranking KPI, which strongly indicates a significantly degraded

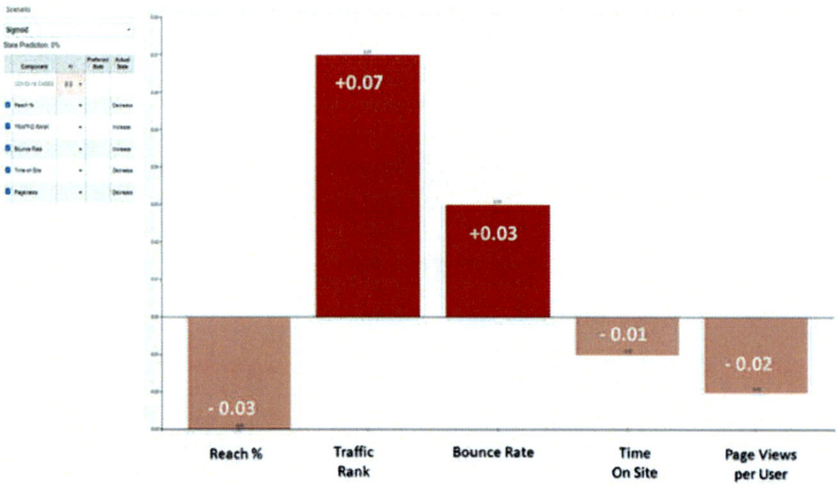

Fig. 4 FCM scenario 3 results

digital branding position during the crisis period, verifying H1. The results of our study confirm the findings of Holland et al. [8] as well as Le and Arcodia [9].

The escalation of the crisis expressed through worldwide confirmed new COVID-19 infections is suggested to significantly correlate with the fluctuation of cruise operators' level of digital branding, confirming H2 but is not correlated to a significant degree with other user engagement metrics. FCM scenario results support these findings and confirm H2 as well.

Authors suggest that in an effort to gain customers' support after travel restrictions are lifted, cruise companies could invest in marketing campaigns that highlight new product development, price reduction, and discount offerings. In addition, the recovery strategy should further include elements that mitigate customers' risk perceptions and address their concerns. To this end, the enhancement of the website's content or development of a blog providing information on risk perceptions and current environmental trends will cover concerns and entice online users back on board.

5 Conclusion

As the cruise industry swifts from response to recovery, the key for the industry's players is to make strategic decisions on building resilience and resetting for a sustainable future. It is therefore essential to initially develop a viable recovery strategy and create an adaptive tool to adjust as the "new normal" emerges. Following this, cruise companies should reset or renew their operations, through service innovations and the development of new pricing models, so as to rescale, reinvent, reduce, or retire specific digital marketing activities for a new reality [32]. As such, finding new ways and ecosystems to reach potential customers and maintain existing ones becomes essential.

Digital marketing optimization activities integrate Web analytics to better understand how recovery strategies influence end sales [33]. Through data-driven monitoring and analysis, cruise companies can evaluate the effectiveness of their recovery strategy, in the direction of altering business operations, in favor of brand optimization. As the future is still largely defined by uncertainty, cruise companies should drive resources to the development of recovery strategies, by using Web analytics as the cornerstone for merging the gap between customers' communication efforts and feedback channels. Therefore, other scientists with similar cases are studying similar problems, using modeling and simulation [34–36].

References

1. Papathanassis, A.: Cruise tourism management: state of the art. Tour. Rev. **72**, 104–119 (2017). https://doi.org/10.1108/TR-01-2017-0003
2. CLIA: 2019 Cruise trends & industry outlook. https://cruising.org/-/media/research-updates/research/clia-2019-state-of-the-industry-presentation-(1).ashx (2019)
3. CLIA: FAQs: the cruise community and COVID-19. https://cruising.org/-/media/Facts-and-Resources/Cruise-Industry-COVID-19-FAQs_August-13-2020 (2020)
4. STATISTA: Percentage change in revenue of leading cruise companies worldwide during the coronavirus (COVID-19) pandemic in 2020 and 2021. https://www.statista.com/statistics/1258281/growth-rate-revenue-leading-cruise-companies-worldwide/. Last accessed 15 Apr 2022
5. Rocklöv, J., Sjödin, H., Wilder-Smith, A.: COVID-19 outbreak on the Diamond Princess cruise ship: estimating the epidemic potential and effectiveness of public health countermeasures. J. Travel Med. **27**, taaa030 (2020). https://doi.org/10.1093/jtm/taaa030
6. U.S. Department of Health and Human Services: Order Under Sections 361 & 365 of the Public Health Service Act and 42 Code (Interstae) and Part 71 (Foreign): No Sail Order and Other Measures Related to Operations. https://www.cdc.gov/quarantine/pdf/signed-manifest-order_031520.pdf (2020)
7. Sahu, A., Naqvi, W.M.: Floating countries and corona pandemic: impact of COVID-19 on stranded Cruise ships. Int. J. Res. Pharmaceut. Sci. **11**, 219–223 (2020). https://doi.org/10.26452/ijrps.v11iSPL1.2702
8. Holland, J., Mazzarol, T., Soutar, G.N., Tapsall, S., Elliott, W.A.: Cruising through a pandemic: the impact of COVID-19 on intentions to cruise. Transp. Res. Interdiscip. Perspect. **9**, 100328 (2021). https://doi.org/10.1016/j.trip.2021.100328
9. Le, T.H., Arcodia, C.: Risk perceptions on cruise ships among young people: concepts, approaches and directions. Int. J. Hosp. Manag. **69**, 102–112 (2018). https://doi.org/10.1016/j.ijhm.2017.09.016
10. Carlsen, J.C., Liburd, J.J.: Developing a research agenda for tourism crisis management, market recovery and communications. J. Travel Tour. Mark. **23**, 265–276 (2008). https://doi.org/10.1300/J073v23n02_20
11. Mann, M., Byun, S.-E.: To retrench or invest? Turnaround strategies during a recessionary time. J. Bus. Res. **80**, 24–34 (2017). https://doi.org/10.1016/j.jbusres.2017.06.018
12. Tikici, M., Omay, E., Derin, N., Seçkin, S.N., Cüreoğlu, M.: Operating turnaround strategies during crisis periods: a research on manufacturing firms. Procedia Soc. Behav. Sci. **24**, 49–60 (2011). https://doi.org/10.1016/j.sbspro.2011.09.046

13. Bhattacharyya, S.S., Malik, A.: Development of an integrated canvas on turnaround strategy. Int. J. Organ. Anal. **28**, 523–554 (2020). https://doi.org/10.1108/IJOA-02-2019-1646
14. Li-Ying, J., Nell, P.C.: Navigating opportunities for innovation and entrepreneurship under COVID-19. Calif. Manag. Rev. (2021). https://cmr.berkeley.edu/2020/06/innovation-entrepreneurship/
15. Pan, T., Shu, F., Kitterlin-Lynch, M., Beckman, E.: Perceptions of cruise travel during the COVID-19 pandemic: market recovery strategies for cruise businesses in North America. Tour. Manag. **85**, 104275 (2021). https://doi.org/10.1016/j.tourman.2020.104275
16. Li, Z., Wang, X., Li, X., Yuen, K.F.: Post COVID-19: health crisis management for the cruise industry. Int. J. Disaster Risk Reduct. **71**, 102792 (2022). https://doi.org/10.1016/j.ijdrr.2022.102792
17. Balci, G., Caliskan, A., Yuen, K.F.: Relational bonding strategies, customer satisfaction, and loyalty in the container shipping market. Int. J. Phys. Distrib. Logist. Manag. **49**, 816–838 (2019). https://doi.org/10.1108/IJPDLM-02-2019-0051
18. da Silva, A.L.R.: An overview of the impact of COVID-19 on the cruise industry with considerations for Florida. Transp. Res. Interdiscip. Perspect. **10**, 100391 (2021). https://doi.org/10.1016/j.trip.2021.100391
19. Han, H., Eom, T., Chung, H., Lee, S., Ryu, H., Kim, W.: Passenger repurchase behaviours in the green cruise line context: exploring the role of quality, image, and physical environment. Sustainability. **11**, 1985 (2019). https://doi.org/10.3390/su11071985
20. Buhusayen, B., Seet, P.-S., Coetzer, A.: Turnaround management of airport service providers operating during COVID-19 restrictions. Sustainability. **12**, 10155 (2020). https://doi.org/10.3390/su122310155
21. Mair, J., Ritchie, B.W., Walters, G.: Towards a research agenda for post-disaster and post-crisis recovery strategies for tourist destinations: a narrative review. Curr. Issue Tour. **19**, 1–26 (2016). https://doi.org/10.1080/13683500.2014.932758
22. Franklyn-Green, L.-G.K., Meredith, N., Ajagunna, I.: The Caribbean cruise industry and COVID-19: what planned strategic options exist for recovery for small island countries? Worldw. Hosp. Tour. Themes. **14**, 162–168 (2022). https://doi.org/10.1108/WHATT-12-2021-0151
23. Muritala, B.A., Hernández-Lara, A.-B., Sánchez-Rebull, M.-V., Perera-Lluna, A.: #CoronavirusCruise: impact and implications of the COVID-19 outbreaks on the perception of cruise tourism. Tour. Manag. Perspect. **41**, 100948 (2022). https://doi.org/10.1016/j.tmp.2022.100948
24. Sakas, D.P., Reklitis, D.P.: Predictive model for estimating the impact of technical issues on consumers interaction in agri-logistics websites. In: Bochtis, D.D., Pearson, S., Lampridi, M., Marinoudi, V., Pardalos, P.M. (eds.) Information and Communication Technologies for Agriculture—Theme IV: Actions, pp. 269–283. Springer International Publishing, Cham (2021). https://doi.org/10.1007/978-3-030-84156-0_14
25. Sakas, D.P., Kamperos, I.D.G., Reklitis, P.: Estimating risk perception effects on courier companies' online customer behavior during a crisis, using crowdsourced data. Sustainability. **13**, 12725 (2021). https://doi.org/10.3390/su132212725
26. Sakas, D.P., Giannakopoulos, N.T., Reklitis, D.P., Dasaklis, T.K.: The effects of cryptocurrency trading websites on airlines' advertisement campaigns. J. Theor. Appl. Electron. Commer. Res. **16**, 3099–3119 (2021). https://doi.org/10.3390/jtaer16070169
27. Sakas, D.P., Kamperos, I.D.G., Reklitis, D.P., Giannakopoulos, N.T., Nasiopoulos, D.K., Terzi, M.C., Kanellos, N.: The effectiveness of centralized payment network advertisements on digital branding during the COVID-19 crisis. Sustainability. **14**, 3616 (2022). https://doi.org/10.3390/su14063616
28. STATISTA: COVID-19 new cases worldwide by day. https://www.statista.com/statistics/1103046/new-coronavirus-covid19-cases-number-worldwide-by-day/. Last accessed 22 Mar 2022

29. Fischhoff, B., Slovic, P., Lichtenstein, S., Read, S., Combs, B.: How safe is safe enough? A psychometric study of attitudes toward technological risks and benefits. Policy Sci. **9**, 127–152 (1978). https://doi.org/10.1007/BF00143739
30. Wikipedia. https://en.wikipedia.org/wiki/List_of_cruise_lines. Last accessed 14 Mar 2022
31. Semrush: Online marketing can be easy. https://www.semrush.com/. Last accessed 21 Mar 2022
32. Dias, Á.L., Silva, R., Patuleia, M., Estêvão, J., González-Rodríguez, M.R.: Selecting lifestyle entrepreneurship recovery strategies: a response to the COVID-19 pandemic. Tour. Hosp. Res. **22**, 115–121 (2022). https://doi.org/10.1177/1467358421990724
33. Chaffey, D., Patron, M.: From web analytics to digital marketing optimization: increasing the commercial value of digital analytics. J. Direct Data Digit. Mark. Pract. **14**, 30–45 (2012). https://doi.org/10.1057/dddmp.2012.20
34. Nasiopoulos, D.K.: Logistics companies reengineering. Change of leadership and human resources planning through management information systems. In: Business Intelligence and Modelling, vol. 6, pp. 207–217. Springer, Cham (2021)
35. Nasiopoulos, D.K.: Web search applications in Logistics companies in the context of evolution of management information systems. In: Business Intelligence and Modelling, vol. 6, pp. 241–254. Springer, Cham (2021)
36. Nasiopoulos, D.K.: Email and E-branding applications in Logistics companies as a field of management information systems. In: Business Intelligence and Modelling, vol. 6, pp. 255–265. Springer, Cham (2021)

The COVID-19 Crisis Effect on Railways' Digital Branding: Risk Management Applications Utilizing Big Data

Damianos P. Sakas, Ioannis Dimitrios G. Kamperos, Marina C. Terzi, and Athanasios Kriemadis

1 Literature Review

1.1 Risk Management in the Railway Sector

According to the CER (Community of European Railway and Infrastructure) Covid Impact Tracker, the passengers' revenue losses of the rail sector in the EE reached the alarming level of 50 billion euros over 2020–2021 [1]. Unfortunately, these major losses in passengers' activities have confirmed, in the worst way, the slow recovery scenario made by UIC (International Union of Railways) in 2020, which had estimated a loss of 52.76 billion euros over the same period [2]. This is a clear alert that the supply chain was not fully aware of its vulnerability to global shocks [3]. The drop in passengers' demand, due to the lockdown and severe transport restrictions, had a direct impact on the economy of the railway sector. However, whether it is a black swan event, like COVID-19, or other unpredictable events in demand, transportation companies have to be prepared to deal with the unexpected [4].

Adjetey-Bahun et al. [5] introduce a simulation-based model as a risk management tool, able to quantify resilience in mass railway transportation systems to rapidly recover from perturbations. Risk management approaches in the railway sector have been extensively researched by the academia, in an attempt to provide

D. P. Sakas · I. D. G. Kamperos (✉) · M. C. Terzi
BICTEVAC LABORATORY Business Information and Communication Technologies in Value Chains Laboratory, Department of Agribusiness and Supply Chain Management, School of Applied Economics and Social Sciences, Agricultural University of Athens, Athina, Greece
e-mail: kamperosdigese@aua.gr

A. Kriemadis
Department of Management Science and Technology, School of Economics and Technology, University of Peloponnese, Tripoli, Greece

D. P. Sakas et al. (eds.), *Computational and Strategic Business Modelling*, Springer Proceedings in Business and Economics,
https://doi.org/10.1007/978-3-031-41371-1_6

insights on risk mitigation tools [6, 7], safety risk management [8], engineering risk [9], sustainability in risk management [10], to name a few. Sasidharan et al.'s [11] paper review on risk management practices in the railway industry elucidates multiple applications on risk management models, examining health, environmental, construction, electricity, information technology, and other issues. The authors further concluded that it is the multifaceted nature of the industry that requires various risk mitigation tools in all stages: design, construction, operation, and maintenance.

1.2 Big Data and the Railway Sector

With digitalization playing a very pivotal role, growing business and increasing traffic are more likely to happen online. Most industries, including rail transport, are experiencing a tidal wave of digital transformation, and thus, they are investing in scalability, accessibility, and, performance to meet and exceed online users' expectations. Therefore, a potential lack of investment in digital technology has no longer place, since most transportation companies heavily rely on their websites to maintain existing customers and attract new ones [12].

With the pandemic still a live threat, transportation companies need to reshape the business ecosystem, with the intention to be characterized by resilience and responsibility. The tendency of the supply chain is to modernize the current technologies and built new digital capabilities [13]. To effectively achieve this purpose, companies should develop a robust framework, which will include a responsive and crisis management strategy. This framework should be technology-led, utilizing platforms that support Web analytics and social media data as a business-as-usual protocol [14].

Specifically for the railway industry, big data has been the point of several pieces of research, investigating operations, maintenance, and safety dimensions [15], since it offers ample opportunities to improve productivity, reliability, velocity, and safety of railroads [16]. Wang and Wang [17] point out the contribution of big data to consumers' marketing aspects. Data exchange activity decodes and simplifies users' online behavior and provides fruitful information on understanding future attitudes based on social features [18].

1.3 Customers' Satisfaction and Social Behavior Within the Railway Sector

Huge amounts of data can be collected, providing great value and big challenges simultaneously, since it offers descriptive and predictive business opportunities in a fast way [19]. Passive crowdsourcing data is proved to be a valuable risk management tool, able to predict users' online behavior [20]. To elevate and leverage business opportunities arising from the increasing data pool and its complexities, big data analysis (BDA) caters to these needs in the long run and analyzes social science phenomena [21].

However, there are aspects of the railway environment that are still in their infancy, regarding BDA [22]. Little research has been conducted on analyzing social behavior within this sector. Passive crowdsourcing has the potential to be used as a tool to collect behavioral information, since it furthers our understanding of each stage of consumers' decision-making process [23, 24]. Thaduri et al. [25] raise concerns regarding customers' satisfaction and railway companies' reputation. However, their research focuses on the advancement of maintenance aspects as a prerequisite for customers' satisfaction. The purpose of this paper is to provide a deep understanding of the impact of COVID-19 on online users' behavior and purchasing attitudes toward railway companies, as an antidote to brand recovery. Capitalizing on data mining techniques for the development of digital marketing strategies, especially to be used as a risk management tool during a novel crisis, will enable the railway ecosystem to recover its brand name and optimize Web traffic and the overall business performance [26].

2 Materials and Methods

2.1 Problem Formulation and Research Hypotheses

Data is important and profitable in the digital era when it is used analytically to enrich and personalize client online experiences and anticipate future online actions. As a result, moving from a corrective to a predictive approach is a huge problem for the transportation industry. Adopting such a strategy will reduce risk perceptions and hence contribute to the railway industry's economic, strategic, and operational growth [20]. By examining Web analytics, the railway business might gain a significant advantage in acquiring crucial behavioral knowledge to properly manage its digital brand. Railways will ensure that their brand thrives in this new digital era by adopting a new unified strategy that is suited to the behavioral key performance indicators (KPIs) accessible. Page views, time on site, reach percentage, and bounce rate are the authors' four behavioral KPIs for this study. Authors also use the number of global daily new COVID-19 infections to track COVID-19 crisis escalation [27].

Behavioral KPIs can highlight the interaction between e-businesses and online consumers, allowing railway organizations to properly measure the effectiveness of marketing campaigns, website structure, and content quality. For organizations to be able to predict customers' behavior during a crisis provides a critical advantage through targeted risk management planning [28–30].

The study's initial goal is to track user engagement metrics, which are indicators of customers' behavior and interactions with the company's domain. During a crisis, this procedure provides crucial insights into Internet users' behavior as well as information to marketers on measures that could result in greater conversion rates.

Two research questions were created in order to promote the purposes of this study:

Hypothesis 1 (H1) *Traffic ranking will be significantly diminished during the COVID period through reach percentage, compared to the pre-COVID period.*

Railways is a special case of the transportation industry, since it is lucking the competition observed in other transportation sectors like aviation. Railways target more "regional" routes, and usually, a connection between two destinations is served between one or very few competitors. Self-isolation and traveling-limiting regulations are expected to negatively affect the digital branding of railway organizations.

Hypothesis 2 (H2) *During the COVID period, COVID-19 cases value fluctuation will be related to the variations in traffic ranking.*

During the pandemic, the number of COVID-19 new infection cases was used as the primary indicator of crisis level of escalation. Customers, as the crisis is escalating and dread for the unknown hazard of COVID-19 is building up, are expected to increase their risk perception for traveling [31] and decrease their online interaction with railway organizations' domains.

2.2 Sample Selection and Data Collection

We created our sample by selecting five of the top ten railway operators in terms of 2018 revenues [32].

We isolated two research periods to investigate the behavior of our selected KPI-related variables. The first period (March 1, 2019, to February 29, 2020) was designated as the "pre-COVID period." March 2, 2020, to February 28, 2021, refers to the first year of the COVID-19 crisis and was defined as the "COVID period." We chose March 1, 2020, as the start date of the pandemic, since most countries had begun enacting border control regulations and self-isolation measures by that date, which we designated as the start date for our investigation.

We collected every KPI value from the sample organizations using online analytics platform big data mining techniques [33] and produced a representative KPI-related variable based on the average value to accomplish more efficient data processing and statistical analysis. This method was used to analyze all of the KPIs. For our research, we used data from the STATISTA database on daily verified COVID-19 cases around the world [27]. These data were processed into an average daily value for each month, forming the study's independent variable, for improved precision and compatibility with the dependent variables (COVID cases).

3 Results

3.1 *Statistical Analysis*

In order to depict any statistically significant differences displayed by our dependent variables, independent samples t-tests were applied between the pre-COVID and the COVID periods datasets. Results indicated statistically significant decreases in reach percentage, time on site, and bounce rate, as well as a significant increase in traffic rank (Table 1).

Pearson correlation coefficient (PCC–r) statistical analysis was applied to our COVID period dataset, to highlight the statistically significant relations between our variables (Table 2). PCC analysis indicated five statistically significant correlations between variables, three at the 0.01 level of significance and two at the 0.05 level (two-tailed).

A total of four statistically significant correlations were spotted, with one of them at the 0.01 level (2-tailed).

3.2 *Development of a Diagnostic Explanatory Model*

In the second level of data analysis, we utilized our variables and the suggested interconnections developed by statistical analysis and created a fuzzy cognitive mapping (Fig. 1) visualizing the proposed underlying neural process.

Table 1 Independent samples t-tests results

		t-Test for equality of means		
Variables	Levene's test for equality of variances significance	Significance (2-tailed)	Mean difference	Standard error difference
Reach %	0.1260	*0.000	0.0124	0.0145
Pageviews	0.5050	0.391	0.0664	0.2237
Bounce rate	0.134	*0.024	0.6930	0.7834
Time on site	0.8880	*0.000	8.8614	−3.1667
Traffic rank	0.0010	*0.000	424.5856	−2396.6771

The symbol "*" indicates a statistically significant result

Table 2 Pearson correlation coefficient (PCC–r) statistical analysis results

Variables			Pearson correlation	Significance (2-tailed)
COVID cases	and	Pageviews	0.664	0.050
COVID cases	and	Bounce rate	−0.647	0.050
Reach %	and	Bounce rate	0.604	0.050
Traffic rank	and	Time on site	0.724	0.010

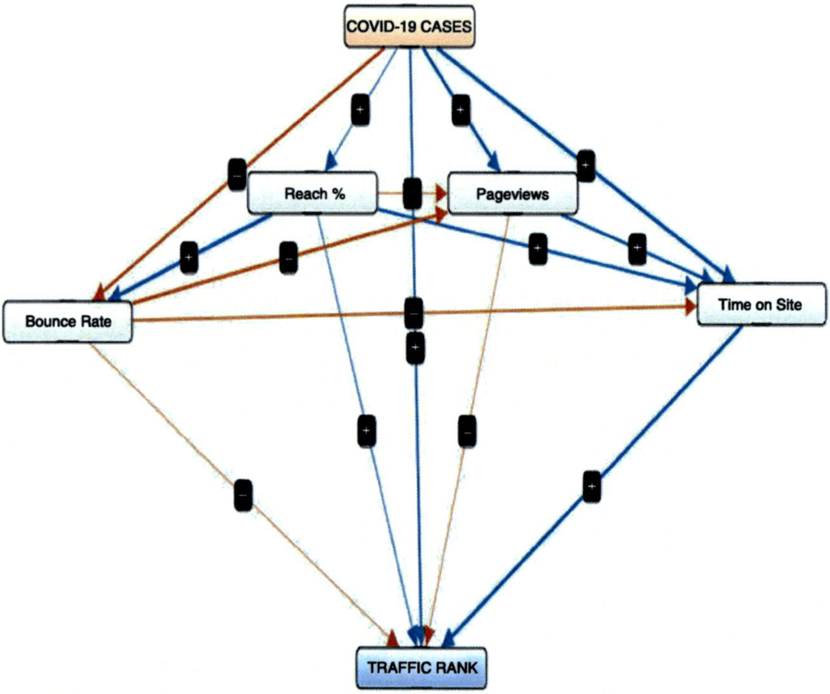

Fig. 1 Fuzzy cognitive mapping of the studied variables

FCMs (fuzzy cognitive mappings) are cause-effect structures, based on human reasoning processes. They can have varying degrees of causation among distinct causal players due to their cloudy nature [34].

To further investigate the interaction between variables in our model, we developed a scenario (Fig. 2) simulating an increase in COVID-19 cases (0.85) and a significant increase (0.75) in page views per user (for example by Web-content development). Results indicated that when both parameters are applied, traffic ranking is affected less by the crisis escalation.

4 Discussion

The purpose of the current study is to investigate the impact of COVID-19 on online users' behavioral attitudes within the railway industry. From March 2020, most countries implemented strict measures of confinement, resulting in a major drop in global mobility. As such, rail traffic has been heavily affected by Covid-19, resulting in financial devastation. The outcomes of this research serve as evidence of this global halt.

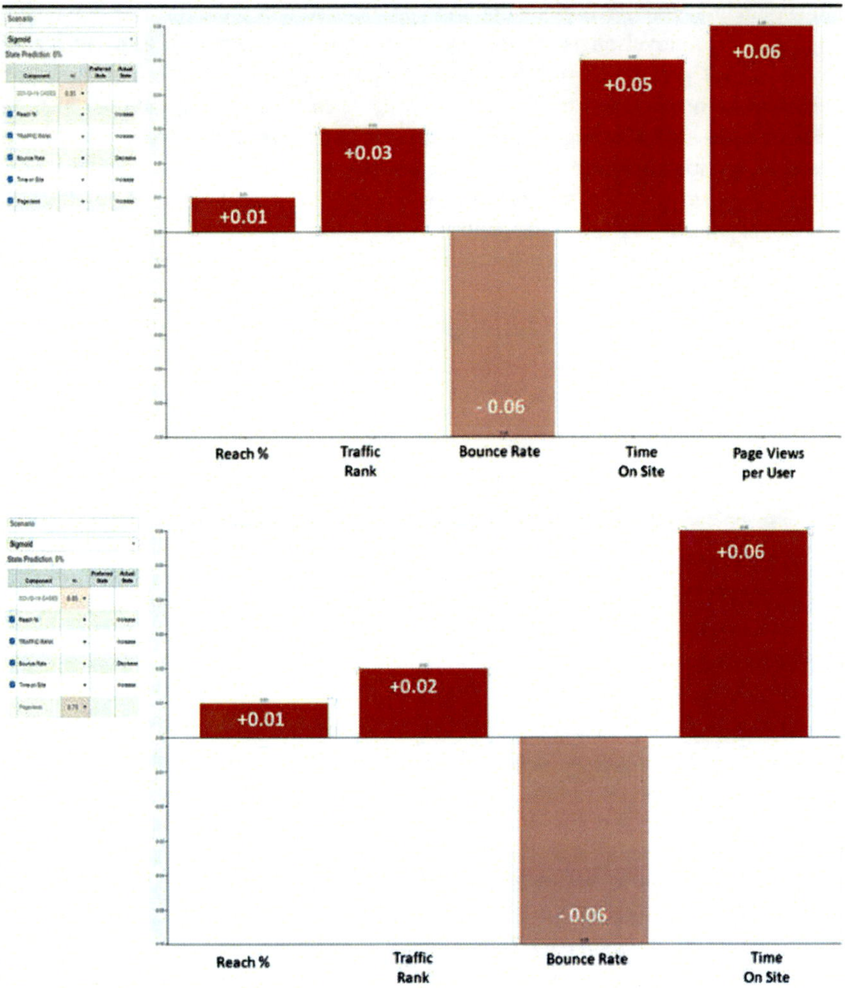

Fig. 2 FCM scenario results indicating a positive effect of page content improvement on domain visibility during a crisis escalation

During the crisis period (COVID period), the reach percentage is significantly reduced, compared to the pre-COVID period. Although there is a statistically significant increase in time on site and, thus, online users tend to interact more with the railways' websites, the traffic ranking is decreased, confirming H1. As online users decrease their searches for railway services, the traffic ranking is being worsened and thus digital branding reaches its lowest levels, compared to the pre-COVID period. Low levels of mobility, in some cases, if any at all, limitation of capacity, and timetable adjustments are only a few of the reasons that influence online users' low interaction with railways' Web domains.

The findings provide further support to previous studies regarding risk perception [31]. During a novel crisis, online users' uncertainty is higher, leading to higher levels of risk perception regarding railway traveling. However, crisis escalation seems not to be related to traffic ranking of the Web domains contradicting H2. As COVID cases are escalating though, online users tend to interact more with railways' websites, by spending more time and viewing more content.

FCM scenarios results suggest that improving site content could be an effective risk management and damage control strategy during a crisis. However, more parameters could be added to the scenarios providing a better perspective to the railways' operators.

5 Conclusion

The Covid-19 pandemic has substantially affected passenger rail traffic, significantly altering customers' travel behavior. The railway ecosystem has experienced COVID-19 in the worst way. Therefore, building a cutting-edge service blueprint, based on lowering consumers' perceived risk, is imperative. Marketers of railway companies should be proactive if they want to stay ahead of the competition. Gaining and maintaining a competitive advantage in the e-marketplace requires the examination of behavioral attitudes, so as to define the indicators that affect consumers' online purchasing process [35].

When the risk is on the table, it has to become the product [36]. Therefore, resources should be allocated to the development of a brand position strategy in order to mitigate risk perception, add value, and differentiate railway services among companies [37]. Since online brand experience is the sum of thoughts, feelings, and reactions, the differentiation should not only focus on technical aspects that eliminate risk, but rather on emotional branding, since it favors brand perceptions and behavioral intentions [38].

The brand position strategy is an urgent that revitalizes brands [39]. One way to emotionally connect with more users and make them feel more involved with the brand is to use social media crowdsourcing effectively. Crowdsourced marketing campaigns usually involve the customers, which further leads to running engaging, interactive campaigns. To reaffirm their status as industry pioneers and tackle the pandemic, companies should invest in product innovations with the contribution of the audience [40]. In this case, behavioral KPIs researched in the present study can serve as an important indicator of customers' online behavior regarding the effectiveness of crowdsourcing marketing.

Once the pandemic is over, the travel behavior is expected to change. The railway industry, in general, and each rail company, in particular, should greatly focus on users' online behavior, not only to track existing trends but also to observe their reactions regarding crowdsourcing marketing. The brand position strategy should work as part of a unified risk mitigation strategy, by incorporating users' social behavior. Analyzing social behaviors through big data could contribute to better conversion rates, in favor of digital brand recovery and performance optimization.

References

1. CER: Latest CER COVID Impact Tracker: railways lose €50 billion since the start of the pandemic. https://www.cer.be/media/press-releases/latest-cer-covid-impact-tracker-railways-lose-%E2%82%AC50-billion-start-pandemic-0. Last accessed 11 Apr 2022
2. UIC: UIC Covid-19 task force management of Covid-19. https://uic.org/IMG/pdf/economic_impact_v2.pdf (2020)
3. Golan, M.S., Jernegan, L.H., Linkov, I.: Trends and applications of resilience analytics in supply chain modeling: systematic literature review in the context of the COVID-19 pandemic. Environ. Syst. Decis. **40**, 222–243 (2020). https://doi.org/10.1007/s10669-020-09777-w
4. Tardivo, A., Carrillo Zanuy, A., Sánchez Martín, C.: COVID-19 impact on transport: a paper from the railways' systems research perspective. Transp. Res. Rec. **2675**, 367–378 (2021). https://doi.org/10.1177/0361198121990674
5. Adjetey-Bahun, K., Birregah, B., Châtelet, E., Planchet, J.-L.: A model to quantify the resilience of mass railway transportation systems. Reliab. Eng. Syst. Saf. **153**, 1–14 (2016). https://doi.org/10.1016/j.ress.2016.03.015
6. Berrado, A.: A framework for risk management in railway sector: application to road-rail level crossings. Open Transp. J. **5**, 34–44 (2011). https://doi.org/10.2174/1874447801105010034
7. Suh, S.D.: Risk management in a large-scale new railway transport system project. IATSS Res. **24**, 53–63 (2000). https://doi.org/10.1016/S0386-1112(14)60029-7
8. Alawad, H., Kaewunruen, S., An, M.: A deep learning approach towards railway safety risk assessment. IEEE Access. **8**, 102811–102832 (2020). https://doi.org/10.1109/ACCESS.2020.2997946
9. Antucheviciene, J., Kou, G., Maliene, V., Vaidogas, E.R.: Mathematical models for dealing with risk in engineering. Math. Probl. Eng. **2016**, 1–3 (2016). https://doi.org/10.1155/2016/2832185
10. Nedeliakova, E., Hudakova, M., Masar, M., Lizbetinova, L., Stasiak-Betlejewska, R., Šulko, P.: Sustainability of railway undertaking services with lean philosophy in risk management—case study. Sustainability. **12**, 5298 (2020). https://doi.org/10.3390/su12135298
11. Sasidharan, M., Burrow, M.P.N., Ghataora, S.G., Torbaghan, M.E.: A review of risk management applications for railways. In: 14th International Conference of Railway Engineering (2017). https://doi.org/10.25084/raileng.2017.0065
12. Sakas, D.P., Reklitis, D.P., Trivellas, P., Vassilakis, C., Terzi, M.C.: The effects of logistics websites' technical factors on the optimization of digital marketing strategies and corporate brand name. Processes. **10**, 892 (2022). https://doi.org/10.3390/pr10050892
13. Jabłoński, A., Jabłoński, M.: Social perspectives in digital business models of railway enterprises. Energies. **13**, 6445 (2020). https://doi.org/10.3390/en13236445
14. Reuter, C., Ludwig, T., Kotthaus, C., Kaufhold, M.-A., von Radziewski, E., Pipek, V.: Big data in a crisis? Creating social media datasets for crisis management research. I-COM. **15**, 249–264 (2016). https://doi.org/10.1515/icom-2016-0036
15. Ghofrani, F., He, Q., Goverde, R.M.P., Liu, X.: Recent applications of big data analytics in railway transportation systems: a survey. Transp. Res. Part C Emerg. Technol. **90**, 226–246 (2018). https://doi.org/10.1016/j.trc.2018.03.010
16. Attoh-Okine, N.: Big data challenges in railway engineering. In: 2014 IEEE International Conference on Big Data (Big Data), pp. 7–9. IEEE, Washington, DC (2014). https://doi.org/10.1109/BigData.2014.7004424

17. Wang, W.Y.C., Wang, Y.: Analytics in the era of big data: the digital transformations and value creation in industrial marketing. Ind. Mark. Manag. **86**, 12–15 (2020). https://doi.org/10.1016/j.indmarman.2020.01.005
18. Cassavia, N., Masciari, E., Pulice, C., Saccà, D.: Discovering user behavioral features to enhance information search on big data. ACM Trans. Interact. Intell. Syst. **7**, 1–33 (2017). https://doi.org/10.1145/2856059
19. Sakas, D.P., Kamperos, I.D.G., Reklitis, D.P., Giannakopoulos, N.T., Nasiopoulos, D.K., Terzi, M.C., Kanellos, N.: The effectiveness of centralized payment network advertisements on digital branding during the COVID-19 crisis. Sustainability. **14**, 3616 (2022). https://doi.org/10.3390/su14063616
20. Sakas, D.P., Kamperos, I.D.G., Reklitis, P.: Estimating risk perception effects on courier companies' online customer behavior during a crisis, using crowdsourced data. Sustainability. **13**, 12725 (2021). https://doi.org/10.3390/su132212725
21. Chang, R.M., Kauffman, R.J., Kwon, Y.: Understanding the paradigm shift to computational social science in the presence of big data. Decis. Support Syst. **63**, 67–80 (2014). https://doi.org/10.1016/j.dss.2013.08.008
22. Mcmahon, P., Zhang, T., Dwight, R.: Requirements for big data adoption for railway asset management. IEEE Access. **8**, 15543–15564 (2020). https://doi.org/10.1109/ACCESS.2020.2967436
23. Hofacker, C.F., Malthouse, E.C., Sultan, F.: Big Data and consumer behavior: imminent opportunities. J. Consum. Mark. **33**, 89–97 (2016). https://doi.org/10.1108/JCM-04-2015-1399
24. Zhang, C., Tan, T.: The impact of big data analysis on consumer behavior. J. Phys. Conf. Ser. **1544**, 012165 (2020). https://doi.org/10.1088/1742-6596/1544/1/012165
25. Thaduri, A., Galar, D., Kumar, U.: Railway assets: a potential domain for big data analytics. Procedia Comput. Sci. **53**, 457–467 (2015). https://doi.org/10.1016/j.procs.2015.07.323
26. Saura, J.R.: Using data sciences in digital marketing: framework, methods, and performance metrics. J. Innov. Knowl. **6**, 92–102 (2021). https://doi.org/10.1016/j.jik.2020.08.001
27. STATISTA: COVID-19 new cases worldwide by day. https://www.statista.com/statistics/1103046/new-coronavirus-covid19-cases-number-worldwide-by-day/. Last accessed 22 Mar 2022
28. Nasiopoulos, D.K.: Logistics companies reengineering. Change of leadership and human resources planning through management information systems. In: Business Intelligence and Modelling, vol. 6, pp. 207–217. Springer, Cham (2021)
29. Nasiopoulos, D.K.: Web search applications in Logistics companies in the context of evolution of management information systems. In: Business Intelligence and Modelling, vol. 6, pp. 241–254. Springer, Cham (2021)
30. Nasiopoulos, D.K.: Email and E-branding applications in Logistics companies as a field of management information systems. In: Business Intelligence and Modelling, vol. 6, pp. 255–265. Springer, Cham (2021)
31. Fischhoff, B., Slovic, P., Lichtenstein, S., Read, S., Combs, B.: How safe is safe enough? A psychometric study of attitudes toward technological risks and benefits. Policy Sci. **9**, 127–152 (1978). https://doi.org/10.1007/BF00143739
32. Railway Technology: The world's biggest railway operators in 2018. https://www.railway-technology.com/analysis/worlds-biggest-railway-operators-2018/. Last accessed 5 Mar 2022
33. Semrush: Online marketing can be easy. https://www.semrush.com/. Last accessed 21 Mar 2022
34. Kosko, B.: Fuzzy cognitive maps. Int. J. Man Mach. Stud. **24**, 65–75 (1986). https://doi.org/10.1016/S0020-7373(86)80040-2
35. Peña-García, N., Gil-Saura, I., Rodríguez-Orejuela, A., Siqueira-Junior, J.R.: Purchase intention and purchase behavior online: a cross-cultural approach. Heliyon. **6**, e04284 (2020). https://doi.org/10.1016/j.heliyon.2020.e04284

36. Rahmi, S., Ilyas, G.B., Tamsah, H., Munir, A.R.: Perceived risk and its role in the influence of brand awareness on purchase intention: study of Shopee users. J. Strategy Bus. **26**, 97–109 (2022). https://doi.org/10.20885/jsb.vol26.iss1.art7
37. Fuchs, C., Diamantopoulos, A.: Evaluating the effectiveness of brand-positioning strategies from a consumer perspective. Eur. J. Mark. **44**, 1763–1786 (2010). https://doi.org/10.1108/03090561011079873
38. Kumar, P., Meng, T., Kabiraj, S.: Effect of crowdsourcing on consumer brand perceptions and behavioral intentions. Bus. Perspect. Res. **7**, 42–58 (2019). https://doi.org/10.1177/2278533718800190
39. Beverland, M.B., Napoli, J., Farrelly, F.: Can all brands innovate in the same way? A typology of brand position and innovation effort*: a brand-innovation typology. J. Prod. Innov. Manag. **27**, 33–48 (2010). https://doi.org/10.1111/j.1540-5885.2009.00698.x
40. Vermicelli, S., Cricelli, L., Grimaldi, M.: How can crowdsourcing help tackle the COVID-19 pandemic? An explorative overview of innovative collaborative practices. R&D Manag. **51**, 183–194 (2021). https://doi.org/10.1111/radm.12443

The Role of Web Page Auditing in Centralized Payment Networks' Digital Marketing Enhancement

Damianos P. Sakas and Nikolaos T. Giannakopoulos

1 Introduction

The digital revolution may give rise to data processing, particularly in the banking industry, which is frequently employed to help the consumer get accessibility to banking goods and services. Digital promotion is any promotion of a commodity via the use of electronic channels and the World Wide Web. A goal of digital marketing is to increase website users and prospective clients as rapidly as possible [1]. Though it is known that innovation as well as the Web are widely accepted in culture, it is not unexpected that digital marketing strategies are indeed the primary option for several businesses.

Digital marketing gained prominence as a result of electronic commerce. The advantage of digital marketing is much more efficient and profitable than conventional marketing, and this seems to be a major reason that firms utilize it [2]. In the latest times, several firms expanded at the same time with the Internet's flourishing, both as a beneficial means of giving information for promotion and as an advertising network to make revenue and improve branding [3]. Organizations develop Web pages to assist their operations, including payments, with many of them being unaware of how successful their Web pages are for acquiring new consumers [4].

Moreover, the influence of digital marketing consumers' data on the branding efficiency of CPN websites has been found valuable [5]. Furthermore, as indicated by an assessment of the website's data [6], visitor activity on CPN domains can primarily impact online ads as well as digital worldwide ranking [7].

D. P. Sakas · N. T. Giannakopoulos (✉)
BICTEVAC LABORATORY: Business Information and Communication Technologies in Value Chains Laboratory, Department of Agribusiness and Supply Chain Management, School of Applied Economics and Social Sciences, Agricultural University of Athens, Athens, Greece
e-mail: n.giannakopoulos@aua.gr

D. P. Sakas et al. (eds.), *Computational and Strategic Business Modelling*,
Springer Proceedings in Business and Economics,
https://doi.org/10.1007/978-3-031-41371-1_7

The paper is organized as follows: The first section is the literature review and related background of the referred topics, like CPNs digital marketing improvement and Web analytics role, as well as settlement of research hypotheses. The second part concerns statistical tool capitalization to extract valuable data for developing the diagnostic and hybrid models in the following stage. Lastly, insights are provided into the importance of website auditing for CPN firms' digital marketing performance.

2 Related Background

With the digitizing of marketing, which serves as one of the transitional mediums frequently utilized by enterprises while monitoring the trend of technology, several organizations are beginning to abandon traditional commerce and transition to contemporary firms utilizing electronic channels. Digitization of promotion, interactions, and payments made a quite easy task, from anywhere across the globe, making it easier for corporate drivers and customers to communicate. Through developing a strategic position between the authority and the online payment operator, there seems to be a substantial interaction in researching and studying creative approaches to the digital payment platform. Additional research conducted by Affandi et al. [8], demonstrated that digitization of branding may enhance revenues for business entrepreneurs.

CPNs seem to be helpful in eliminating knowledge imbalance. Initially, CPNs might assist financing companies in improving affordability and consistency, as well as reducing communication barriers among banking institutions and SMEs [9]. This can assist banks in improving their capacity to handle risk data and lowering operational expenses [10]. Furthermore, Jaki and Marin [11] demonstrated that CPNs are a powerful driver in promoting loan availability to SMEs. Lu [12] indicated that CPNs' services and option funding are successful in overcoming the financing deficit for SMEs, while Baber [13] displayed that CPNs enhance SMEs' chance of borrowing with low rates.

Big data might provide meaningful perspectives to a broad array of firms, including government entities, CPNs, and fundraising [14]. Big data utility and analysis endorses advancement in the CPN and Fintech market by turning operations more dynamic, platforms easier to use and promoting innovative approaches [15, 16].

Optimizing Web page functionality is an important strategy that firms may employ to boost online visibility, engage and keep new visitors, and provide precise and trustworthy content to a wider readership. User experience extends past surface-level design; that generally relates [17] to a package's utility, which encompasses characteristics including browsing or user-experienced difficulties on a Web page [18]. This covers the issue of how accessible or appealing a Web page is, both of which might impact the number of visitors who interact with the Web page.

Usability additionally considers user involvement and web page's capacity to meet other goals. Whenever people are unable to readily browse and utilize a Web page, they are less willing to keep accessing it as a resource of knowledge. Greater usability, on the other hand, can increase a platform's appeal. As a result, Web pages are increasingly required to adapt to customer expectations, wishes, and demands [19, 20]. For optimizing Web page usability, some sectors have set defined rules for ease of access, information, advertising, and software [21].

Having analyzed the above factors, a research gap is noted regarding CPN firms' individual website issues' impact on their digital marketing strategy efficiency. Each website issue should be addressed individually and in total compared with representative variables of digital marketing strategy efficacy variables, such as organic traffic and its subfactor branded traffic. Hence, the following research hypotheses are formed to answer the above reflections:

H1: "The more intense the website problems the more CPNs' brand is weakened."
H2: "Presentation, size and links of a website affect the importance of CPNs' onsite issues."
H3: "CPNs' brand power and website issues' importance variate their digital marketing performance."

3 Methodology

To shape the methodological framework of this paper, the authors invoked significant research tools. These refer to performing a diagnostic model combined with hybrid modeling deployment. As inputs of these models, the coefficients of the regression analyses are used. In this way, the results of the models depict the current situation of CPNs' website performance and its effect on the efficacy of digital marketing strategies. To do so, the exploratory and diagnostic model is developed according to Mpelogianni and Groumpos [22] research, to provide a cognitive mind map of the factors of the framework and their observed linking, by utilizing correlations and regression coefficients. Then, hybrid modeling takes place, by performing both agent-based and dynamic models to estimate the variation of the factors during a 100-day simulation period [23]. In this way, results of simulation can be explained and shape an analytical context for CPN firms to follow, regarding:

- Granting their digital marketers, the needful insights to increase their marketing strategies' efficiency and attract more potential customers at their Web pages
- Accord and collect the indispensable big data analytics to decision-makers to estimate Web page issues effect on their digital marketing strategies and set action courses

For paper's purposes, the authors extracted website analytic data from the top 5 leading firms in centralized payment networks market [24]. These data refer to daily observation and assortment of their Web page traffic and issues, capitalizing website decision support platforms [25]. Gathered data are analyzed in the coming sections, as described at the beginning of the methodological framework.

4 Results

In this section of the paper, the outputs of the regression analysis are shown. Table 1 shows the outcomes of the three performed linear regressions with dependent variables branded traffic, website issues, and organic traffic. As independent variables, the Web page issues were used, which sum up a number of potential other factors,

Table 1 Centralized payment networks' regressions

Variables	Standardized coefficient	R^2	F	p-value
Dependent: Branded traffic				
Constant	–	0.977	9.458	0.099
Web page issues	7.618			0.056
Size of JS_CSS	6.237			0.063
Number of JS_CSS	−12.632			0.141
Raw internal link rate	−1.000			0.246
Crawl depth	0.795			0.676
Page load time	−1.841			0.200
Incoming internal links	0.834			0.570
Outcoming internal links	1.631			0.059
Outcoming external links	0.924			0.439
Dependent: Web page issues				
Constant	–	1.00	629.678	0.002**
Branded traffic	0.117			0.056
Size of JS_CSS	−0.797			0.034*
Number of JS_CSS	1.749			0.040*
Raw internal link rate	0.110			0.331
Crawl depth	−0.154			0.495
Page load time	0.192			0.325
Incoming internal links	−0.063			0.737
Outcoming internal links	−0.205			0.045*
Outcoming external links	−0.130			0.363
Dependent: Organic traffic				
Constant	–	0.999	81.296	0.086
Branded traffic	−0.291			0.427
Size of JS_CSS	3.706			0.251
Number of JS_CSS	−6.314			0.315
Raw internal link rate	−0.512			0.344
Crawl depth	−0.888			0.362
Page load time	−0.400			0.589
Incoming internal links	1.436			0.193
Outcoming internal links	2.174			0.116
Outcoming external links	0.739			0.303
Web page issues	3.190			0.337

* and ** indicate statistical significance at the 95% and 99% levels, respectively

like the size and number of JavaScript (JS) and Cascading Style Sheets (CSS), the Web page load time and crawl depth, the incoming and outcoming internal and external links, and also the raw internal link rate. All these factors form the most common Web page issues encountered.

Table 1 shows that CPNs' branded traffic is marginally insignificantly impacted by the total of Web page issues, the size of JS and CSS, and the outcoming internal links, due to regression's p-value $= 0.099 > a = 0.05$ significance level and each issues' p-value being also above the significance level of 0.05. Hence, the first research hypothesis (H1) is marginally rejected, meaning that CPNs' brand does not weaken when website problems intensify.

Next, the regression of Web page issues is produced, which is verified overall with $R^2 = 1.00$ and p-value $= 0.002 < a = 0.01$ significance level. At least three independent variables are flagged as significant (the size and number of JS and CSS and the outcoming internal links), with p-values below 0.05 level of significance. So, the second research hypothesis (H2) is verified and onsite issues do get affected by the size, presentation, and links of a Web page.

At last, organic traffic's linear regression is neither verified overall, despite satisfactory $R^2 = 0.999$, with p-value $= 0.086 > a = 0.05$ significance level. Additionally, none of the Web page factors pose significant effect on organic traffic variable (p-values over 0.05 level of significance). Hence, the third hypothesis (H3) is rejected, with website issues' being unable to impact CPNs' digital marketing performance, expressed by their organic traffic.

5 Diagnostic Model

Development of the diagnostic model of the research utilizes the fuzzy cognitive mapping (FCM) tool, offered from the MentalModeler website application [26]. FCMs are fuzzy-graph constructs that are used to describe inductive inference. Because of such fuzzy sets, they can have murky levels of causation among fuzzy causative elements. This network framework provides methodical causative dispersion, particularly forward and backward stringing, as well as the expansion of information sources by linking disparate factors [22]. As illustrated in Fig. 1, the strong positive correlations (blue bold arrows) of organic traffic, Web page issues, and branded traffic are discerned. On the other hand, negative correlations are represented by the red-colored arrows.

6 Hybrid Model

Hybrid modeling consists of both agent-based analysis and a dynamic modeling process. The model's starting point is set as the potential Web visitor statechart, giving also start to the dynamic model of each website issue. The overall process

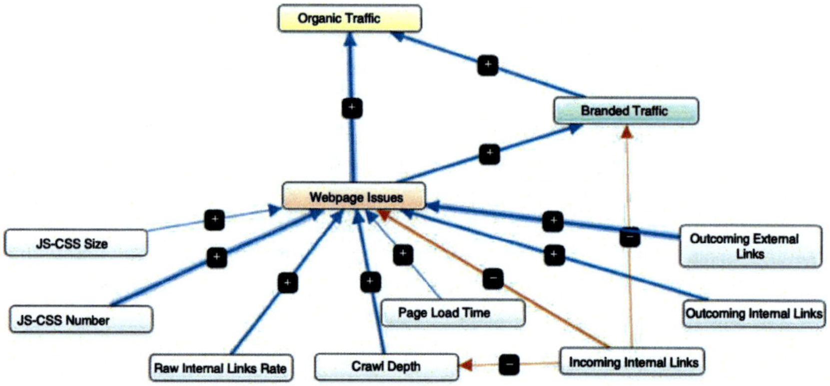

Fig. 1 Fuzzy cognitive mapping of CPNs website issues

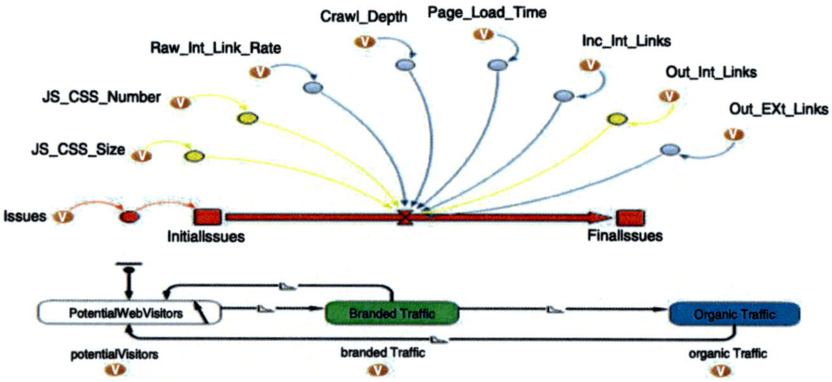

Fig. 2 Hybrid modeling of CPNs' website issues and traffic

measures CPN web page issues' variation, from eight factors commonly affecting a website to effects on their organic and branded traffic attraction. Thus, each website factor is made to follow the normal distribution and to affect the total Web page issues of CPNs, while at the same time, visitors enter these Web pages and interact with their content, as depicted by the statecharts' links [23]. Through running the model shown at Fig. 2, the fluctuations of Web page issues and branded and organic traffic are produced.

The execution of the hybrid model, as seen in Fig. 3, results in the fact that the website issues variation does not explain that of organic traffic, since these variables appear to have different highs and lows overall. On the contrary, branded traffic's movement seems to shortly follow that of website issues, with highly connected variations over the model's period. These results suggest that CPNs' organic traffic, as a representative variable of digital marketing's performance, does not variate in coordination with the issues appearing on their Web pages.

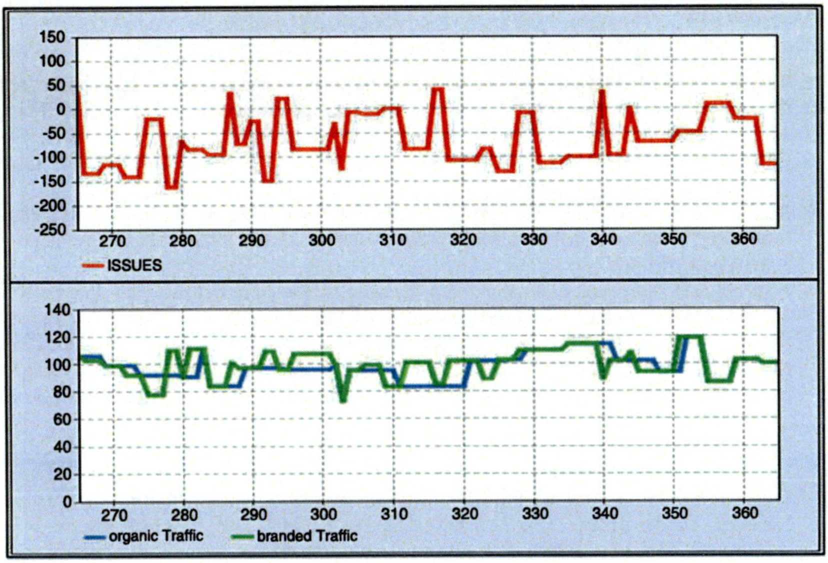

Fig. 3 Model's issues and traffic outcomes in 100 days

7 Conclusions

Throughout this paper, the aim of authors focused on developing a data-driven context for modeling CPNs' digital marketing efficiency via analyzing their Web pages' issues. The fact that the sum of Web page issues was not statistically causing a significant impact on organic and branded traffic was not endorsed by the simulation process. Therefore, it was indicated that branded traffic does follow the variation of Web page issues over 100 days' simulation, thus creating a strong connection between them. Organic traffic was yet to be impacted by the analyzed Web page issues. Other scientists with similar cases are studying similar problems, using modeling and simulation [27–29].

8 Practical Implications

The implications applied in real-time situations present the need for CPN firms to further optimize their Web pages to reduce the observed issues for the visiting customers, as the analyzed factors presented in the paper's results section. In this way, mostly their branded, and lesser their organic traffic can be increased, thus improving their digital marketing efficiency.

References

1. Ritz, W., Wolf, M., McQuitty, S.: Digital marketing adoption and success for small businesses: the application of the do-it-yourself and technology acceptance models. J. Res. Interact. Mark. **13**(2), 179–203 (2019). https://doi.org/10.1108/JRIM-04-2018-0062
2. Chaffey, D., Ellis-Chadwick, F.: Digital Marketing, 7th edn. Pearson, Harlow (2019) ISBN 978-1-292-24157-9
3. Sakas, D., Giannakopoulos, N.: Harvesting crowdsourcing platforms' traffic in favour of air forwarders' brand name and sustainability. Sustainability. **13**(15), 8222 (2021). https://doi.org/10.3390/su13158222
4. Phippen, A.D., Sheppard, L., Furnell, S.: A practical evaluation of web analytics. Internet Res. **14**, 284–293 (2004). https://doi.org/10.1108/10662240410555306
5. Sakas, D.P., Kamperos, I.D.G., Reklitis, D.P., Giannakopoulos, N.T., Nasiopoulos, D.K., Terzi, M.C., Kanellos, N.: The effectiveness of centralized payment network advertisements on digital branding during the COVID-19 crisis. Sustainability. **14**, 3616 (2022). https://doi.org/10.3390/su14063616
6. Sakas, D.P., Giannakopoulos, N.T., Reklitis, D.P., Dasaklis, T.K.: The effects of cryptocurrency trading websites on airlines advertisement campaigns. J. Theor. Appl. Electron. Commer. Res. **16**(7), 3099–3119 (2021). https://doi.org/10.3390/jtaer16070169
7. Kanellos, N. by Sakas, D.P., Kamperos, I.D.G., Reklitis, D.P., Giannakopoulos, N.T., Nasiopoulos, D.K., Terzi, M.C.: The effectiveness of centralized payment network advertisements on digital branding during the COVID-19 crisis. Sustainability. **14**(6), 3616 (2022)
8. Affandi, A., Sarwani, A.S., Erlangga, H., Siagian, A.O., Purwanto, A., Effendy, A.A., Juhaeri, G.: Optimization of MSMEs empowerment in facing competition in the global market during the COVID-19 pandemic time. Syst. Rev. Pharm. **11**(11), 1506–1515 (2020)
9. Sánchez, J.M.: The information technology revolution and the unsecured credit market. Econ. Inq. **56**(2), 914–930 (2018). https://doi.org/10.1111/ecin.12519
10. Livshits, I., Mac Gee, J.C., Tertilt, M.: The democratization of credit and the rise in consumer bankruptcies. Rev. Econ. Stud. **83**(4), 1673–1710 (2016). https://doi.org/10.1093/restud/rdw011
11. Jakšič, M., Marinč, M.: Relationship banking and information technology: the role of artificial intelligence and FinTech. Risk Manag. **21**(1), 1–18 (2019). https://doi.org/10.1057/s41283-018-0039-y
12. Lu, L.: Promoting SME finance in the context of the fintech revolution: a case study of the UK's practice and regulation. Bank. Finance Law Rev. **31**, 7–343 (2018) https://ssrn.com/abstract=3144767
13. Baber, H.: FinTech, crowdfunding and customer retention in Islamic banks. Vision. **24**(3), 260–268 (2020). https://doi.org/10.1177/0972262919869765
14. Xu, Z., Liu, Y., Yen, N.Y., Mei, L., Luo, X., Wei, X., Hu, C.: Crowdsourcing based description of urban emergency events using social media big data. IEEE Trans. Cloud Comput. **8**, 387–397 (2020). https://doi.org/10.1109/TCC.2016.2517638
15. Meng, S., He, X., Tian, X.: Research on Fintech development issues based on embedded cloud computing and big data analysis. Microprocess. Microsyst. **83**, 103977 (2021). https://doi.org/10.1016/j.micpro.2021.103977
16. Lee, I., Shin, Y.J.: Fintech: ecosystem, business models, investment decisions, and challenges. Bus. Horiz. **61**, 35–46 (2018). https://doi.org/10.1016/j.bushor.2017.09.003
17. Griffiths, K.M., Christensen, H.: Website quality indicators for consumers. J. Med. Internet Res. **7**(5), 55 (2005). https://doi.org/10.2196/jmir.7.5.e55
18. Otis, M., Zhu, J., Mustafa-Kutana, S.N., Bernier, A.V., Shum, J.M., Dupre, J.A.S., et al.: Testing usability and feasibility of a mobile educator tool for pediatric diabetes self-management: mixed methods pilot study. JMIR Form. Res. **4**(5), 16262 (2020). https://doi.org/10.2196/16262

19. Mittler, J.N., Volmar, K.M., Shaw, B.W., Christianson, J.B., Scanlon, D.P.: Using websites to engage consumers in managing their health and healthcare. Am. J. Manag. Care. **18**(6), 177–184 (2012)
20. Liang, C., Chen, H.: A study of the impacts of website quality on customer relationship performance. Total Qual. Manag. Bus. Excell. **20**(9), 971–988 (2009). https://doi.org/10.1080/14783360903181784
21. Oermann, M.H., McInerney, S.M.: An evaluation of sepsis Web sites for patient and family education. Plast. Surg. Nurs. **27**(4), 192–196 (2007). https://doi.org/10.1097/01.PSN.0000306184.95812.8a
22. Mpelogianni, V., Groumpos, P.P.: Re-approaching fuzzy cognitive maps to increase the knowledge of a system. AI & Soc. **33**, 175–188 (2018). https://doi.org/10.1007/s00146-018-0813-0
23. Dehghani, M., Trojovský, P.: Hybrid leader based optimization: a new stochastic optimization algorithm for solving optimization applications. Sci. Rep. **12**, 5549 (2022). https://doi.org/10.1038/s41598-022-09514-0
24. Devathon: Top 10 payment gateways around the world (2022, March 16). https://devathon.com/top-10-paymentgateways/
25. Semrush. (2022, March 16). https://www.semrush.com/
26. MentalModeler. (2022, March 16). https://dev.mentalmodeler.com/
27. Nasiopoulos, D.K.: The role of digital marketing in the development of a distribution and logistics network of information technology companies. In: Business Intelligence and Modelling, vol. 6, pp. 267–276. Springer, Cham (2021)
28. Dimitrios, N.K., Sakas, D.P., Vlachos, D.S.: The contribution of dynamic simulation model of depiction of knowledge, in the leading process of high technology companies. Key Eng. Mater. **543**, 406–409 (2013)
29. Nasiopoulos, D.K., Sakas, D.P., Masselos, K.: Open Source Web Applications. How it spread through the internet and their contribution to education. In: 4th Symposium on Business and Management and Dynamic Simulation Models Supporting Management Strategies (2011)

Effects of Big Data and Web Analytics on the Optimization of Startups' Branding and Market Share

Damianos P. Sakas, Dimitrios P. Reklitis, and Panagiotis Trivellas

1 Introduction and Hypotheses Developments

According to prior studies, digital marketing may help startups grow, establish, and expand while also providing valuable methods to acquire a bigger market share [1, 2]. Additionally, digital marketing can provide an added value on the optimization of corporate visibility and brand name [3–12]. Furthermore, approximately, 40% of all startups lack a website, while 60% of those who do have a website lack a dedicated marketing team [13]. This unawareness may also explain why over 80% of startups bankrupt within the first months of conducting business [14].

Big data and web analytics can play a crucial role in the optimization of digital marketing strategies, since the extraction and their analysis are purified from potential cognitive biases that might affect the marketer's decisions [15–17]. Our research is focused on big data predictive analytics and their relationship with web analytics for startups [18–20]. More precisely, marketers and decision-makers may use this type of analytics to collect historical data from startup websites and then develop a

D. P. Sakas · D. P. Reklitis (✉)
BICTEVAC LABORATORY Business Information and Communication Technologies in Value Chains laboratory, Department of Agribusiness and Supply Chain Management, School of Applied Economics and Social Sciences, Agricultural University of Athens, Athina, Greece
e-mail: d.sakas@aua.gr; drekleitis@aua.gr

P. Trivellas
Organizational Innovation and Management Systems, Department of Agribusiness and Supply Chain Management, School of Applied Economics and Social Sciences, Agricultural University of Athens, Athens, Greece
e-mail: ptrivel@aua.gr

D. P. Sakas et al. (eds.), *Computational and Strategic Business Modelling*, Springer Proceedings in Business and Economics,
https://doi.org/10.1007/978-3-031-41371-1_8

Table 1 Web analytics KPIs

Key performance indicators	Description of the key performance indicators
World ranking	Refers to the website global rank of a website on a specific day [21]
Organic traffic cost	Refers to an estimated amount of paid organic traffic in correlation to the competitors [21]
Paid traffic	Refers to an estimated amount of the total paid traffic in correlation to the competitors [21]
Social media paid	Refers to an estimated number of social media paid advertisements in correlation to the competitors [21]
Average duration	Refers to the average time that a customer spends on a website [21]
Total visitors	Refers to the number of total visitors that enter in a startup website [21]
Pages per visit	Refers to the total number of the pages that are viewed by a visitor every time they enter the website [21]

model to analyze the relationships between the key performance indicators (KPIs) [18–20]. Table 1 presents the big data KPIs that have been extracted and analyzed for the purpose of this research.

After defining the extracted KPIs, it is crucial to formulate the research hypothesis. Those hypotheses are designed in a way to help the startups' marketers and decision-makers to understand the usage of those KPIs and incorporate them into their holistic strategy.

Hypothesis 1 The "organic traffic cost" affects the "world ranking" variable through their "paid traffic" and the "social media paid" metrics.

Hypothesis 2 "Pages per visits" affects the "world ranking" variable through their "total visitors" and the "average duration" metrics.

2 Methodology

The authors used a methodology that included quantitative elements of consumers' digital path by investigating their behavior using big data and web analytics. To start, raw big data was collected daily from five large startup organizations over a span of 6 months [22] and statistically processed, using Pearson's correlation and regressions. After that, the statistical study's findings were utilized to create a paid campaign and visibility optimization scenarios that could be followed by startup marketers at a minimal cost and yield high returns. This approach was used because the raw collected big data were not impacted by any possible cognitive bias [15–17].

3 Results

3.1 Statistical Analysis

Tables 2 and 3 present the findings of the H1. As can be observed, all correlations are significant. More specifically, negative correlations found between the brand name (world ranking) and the organic traffic cost, paid traffic, and social media paid advertisements with $\boldsymbol{\rho = -0.587^{**}}$, $\boldsymbol{\rho = -0.564^{**}}$, and $\boldsymbol{\rho = -0.769^{**}}$, respectively. These findings highlight that the paid advertisements provide an added value to the brand name, since the lower the ranking, the better the brand name; number 1 in rankings is better than number 100. Furthermore, the regression analysis, as presented in Table 2, illustrates that for every 1% increase in brand name, the organic traffic cost increases by 17.9%, the total paid traffic decreases by 3.8%, and the social media paid advertisements decrease by 98.7%. The findings are significant with $p < 0.005$ and suggests to startup managers to invest in social media ads, because this investment will attract much more users to their websites than using other type of advertisements.

Tables 4 and 5 present the findings of the H2. As can be shown, negative correlations found between the brand name (world ranking) and the pages per visits and total visitors with $\boldsymbol{\rho = -0.251}$ and $\boldsymbol{\rho = -0.751^{**}}$. Furthermore, a significant positive correlation can be observed between the world ranking and the average

Table 2 H1 correlations

H1	World ranking	Organic traffic cost	Paid traffic	Social media paid
World ranking	1			
Organic traffic cost	−0.578**	1		
Paid traffic	−0.564**	0.548**	1	
Social media paid	−0.769**	0.859**	0.733**	1

**Correlation is significant at the 0.01 level (1-tailed)

Table 3 H1 regression

Variables	Standardized coefficient	R^2	F	p Value
Constant (world ranking)	–	0.614	16.435	<0.001
Organic traffic cost	0.179			0.368
Paid traffic	−0.038			0.810
Social media paid	−0.897			<0.001

Table 4 H2 correlations

H1	World ranking	Average duration	Pages per visit	Visitors
World ranking	1			
Average duration	0.463**	1		
Pages per visit	−0.251	0.175	1	
Visitors	−0.751**	−0.278	0.148	1

**Correlation is significant at the 0.01 level (1-tailed)

Table 5 Web analytics KPIs

Variables	Standardized coefficient	R^2	F	p Value
Constant (world ranking)	–	0.678	21.750	<0.001
Average duration	0.325			0.005
Pages per visit	−0.215			0.051
Visitors	−0.629			<0.001

duration with $\rho = \mathbf{0.463}^{**}$. These findings suggest that the content of the website needs to be rich in order to keep the customer on the website longer for optimizing the brand name. Furthermore, the regression analysis, as presented in Table 5, illustrates that for every 1% increase in brand name, the average duration increases by 32.5%, the pages per visits decreases by 21.5%, and the total visitors decrease by 62.9%. The findings are significant with $p < 0.005$ and suggests to startup decision-makers to make their website user-friendly in order to optimize visibility.

3.2 *FCM*

An FCM is a scaled kind of mind mapping in which stable models are created by specifying basic system properties like positive and negative Pearson's correlations between the examined metrics and the impact of connection that a metric might produce to others. FCM is often used to illustrate the cause-and-effect relationships of a system on a macro-level analysis on the (−1,1) scale [23]. Additionally, FCMs may be used to describe a system that is impacted by a large number of metrics and to map the Pearson's correlations between the metrics as well as the system [24]. Previous studies propose using a fuzzy cognitive map-based web analytics data for evaluating a digital marketing strategy. The creation of FCM will assist firms in implementing efficient digital advertising campaigns [10] with the goal of growing their digital brand name.

Figure 1 illustrates the creation of the FCM by incorporating the analyzed big data KPIs and their interconnections. Figure 2 illustrates the paid advertisement optimization scenario. More specifically, as verified in the statistical analysis decision-makers need to invest much more in social media advertisements, since this kind of investment will provide very fast a return on the corporate brand name. This is crucial for startups since the vast majority of them fail to survive the first year of operation [13, 14]. More specifically, to increase the corporate brand name by 10%, marketers need to increase their investments in social media ads by 5% and decrease the general paid advertisements on other websites by 2%.

Figure 3 illustrates a scenario for optimizing the brand name by increasing the website's visibility. As verified previously from statistical analysis, startup' website developers need to develop their websites with more rich content in order to increase the average time on site. This will promote visibility and brand name. More specifically, to increase the brand name by 10%, the website user must remain on the website for 39% more time.

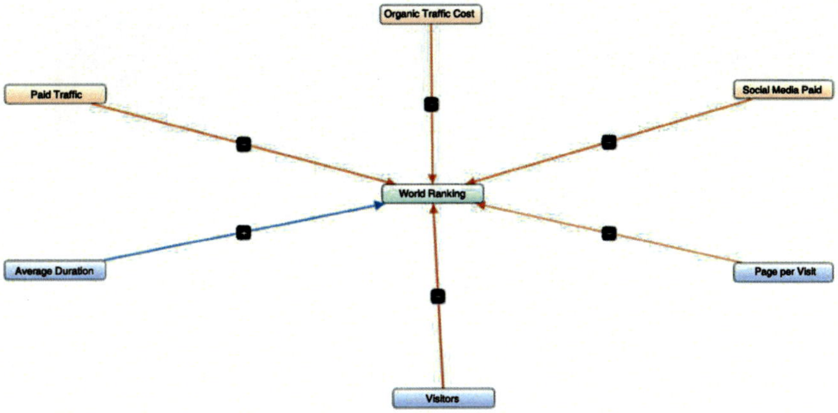

Fig. 1 Fuzzy cognitive map system

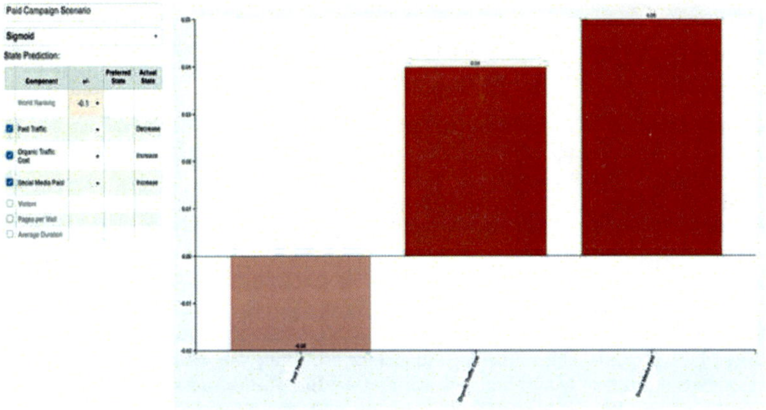

Fig. 2 Paid advertisement optimization scenario

4 Conclusions

As discussed, the majority of the startups bankrupt in the very first months of business activity [13, 14]. Our research provides valuable suggestions to marketers and decision-makers in order to optimize their startup's brand name. First, they need to invest in social media ads much more than other types of advertisements. Additionally, they need to develop their websites in a user-friendly manner to extend the average time that a user spends on their websites. Finally, further research that will incorporate big data with knowledge management and corporate social responsibility would be useful for the optimizations of the corporate brand name [25, 26].

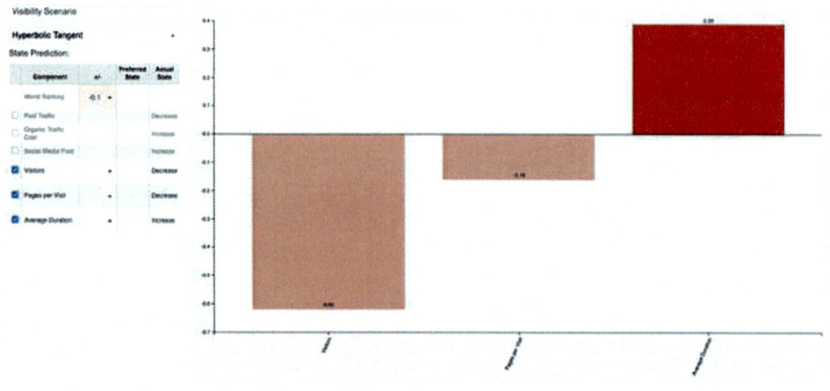

Fig. 3 Scenario for optimizing brand name by increasing the website' visibility

References

1. Patel, C.P., Chugan, P.K.: Digital inbound marketing to drive the success of startups. Int. J. Fam. Bus. Manag. **2**, 1–6 (2018). https://doi.org/10.15226/2577-7815/2/1/00111
2. Teixeira, S., Martins, J., Branco, F., Gonçalves, R., Au-Yong-Oliveira, M., Moreira, F.: A theoretical analysis of digital marketing adoption by startups. In: Trends and Applications in Software Engineering, pp. 94–105. Springer International Publishing, Cham (2018). https://doi.org/10.1007/978-3-319-69341-5_9
3. Sakas, D.P., Reklitis, D.P.: The impact of organic traffic of crowdsourcing platforms on airlines' website traffic and user engagement. Sustainability. **13**, 8850 (2021). https://doi.org/10.3390/su13168850
4. Sakas, D.P., Giannakopoulos, N.T., Reklitis, D.P., Dasaklis, T.K.: The effects of cryptocurrency trading websites on airlines' advertisement campaigns. J. Theor. Appl. Electron. Commer. Res. **16**, 3099–3119 (2021). https://doi.org/10.3390/jtaer16070169
5. Sakas, D.P., Reklitis, D.P.: Predictive model for estimating the impact of technical issues on consumers interaction in agri-logistics websites. In: Information and Communication Technologies for Agriculture—Theme IV: Actions, pp. 269–283. Springer International Publishing, Cham (2021). https://doi.org/10.1007/978-3-030-84156-0_14
6. Sakas, D.P., Kamperos, I.D.G., Reklitis, P.: Estimating risk perception effects on courier companies' online customer behavior during a crisis, using crowdsourced data. Sustainability. **13**, 12725 (2021). https://doi.org/10.3390/su132212725
7. Nasiopoulos, D.K., Sakas, D.P., Reklitis, P.: The interaction of users' behavior to the development of a distribution and logistics network of the information technology companies. In: Business Intelligence and Modelling, pp. 219–227. Springer International Publishing, Cham (2021). https://doi.org/10.1007/978-3-030-57065-1_23
8. Nasiopoulos, D.K., Sakas, D.P., Trivellas, P.: The role of digital marketing in the development of a distribution and logistics network of information technology companies. In: Business Intelligence and Modelling, pp. 267–276. Springer International Publishing, Cham (2021). https://doi.org/10.1007/978-3-030-57065-1_27
9. Sakas, D.P., Nasiopoulos, D.K., Reklitis, P.: Modeling and simulation of the strategic use of marketing in search engines for the business success of high technology companies. In: Strategic Innovative Marketing, pp. 217–226. Springer International Publishing, Cham (2019). https://doi.org/10.1007/978-3-030-16099-9_27

10. Sakas, D.P., Reklitis, D.P., Trivellas, P., Vassilakis, C., Terzi, M.C.: The effects of logistics websites' technical factors on the optimization of digital marketing strategies and corporate brand name. Processes. **10**, 892 (2022). https://doi.org/10.3390/pr10050892
11. Sakas, D.P., Reklitis, D.P., Giannakopoulos, N.T., Trivellas, P.: The influence of websites user engagement on the development of digital competitive advantage and digital brand name in logistics startups. European Research on Management and Business Economics. **29**, 100221 (2023). https://doi.org/10.1016/j.iedeen.2023.100221
12. Sakas, D.P., Kamperos, I.D.G., Reklitis, D.P., Giannakopoulos, N.T., Nasiopoulos, D.K., Terzi, M.C., Kanellos, N.: The effectiveness of centralized payment network advertisements on digital branding during the COVID-19 crisis. Sustainability. **14**, 3616 (2022). https://doi.org/10.3390/su14063616
13. Fundera.com: Small business marketing statistics and trends. https://www.fundera.com/resources/small-business-marketing-statistics (2021). Last accessed 20 June 2022
14. Maital, S., Barzani, E.: Why startups fail: a survey of empirical studies. https://www.neaman.org.il/EN/Files/Report_Why%20Startups%20Fail.pdf (2021). Last accessed 20 June 2022
15. Phillips-Wren, G., Power, D.J., Mora, M.: Cognitive bias, decision styles, and risk attitudes in decision making and DSS. J. Decis. Syst. **28**, 63–66 (2019). https://doi.org/10.1080/12460125.2019.1646509
16. Merendino, A., Dibb, S., Meadows, M., Quinn, L., Wilson, D., Simkin, L., Canhoto, A.: Big data, big decisions: the impact of big data on board level decision-making. J. Bus. Res. **93**, 67–78 (2018). https://doi.org/10.1016/j.jbusres.2018.08.029
17. Acciarini, C., Brunetta, F., Boccardelli, P.: Cognitive biases and decision-making strategies in times of change: a systematic literature review. Manag. Decis. **59**, 638–652 (2020). https://doi.org/10.1108/MD-07-2019-1006
18. Chantzaras, A., Dimitrios, N.K., Vlachos, D.S.: Mobile commerce and success factors. Simulation and modeling of the problem. In: Strategic Innovative Marketing, pp. 349–355. Springer International Publishing, Cham (2017). https://doi.org/10.1007/978-3-319-56288-9_46
19. Chaffey, D., Ellis-Chadwick, F.: Digital Marketing: Strategy, Implementation & Practice. Pearson, Harlow (2019)
20. Alexandros, N.K., Damianos, S.P., Dimitrios, N.K., Dimitrios, V.S.: Modelling the process of a web-based collaboration tool development. In: Strategic Innovative Marketing, pp. 383–389. Springer International Publishing, Cham (2017). https://doi.org/10.1007/978-3-319-56288-9_51
21. Semrush: Traffic analytics overview report. https://www.semrush.com/kb/895-traffic-analytics-overview-report. Last accessed 4 Apr 2022
22. Statista: Global startups – statistics & facts. https://www.statista.com/topics/4733/startups-worldwide/#dossierContents__outerWrapper. Last accessed 20 June 2022
23. Salmeron, J.L.: Supporting decision makers with fuzzy cognitive maps. Res. Technol. Manag. **52**, 53–59 (2009). https://doi.org/10.1080/08956308.2009.11657569
24. Giabbanelli, P.J., Gray, S.A., Aminpour, P.: Combining fuzzy cognitive maps with agent-based modeling: frameworks and pitfalls of a powerful hybrid modeling approach to understand human-environment interactions. Environ. Model. Softw. **95**, 320–325 (2017). https://doi.org/10.1016/j.envsoft.2017.06.040
25. Dimitrios, R.: Cloud-based knowledge management in Greek SME's. http://www.diva-portal.org/smash/record.jsf?pid=diva2:1261304 (2018)
26. Reklitis, P., Trivellas, P., Mantzaris, I., Mantzari, E., Reklitis, D.: Employee perceptions of corporate social responsibility activities and work-related attitudes: the case of a Greek Management Services Organization. In: Gal, G., Akisik, O., Wooldridge, W. (eds.) Sustainability and Social Responsibility: Regulation and Reporting, pp. 225–240. Springer Singapore, Singapore (2018). https://doi.org/10.1007/978-981-10-4502-8_10

Competitive Differentiation in Airlines' Mobile Applications with the Assistance of Eye-Tracking Research

Damianos P. Sakas, Dimitrios P. Reklitis, and Panagiotis Trivellas

1 Introduction and Research Hypotheses

It is vital for airlines to use digital marketing techniques in order to accomplish competitive differentiation [1]. A company's continual relationship with its customers is built with the assistance of electronic commerce [2]. In order to obtain more attention, companies promote their products through their mobile applications and their web pages [3]. As a consequence of that, airline companies could gain market share and accomplish competitive advantage through the correct usage of their platforms [4].

Social media advertising is among the most essential and effective ways to acquire new customers since it is easier to establish brand recognition on social media [5–11]. Additionally, by using social media marketing, airline companies may increase their brand name while simultaneously reducing misunderstandings that hurt their brand [12]. Despite that the amount of followers for airline companies has grown, user involvement has remained at low levels [12]. This might be averted if the marketing was "driven by ad content," implying that customers attach great importance to ad content, especially if the ad can solve a specific issue or meet a specific customer need, such as booking a ticket as fast as possible [13].

D. P. Sakas · D. P. Reklitis (✉)
Department of Agribusiness and Supply Chain Management, BICTEVAC LABORATORY Business Information and Communication Technologies in Value Chains laboratory, School of Applied Economics and Social Sciences, Agricultural University of Athens, Athens, Greece
e-mail: d.sakas@aua.gr; drekleitis@aua.gr

P. Trivellas
Department of Agribusiness and Supply Chain Management, Organizational Innovation and Management Systems, School of Applied Economics and Social Sciences, Agricultural University of Athens, Athens, Greece
e-mail: ptrivel@aua.gr

D. P. Sakas et al. (eds.), *Computational and Strategic Business Modelling*, Springer Proceedings in Business and Economics,
https://doi.org/10.1007/978-3-031-41371-1_9

Table 1 KPIs

Web analytics KPISs	
Social media traffic	The metric social media traffic illustrates the redirections from social media sites such as Facebook, Twitter, and YouTube to the corporate mobile app [15].
Average time on site	This KPI illustrates a median estimate of how much time each visitor stays on a mobile app [15]
Bounce rate	This metric presents the percentage of users that exit after seeing only one page in the mobile app [15]
Pages per visit	This metric illustrates an average of the total pages a visitor views on the mobile app in a single session [15]
Total visits	This metric presents the amount of total mobile app visitors for the given period [15]

In order to examine the possible factors that play a role in the digital marketing strategy, five big data key performance indicators [14] have been extracted and used in the formulation of the research hypothesis (Table 1).

Since the study of the digital marketing key performance indicators is critical for the acquisition of competitive differentiation of airlines, those KPIs were utilized in the formation of the hypothesis. Three key hypotheses have been created in the process of selecting the best digital marketing approach for airline firms to use. These study hypotheses are important for decision-makers since the findings might provide useful information about visitors' activity on a mobile app.

Hypothesis 1 The "Pages per Visits" affects the "Social Media Traffic" variable through their "Bounce Rate" metric.

Hypothesis 2 "Pages per Visits" affects the "Average Time on Site" variable through their "Total Visits" metric.

Hypothesis 3 The eye-tracking and facial expression research provide an added value to airline companies to achieve competitive differentiation from the examination of user engagement in their mobile apps.

2 Methodology

In this study, we implemented an approach that includes statistical analysis, Fuzzy Cognitive Maps, and eye-tracking research in order to assess the consumers' mobile behavior. At first, behavioral big data was collected each and every day from four world-leading Airline mobile apps over 9 months. Next, those extracted data, are processed statistically and produce the Pearson's correlations for the two main hypotheses. Following that, the correlations were utilized to create a mobile engagement optimization scenario that could be implemented by airline companies. Finally, an eye-tracking experiment was conducted with the assistance of eight participants.

3 Results

3.1 *Statistical Analysis*

Tables 2 and 3 present the results of the Pearsons' correlations.

As for the H1, the positive correlation between social media traffic and the bounce rate; with $\rho = 0.577^{*}$, indicates that the majority of the visitors that enter mobile apps from the social media advertisements entered accidentally or couldn't find what they needed on time and exited from the app fast. The interesting negative correlation between the social media traffic and the total visits; with $\rho = -0.520^{*}$, illustrates that the vast majority of the visitors to the mobile apps came from other sources such as Google or search engines.

As for the H2, the positive correlation between the average time on site and the pages per visit; with $\rho = 0.911^{**}$, illustrates that there is a high connection between those two behavioral factors. The negative correlation between the pages per visit and the total visits; with $\rho = -0.989^{**}$, indicates that mobile app users aim to book a ticket as fast as possible without redirecting to a lot of pages.

3.2 *Fuzzy Cognitive Map*

The correlation is incorporated into the Fuzzy Cognitive Map (FCM) in order to illustrate the macro analysis between the Key Performance Indicators (Fig. 1) [16].

We created a mobile engagement optimization scenario after applying Pearson's correlations to the FCM (Fig. 2). More specifically, in order to optimize mobile engagement by attracting more visitors by 33%, airline companies need to invest more in social media advertisements for mobile apps.

Table 2 H1 Pearsons' correlation

H1	Social Media Traffic	Total visits	Bounce rate
Social media traffic	1		
Total visits	-0.520^{*}	1	
Bounce rate	0.577^{*}	-0.976^{**}	1

** Correlation is significant at the 0.01 level (1-tailed), *Correlation is significant at the 0.05 level (1-tailed)

Table 3 H2 Correlations

H2	Average time on site	Pages per visits	Total Visits
Average time on site	1		
Pages per visit	0.911^{**}	1	
Total visits	0.867^{**}	-0.989^{**}	1

** Correlation is significant at the 0.01 level (1-tailed)

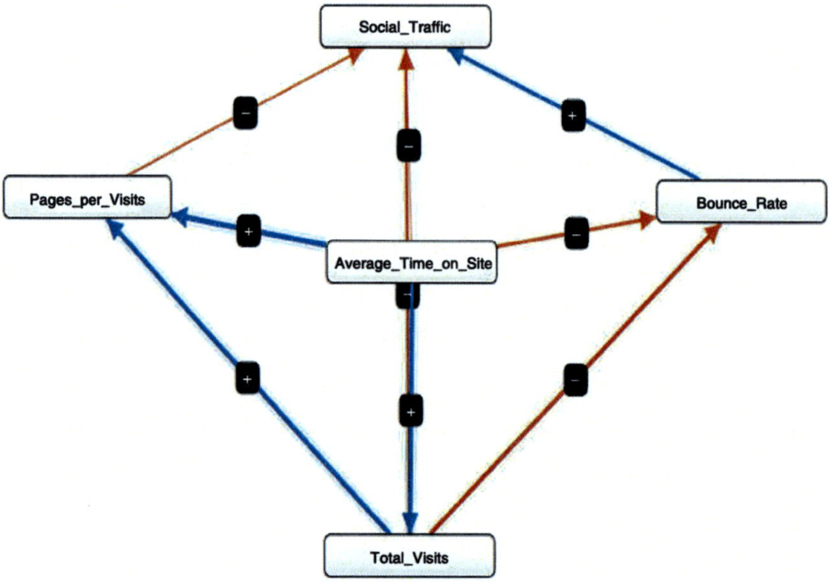

Fig. 1 FCM

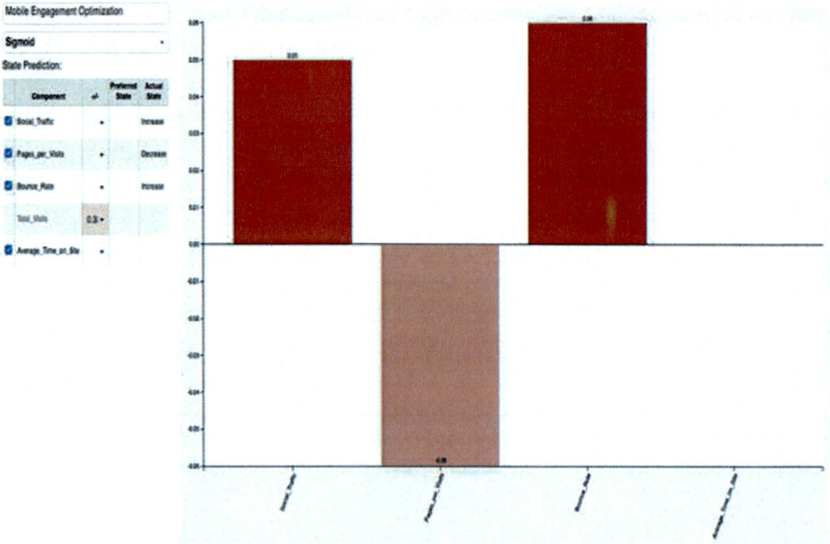

Fig. 2 Mobile engagement optimization scenario

3.3 Neuromarketing Analysis

Because of the strengths of neuromarketing analysis, this study has included the involvement of eight participants who had prior experience with airline mobile applications. All neuroethics criteria were considered in the research, including "Transparency and impartiality" and "Subjects' Consent to Participate in the Study" [17]. The applicants agreed to take part in the study because they wanted to learn more about mobile engagement and involvement, as well as evaluate the success of various digital marketing techniques. The smartphone application "UXReality" was utilized to determine the user's facial expressions and eye-tracking for this neuromarketing research. The users were provided with a weblink and instructions. The participants were asked to book a ticket using each airline's mobile app.

Figure 3 illustrates the heatmap, while Fig. 4 displays a snapshot of the process. The core focus of this neuromarketing analysis is to gather the metrics "AVG Duration" and "Involvement", to verify the results of the FCM and the statistical analysis. The results, shown in Fig. 4, reveal that while the participants were able to finish the assigned task, their engagement "Involvement" with the digital platform was low. In the center of Fig. 3, a heat map is displayed, which represents the area in which the user's attention was caught. Furthermore, as shown on the right-hand side

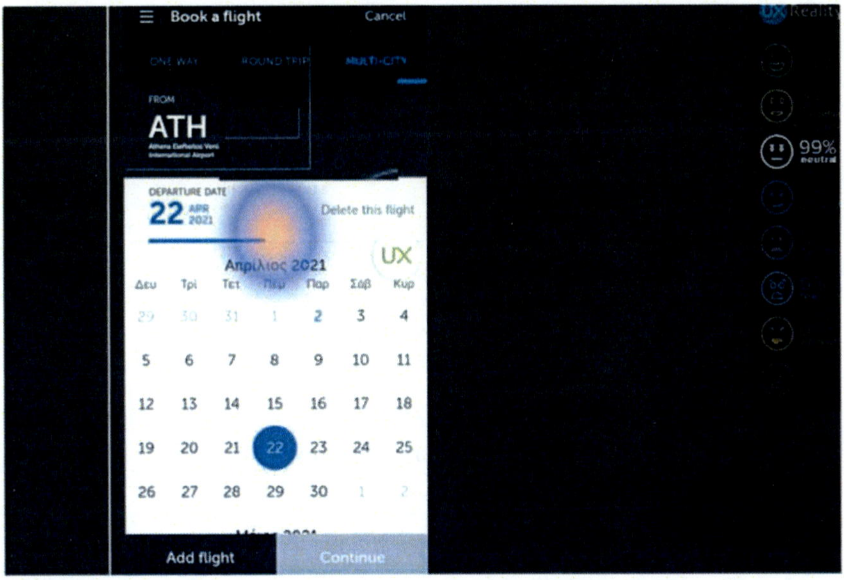

Fig. 3 Heatmap

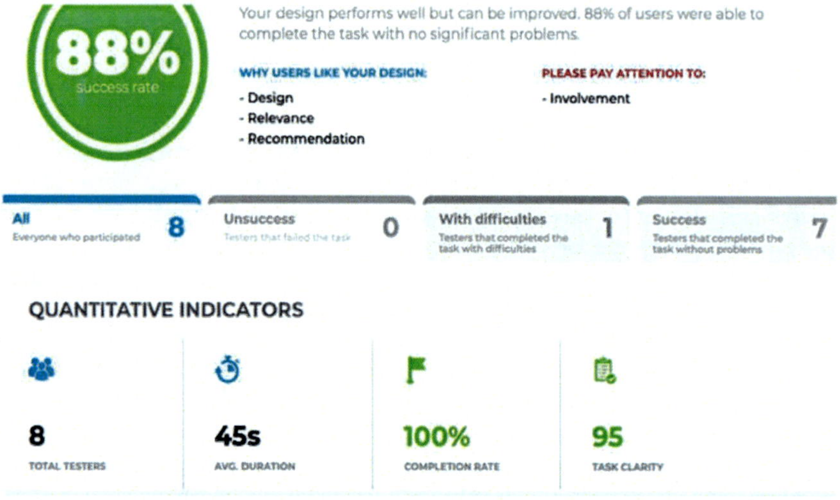

Fig. 4 Eye-tracking results

of Fig. 3, the user remains "neutral"; the heatmap does not record any depiction of strong emotion and focuses, as was anticipated, on the selection of the fictional departure date. Figure 4 presents the assigned tasks, and the results reveal that the task clarity obtained a high marking of 95%, while the completion rate of the assigned tasks ranked at 100%. On average, the participants completed the assigned tasks in 45 seconds. These findings reveal that although the assigned tasks contained clear instructions, the user's involvement remained low.

4 Conclusion and Limitations

The study suggests to that airlines' companies develop their mobile apps with clear instructions on how to book a ticket in order to accomplish competitive differentiation as well as invest in social media advertisements. The limitation of the study can be identified in the number of the examined airlines and the number of participants in the experiment. Finally, future research on the possible benefits of the integration of knowledge management techniques in those platforms could be useful. Finally, future research on the possible benefits of the integration of knowledge management techniques in those platforms and their effects on corporate social responsibility could be useful [18, 19].

References

1. Kerdpitak, C., Pungnirund, B., Hotrawaisaya, C., Jariyachamsit, S., Yen, W.-H., Chantranon, S.: Effect of competitive advantage, digital marketing to supply chain management on tourism business performance in Thailand. Uncertain Supply Chain Manag. **10**, 721–728 (2022). https://doi.org/10.5267/j.uscm.2022.5.003
2. Sakas, D.P., Reklitis, D.P.: The Impact of Organic Traffic of Crowdsourcing Platforms on Airlines' Website Traffic and User Engagement. Sustainability. **13**, 8850 (2021). https://doi.org/10.3390/su13168850
3. Sakas, D.P., Giannakopoulos, N.T.: Big Data Contribution in Desktop and Mobile Devices Comparison, Regarding Airlines' Digital Brand Name Effect. Big Data Cogn Comput. **5**, 48 (2021). https://doi.org/10.3390/bdcc5040048
4. Karaağaoğlu, N., Çiçek, M.: An evaluation of digital marketing applications in airline sector. Human Sci. **16**, 606–619 (2019)
5. Sakas, D.P., Giannakopoulos, N.T., Reklitis, D.P., Dasaklis, T.K.: The Effects of Cryptocurrency Trading Websites on Airlines' Advertisement Campaigns. J. Theor. Appl. Electron. Commer. Res. **16**, 3099–3119 (2021). https://doi.org/10.3390/jtaer16070169
6. Sakas, D.P., Reklitis, D.P.: Predictive model for estimating the impact of technical issues on consumers interaction in agri-logistics websites. In: Information and Communication Technologies for Agriculture—Theme IV: Actions, pp. 269–283. Springer International Publishing, Cham (2021). https://doi.org/10.1007/978-3-030-84156-0_14
7. Sakas, D.P., Kamperos, I.D.G., Reklitis, D.P., Giannakopoulos, N.T., Nasiopoulos, D.K., Terzi, M.C., Kanellos, N.: The effectiveness of centralized payment network advertisements on digital branding during the COVID-19 crisis. Sustainability. **14**, 3616 (2022). https://doi.org/10.3390/su14063616
8. Sakas, D.P., Giannakopoulos, N.T.: Harvesting crowdsourcing platforms' traffic in favour of air forwarders' brand name and sustainability. Sustainability. **13**, 8222 (2021). https://doi.org/10.3390/su13158222
9. Sakas, D.P., Kamperos, I.D.G., Reklitis, P.: Estimating risk perception effects on courier companies' online customer behavior during a crisis, using crowdsourced data. Sustainability. **13**, 12725 (2021). https://doi.org/10.3390/su132212725
10. Nasiopoulos, D.K., Sakas, D.P., Reklitis, P.: The interaction of users' behavior to the development of a distribution and logistics network of the information technology companies. In: Business Intelligence and Modelling, pp. 219–227. Springer International Publishing, Cham (2021). https://doi.org/10.1007/978-3-030-57065-1_23
11. Sakas, D.P., Reklitis, D.P., Trivellas, P., Vassilakis, C., Terzi, M.C.: The effects of logistics websites' technical factors on the optimization of digital marketing strategies and corporate brand name. Processes. **10**, 892 (2022). https://doi.org/10.3390/pr10050892
12. Grieve, R., Witteveen, K., Tolan, G.A.: Social media as a tool for data collection: examining equivalence of socially value-laden constructs. Curr. Psychol. **33**, 532–544 (2014). https://doi.org/10.1007/s12144-014-9227-4
13. Bühler, J., Lauritzen, M., Bick, M.: Download limit exceeded, http://citeseerx.ist.psu.edu/viewdoc/download?doi=10.1.1.962.9801&rep=rep1&type=pdf. Last accessed 12 June 2022.
14. Fitz-Gibbon, C.T.: Performance Indicators. Multilingual Matters (1990)
15. Semrush: Traffic Analytics Overview Report, https://www.semrush.com/kb/895-traffic-analytics-overview-report. Last accessed 4 Apr 2022.
16. Giabbanelli, P.J., Gray, S.A., Aminpour, P.: Combining fuzzy cognitive maps with agent-based modeling: Frameworks and pitfalls of a powerful hybrid modeling approach to understand human-environment interactions. Environ. Model Softw. **95**, 320–325 (2017). https://doi.org/10.1016/j.envsoft.2017.06.040
17. Santos, Oliveira, Rocha: Eye tracking in neuromarketing: a research agenda for marketing studies. Int. J. High Risk Behav. Addict. (2015)

18. Reklitis, P., Trivellas, P., Mantzaris, I., Mantzari, E., Reklitis, D.: Employee perceptions of corporate social responsibility activities and work-related attitudes: the case of a greek management services organization. In: Gal, G., Akisik, O., Wooldridge, W. (eds.) Sustainability and Social Responsibility: Regulation and Reporting, pp. 225–240. Springer Singapore, Singapore (2018). https://doi.org/10.1007/978-981-10-4502-8_10
19. Dimitrios, R.: Cloud-based Knowledge Management in Greek SME's, http://www.diva-portal.org/smash/record.jsf?pid=diva2:1261304, (2018).

Digital Marketing Strategy for Competitive Advantage Acquisition Through Neuromarketing in the Logistics Sector

Damianos P. Sakas, Dimitrios P. Reklitis, and Panagiotis Trivellas

1 Introduction and Research Hypothesis

Digital marketing can be defined as the promotion of companies via the use of the internet in order to interact with potential clients [1]. In order to reach those customers, companies must use all those digital strategies and techniques that are available as well as develop various mechanisms to evaluate the results [2, 3]. According to previous research, implementing an efficient digital marketing strategy is crucial for optimizing the corporate brand name, profit, engagement, and user experience [4–13]. More specifically, logistics companies need to invest in digital marketing since is one of the key elements in the creation of competitive advantage [14]. Another crucial aspect of digital competitive advantage is the incorporation of big data in digital marketing strategy [14]. This can be beneficial since the great number of those data can provide a clear picture of the logistics of digital marketing campaign results in terms of visibility and user engagement [1, 4, 6].

Furthermore, the adoption of a neuromarketing analysis could be valuable in order to identify users' engagement with logistics companies' websites. An eye-tracking analysis is only but a minor sub-part of wider neuromarketing analysis techniques. Increasingly, researchers rely on eye-tracking analysis to draw

D. P. Sakas · D. P. Reklitis (✉)
Department of Agribusiness and Supply Chain Management, BICTEVAC LABORATORY Business Information and Communication Technologies in Value Chains laboratory, School of Applied Economics and Social Sciences, Agricultural University of Athens, Athena, Greece
e-mail: d.sakas@aua.gr; drekleitis@aua.gr

P. Trivellas
Department of Agribusiness and Supply Chain Management, Organizational Innovation and Management Systems, School of Applied Economics and Social Sciences, Agricultural University of Athens, Athens, Greece
e-mail: ptrivel@aua.gr

D. P. Sakas et al. (eds.), *Computational and Strategic Business Modelling*,
Springer Proceedings in Business and Economics,
https://doi.org/10.1007/978-3-031-41371-1_10

Table 1 Web analytics KPIs

KPIs	Description of the KPIs
Organic traffic	Organic traffic refers to visitors who enter the webpage without clicking or being redirected from a paid advertisement [16].
Average visits duration	This key performance indicator monitors the amount of time a user spends on a webpage [16].
Bounce rate	The KPI bounce rate is generated when a customer visits a website and then leaves without viewing anything else [16].
Pages per visit	This KPI illustrates the number of pages that a user accesses on a website [16].
Total visitors	This KPI measures the daily total number of people who visit a logistics website [16].

conclusions from both corporate websites and industries as a whole [15]. Eye-tracking appears to be a useful method that proves to be valid when used in combination with facial expression detection [15], as it reveal the emotional state of the user. For these reasons, a multitude of software and hardware can be used to encourage the use of eye-tracking analysis. For the purposes of this study, eye-tracking analysis was chosen as it enables the generation of outputs from user engagement metrics and in particular users' emotional depiction and Involvement variables. But which are those involvement variables? For this research, the authors gathered from web analytics platforms the following Key Performance Indicators (KPIs) as depicted in Table 1.

Those KPIs have been used in the formulation of the research hypothesis since the analysis of the digital KPIs is crucial for the corporate digital marketing strategy. In the process of identifying the optimum digital marketing strategy that can be adopted by the logistics companies, three main research hypotheses have been formulated. These research hypotheses are crucial for marketers and website developers since their results could produce valuable insights into the users' conscious and subconscious behavior on a website.

Hypothesis 1 The "Average Visits Duration" affects the "Organic Traffic" variable through their "Bounce Rate" metric.

Hypothesis 2 "Pages per Visits" affects the "Total Visitors" variable through their "Bounce Rate" metric.

Hypothesis 3 The eye-tracking research provides an added value to digital marketing strategy in logistics companies through the assessment of user engagement.

2 Methodology

In this chapter, the authors followed a methodology that incorporates both quantitative and qualitative aspects of users' digital behavior. First, raw behavioral big data were gathered from three logistics leading companies, daily, over a 12-month period

[17] and analyzed statistically including correlation and regressions. Followingly, the results of the statistical analysis have been used to develop a visibility and SEO optimization scenario which can be adopted from the logistics with low cost and could provide high results. Finally, to confirm the results obtained from the web analytics quantitative analysis, a qualitative analysis has been conducted with the use of eye-tracking neuromarketing analysis.

Moreover, in order to identify user engagement within logistics company websites, a neuromarketing eye-tracking analysis was conducted with the assistance of 16 adults. Eye-tracking was selected for the purposes of this research, as this form of analysis helps identify the areas of highest fixations within a website [18]. When individuals click on a website their eyes immediately focus on information that attracts their attention and will inherently display emotions during this observation process. Research has shown that emotional pictures and explicit context is more likely to attract the attention of a website visitor, who has also displayed more positive emotions during this type of exposure [19]. Before taking part, each participant provided written informed permission.

3 Results

3.1 *Statistical Analysis*

The results of the first hypothesis are presented in Tables 2 and 3. Table 2 depicts the correlations between the examined metrics and more specifically, significant correlations can be observed between the bounce rate, the organic traffic, and the average visit duration with $\rho = 0.223^{**}$, and $\rho = 0.608^{**}$ respectively. Additionally, a significant negative correlation can be observed between the organic traffic and the average visit duration with $\rho = -0.207^{**}$. Table 3 illustrates the regression analysis for the H1. The results are significant ($p < 5\%$) and illustrate a 1% increase in organic traffic, followed by a 54.3% decrease in average visit duration and a 55.4%

Table 2 H1 correlations

H1	Organic traffic	Average visits duration	Bounce rate
Organic traffic	1		
Average visits duration	-0.207^{**}	1	
Bounce rate	0.223^{**}	0.608^{**}	1

**Correlation is significant at the 0.01 level (1-tailed)

Table 3 H1 regression

Variables	Standardized coefficient	R^2	F	p value
Constant (organic traffic)	–	0.236	55.140	<0.001
Average visits duration	−0.543			<0.001
Bounce rate	0.554			<0.001

Table 4 H2 correlations

H2	Total visitors	Pages per visits	Bounce rate
Total visitors	1		
Pages per visit	−0.050	1	
Bounce rate	0.197**	−0.886**	1

**Correlation is significant at the 0.01 level (1-tailed)

Table 5 Web analytics KPIs

Variables	Standardized coefficient	R^2	F	p-value
Constant (total visitors)	–	0.111	22.252	<0.001
Pages per visits	0.580			<0.001
Bounce rate	0.711			<0.001

increase in bounce rate. The findings suggest that when the organic traffic increases, the average visit duration decreases which illustrates that either some visitors enter the website accidentally or complete their query very fast.

Tables 4 and 5 illustrate the results of the H2. Table 4 illustrates the Pearson correlations. More specifically, significant correlations can be found among the bounce rate, the pages per visit, and total visitors with $\rho = -0.886^{**}$ and $\rho = 0.197^{**}$, respectively. Additionally, a negative correlation was observed between the total visitors and pages per visit with $\rho = -0.050$. Table 5 presents the regression analysis for the second hypothesis. The results are significant ($p < 5\%$) and suggest a 1% increase in total visitors, followed by a 58.8% increase in pages per visit and 71.1% increase in bounce rate. The results suggest that when the total number of visitors increases, the number of pages per visit decreases which illustrates that the logistics websites to be effective need to be developed with fewer pages but with clear instructions.

3.2 Formulation of Fuzzy Cognitive Map

The above statistical analysis has been used to create a macro scale fuzzy cognitive map (Fig. 1). The fuzzy cognitive map (FCM) illustrates the cause-and-effect relationship between the KPIs using the (−1,1) range [20].

After incorporating the statistical analysis into the FCM, the authors present a visibility and SEO optimization scenario (Fig. 2). More specifically, if the logistics companies wish to improve their website visibility by 20%, companies need to improve their website development and enrich it with more keywords in order to elevate search engines. Furthermore, the websites need to be developed with fewer web pages because this creates less frustration for the user.

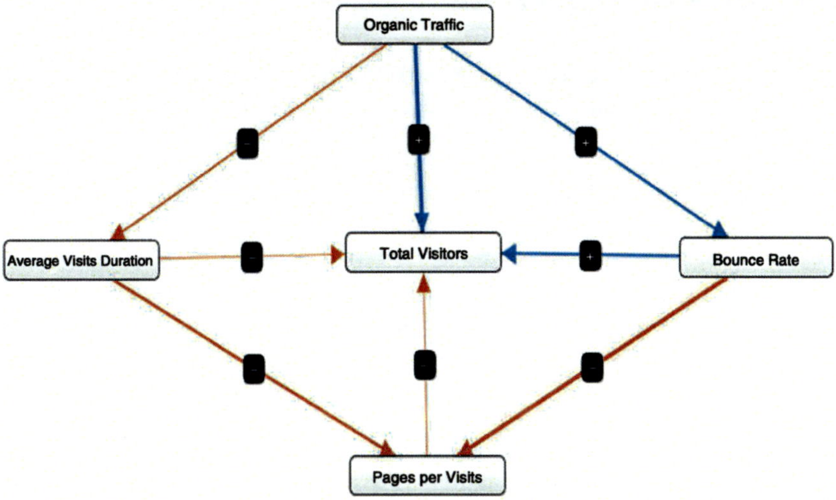

Fig. 1 Fuzzy cognitive map

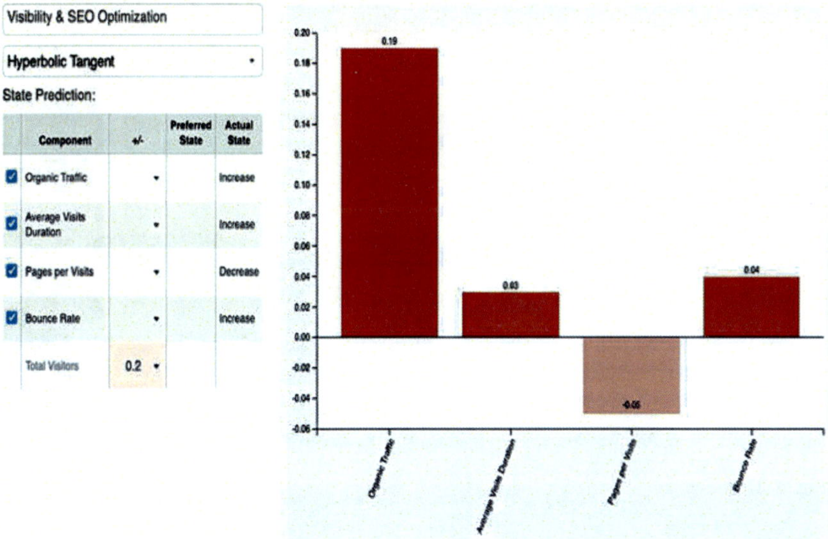

Fig. 2 Visibility and SEO scenario

3.3 Neuromarketing Analysis

Due to the benefits provided by neuromarketing research, this study integrated the participation of 16 adults who had experience using logistics platforms. The study took into consideration all the neuroethics parameters, such as "Transparency and objectivity" and "Consent of subjects to participate in the study" [21]. The candidates volunteered to participate in the study to gather data about user engagement and participation and to consider the effectiveness of potential digital marketing strategies. To carry out this neuromarketing analysis, the mobile application "UXReality" was used to identify the facial expressions and eye-tracking of the user.

The users were provided with a weblink and instructions. The participants were asked to access a logistics website where they were asked to assess the efficacy of the booking process used on the website. Figure 4 illustrates the relevant results for H3, while Fig. 3 displays a snapshot of the process. The core focus of this neuromarketing analysis is to gather the metric "Involvement" and the variety of experienced emotions during the process. The results, shown in Fig. 4, reveal that while the participants were able to finish the assigned task, their "Involvement" with the digital platform was low. On the right-hand side of Fig. 3 a heat map is displayed, which represents the area in which the user's attention was caught. Additionally, a snapshot of the current emotion is illustrated at the same time. Furthermore, as shown on the right-hand side of Fig. 3, the user's status is "anger"; when tries to find

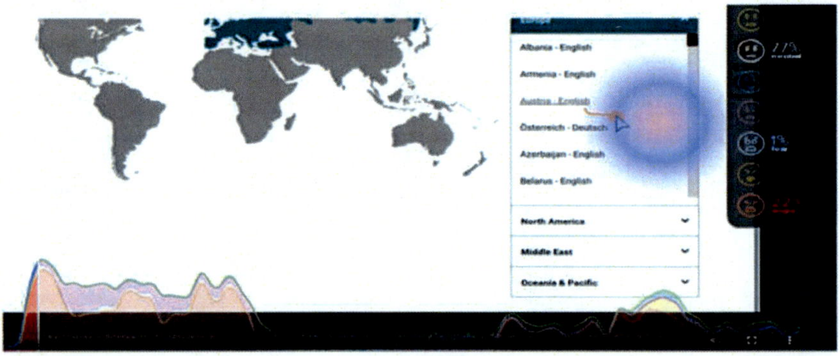

Fig. 3 Heatmap

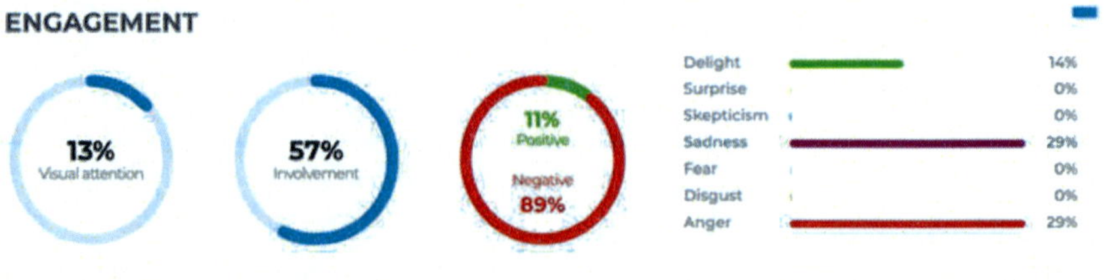

Fig. 4 Eye-tracking results

the correct webpage to search for a booking and cannot find it. Figure 4 presents the assigned tasks, and the results reveal that the "visual attention" obtained a high marking of 13%, while the "Involvement" obtained 57% and the study concluded with the average total emotional state of the participants in 89% negative.

These findings reveal that although the assigned tasks contained clear instructions, the user's "Involvement" remained low. Additionally, the longer it takes to search for a parcel creates sadness and anger. To optimize the SEO strategy logistics companies must make the websites with fewer pages with big and clear instructions. This result also highlights the findings of previous research which suggest creating easy-to-use logistics websites in order to increase user engagement [14].

4 Conclusion and Limitations

In order to improve digital marketing strategy, logistics companies must develop their websites with fewer web pages as well as have to provide clear instructions on the main pages. The limitation of this research lies in the fact that 3pl companies used. Further research on other sectors could provide a more holistic view [22–24].

References

1. Chaffey, D., Ellis-Chadwick, F.: Digital Marketing: Strategy, Implementation & Practice. Pearson (2019)
2. Olson, E.M., Olson, K.M., Czaplewski, A.J., Key, T.M.: Business strategy and the management of digital marketing. Bus. Horiz. **64**, 285–293 (2021). https://doi.org/10.1016/j.bushor.2020.12.004
3. Zhu, G., Gao, X.: Precision retail marketing strategy based on digital marketing model. Sci J Business Manag. **7**, 33–37 (2019)
4. Sakas, D.P., Reklitis, D.P.: The impact of organic traffic of crowdsourcing platforms on airlines' Website traffic and user engagement. Sustainability. **13**, 8850 (2021). https://doi.org/10.3390/su13168850
5. Sakas, D.P., Giannakopoulos, N.T., Reklitis, D.P., Dasaklis, T.K.: The effects of cryptocurrency trading websites on airlines' advertisement campaigns. J. Theor. Appl. Electron. Commer. Res. **16**, 3099–3119 (2021). https://doi.org/10.3390/jtaer16070169
6. Sakas, D.P., Reklitis, D.P.: Predictive model for estimating the impact of technical issues on consumers interaction in agri-logistics websites. In: Information and Communication Technologies for Agriculture—Theme IV: Actions, pp. 269–283. Springer International Publishing, Cham (2021). https://doi.org/10.1007/978-3-030-84156-0_14
7. Sakas, D.P., Kamperos, I.D.G., Reklitis, D.P., Giannakopoulos, N.T., Nasiopoulos, D.K., Terzi, M.C., Kanellos, N.: The effectiveness of centralized payment network advertisements on digital branding during the COVID-19 crisis. Sustainability. **14**, 3616 (2022). https://doi.org/10.3390/su14063616
8. Sakas, D.P., Reklitis, D.P., Terzi, M.C., Glaveli, N.: Growth of digital brand name through customer satisfaction with big data analytics in the hospitality sector after the COVID-19 crisis. Int J Inf Manag Data Insights. **3**, 100190 (2023). https://doi.org/10.1016/j.jjimei.2023.100190

9. Sakas, D.P., Reklitis, D.P., Terzi, M.C.: Leading logistics Firms' Re-Engineering through the Optimization of the Customer's Social Media and Website Activity. Electronics. **12**, 2443 (2023). https://doi.org/10.3390/electronics12112443
10. Sakas, D.P., Kamperos, I.D.G., Reklitis, P.: Estimating risk perception effects on courier companies' online customer behavior during a crisis, using crowdsourced data. Sustainability. **13**, 12725 (2021). https://doi.org/10.3390/su132212725
11. Nasiopoulos, D.K., Sakas, D.P., Reklitis, P.: The interaction of users' behavior to the development of a distribution and logistics network of the information technology companies. In: Business Intelligence and Modelling, pp. 219–227. Springer International Publishing, Cham (2021). https://doi.org/10.1007/978-3-030-57065-1_23
12. Nasiopoulos, D.K., Sakas, D.P., Trivellas, P.: The role of digital marketing in the development of a distribution and logistics network of information technology companies. In: Business Intelligence and Modelling, pp. 267–276. Springer International Publishing, Cham (2021). https://doi.org/10.1007/978-3-030-57065-1_27
13. Sakas, D.P., Nasiopoulos, D.K., Reklitis, P.: Modeling and simulation of the strategic use of marketing in search engines for the business success of high technology companies. In: Strategic Innovative Marketing, pp. 217–226. Springer International Publishing, Cham (2019). https://doi.org/10.1007/978-3-030-16099-9_27
14. Sakas, D.P., Reklitis, D.P., Trivellas, P., Vassilakis, C., Terzi, M.C.: The effects of logistics websites' technical factors on the optimization of digital marketing strategies and corporate brand name. Processes. **10**, 892 (2022). https://doi.org/10.3390/pr10050892
15. Białowąs, S., Szyszka, A.: Eye-tracking in Marketing Research. Managing Economic Innovations--Methods and Instruments. 91 (2019).
16. Semrush: Traffic Analytics Overview Report, https://www.semrush.com/kb/895-traffic-analytics-overview-report. Last accessed 4 Apr 2022
17. Statista: Logistics companies worldwide in 2020, based on logistics revenue, https://www.statista.com/statistics/270168/logistics-companies-worldwide-by-revenue/. Last accessed 4 Dec 2022
18. Oliveira, P.M., Guerreiro, J., Rita, P.: Neuroscience research in consumer behavior: A review and future research agenda. Int. J. Consum. Stud. (2022). https://doi.org/10.1111/ijcs.12800
19. Azimzade, M., Nobakht, F., Aminiroshan, Z., Sotoode, M.S.: Employing eye tracking in quantifying and qualifying visual attention of web site viewers (physical education faculties). J Adv Sport Technol. **4**, 9–19 (2020)
20. Salmeron, J.L.: Supporting decision makers with fuzzy cognitive maps. Res. Technol. Manag. **52**, 53–59 (2009). https://doi.org/10.1080/08956308.2009.11657569
21. Bercea Olteanu, M.D.: Neuroethics and responsibility in conducting neuromarketing research. Neuroethics. **8**, 191–202 (2015). https://doi.org/10.1007/s12152-014-9227-y
22. Dimitrios, R.: Cloud-based Knowledge Management in Greek SME's, http://www.diva-portal.org/smash/record.jsf?pid=diva2:1261304 (2018).
23. Reklitis, P., Trivellas, P., Mantzaris, I., Mantzari, E., Reklitis, D.: Employee perceptions of corporate social responsibility activities and work-related attitudes: The case of a greek management services organization. In: Gal, G., Akisik, O., Wooldridge, W. (eds.) Sustainability and Social Responsibility: Regulation and Reporting, pp. 225–240. Springer Singapore, Singapore (2018). https://doi.org/10.1007/978-981-10-4502-8_10
24. Chantzaras, A., Dimitrios, N.K., Vlachos, D.S.: Mobile commerce and success factors. Simulation and modeling of the problem. In: Strategic Innovative Marketing, pp. 349–355. Springer International Publishing, Cham (2017). https://doi.org/10.1007/978-3-319-56288-9_46

Social Media Analytics for Customer Satisfaction Based on User Engagement and Interactions in the Tourism Industry

Damianos P. Sakas, Dimitrios P. Reklitis, and Panagiotis Trivellas

1 Introduction and Hypotheses Developments

When travelers are making vacation plans and making hotel reservations, they frequently look for locations they have never stayed at before [1]. Gathering hotel information might help them to make better booking decisions [2–4]. Word of mouth (WOM) from friends has historically been used to lower risks, and as the web has gone mainstream, WOM is starting to become more significant [5].

Owing to the difficulty of the customer experience, determining customer satisfaction is a basic task for hotel management. Studies have shown that there is a discrepancy between hotel administrators' ideas of what constitutes customer satisfaction and what precisely clients think is important when they reserve and review their hotel accommodations in the hospitality sector [6, 7].

This study field is difficult to understand using conventional survey methodologies. Therefore, it is critical to create new metrics and research methodologies to investigate the variables influencing consumer satisfaction. Utilizing big data and web analytic (WA) sources and cutting-edge research techniques to learn more about customer satisfaction is an intriguing area for future study [8, 9].

D. P. Sakas · D. P. Reklitis (✉)
Department of Agribusiness and Supply Chain Management, BICTEVAC LABORATORY Business Information and Communication Technologies in Value Chains laboratory, School of Applied Economics and Social Sciences, Agricultural University of Athens, Athena, Greece
e-mail: d.sakas@aua.gr; drekleitis@aua.gr

P. Trivellas
Department of Agribusiness and Supply Chain Management, Organizational Innovation and Management Systems, School of Applied Economics and Social Sciences, Agricultural University of Athens, Athens, Greece
e-mail: ptrivel@aua.gr

D. P. Sakas et al. (eds.), *Computational and Strategic Business Modelling*, Springer Proceedings in Business and Economics,
https://doi.org/10.1007/978-3-031-41371-1_11

Table 1 Web analytics

WA	Description of the WA
Page performance index	"The Page Performance Index (PPI) is a combination of engagement and growth" [14]
Follower growth	This analytic presents the percentage of followers growth on social media per month [14]
Number of likes	This metric presents the total likes of the social media page [14].
Number of friends	This metric presents the total number of friends on the social media page [14].
Engagement	Refers to the total interactivity of the customer with social media, including posts, likes, reactions, and comments [14].
Post interaction	Refers to the number of interactions per post [14].

Numerous research on the topic has been conducted since social media has emerged as a crucial asset in the tourism sector for digital marketing and digital advertising [9, 10]. Before choosing where should stay, the potential clients, utilize online rating sites to do research and interact with corporate social media [11]. It is very important nowadays for potential customers to interact with corporate social media in order to book an apartment or even to evaluate their experience [12, 13]. For the purpose of this study, we gathered the following web analytics that will help us to indirectly evaluate the customer experience (Table 1).

It is essential to develop the study hypothesis after establishing the retrieved web analytics. The two hypotheses are created in a way that will assist the marketers in comprehending how to use those KPIs and incorporating them into their overall plan and will provide an indirect picture of customer satisfaction.

Hypothesis 1 The "Page Performance Index" affects the "Follower Growth" variable through the "Number of Likes" metric.

Hypothesis 2 The "Number of Friends" affects the "Page Performance Index" variable through their "Engagement" and the "Post Interaction" metrics.

2 Methodology

The authors' research involved analyzing big data and web analytics to look at the quantitative aspects of customer satisfaction. Over the course of six months, daily big data was collected from eight world-leading hospitality social media [15]. This data was then statistically analyzed using correlation and regressions. The results of the statistical investigation were then used to develop a fuzzy cognitive map and two optimization scenarios. This method was utilized since there was no cognitive bias present in the big data that had been obtained in its unprocessed form [16].

3 Results

3.1 Statistical Analysis

Tables 2 and 3 illustrate the findings of the H1. We can observe two significant correlations between the follower growth and the performance index as well as with the number of likes with $\rho = 0.270^{*}$ and $\rho = 0.286^{*}$, respectively. That highlights that when the customers interact with the hospitality websites, they gain more followers and the brand name increases [17, 18]. Additionally, the regression suggests that for every 1% increase in followers, the page performance index, and the number of likes increase by 27.5% and 25.9%, respectively. These significant results, with $p < 0.005$ suggest to social media administrators to produce more posts per day to increase interactivity and as a consequence, followers and customer satisfaction will be increased.

Tables 4 and 5 illustrate the findings of the H2. We can observe two significant correlations between the performance index with the number of fans as well as with the post-interaction with $\rho = -0.633^{**}$ and $\rho = 0.496^{**}$, respectively. That highlights that when the customers interact with the hospitality websites, they don't gain more fans but the interaction with the post increases. Additionally, the regression suggests that for every 1% increase in page performance index, the

Table 2 H1 correlations

H1	Follower growth	Page performance Index	Number of likes
Follower growth	1		
Page performance index	0.270*	1	
Number of likes	0.286*	0.042	1

*Correlation is significant at the 0.05 level (2-tailed)

Table 3 H1 Regression

Variables	Standardized coefficient	R^2	F	p value
Constant (follower growth)	–	0.149	5.757	0.005
Page performance index	0.275			0.018
Number of likes	0.259			0.026

Table 4 H2 correlations

H1	Page performance Index	Number of fans	Engagement	Post-interaction
Page performance index	1			
Number of fans	−0.633**	1		
Engagement	−0.093	−0.105	1	
Post-interaction	0.496**	0.062	−0.157	1

**Correlation is significant at the 0.01 level (1-tailed)

Table 5 Web analytics KPIs

Variables	Standardized coefficient	R^2	F	p-value
Constant (page performance index)	–	0.519	23.397	<0.001
Engagement	0.035			0.686
Number of fans	−0.543			<0.001
Post-interaction	0.359			<0.001

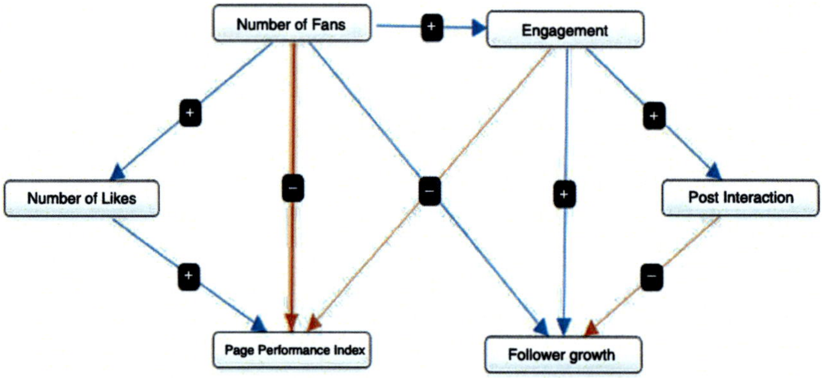

Fig. 1 Fuzzy cognitive map system

number of fans decreases by 54.3%, and engagement increases by 3.5%. These significant results, with $p < 0.005$, suggest that to increase customer interactivity and satisfaction there is a necessity to upload engaging content regularly without worrying about the number of total fans.

3.2 FCM

A fuzzy cognitive map is a scaled form of mind mapping where stable models are built by defining fundamental system attributes such as positive and negative correlations between the studied metrics and the influence of connectivity that a metric may have on others. The (−1,1) scale macro-level study of a system's cause-and-effect interactions is frequently illustrated using FCM [19]. These models might be used to explain a system that is affected by a lot of measurements [19]. Prior research suggests using web analytics data based on fuzzy cognitive maps to assess a digital marketing plan. The development of FCM will help businesses run effective digital marketing campaigns [20] to enhance their online reputation.

The development of the FCM using the examined big data analytics and their relationships is shown in Fig. 1. The scenario for fans optimization is shown in Fig. 2. To improve the number of fans and indirect customer satisfaction,

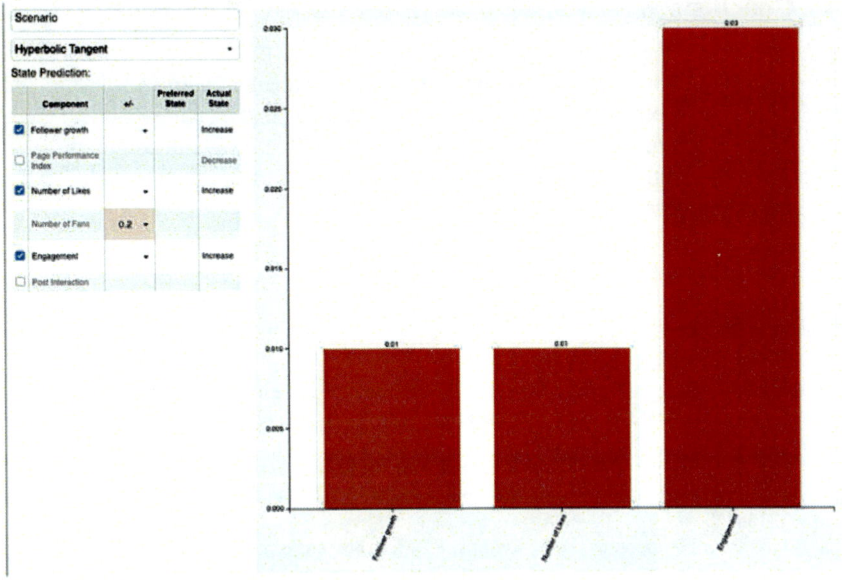

Fig. 2 Fans optimization scenario

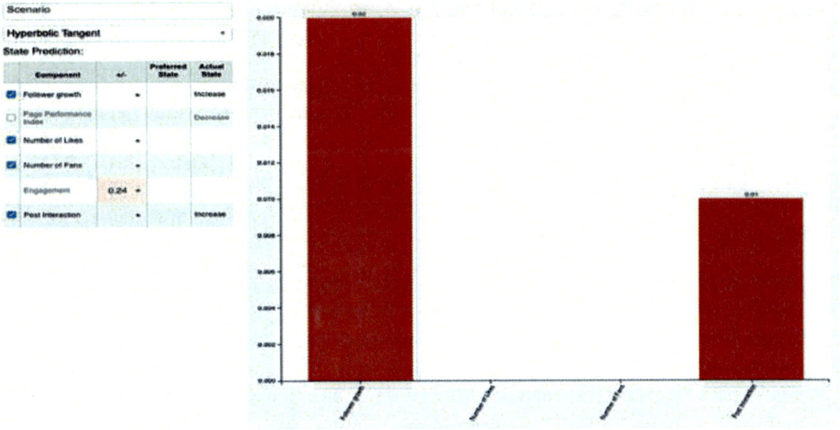

Fig. 3 Scenario for optimizing engagement and interactivity

decision-makers especially need to spend much more on social media marketing and uploading material often [1, 5]. More specifically, to increase the number of fans by 20%, the social media administrator must increase user engagement and interaction by 3%. The scenario for optimizing engagement and interactivity is shown in Fig. 3. To improve engagement by 24% the social media administrators, must provoke the post-interaction to be increased by only 1%.

4 Conclusions

As discussed, a company must optimize the user engagement with its websites and social media to promote indirectly customer satisfaction and increase its brand name [21, 22]. Finally, further research that will incorporate big data with corporate social responsibility would be useful for the tourism industry to investigate the potential benefits of customer satisfaction [23].

References

1. Kim, Y.-J., Kim, H.-S.: The impact of hotel customer experience on customer satisfaction through online reviews. Sustain. Sci. Pract. Policy. **14**, 848 (2022). https://doi.org/10.3390/su14020848
2. Sakas, D.P., Reklitis, D.P.: The impact of organic traffic of crowdsourcing platforms on airlines' website traffic and user engagement. Sustainability. **13**, 8850 (2021). https://doi.org/10.3390/su13168850
3. Sakas, D.P., Nasiopoulos, D.K., Reklitis, P.: Modeling and simulation of the strategic use of marketing in search engines for the business success of high technology companies. In: Strategic Innovative Marketing, pp. 217–226. Springer International Publishing, Cham (2019). https://doi.org/10.1007/978-3-030-16099-9_27
4. Nasiopoulos, D.K., Sakas, D.P., Reklitis, P.: The interaction of users' behavior to the development of a distribution and logistics network of the information technology companies. In: Business Intelligence and Modelling, pp. 219–227. Springer International Publishing, Cham (2021). https://doi.org/10.1007/978-3-030-57065-1_23
5. Park, J., Hyun, H., Thavisay, T.: A study of antecedents and outcomes of social media WOM towards luxury brand purchase intention. J. Retail. Consum. Serv. **58**, 102272 (2021). https://doi.org/10.1016/j.jretconser.2020.102272
6. Sakas, D.P., Kamperos, I.D.G., Reklitis, D.P., Giannakopoulos, N.T., Nasiopoulos, D.K., Terzi, M.C., Kanellos, N.: The effectiveness of centralized payment network advertisements on digital branding during the COVID-19 crisis. Sustainability. **14**, 3616 (2022). https://doi.org/10.3390/su14063616
7. Vlachvei, A., Notta, O., Koronaki, E.: Effects of content characteristics on stages of customer engagement in social media: Investigating European wine brands. Interact. J. Med. Res. (2021)
8. Sakas, D.P., Giannakopoulos, N.T., Reklitis, D.P., Dasaklis, T.K.: The effects of cryptocurrency trading websites on airlines' advertisement campaigns. J. Theor. Appl. Electron. Commer. Res. **16**, 3099–3119 (2021). https://doi.org/10.3390/jtaer16070169
9. Del Vecchio, P., Mele, G., Ndou, V., Secundo, G.: Open innovation and social big data for sustainability: Evidence from the tourism industry. Sustain. Sci. Pract. Policy. **10**, 3215 (2018). https://doi.org/10.3390/su10093215
10. Sakas, D.P., Reklitis, D.P., Trivellas, P., Vassilakis, C., Terzi, M.C.: The effects of logistics websites' technical factors on the optimization of digital marketing strategies and corporate brand name. Processes. **10**, 892 (2022). https://doi.org/10.3390/pr10050892
11. Mariani, M., Di Fatta, G., Di Felice, M.: Understanding customer satisfaction with services by leveraging big data: The role of services attributes and consumers' cultural background. IEEE Access. **7**, 8195–8208 (2019). https://doi.org/10.1109/access.2018.2887300
12. Ramanathan, U., Williams, N.L., Zhang, M., Sa-nguanjin, P., Garza-Reyes, J.A., Borges, L.A.: A new perspective of E-trust in the era of social media: Insights from customer satisfaction data. IEEE Trans. Eng. Manag. **69**, 1417–1431 (2022). https://doi.org/10.1109/tem.2020.2985379

13. Lee, M., (Ally), L.S., Koh, Y: Multisensory experience for enhancing hotel guest experience: Empirical evidence from big data analytics. Int. J. Contemp. Hosp. Manag. **31**, 4313–4337 (2019). https://doi.org/10.1108/IJCHM-03-2018-0263
14. Fanpagekarma.com: Social Media KPIs explained., https://blog.fanpagekarma.com/2014/07/01/social-media-kpis-explained/. Last accessed 20 June 2022
15. Statista Research Department: Most visited travel and tourism websites worldwide as of May 2022., https://www.statista.com/statistics/1215457/most-visited-travel-and-tourism-websites-worldwide/. Last accessed 7 Jul 2022
16. Paxton, A., Griffiths, T.L.: Finding the traces of behavioral and cognitive processes in big data and naturally occurring datasets. Behav. Res. Methods. **49**, 1630–1638 (2017). https://doi.org/10.3758/s13428-017-0874-x
17. Chantzaras, A., Dimitrios, N.K., Vlachos, D.S.: Mobile commerce and success factors. Simulation and modeling of the problem. In: Strategic Innovative Marketing, pp. 349–355. Springer International Publishing, Cham (2017). https://doi.org/10.1007/978-3-319-56288-9_46
18. Tasci, A.D.A.: Added value of a destination brand name calculated by crimmins's method. Tour. Anal. **21**, 669–673 (2016). https://doi.org/10.3727/108354216X14713487283327
19. Sakas, D.P., Kamperos, I.D.G., Reklitis, P.: Estimating risk perception effects on courier companies' online customer behavior during a crisis, using crowdsourced data. Sustainability. **13**, 12725 (2021). https://doi.org/10.3390/su132212725
20. Sakas, D.P., Reklitis, D.P.: Predictive model for estimating the impact of technical issues on consumers interaction in agri-logistics websites. In: Information and Communication Technologies for Agriculture—Theme IV: Actions, pp. 269–283. Springer International Publishing, Cham (2021). https://doi.org/10.1007/978-3-030-84156-0_14
21. Nasiopoulos, D.K., Sakas, D.P., Trivellas, P.: The role of digital marketing in the development of a distribution and logistics network of information technology companies. In: Business Intelligence and Modelling, pp. 267–276. Springer International Publishing, Cham (2021). https://doi.org/10.1007/978-3-030-57065-1_27
22. Sakas, D.P., Reklitis, D.P., Terzi, M.C.: Leading Logistics Firms' Re-Engineering through the Optimization of the Customer's Social Media and Website Activity. Electronics. **12**, 2443 (2023)
23. Reklitis, P., Trivellas, P., Mantzaris, I., Mantzari, E., Reklitis, D.: Employee perceptions of corporate social responsibility activities and work-related attitudes: The case of a greek management services organization. In: Gal, G., Akisik, O., Wooldridge, W. (eds.) Sustainability and Social Responsibility: Regulation and Re-porting, pp. 225–240. Springer Singapore, Singapore (2018). https://doi.org/10.1007/978-981-10-4502-8_10

Part II
Decision Support Systems (DSS) and Business Modelling

Dimitrios K. Nasiopoulos

Description: Management Information Systems, Decision Support Systems, and Simulation modeling provide a powerful methodology for advancing theory and research on complex behaviors and systems. This session aims to promote the understanding of simulation methodology and to develop an appreciation of its potential contributions to management theory by describing the nature of simulations, its attractions, and its special problems, as well as some uses of computational modeling in management research.

D. K. Nasiopoulos
Agricultural University of Athens, Athens, Greece

Business Strategy to Increase Profit and Customer Base Using Eco-Advertising, Modeling and Optimization

Dimitrios M. Mastrakoulis, Dimitrios K. Nasiopoulos, Dimitrios A. Arvanitidis, Elli C. H. Zara, Panagiotis Trivellas, and Catherine Marinagi

1 Introduction

The purpose is to find out how useful and beneficial it is for the company to allocate all its resources to the field of eco-advertising as well as the impact it will have on consumers. In this particular investment, what the company seeks is the competitive advantage over other companies and the recognition of the product. The time given to the company to evaluate these results is 12 months [4].

Moreover, Staff members of high-tech companies with high levels of technical knowledge led to increased financial and economic performance, in comparison with companies, in the same industrial sector, employing less technically qualified staff

D. M. Mastrakoulis
Department of Informatics and Telecommunications, University of Thessaly, Lamia, Greece
e-mail: dmastrakoulis@aua.gr

D. K. Nasiopoulos
Department of Agribusiness and Supply Chain Management, BICTEVAC LABORATORY Business Information and Communication Technologies in Value Chains laboratory, School of Applied Economics and Social Sciences, Agricultural University of Athens, Athens, Greece

D. A. Arvanitidis (✉) · E. C. H. Zara
Department of Informatics and Telecommunications, University of Peloponnese, Tripolis, Greece
e-mail: darvanitidis@aua.gr

P. Trivellas
Department of Agribusiness and Supply Chain Management, Organizational Innovation and Management Systems, School of Applied Economics and Social Sciences, Agricultural University of Athens, Athens, Greece

C. Marinagi
Department of Agribusiness and Supply Chain Management, School of Applied Economics and Social Sciences, Agricultural University of Athens, Athens, Greece

D. P. Sakas et al. (eds.), *Computational and Strategic Business Modelling*, Springer Proceedings in Business and Economics,
https://doi.org/10.1007/978-3-031-41371-1_12

[30]. A significant relation among major capital structure and corporate governance parameters and firm performance of 10 leading energy sector companies traded in the NYSE. Furthermore, pecking order and agency cost theories play an important role in the financing of these firms, while static trade and irrelevance theory find no support [31].

This research will analyze whether the eco-advertising factor can directly and to a large extent influence the success of a product in the market. The goal is to identify and analyze how eco-advertising as one of the tools of digital marketing will determine the relationship between eco-innovation and product success (product recognition and competitive advantage).

1.1 Ecological Advertising

The concept of eco-green advertising began in the 1970s due to the concerns of consumers about the environmental damage caused by the promotion of products, services, ideas, etc. [1] The number of green ads has increased almost tenfold in the last 20 years and is a strategy used by businesses to show that they are interested in environmental issues.

2 Description of Hypothesis and Modeling Process (Table 1)

Table 1 Description of the factors applied

Analysis of the main factors:	
Social Media	A key factor in the process is social media, in addition to the direct communication that is offered as a service. The purpose of the business is to attract and expand online communication with potential customers. Immediately after, the parameters that have been set for the proper operation of the factor are applied.
Google Ads	An additional key factor process is Google Ads, which also play a very important role and contribution to the overall business development and product promotion. Google ads enable businesses to promote their products or services through search campaigns (text campaigns) through text ads, Google display network (display network campaigns) and YouTube (video display campaigns), a forementioned platforms, the advantage of advertising in Google Ads is the ability to target advertising to specific audiences, based on the goals and needs of each business. Immediately after, the parameters that have been set for the proper operation of the factor are applied.
Email	A last key factor in the process is email marketing, which is the digital version of the well-known correspondence. This way the details become measurable (how many opened and read the emails, what date and time, and where they clicked), the upcoming message can be customized each time, making it more attractive for optimization of the results of the respective campaigns. Immediately after, the parameters that have been set for the proper operation of the factor are applied.

3 Hypothesis

H1: *CA > 0, A < 0/Existence of social_media and google_ads, makes email don't have the same acceptance from the customer.*

The effectiveness of the email has been reduced due to the existence of social_media and google_ads.

H2: *CA > 0, A > 0/Email gets customers' trust.*

Acceptance of email rises and gets positive values.

H3: *CA > 0, A > 0/Google platform crashes and Instagram announces user interface problems.*

The customers turn to Facebook and twitter as an alternan solution getting an equal percentage of users (+20%).

H4: *CA > 0, A > 0/Facebook server malfunctions and google loses half of its users.*

Instagram and Twitter gain more power and google_ads lose half of its power.

H5: *CA > 0, A > 0, influencers < 0/All mail servers broke down because of a cyber-attack and Instagram's owner resigned.*

The factor of email is excluded from the system's equation and Instagram loses 60% of its influence.

4 Methodology

The aim of the system under consideration is to find out how useful and beneficial it is for the company to allocate all its resources to the field of eco-advertising as well as the impact it will have on consumers.

> *The business is considered to succeed when the resources spent on eco-advertising will bring positive results both in terms of product recognition and competitive advantage.*

This practically means for the system, that after the initial stock that is provided and constitutes 100% of the resources, those that feed the stocks, must lead it to its maximum possible value, which implies the ever-increasing value of the final value. The purpose of the whole project is the non-interruption of the process and at the same time the overcoming of the sum of the other opposite measure of parameters and factors that are applied. To achieve this, percentage changes had to be made. Time is referred as time units (tu).

Initially, 100% of the processing is placed in Green_Adv in the context of perfectly competitive. This practically means that the resources that supply the following stocks must lead it to its maximum possible value, which implies the ever-increasing value of the final amount.

The first element being used is Green_Adv. Then, if this stock is analyzed and goes through the factors that exist in the model, it will end up in A and CA. It is pointed out that the examined operating time of the system is from $t = 0$ to $t = 12$.

Initially, 100% of the resources that will affect the system during the starting process at time $t = 0$ are placed in Green_Adv. During the process of the system, other parameters will act because of the production of extra factors such as the following:

Facebook Facebook has over 2 billion monthly users worldwide with 58 min average daily usage and is one of the most popular and important social networks, which makes it the ultimate tool to attract the right users, based on the common goal of businesses.

Instagram Instagram, with over 800 million monthly users and 53 min of average daily usage worldwide, has grown significantly in recent years. This platform focuses mainly on professionals and businesses that base their communication on image/video while following a different approach from that of Facebook.

Twitter Twitter, with over 330 million monthly users worldwide, is an important tool for businesses to convey their messages quickly and instantly to their audience. Twitter's platform allows for a small content so that the messages posted by businesses are clear and highly targeted.

Text Ads Search Campaigns through text ads, i.e., search campaigns are ads that have the form of text. In other words, Google enables businesses to define the keywords for the product or service they offer, because of which they appear at the top of its search results.

Google Display Network GDN essentially offers the ability to display advertising messages on various sites, without any communication or agreement with them. It can be used mainly to promote offers from sites that have an e-shop, as well as to increase their brand awareness.

YouTube It's one of the most popular websites in the world, after Google. Businesses through advertising have the potential to attract new customers by encouraging them to either watch more videos and learn more about the product or lead them to buy it. The information in the videos is preferred by the consumers because the combination of image and sound is better reflected in the consciousness of the customers.

Analytics The Email Click and Email Open Rate indicators show us the total number of users who opened the sent emails, in relation to the total number of recipients. The higher the index, the more successful the recipient list as well as the upcoming email content. In other words, the company can make decisions based on statistical performance results (data-driven decisions).

Campaigns Through email campaigns, the business can achieve a lot. Initially to introduce a new product to the public, to promote a special edition of the best seller product, to reward the public with a discount. The most important thing about these campaigns is that it notify the customers about the offers that are valid for a certain period.

Newsletter One of the most effective and cost-effective methods used in internet marketing for product advertising campaign is the newsletter. These are newsletters, which contain news, offers, announcements, tips, or general information and are sent via email. It is not just an ad text, but it is a top tool for internet marketing and green advertising.

After analyzing the key factors and sub-factors of Green_Adv an **independent factor** will be examined, to see if it will affect the results of the model. Specifically, one of the two factors that will lead to the success or failure of this product of recognition.

Due to the increasing use of social media and especially Instagram, people from various sectors promote a product to influence the consumer habits of users. In other words, it was about the influencers, who have concluded an agreement with the companies to present in the most representative way the values of the brand, as well as the usefulness of the product.

It is usually observed that if they finally connect with the product, then their followers will surely do the same, contributing positively to the growth of the brand. To note at this point, that a business is considered a success when the resources spent on eco-advertising will bring positive results both in terms of product recognition and competitive advantage.

5 Hypothesis Trials and Results

Initially, 100% of green advertising resources (green advertising) should be divided into three factors which are social media, advertising through the largest information retrieval company Google (google_ads) and email as shown in Table 2. The distribution of the initial percentage is chosen to be as it is shown in Table 2. This is achieved by using a converter on the respective flow from the Green_Adv to each factor. Finally, the effect of influencers is applied as mentioned above. After researches, the following results are obtained regarding the behavior of the system. The green color in the following diagram indicates a positive to the system parameter, and red indicates the negative one.

Social_media

In this case, 55% of the initial percentage goes to social_media. From sm, 50% of the resources are selected to be distributed on Facebook via sm2F, 40% on Instagram via sm2I and the remaining 10% on Twitter via sm2T.

Table 2 Quotation of variables

Variables	Percentage (%)
Social_media (sm)	55%
Google_ads (ga)	30%
Email (e)	(−)15%

google_ads

The next 30% of the percentage goes to ***google_ads*** in the same way as above. Flows distribute resources in the same way to Text_Ads, Google_Display_Network, and YouTube. A total of 40% is distributed through ga2TA in search campaigns via text ads and 10% in network advertising campaigns.

Email

The next 15% of Green_Adv resources go to email after the percentage has been placed on Green_Adv in the GA2e flow. A total of 45% of the email resources are available to the company in the analysis of data.

And about the remaining parameters:

- All of Facebook's resources are driven by the F2CA flow into one of the two factors that will determine the cost-effectiveness of competitive advantage (CA).
- 50% of Instagram resources are directed to the influencer factor. 85% of Twitter resources are directed to competitive advantage through T2CA streaming.
- All Text_ads resources will be routed to the CA through the TA2CA stream. 25% of GDN to -A- via GDN2A and 10% of YouTube Y2i resources to the influencers. YouTube is a key product promotion tool for influencers.
- 30% in the campaigns carried out by the company via mail mainly concern offers and discounts and 25% in the newsletter, this is achieved by the corresponding functions selected on each stream (e2A, e2C, e2N) with the percentage on the email.
- 95% of the data analysis resulting from the sending of emails is promoted to the competitive advantage through A2CA. will have a negative sign. Finally, 5% of the N2i newsletter flows to the influencers.
- The influencers factor gathers all the resources from the N2i, Y2i streams and 50% of the I2i resources and from there 75% through i2A leads to the recognition. The total resources accumulated in CA and A feedback the model.
- The resources are on the vertical axis and the twelve months of the year (tu) are on the horizontal axis. Graphically shows the course of Green_Adv, CA, A and influencers.

As it will be shown below, the system does not become viable, which is verified in the last stage of the process.

6 Modeling and Simulation

The system from the beginning of its operation and although the initial stocks that feed the positive route are at almost 95% of the initial percentage, it never manages to get positive for the A stock for 0–12tu. The consequence of this is the cancellation of the process since it is necessary for all three factors (CA, A, influencers) to be positive (Fig. 1).

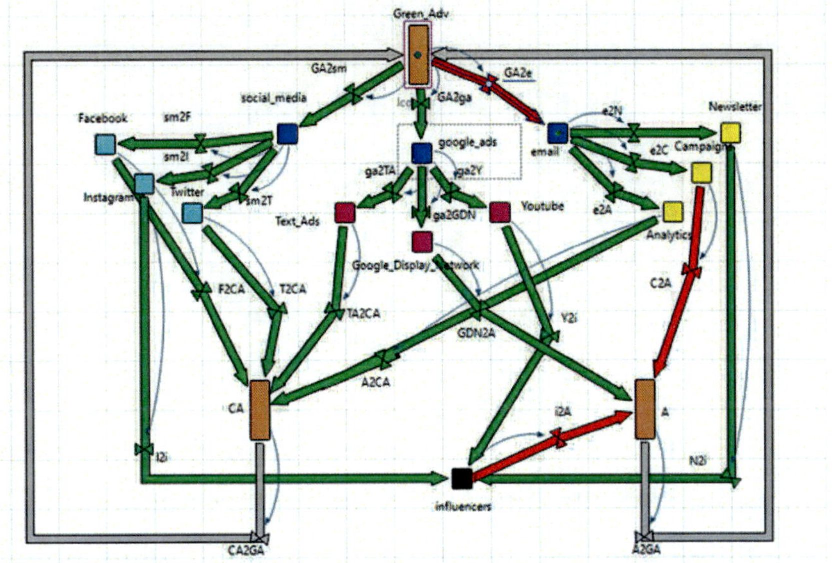

Fig. 1 Simulation model

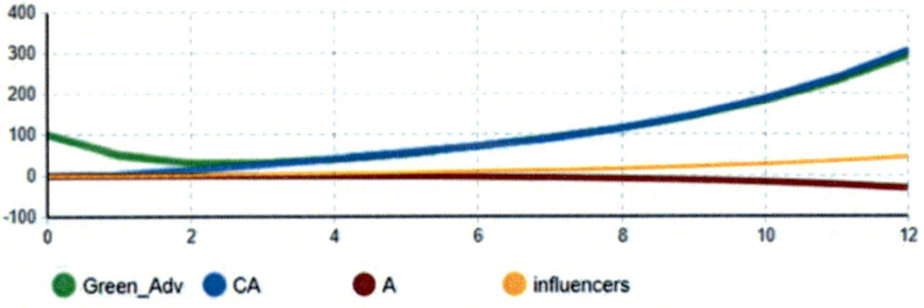

Fig. 2 Diagram during simulation phase

Table 3 Quotation of variables in optimization

Variables	Percentage (%)
Social_media (sm)	57.28%
Google_ads (ga)	37.92%
Email (e)	4.8%

The above configuration of the model will result in the following results (Fig. 2): It is observed that although **CA** takes a positive value and equals to **1037*Green_Adv** at $t =$ 12tu the **A** stock never takes a positive value and equals to **−0.1046*Green_Adv**. Consequently, the system will never manage to work (Table 3; Fig. 4).

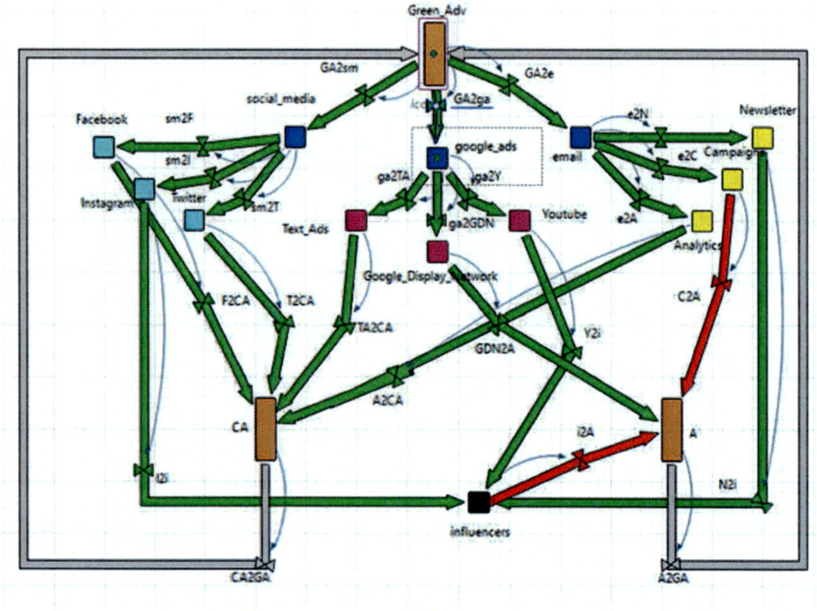

Fig. 3 Optimization model

7 Optimization (Fig. 3)

The goal of the company is through Green_Adv to gain a competitive advantage and recognition. The company aims at the maximum possible competitive advantage and maximum recognition. This practically, in the model, means that after the initial factor invested and is 100% of the resources, the resources that feedback the GA should lead it to its maximum possible value, as the higher the GA the bigger the CA-A. To achieve this goal modifications had to be made to the model values shown in the image below. Toward that direction, the resource percentages will vary. After many simulations, the result is as the following table:

The maximum possible profit is achieved if the percentage of resources flowing from GA to social_media, google_ads, and email is differentiated. 57.28% should be available on social_media, ga 37.92% and email 4.8%. It is observed that in the fourth month, after the optimization GA also increased the key factors CA and A. As the GA increased both the financial profit of the company and the success of the product increased. An important factor that positively affected the recognition is the influencers factor due to the positive impact 25% of the resources are directed to A through i2A (Tables 4, 5, 6, 7, 8, 9).

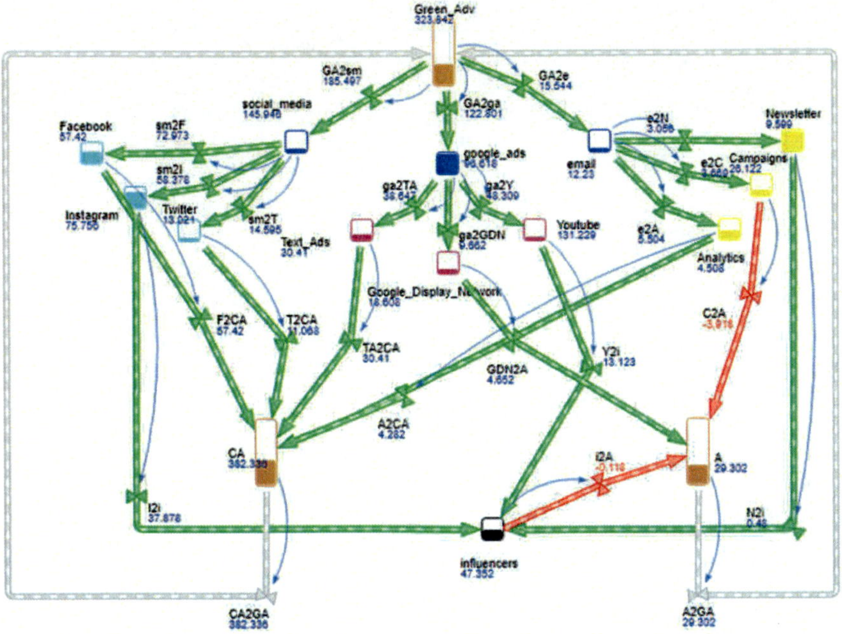

Fig. 4 Diagram during optimization phase

Table 4 Explanation of flows and factors

Variables	Type	Explanation
sm2F	social_media*50%	Flow to Facebook (F)
sm2I	social_media*40%	Flow to Instagram (I)
sm2T	social_media*10%	Flow to Twitter (T)

Table 5 Explanation of flows and factors

Variables	Type	Explanation
ga2TA	google_ads*40%	Flow to Text_Ads (ta)
ga2GDN	google_ads*10%	Flow to Google_Display_Network (GDN)
ga2Y	google_ads*50%	Flow to Youtube (Y)

Table 6 Explanation of flows and factors

Variables	Type	Explanation
e2A	Email*45%	Flow to analytics (A)
e2C	Email*30%	Flow to campaigns (C)
e2N	Email*25%	Flow to newsletter (N)

Table 7 Explanation of flows and factors

Variables	Type	Explanation
I2i	Instagram*50%	Flows to influencers (i) from Instagram
Y2i	Youtube*10%	Flows to influencers (i) from Youtube
N2i	Newsletter*5%	Flows to influencers (i) from newsletter

Table 8 Explanation of flows and factors

Variables	Type	Explanation
F2CA	Facebook*100%	Flows to CA from Facebook
TA2CA	100%	Flows to CA from Text_ads
A2CA	Analytics* 95%	Flows to CA from analytics
T2CA	Twitter* 85%	Flows to CA from twitter
GDN2A	Google_Display_Network*25%	Flows to A from GDN
C2A	– Campaigns*15%	Flows to A from campaigns
i2A	– Influencers*0.25%	Flows to A from influencers
CA2GA	CA*100%	Green_Adv(GA) from CA
A2GA	A*100%	Green_Adv(GA) from A

Table 9 Explanation of flows and factors

Variables	Type	Explanation
social_media (sm)	74.81%	Green_adv
Email (e)	25.19%	Green_adv
sm2F	social_media*70/100	Flow to Facebook (F)
sm2I	social_media*0%	Flow to Instagram (I)
sm2T	social_media*30%	Flow to twitter (T)
Y2i	Youtube*52.5%	Flows to influencers (i) from Youtube
N2i	Newsletter*47.5%	Flows to influencers (i) from newsletter

More specifically:

(a) The following factors affect social media:
(b) The following factors affect google ads:
(c) The following factors affect email:
(d) The following factors affect influencers:
(e) The following factors are from and to CA and A:

It is observed that CA & A at **t = 0tu** begin from the initial value of 100% but then at the output, and at the same time, they both obtain a positive value as well. At t = 12tu the **CA≈1,18*Green_Adv & A ≈ 0.0904*Green_Adv** approximately. A result that, compared to modeling, shows a tremendous change (Fig. 5).

After the change in the distribution of resources, the model shows an increase in competitive advantage and recognition. The model is consequently fed back. Therefore, the model after optimization is much more efficient.

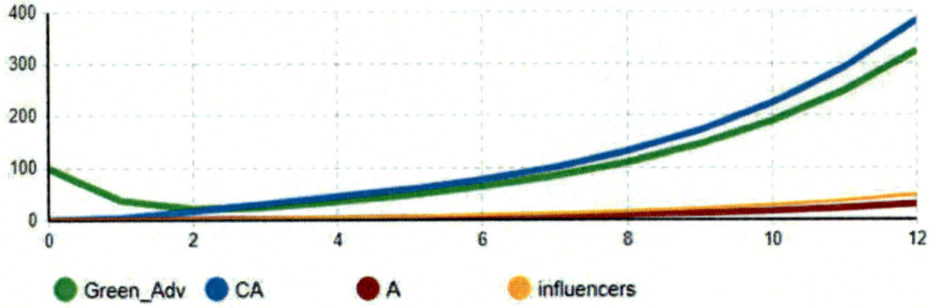

Fig. 5 Diagram during optimization phase

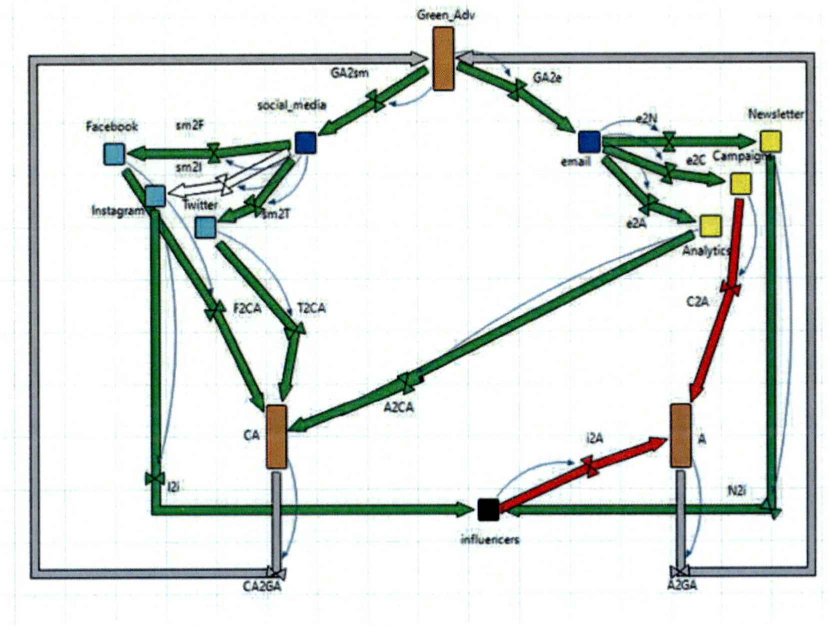

Fig. 6 Optimization model

(a) However, to achieve the examination of other scenarios, the viability of the system will be examined if it is considered that Google platform crashes and Instagram announces user interface problems. Based on this assumption, below the behavior of the system is listed (Fig. 6).

The parameter percentages that are applied to the above are:

The following diagram clarifies the behavior of the system (Fig. 7).

It is observed that at **t = 12tu** the **CA≈1.264*Green_Adv and A ≈ 0.0988*Green_Adv** approximately, which means that the initial scope is succeeded (Table 10).

Table 10 Explanation of flows and factors

Variables	Type	Explanation
social_media (sm)	52.39%	Green_adv
google_ads (ga)	28.47%	Green_adv
Email (e)	19.14%	Green_adv
sm2F	social_media*0%	Flow to Facebook (F)
sm2I	social_media*65%	Flow to Instagram (I)
sm2T	social_media*35%	Flow to Twitter (T)
ga2TA	google_ads*20%	Flow to Text_Ads (ta)
ga2GDN	google_ads*5%	Flow to Google_Display_Network (GDN)
ga2Y	google_ads*25%	Flow to Youtube (Y)

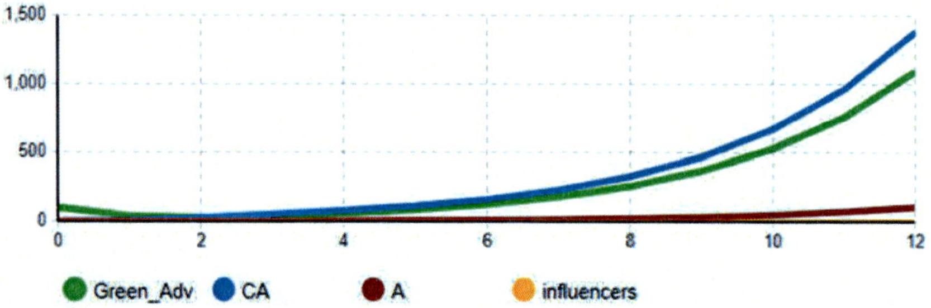

Fig. 7 Diagram during optimization phase

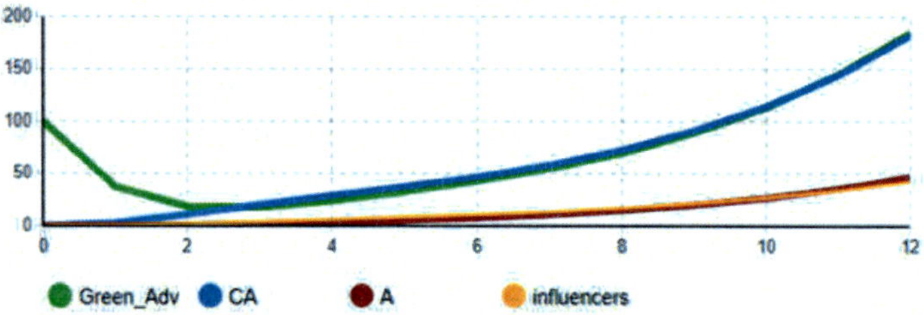

Fig. 8 Resources diagrams during optimization phase

(b) Additionally, the viability of the system will be examined if it is considered that Facebook server malfunctions and Google loses half of its users. Based on this assumption, below the behavior of the system is listed. The parameter percentages that are applied to the above are:

The following diagram clarifies the behavior of the system (Fig. 8).

It is observed that at **t = 12tu** the **CA≈0.9879*Green_Adv and A ≈ 0.2551*Green_Adv** approximately, which means that the initial scope is succeeded.

Table 11 Explanation of flows and factors

Variables	Percentage (%)	Explanation
social_media (sm)	57.28%	Green_adv
google_ads (ga)	37.92%	Green_adv
Email (e)	4.8%	Green_adv

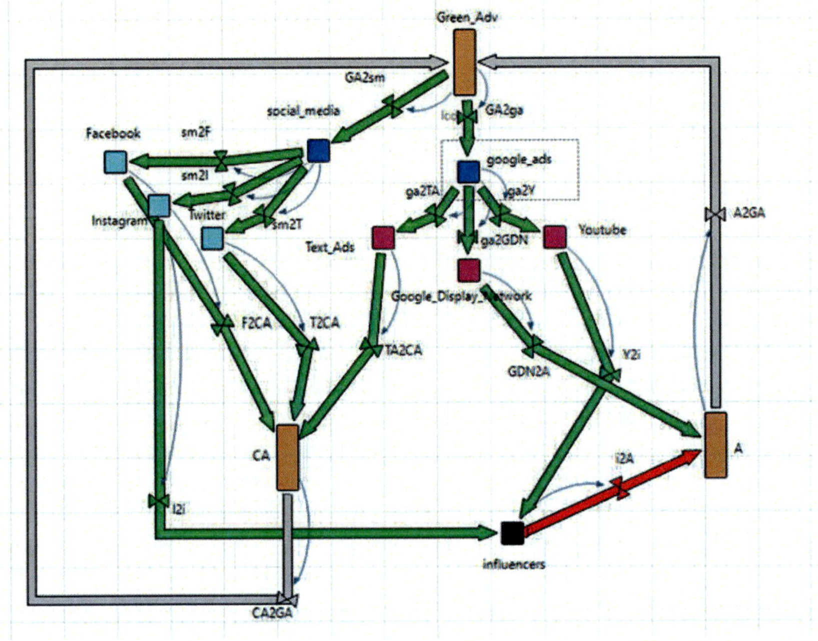

Fig. 9 Optimization model

(c) Closing, the viability of the system will be examined if it is considered that all mail servers broke down because of a cyber-attack and Instagram's owner resigned. Based on this assumption, below the behavior of the system is listed (Table 11; Fig. 9).

The parameter percentages that are applied to the above are:
As a result, influencers = Y2i*25/100 – I2i*60/100.
The following diagram clarifies the behavior of the system (Fig. 10).

It is observed that at **t** = **12tu** the **CA≈1.182*Green_Adv and A ≈ 0.0792*Green_Adv** approximately, but **influencers = −0.1714*Green_Adv** which means that the initial scope is succeeded. Therefore, the model is considered to succeed in the process.

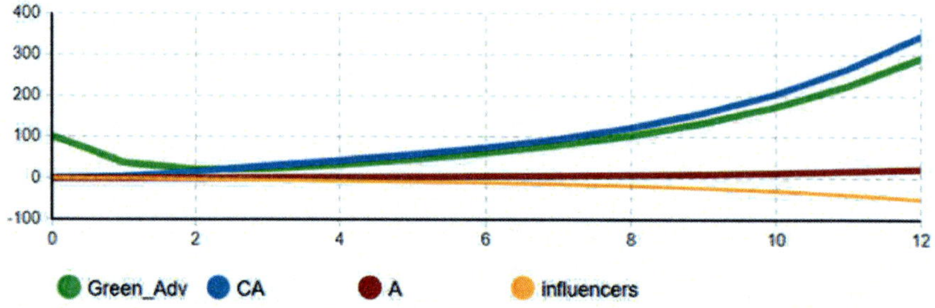

Fig. 10 Resources diagrams during optimization phase

8 Conclusion

This research is part of a research project examining and optimizing how useful and beneficial it is for the company to allocate all its resources to the field of eco-advertising as well as the impact it will have on consumers. The target of the company is to have a return of at least the initial return of investment (ROI) resources. But to be as profitable as possible, the company's investment must be sought in an optimal version of the model with a different sharing of resources. After a study and several tests, the resources were distributed so that the company with the same financial investment had the maximum possible monetary gains. Optimization played a prominent role in this. It is observed that the GA starts from the initial value of 100%, then decreases for a short time due to the investment and after the fourth month increases more especially after the optimization than before. So, it brings more money to the company. After the second month, there is an increase in CA and A after optimization, where this is particularly important because they are the two main factors that will lead to the optimal increase in GA. Therefore, the Post-optimization model is more efficient.

At the bottom line, as digital marketing in general and green advertising and the media used evolve day by day, companies are called upon to adapt to change to remain competitive [15, 16]. Year after year, more and more consumers are turning to products and businesses that are transparent about their impact on society and their ecological footprint. Eco-advertising as an aspect of eco-innovation and R&D orientation [17] can be a key and determining factor for the success of a product in the market.

Across this line of reasoning, during pandemic and economic crises, a "place-based" typology of stakeholders has been put forth [18], which synthesize and enrich human activity and well-being with several factors, such as natural environment, remoteness, connectivity, and eco-efficiency, in order to establish sustainable development in challenging destinations such as tourist attractions [19–24]. Additionally, future research in CSR and advertising could be beneficial for the environment [25–29].

References

1. Arminda Maria Finisterra do Paço & Rosa Reis: Factors affecting skepticism toward green advertising. J. Advert. **41**, 147–155 (2013)
2. D'Souza, C., Taghian, M.: Green advertising effects on attitude and choice of advertising themes. Asia Pac. J. Mark. Logist. **17**, 51–66 (2005)
3. Pujari, D.: Eco-innovation and new product development: understanding the influences on market performance. Technovation. **26**, 76–85 (2006)
4. Haytko, D.L., Matulich, E.: Green advertising and environmentally responsible consumer behaviors: Linkages examined. J Manag Market Res. **1**, 1–20 (2008)
5. Iyer, E., Banerjee, B.: Anatomy of green advertising. Assoc Consum Res. **20**, 494–501 (1993)
6. Diana, H.L., Erika, M.: Green advertising and environmentally responsible consumer behaviors: Linkages examined. J Manag Market Res. **1**, 2–11 (2008)
7. Iyer, E., Banerjee: Anatomy of green advertising. Adv Consumer Res, 494–501 (1993)
8. Murin, I., Marková, I., Zelený, J., Jaďuďová, J.: Green marketing as a tool influencing consumerś behavior: Slovak case study of regional mark preference. Procedia Economic Finance. **34**, 260–267 (2015)
9. de Medeiros, J.F., Ribeiro, J.L.D., Cortimiglia, M.N.: Success factors for environmentally sustainable product innovation: A systematic literature review. J. Clean. Prod. **65**, 76–86 (2014)
10. Karna, J., Juslin, H., Ahonen, V., Hansen, E.: Green advertising: Greenwash or a true reflection of marketing strategies? Greener Manag. Int. **33**, 59–70 (2001)
11. Leonidou, L.C., Leonidou, C.N., Palihawadana, D., Hultman, M.: Evaluating the green advertising practices of international firms: a trend analysis. Int Market Rev. **6**, 6–33 (2011)
12. Ángeles, P., Xavier, V.: Key business factors for eco-innovation: an overview of recent firm-level empirical studies. Cuadernos de Gesti6n. **12**, 73–103 (2012)
13. Golińska, P., Kawa, A.: Technology management for sustainable production and logistics, pp. 173–222. Springer (2015)
14. Kao, T.F., Du, Z.: A study on the influence of green advertising design and environmental emotion on advertising effect. J. Clean. Prod. **242** (2020)
15. Sakas, D.P., Reklitis, D.P., Trivellas, P., Vassilakis, C., Terzi, M.C.: The effects of logistics websites' technical factors on the optimization of digital marketing strategies and corporate brand name. Processes. **10**(5), 892 (2022)
16. Reklitis, P., Trivellas, P.: Performance implications of aligning generic strategies with the business environment. Int J Manag Decis Mak. **3**(3–4), 319–336 (2002)
17. Trivellas, P.: Investigating the impact of research and development strategy on firm performance. Key Eng Mater. **495**, 306–309 (2012)
18. Trivellas, P., Anastasopoulou, A., Malindretos, G., Reklitis, P., Sakas, D.P.: A place based analysis of stakeholders' advancing sustainability in remote and isolated communities: The case of North Aegean Islands. IOP Conf. Ser.: Earth Environ. Sci. **899**(1), 12044 (2021)
19. Trivellas, P., Kakkos, N., Vasiliadis, L., Belias, D.: Sustainability, Social Marketing and Host Attitudes About Tourism in the City of Larissa, pp. 89–95. Springer Proceedings in Business and Economics (2017)
20. Trivellas, P., Kakkos, N., Vasiliadis, L., Belias, D.: Residents' Perceptions Toward Cultural, Social and Economic Benefits and Costs of Tourism Industry: An Empirical Survey, pp. 171–181. Springer Proceedings in Business and Economics (2016)
21. Folinas, D., Fotiadis, T.: Marketing and Supply Chain Management: A Systemic Approach, 1st edn. Routledge (2017)
22. Fotiadis, T.: Strategic Marketing for High Technology Products: An Integrated Approach (Routledge Studies in Innovation, Organizations and Technology), 1st edn. Routledge (2018)
23. Fotiadis, T., Folinas, D., Vasileiou, K., Konstantoglou, A.: Marketing and the Customer Value Chain: Integrating Marketing and Supply Chain Management 1st Edition, Kindle Edition, 1st edn. Routledge (2022)

24. Katsikari, C., Hatzithomas, L., Fotiadis, T., Folinas, D.: Push and pull travel motivation: segmentation of the Greek market for social media marketing in tourism. Sustainability. **12**(11), 4770 (2020)
25. Sakas, D.P., Giannakopoulos, N.T., Reklitis, D.P., Dasaklis, T.K.: The Effects of Cryptocurrency Trading Websites on Airlines' Advertisement Campaigns. J. Theor. Appl. Electron. Commer. Res. **16**, 3099–3119 (2021). https://doi.org/10.3390/jtaer16070169
26. Sakas, D.P., Reklitis, D.P.: The Impact of Organic Traffic of Crowdsourcing Platforms on Airlines' Website Traffic and User Engagement. Sustainability. **13**, 8850 (2021). https://doi.org/10.3390/su13168850
27. Sakas, D.P., Reklitis, D.P.: Predictive model for estimating the impact of technical issues on consumers interaction in agri-logistics websites. In: Information and Communication Technologies for Agriculture—Theme IV: Actions, pp. 269–283. Springer International Publishing, Cham (2021). https://doi.org/10.1007/978-3-030-84156-0_14
28. Reklitis, P., Trivellas, P., Mantzaris, I., Mantzari, E., Reklitis, D.: Employee Perceptions of Corporate Social Responsibility Activities and Work-Related Attitudes: The Case of a Greek Management Services Organization. In: Gal, G., Akisik, O., Wooldridge, W. (eds.) Sustainability and Social Responsibility: Regulation and Re-porting, pp. 225–240. Springer, Singapore (2018). https://doi.org/10.1007/978-981-10-4502-8_10
29. Sakas, D.P., Kamperos, I.D.G., Reklitis, D.P., Giannakopoulos, N.T., Nasiopoulos, D.K., Terzi, M.C., Kanellos, N.: The effectiveness of centralized payment network advertisements on digital branding during the COVID-19 crisis. Sustainability. **14**, 3616 (2022). https://doi.org/10.3390/su14063616
30. Toudas, K., Kanellos, N.: Economic and accounting performance of greek innovative firms through knowledge-based entrepreneurship. J Account Taxation. **1492**, 150–160 (2022). https://doi.org/10.5897/JAT2022.0530
31. Georgakopoulos, G., Toudas, K., Poutos, E., Kounadeas, T., Tsavalias, S.: Capital structure, corporate governance, equity ownership and their impact on firms' profitability and effectiveness in the energy sector. Energies. **15**(3625), 1–10 (2022). https://doi.org/10.3390/en15103625

Increasing the Use of Social Networks by Businesses and Their Crucial Role in the Creation of Online Stores, Modeling and Optimization

Dimitrios M. Mastrakoulis, Dimitrios K. Nasiopoulos, Dimitrios A. Arvanitidis, Elli C. H. Zara, and Catherine Marinagi

1 Introduction

The purpose is to find out how useful and beneficial it is for the company to allocate all its resources to the online store as well as the impact it will have on consumers [1]. In this particular investment, what the company seeks is the competitive advantage over other companies and the recognition of the product. The time given to the company to evaluate these results is 12 months.

Moreover, staff members of high-tech companies with high levels of technical knowledge led to increased financial and economic performance, in comparison with companies, in the same industrial sector, employing less technically qualified staff [2]. A significant relation among major capital structure and corporate governance parameters and firm performance of 10 leading energy sector companies traded in the NYSE. Furthermore, pecking order and agency cost theories play an important

D. M. Mastrakoulis
Department of Informatics and Telecommunications, University of Thessaly, Lamia, Greece
e-mail: dmastrakoulis@aua.gr

D. K. Nasiopoulos
Department of Agribusiness and Supply Chain Management, BICTEVAC LABORATORY Business Information and Communication Technologies in Value Chains laboratory, School of Applied Economics and Social Sciences, Agricultural University of Athens, Athens, Greece

D. A. Arvanitidis (✉) · E. C. H. Zara
Department of Informatics and Telecommunications, University of Peloponnese, Tripolis, Greece
e-mail: darvanitidis@aua.gr

C. Marinagi
Department of Agribusiness and Supply Chain Management, School of Applied Economics and Social Sciences, Agricultural University of Athens, Athens, Greece

D. P. Sakas et al. (eds.), *Computational and Strategic Business Modelling*, Springer Proceedings in Business and Economics,
https://doi.org/10.1007/978-3-031-41371-1_13

role in the financing of these firms, while static trade and irrelevance theory find no support [3].

This research will analyze whether the e-shop factor can directly and to a large extent influence the success and recognition of a product in the market. The main goal is to identify and analyze how the online store (e-shop) as one of the tools of digital marketing will determine the relationship between eco-innovation and product success (competitive advantage and recognition) [4–8].

1.1 E-Shop

E-shops are constantly growing and enable each customer to search for the item they want, compare it (video images), and see the current offers [9]. In addition, it can have an image of the time of preparation, shipping, and receipt of its order. In addition, they enable the customer to order specialized items from other cities even abroad. The prices in the online stores are usually cheaper, while the payment method (credit card, cash on delivery, and Paypal) of the order facilitates the acquisition of the products.

2 Description of Hypothesis and Modeling Process (Table 1)

Table 1 Description of the factors applied

Analysis of the main factors	
Website	A key factor in the process is the website. The companies create a website, which contains detailed information about the products and services they offer. The purpose of the website through the presentation of the good is to make the browsers and the existing customers buy the product again or one of the other products. Immediately after, the parameters that have been set for the proper operation of the factor are applied.
Advertising	An additional key factor process is advertising. There are many ways in which the e-shop can be advertised and promoted; however, the ideal approach is to choose the most suitable methods that suit the respective online store. The purpose, however, is to develop and increase e-shop sales. Immediately after, the parameters that have been set for the proper operation of the factor are applied.
Cooperations	A last key factor in the process is cooperation. An important part, perhaps even before the construction of the online store begins, is the collaborations that need to be done. Especially those that are developed with the transport companies (courier), with the banking institutions, and with lawyers and accountants. Immediately after, the parameters that have been set for the proper operation of the factor are applied.

3 Hypothesis

H1: $CA > 0$, $A < 0$, e-bay<0/ *The cyber attack reduces by 50% the functionality of the e-shop.*

The e-shop page is under cyber attack with the result that the site displays 50% of the products and makes it impossible to communicate with the banking institutions.

H2: $CA > 0$, $A > 0$, e-bay>0/ *e-bay-marketplaces are down 30%.*

The facilities of the e-bay server are occupied by terrorists so 30% of the auctions do not take place.

H3: $CA > 0$, $A > 0$, e-bay>0/ *Social media loses 47% of its power.*

Due to the emergence of a global movement with complaints about the violation of the privacy of social media users, 47% of all users are affected and abstained.

H4: $CA > 0$, $A > 0$, e-bay>0/ *The marketplace stops working.*

The institution of the global marketplace is deactivated due to a decrease in the absorption of transactions.

H5: $CA > 0$, $A > 0$, e-bay>0/ *95% failure to deliver products.*

Transport companies in Europe due to a war between three countries receive the goods seven days late and fail to deliver 95% of the products on time.

4 Methodology

The aim of the system under consideration is to identify and analyze how the online store (e-shop) as one of the tools of digital marketing will determine the relationship between ecological innovation and product success (competitive advantage and recognition).

The purpose is to find out how useful and beneficial it is for the company to allocate all its resources to the online store as well as the impact it will have on consumers. In this particular investment, what the company seeks is the competitive advantage over other companies and the recognition of the product.

This practically means for the system that after the initial stock is provided and constitutes 100% of the resources, the resources that feed the stocks, must lead it to its maximum possible value, which implies the ever-increasing value of the final value. The purpose of the whole project is the non-interruption of the process and at the same time the overcoming of the sum of the other opposite measure of parameters and factors that are applied. To achieve this goal, percentage changes had to be made. Time is referred to as time units (tu).

Initially, 100% of the processing is placed in e-shop in the context of perfectly competitive. This practically means after the resources that supply the following stocks must lead it to its maximum possible value, which implies the ever-increasing value of the final amount.

The first element being used is e-shop. Then, if this stock is analyzed and goes through the factors that exist in the model, it will end up in A and CA. It is pointed out that the examined operating time of the system is from $t = 0$ to $t = 12$.

Initially, 100% of the resources that will affect the system during the starting process at time $t = 0$ is placed in e-shop. During the process of the system, other parameters will act because of the production of extra factors such as the following:

Visual Approach The visual approach or graphic representation of a website, in essence, concerns the graphic image of the e-shop and the way in which the page is displayed on the internet.

CMS The CMS concerns the management part of the e-shop. It is the software that helps the user to manage his website and to present it properly on the internet. What a CMS seeks is to automate some repetitive processes and to facilitate others so that no technical knowledge is required to implement them.

Technical Support After the completion of the website and from time to time, the website needs to be periodically improved and maintained. The site should be regularly maintained and constantly monitored to avoid errors or even malicious practices.

Marketplaces The marketplace is a large e-shop that hosts many merchants. The final consumer basket may consist of products received from different merchants/companies.

Google Ads E-shop advertising can be done either through Google Dynamic Search Ads where it is an easy way to find people who are looking for exactly. Shoppers look for it through Google Shopping Ads where the products appear at the top of the search results along with images, prices, and titles, enabling the online store to make direct sales.

Social Media Social media is a tool that enhances the recognition of the online store and increases the revenue and sales of the e-shop [10].

Transport Company In case the company does not have its own means of transporting products, one of the most important collaborations is with a transport company. There are many shipping options, and special attention should be paid to both the cost and the services offered by each of them (delivery time, shipping areas, and delivery cost).

Banking Institutions Another important cooperation of the e-shop is with the banking institution. For these transactions, the banking institutions provide their services for a fee. This charge is usually a fixed amount plus a percentage of the total amount.

Lawyer—Accountant The cooperation with a lawyer should be done early, before the construction of the e-shop as there are issues that need to be considered and addressed (company articles of association, amendments, GDPR). The accountant on the other hand gives advice on tax issues, business, and insurance. In addition, he is the one who deals with the observance of tax obligations, and not only.

E-bay is a shopping website that is best known for consumer auctions and sales to consumers but is also extremely popular with online marketers where they use it as a sales channel [11].

E-shops are constantly growing and can provide a wide variety of products at low and competitive prices. They enable the respective customer to have an idea of the time of preparation, sending, and receipt of his order. In addition to products that can be found in stores in the area, e-shops allow the customer to order specialized items from other cities even abroad. The prices in the online stores are usually cheaper, while the payment method (credit card, cash on delivery and Paypal) of the order facilitates their acquisition. The online stores manage to obtain the lowest prices on products since they remain "open" 24 h a day, 365 days a year, do not have high rental outlets, and do not need specific specifications as a conventional store would need to be considered successful. The size of the staff varies depending on the needs of each company.

5 Hypothesis Trials and Results

Initially, 100% of e-shop resources should be divided into three factors which are website, advertising, and collaborations as shown in the Figure below. The distribution of the initial percentage is chosen to be as it is shown in Table 2. This is achieved by using a converter on the respective flow from the e-shop to each factor. After researches, the following results are obtained regarding the behavior of the system. The green color in the following diagram indicates a positive to the system parameter and red indicates the negative one.

Website
For example, 29.3% of the e-shop 's resources go to the website since the es2w stream has the percentage set to es. The website selects 50% of the resources to be distributed to g_i (graphic_illustration) via the w2gi stream, 30% to the CMS via w2CMS, and the remaining 20% to technical support via the w2ta stream.

Advertising
The next 25.4% of the e_shop goes to a (advertising) In the same way as previously explained. Through streams, resources are distributed in the same way to marketplaces, google_ads, and social_media. 30% is distributed through a2m to marketplaces, 30% to ads via Google (google_ads) via a2ga and the remaining 40% to ads via social_media with the flow a2sm.

Table 2 Quotation of variables

Variables	Percentage (%)
Website(w)	29.3%
Advertising(a)	25.4%
Collaborations(c)	45.3%

Collaborations

The remaining 45.3% of the e_shop goes to the collaborations that will be made to properly build and operate the e_shop. From the resources of collaborations, the company has 10% in the cooperation with the transport companies. 45% in collaborations with banking institutions and the remaining 45% in collaborations with lawyers-accountants. This is achieved with the corresponding functions selected on the respective flow (c2c, c2b, c2l_a) with the percentage on the e_shop.

And about the remaining parameters:

- 15% of marketplace resources along with 25% of google_ads resources will be redirected to the external eBay agent via m2s and ga2s streams respectively. 95% of sm resources will be redirected to -A- via sm2A.
- All g_i and CMS resources are routed through the gi2A and CMS2A stream to one of the two factors that will determine the success of the cost, in recognition (A). ta2CA flow.
- 25% of the resources spent on lawyers-accountants together with 85% of the resources in banking institutions are promoted to product recognition through b2A. A percentage with a negative sign is driven through the c2s flow to the eBay sub-factor.
- The e-bay factor gathers all the resources from the flows ga2s, m2s, and c2s and from there 50% through eb2CA leads to the competitive advantage. All the resources gathered in CA and A feeds back the model.

As it will be shown below, the system does not become viable, which is verified in the last stage of the process. As mentioned above a business success is considered when the resources spent to create online stores (e-shop), will bring positive results in terms of product recognition and competitive advantage. The results from the operation of the model are shown in the following graph. The resources are on the vertical axis and the twelve months of the year are on the horizontal axis. Graphically shows the trajectory of e_shop, CA, A, and e-bay.

6 Modelling and Simulation

The system from the beginning of its operation and although the initial stocks that feed the positive route is 100% of the initial percentage, it never manages to get positive for the A, e_shop, and e_bay stock for 0-12tu. The consequence of this is the cancellation of the process since it is necessary all three factors (CA, A, ebay) to be positive (Fig. 1).

The above configuration of the model will result in the following results (Fig. 2):

It is observed that although **CA** takes a positive value it equals to − **0.0386*e_shop** at t = 12tu, the **A** stock never takes a positive value and equals to **1.202*e_shop**. Consequently, the system will never manage to work.

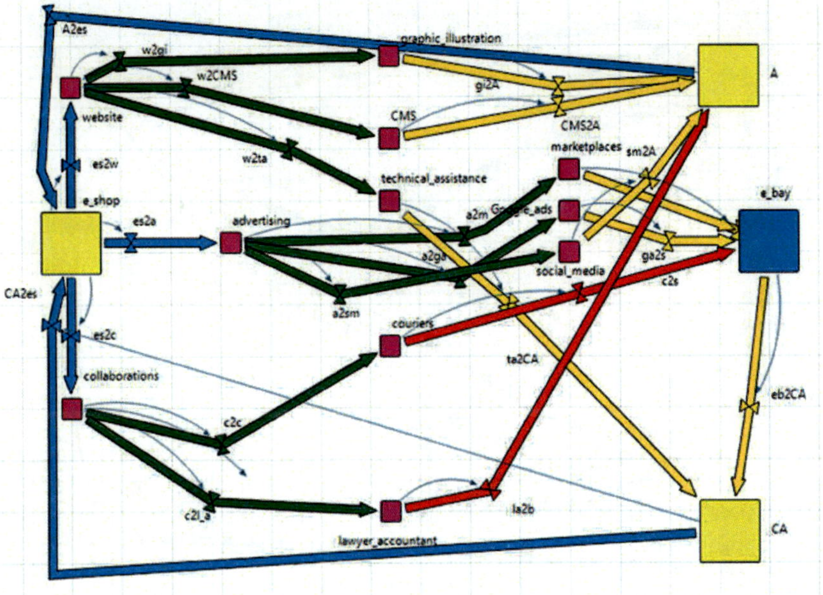

Fig. 1 Simulation model

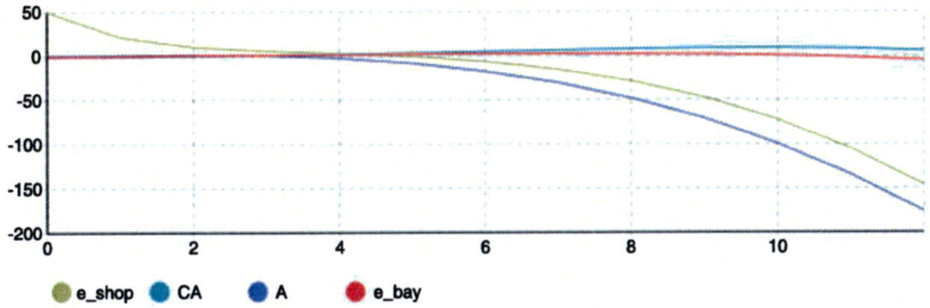

Fig. 2 Diagram during simulation phase

7 Optimization (Fig. 3)

The goal of the company is through the e_shop to gain a competitive advantage and recognition. The company aims at the maximum possible competitive advantage and maximum recognition. This practically, in the model, means that after the initial capital invested and which is 100% of the resources, the resources that return the es must lead it to its maximum possible value, as the larger the es, the larger the CA-A (Table 3; Fig. 4).

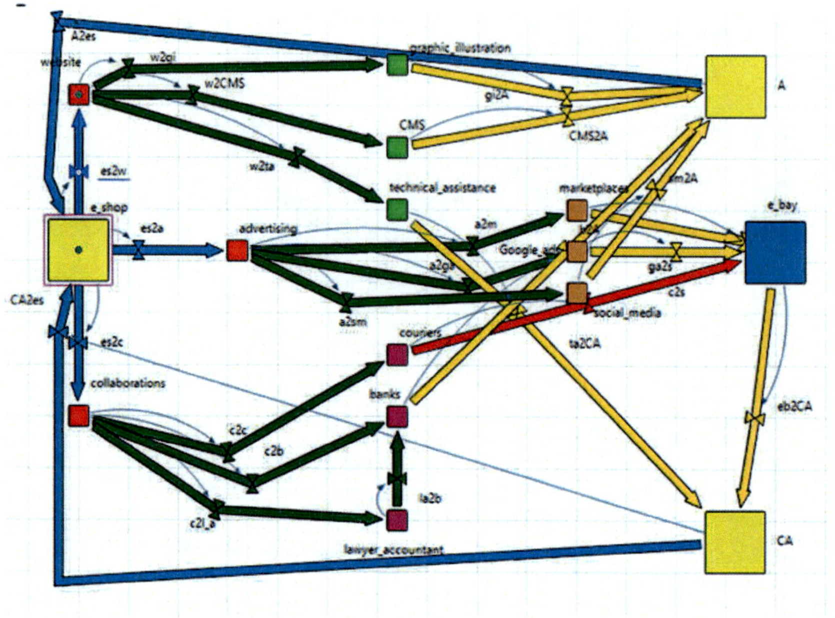

Fig. 3 Optimization model

Table 3 Quotation of variables in optimization

Variables	Percentage (%)
Website(w)	31.7%
Advertising(a)	29.1%
Collaborations(c)	39.2%

Towards that direction, the resource percentages will vary. After many simulations, the result is as the following table:

The maximum possible profit is achieved if the percentage of resources flowing from e_shop to website, advertising, and collaborations is differentiated. 31.7% should be available on website, advertising 29.1%, and collaborations 39.2%. It is observed that the in fourth month, after the increase of the competitive advantage CA, the resources of the e_shop also increased. As the e_shop grew, so did the business profit of the business and the success of the product. An important factor that had a positive effect is the eBay factor which, due to the positive effect it had on the optimization of the model, had a positive effect both in terms of increasing the competitive advantage and in terms of increasing the total resources that feed back the model.

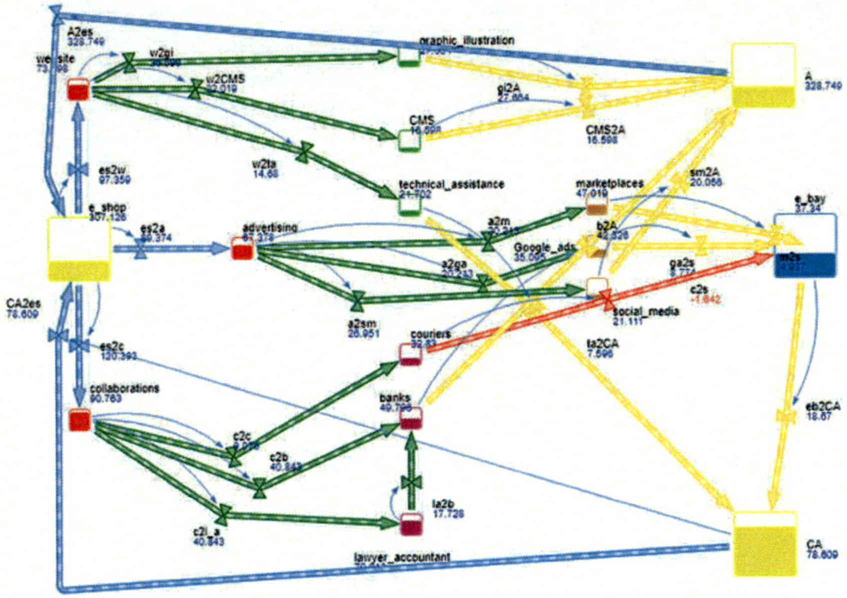

Fig. 4 Diagram during the optimization phase

Table 4 Explanation of flows and factors

Variables	Type	Explanation
w2gi	Website*50%	Flow from website to graphic_illustration
w2CMS	Website*30%	Flow from website to CMS
w2ta	Website*20%	Flow from website to technical_assistance

More specifically:

(a) The following factors affect the website (Table 4):

(b) The following factors affect advertising (Table 5):

(c) The following factors collaborations (Table 6):

(d) The following factors affect e_bay (Table 7):

(e) The following factors are from and to CA and A (Table 8):

It is observed that CA & A at **t = 0tu** begin from the initial value of 100% but then at the output, and at the same time, they both obtain a positive value as well. At t = 12tu the **CA = e_shop & A ≈ 4182*e_shop** approximately. A result that, compared to modeling, shows a tremendous change (Fig. 5).

Table 5 Explanation of flows and factors

Variables	Type	Explanation
a2m	Advertising*30%	Flow from advertising to marketplaces
a2ga	Advertising*30%	Flow from advertising to Google_ads
a2sm	Advertising*40%	Flow from advertising to social_media

Table 6 Explanation of flows and factors

Variables	Type	Explanation
c2c	Collaborations*10%	Flow from collaborations to courier
c2b	Collaborations*45%	Flow from collaborations to banks
c2la	Collaborations*45%	Flow from collaborations to lawyer_accountant

Table 7 Explanation of flows and factors

Variables	Type	Explanation
m2s	Marketplaces*15/100*70/100	Flow from marketplaces to ebay
ga2s	Google_ads*25/100	Flow from Google_Ads to ebay
c2s	-Couriers*5/100	Flow from couriers to ebay

Table 8 Explanation of flows and factors

Variables	Type	Explanation
ta2CA	technical_assistance*35/100	Flow to CA
eb2CA	e_bay*50/100	Flow to CA
gi2A	graphic_illustration	Flow to A
CMS2A	CMS	Flow to A
sm2A	social_media*95/100	Flow to A
b2A	Banks*85/100	Flow to A

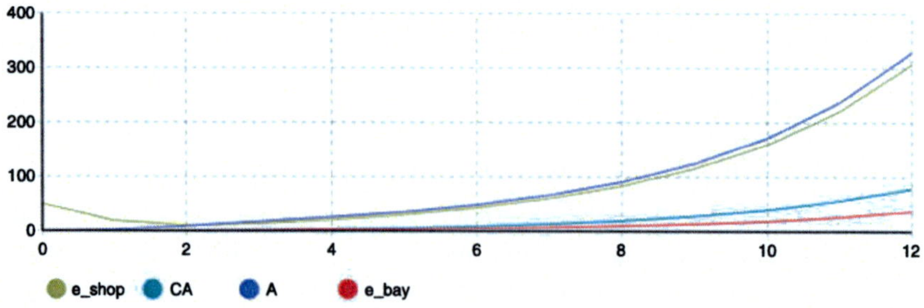

Fig. 5 Diagram during optimization phase

Table 9 Quotation of variables in optimization

Variables	Percentage (%)
Website(w)	19.48%
Advertising(a)	14.71%
Collaborations(c)	65.81%

Table 10 Explanation of flows and factors

Variables	Type	Explanation
sm2A	Social_media*95/100*53/100	Flow from social_media to A

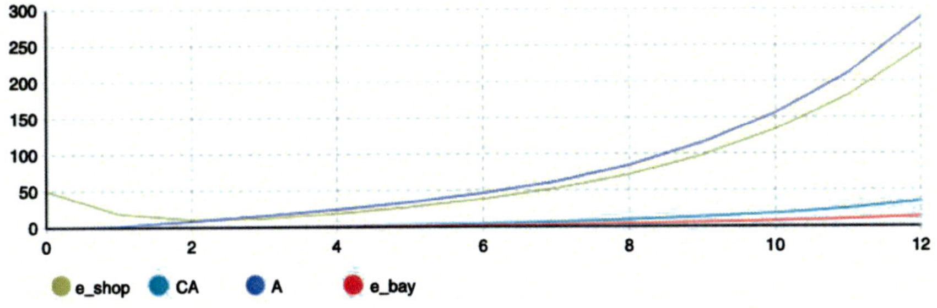

Fig. 6 Diagram during optimization phase

After the change in the distribution of resources, the model shows an increase in competitive advantage and recognition. The model is consequently fed back. Therefore, the model after optimization is much more efficient.

(a) However, to achieve the examination of other scenarios, the viability of the system will be examined if it is considered that social media loses 47% of its power. Based on this assumption, below the behavior of the system is listed.

The parameter percentages that are applied to the above are (Tables 9 and 10):

The following diagram clarifies the behavior of the system (Fig. 6).

It is observed that at **t = 12tu** the **CA = e_shop** and **A ≈ 8.4623*e_shop** approximately, which means that the initial scope is succeeded.

(b) Additionally, the viability of the system will be examined if it is considered that the institution of the global marketplace is deactivated. Based on this assumption, below the behavior of the system is listed (Fig. 7).

The parameter percentages that are applied to the above are (Table 11):

The following diagram clarifies the behavior of the system (Fig. 8).

It is observed that at **t = 12tu** the **CA = e_shop and A ≈ 6.2253* e_shop** approximately, which means that the initial scope is succeeded.

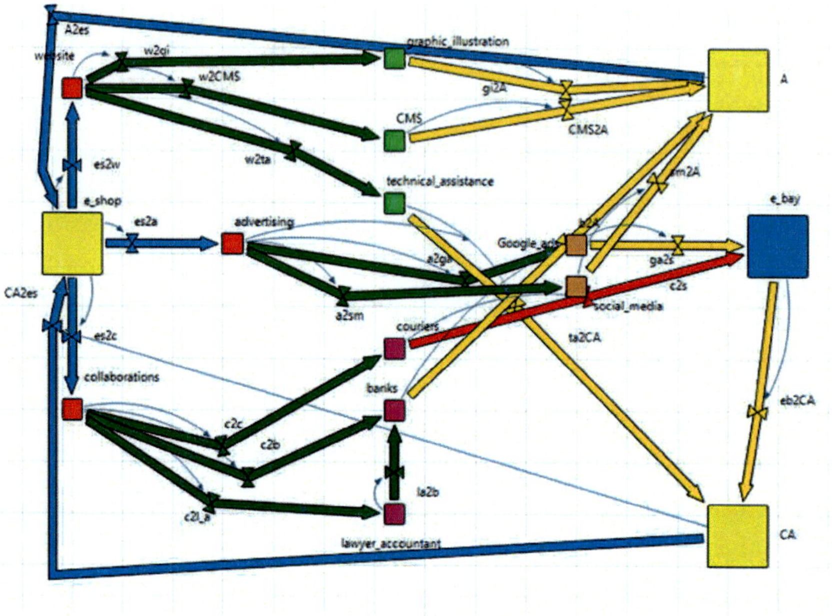

Fig. 7 Diagram during optimization phase

Table 11 Quotation of variables in optimization

Variables	Percentage (%)
Website(w)	42.13%
Advertising(a)	17.39%
Collaborations(c)	40.48%

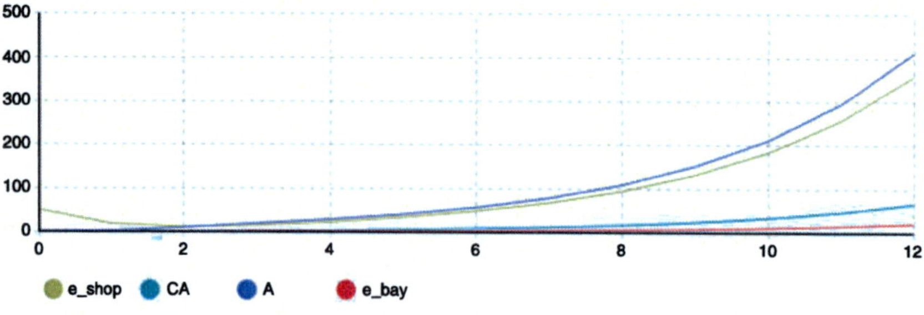

Fig. 8 Resources diagrams during optimization phase

Table 12 Explanation of flows and factors

Variables	Percentage (%)
Social_media (sm)	14.91%
Google_ads (ga)	47.25%
Email (e)	37.24%

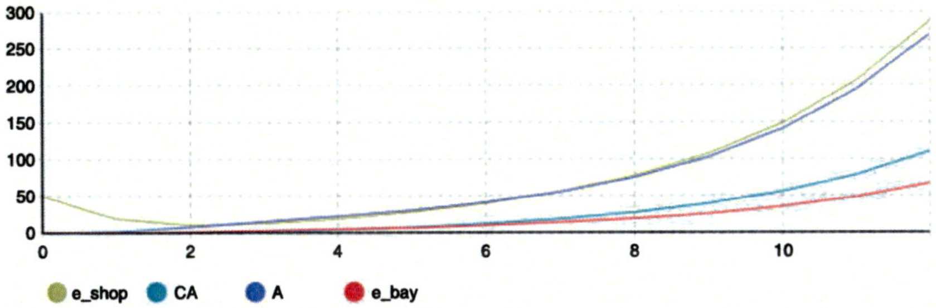

Fig. 9 Resources diagrams during optimization phase

(c) In conclusion for another scenario, the viability of the system will be examined if it is considered that couriers fail to deliver products for a percentage of 95%.

The parameter percentages that are applied to the above are (Table 12):

The following diagram clarifies the behavior of the system (Fig. 9).

It is observed that at **t = 12tu** the **CA = e_shop and A ≈ 2.4479*e_shop** approximately which means that the initial scope is succeeded. Therefore, the model is considered to succeed in the process.

8 Conclusion

This research is part of a research project examining and optimizing how useful and beneficial it is for businesses to maintain an e-shop. Businesses need to use the proper marketing and strategies to cope with the competitive environment and maximize their sales profits and product recognition [12–14]. The rapid changes that are taking place in the IT sector, shape and bring a series of changes that affect commercial companies. E-commerce is one of these changes since it is a modern method of promoting products and services with rapid development in the field of internet [15–17].

The speed and ease with which trading can take place, rank e-commerce high in the preferences and activities of internet users. As it has been analyzed above, through the appropriate website, the right e-shop advertising, and the right collaborations before the creation and during its operation will determine the success of the product and will bring the desired results for the company. In conclusion, as shown by the modeling, the use of this means of ecological innovation—creation of an online store—for both a newly established company and an existing one, is now necessary.

Following this argumentation, pandemic and economic crises enforce a "place-based" typology of stakeholders [18], which promote eco-innovation and sustainable development in challenging locations and tourist destinations [19–21], coupling human activity and well-being to several factors, such as digitalization, natural environment, connectivity, and eco-efficiency. Furthermore, future research on corporate social responsibility should be considered with the assistance of digital marketing in order to ensure the survivability of the e-shops [22–26].

References

1. Toudas, K., Kanellos, N.: Economic and Accounting Performance of Greek innovative firms through Knowledge-based entrepreneurship. J Account Taxation. **1492**, 150–160 (2022). https://doi.org/10.5897/JAT2022.0530
2. Georgakopoulos, G., Toudas, K., Poutos, E., Kounadeas, T., Tsavalias, S.: Capital structure, corporate governance, equity ownership and their impact on firms' profitability and effectiveness in the energy sector. Energies. **15**(3625), 1–10 (2022). https://doi.org/10.3390/en15103625
3. Pujari, D.: Eco-innovation and new product development: understanding the influences on market performance. Technovation. **26**, 76–85 (2006)
4. Khosrow-Pour: Strategic E-Commerce Systems and Tools for Competing in the Digital Marketplace, p. 2. IRMA (2015)
5. Krupcała, K., Januszewski, A.: Website and e-shop Development as an e business Teaching Programme Innovation in Management Education, vol. 176, pp. 2476–2486. Procedia Computer Science (2020)
6. Aghion, P., Tirole, J.: The management of innovation. Quart J Econ. **4**, 1185–1209 (1994)
7. Niranjanamurthy, M., Kavyashree, N., Mr Jagannath, S., Chahar, D.: Analysis of e-commerce and m-commerce: advantages, limitations and security issues. Int. J. Comput. Commun. Eng. **2**, 2360–2370 (2013)
8. Bellal, M., Fahima Nader: E-shop: A Collaborative Learning Activity, pp. 214–218. Procedia Social and Behavioral Sciences (2014)
9. Resnick, P., Zeckhauser, R., Swanson, J., Lockwood, K.: The value of reputation on eBay: a controlled experiment. Exp Econ. **9**, 79–101 (2006)
10. Sakas, D.P., Reklitis, D.P., Trivellas, P., Vassilakis, C., Terzi, M.C.: The effects of logistics websites' technical factors on the optimization of digital marketing strategies and corporate brand name. Processes. **10**(5), 892 (2022)
11. Reklitis, P., Trivellas, P.: Performance implications of aligning generic strategies with the business environment. Int J Manag Decis Mak. **3**(3-4), 319–336 (2002)
12. Trivellas, P.: Investigating the impact of research and development strategy on firm performance. Key Eng Mater. **495**, 306–309 (2012)
13. Christianna, T.: Business plan of e-shop development,1 66, (2019)
14. The Chartered Institute of Marketing: Marketing is the management process responsible for and satisfying customer, 211, (2015)
15. Hatzithomas, L., Fotiadis, T.A., Coudounaris, D.N.: Standardization, adaptation, and personalization of international corporate social media communications. Psychol. Mark. **33**(12), 1098–1105 (2016)
16. Trivellas, P., Anastasopoulou, A., Malindretos, G., Reklitis, P., Sakas, D.P.: A place based analysis of stakeholders' advancing sustainability in remote and isolated communities: The case of North Aegean Islands. IOP Conf. Ser.: Earth Environ. Sci. **899**(1), 012044 (2021)

17. Trivellas, P., Kakkos, N., Vasiliadis, L., Belias, D.: Sustainability, Social Marketing and Host Attitudes About Tourism in the City of Larissa, pp. 89–95. Springer Proceedings in Business and Economics (2017)
18. Trivellas, P., Kakkos, N., Vasiliadis, L., Belias, D.: Residents' Perceptions Toward Cultural, Social and Economic Benefits and Costs of Tourism Industry: An Empirical Survey, pp. 171–181. Springer Proceedings in Business and Economics (2016)
19. Katsikari, C., Hatzithomas, L., Fotiadis, T., Folinas, D.: Push and pull travel motivation: segmentation of the Greek market for social media marketing in tourism. Sustainability. **12**(11), 4770 (2020)
20. Sakas, D.P., Reklitis, D.P.: The Impact of Organic Traffic of Crowdsourcing Platforms on Airlines' Website Traffic and User Engagement. Sustainability. **13**, 8850 (2021). https://doi.org/10.3390/su13168850
21. Reklitis, P., Trivellas, P., Mantzaris, I., Mantzari, E., Reklitis, D.: Employee perceptions of corporate social responsibility activities and work-related attitudes: the case of a greek management services organization. In: Gal, G., Akisik, O., Wooldridge, W. (eds.) Sustainability and Social Responsibility: Regulation and Re-porting, pp. 225–240. Springer Singapore, Singapore (2018). https://doi.org/10.1007/978-981-10-4502-8_10
22. Sakas, D.P., Giannakopoulos, N.T., Reklitis, D.P., Dasaklis, T.K.: The effects of cryptocurrency trading websites on airlines' advertisement campaigns. J. Theor. Appl. Electron. Commer. Res. **16**, 3099–3119 (2021). https://doi.org/10.3390/jtaer16070169
23. Sakas, D.P., Kamperos, I.D.G., Reklitis, D.P., Giannakopoulos, N.T., Nasiopoulos, D.K., Terzi, M.C., Kanellos, N.: The effectiveness of centralized payment network advertisements on digital branding during the COVID-19 crisis. Sustainability. **14**, 3616 (2022). https://doi.org/10.3390/su14063616
24. Sakas, D.P., Reklitis, D.P.: Predictive model for estimating the impact of technical issues on consumers interaction in agri-logistics websites. In: Information and Communication Technologies for Agriculture—Theme IV: Actions, pp. 269–283. Springer International Publishing, Cham (2021). https://doi.org/10.1007/978-3-030-84156-0_14
25. Horch, A., Wohlfrom, A., Weisbecker, A.: An E-Shop Analysis with a Focus on Product Data Extraction, pp. 61–72. Springer (2017)
26. Koumparoulis, D.N.: PEST Analysis: The case of E-shop. Int. J. Econ. Manag. Soc. Sci. **2**, 31–36 (2013)

Social Networks and User Interaction as a Determinant of Business Viability, Modeling, and Optimization

Dimitrios M. Mastrakoulis, Dimitrios K. Nasiopoulos, Dimitrios A. Arvanitidis, Elli C. H. Zara, Panagiotis Trivellas, and Catherine Marinagi

1 Introduction

The purpose of this research is to find out how useful and beneficial it is for the company to allocate all its resources to the social media sector as well as the impact it will have on consumers. In this particular investment, what the company seeks is a competitive advantage over other companies and the recognition of the product. The time given to the company to evaluate these results is 12 months. At first glance and on a highly theoretical basis, it is believed that the costs that the company will spend on social media will have a positive impact and consequently will determine the success of the product [1–6].

D. M. Mastrakoulis
Department of Informatics and Telecommunications, University of Thessaly, Lamia, Greece
e-mail: dmastrakoulis@aua.gr

D. K. Nasiopoulos
Department of Agribusiness and Supply Chain Management, BICTEVAC LABORATORY Business Information and Communication Technologies in Value Chains laboratory, School of Applied Economics and Social Sciences, Agricultural University of Athens, Athens, Greece

D. A. Arvanitidis (✉) · E. C. H. Zara
Department of Informatics and Telecommunications, University of Peloponnese, Tripolis, Greece
e-mail: darvanitidis@aua.gr

P. Trivellas
Department of Agribusiness and Supply Chain Management, Organizational Innovation and Management Systems, School of Applied Economics and Social Sciences, Agricultural University of Athens, Athens, Greece

C. Marinagi
Department of Agribusiness and Supply Chain Management, School of Applied Economics and Social Sciences, Agricultural University of Athens, Athens, Greece

D. P. Sakas et al. (eds.), *Computational and Strategic Business Modelling*, Springer Proceedings in Business and Economics,
https://doi.org/10.1007/978-3-031-41371-1_14

Moreover, staff members of high-tech companies with high levels of technical knowledge led to increased financial and economic performance, in comparison with companies, in the same industrial sector, employing less technically qualified staff [7]. A significant relation among major capital structure and corporate governance parameters and firm performance of 10 leading energy sector companies traded in the NYSE. Furthermore, pecking order and agency cost theories play an important role in the financing of these firms, while static trade and irrelevance theory find no support [8].

This research will analyze whether the social media factor can directly and to a large extent, influence the success of a product in the market. The goal is to identify and analyze how social media as one of the tools of digital marketing will determine the relationship between eco-innovation and product success (product recognition and competitive advantage) [9].

1.1 Social Media

The phenomenon of social networks is not something new, what has changed is their scope and penetration into the daily lives of users.

The modern way of life and the need of people for communication have made social networks very popular with the result that the development of these networks is growing day by day. Their ever-increasing usage rates prove their consolidation [1–6].

2 Description of Hypothesis and Modeling Process (Table 1)

Table 1 Description of the factors applied

Analysis of the main factors:	
Facebook	A key factor in the process is social media, in addition to the direct communication that is offered as a service. Facebook can be used effectively by businesses to spread a message and at the same time is a feedback tool offering important information about both consumer behavior and their reviews of the products and services they offer. The presence of a company on Facebook through a positive and attractive appearance, can significantly enhance its social image. Immediately after, the parameters that have been set for the proper operation of the factor are applied.
YouTube	An additional key factor process is YouTube. Most users use YouTube to listen to music, while a slightly smaller percentage use it to watch everyday videos. YouTube offers businesses the opportunity to create their own channel and advertise their product through it. Immediately after, the parameters that have been set for the proper operation of the factor are applied [2, 10, 11].
Twitter	The last key factor in the process is Twitter. This platform allows users to read and send short messages called tweets. Twitter, like other social media, facilitates the way businesses communicate and connect with the rest of the world as they have the potential to reach a huge market without geographical restrictions. Immediately after, the parameters that have been set for the proper operation of the factor are applied.

3 Hypothesis

H1: CA > 0, A < 0/Hackers affect Facebook and personalized ads post obscene content resulting in 6% of users deleting their account.

Facebook's impact is reduced by 48% resulting to video marketing turn to negative axis.

H2: CA > 0, A > 0/*Facebook's Content marketing is being replaced by another application that is not frequently updated with business information.*

Thirty-eight percent of businesses stop sending information.

H3: CA > 0, A > 0/ *Twitter's programmatic returns to the previous state, targeting only a small portion of customers, with a lot of money.*

Twenty-four percent of users delete their account.

H4: CA > 0, A > 0/ *Twitter stops offering discount coupons from small businesses to users.*

Eighteen percent of users and 14% of businesses stop using Twitter and the traffic becomes a negative factor.

H5: *CA < 0, A < 0, youtubers > 0, social_media < 0/All mail servers broke down because of a cyber-attack and YouTube's owner resigned.*

The factor of Metrics is excluded from the system's equation and so YouTube loses 60% of its youtubers.

4 Methodology

The aim of the system under consideration is considered when the resources spent on the creation and support of social media will bring positive results both in terms of product recognition and competitive advantage. Additionally significant purpose of businesses is to attract new customers, to strengthen their relationship with existing ones and to convey the information they want. This practically means for the system, that after the initial stock is provided and constitutes 100% of the resources, the resources that feed the stocks, must lead it to its maximum possible value, which implies the ever-increasing value of the final value. The purpose of the whole project is the non-interruption of the process and at the same time the overcoming of the sum of the other opposite measure of parameters and factors that are applied. To achieve this goal, percentage changes had to be made. Time is referred to as time units (tu).

Initially, 100% of the processing is placed in **social_media** in the context of perfectly competitive. This practically that after the resources that supply the following stocks must lead it to its maximum possible value, which implies the ever-increasing value of the final amount.

The first element that will start the model with, is **social_media**. Then, if this stock is analyzed and goes through the factors that exist in the model, it will end up in A and CA. It is pointed out that the examined operating time of the system is from $t = 0$ to $t = 12$.

Initially, 100% of the resources that will affect the system during the starting process at time $t = 0$ is placed in **social_media**. During the process of the system, other parameters will act because of the production of extra factors such as the following:

Marketplace In addition to the many features that Facebook offers (messages, notifications, status updates, and event updates) a very important part is the Marketplace where in addition to the fact that users can post their own ads for product sales (in their area or with the ability to send), companies and their brands can follow. The Marketplace is used by businesses mainly to advertise their products, reach new people and as a sales channel as they enable customers to make purchases without leaving the company's Facebook page [12].

Personalized Ads In personalized ads the communication with the customer becomes more personal, making him feel familiar with the brand and gaining a place in his shopping choices. Targeted advertising always makes a difference in a business. The campaigns find their target and the data collected boosts sales. Personalized ads on Facebook, that is, those that use a user's psychological profile and not just his demographic characteristics (age and gender), significantly increase the likelihood that users will click on them and buy products [11].

Video Marketing Video marketing is a powerful tool in the hands of businesses that can inform, entertain, and attract the public. To date, most professionals who use video as part of their strategy plan to continue using it, and a large percentage plan to invest further in future video campaigns. Using videos, businesses can engage their audience in a creative way, share key messages, promote a new product, and increase brand recognition [11].

Content Marketing Content marketing is the art of communicating with customers and creating opportunities without selling. The purpose is to provide information that makes the buyer smarter. In other words, companies on their part provide consistent, continuous, and valuable information to the buyers, and they, in turn, reward them with their loyalty and loyalty.

Content marketing can take many forms, the most common of which are like a blog post, videos, infographics, PDFs, eBooks, social media posts, and interviews.

Homepage Ads In addition to the two main categories of ads (TrueView Video and Display Ads) that have been available on YouTube in recent years, there is also the possibility of advertising on the main YouTube page.

Metrics – YouTube Analytics (YouTube Insight) This is a free analysis tool (metrics and reporting tool) which is offered to all YouTube users who have an account and allows them to see statistics for the audience that watches their videos. This tool alone can't help to increase the views of a video, but with the correct analysis of the statistics it provides, the chances of maximizing the views and consequently the revenue increase.

Programmatic A big innovation in the field of digital marketing is programmatic buying, a specially designed software, which has revolutionized the way in which advertising space is purchased from websites. Automates the purchase, placement, and optimization of business advertising using AI (artificial intelligence) and machine learning technologies.

Traffic One of the key metrics that indicate signs of success is traffic to the social site. The more customers who visit it, the more chances it has for sales.

Convert The main goal of businesses using social media is to convert visitors into leads, which to achieve this, their personal data/contact information must have been collected.

5 Hypothesis Trials and Results

Initially, 100% of social_media resources should be shared among the three factors which are Facebook, YouTube, and Twitter. The distribution of the initial percentage is chosen to be as it shown in Table 2. This is achieved by using a converter on the respective flow from the social_media to each factor. Finally, the effect of influencers is applied as mentioned above. After research the following results are obtained regarding the behavior of the system. The green color in the following diagram indicates a positive to the system parameter and red indicates the negative one.

Facebook

In this case, 38.42% of the social_media resources go to Facebook since in the sm2F stream the percentage on sm has been placed. Facebook selects 25% of the resources to be distributed in the marketplace via F2m streaming, 35% to personalized_ads (pa) via F2pa and the remaining 40% to video_marketing (vm) via F2vm streaming.

YouTube

The next 35.31% of social_media goes to YouTube in the same way as previously explained. Through streams, resources are distributed in the same way to content_marketing, Homepage_ads, and metrics; 30% is distributed through Y2cm to content_marketing, 30% to Homepage_ads through Y2Ha, and the remaining 40% to metrics with m2p flow.

Table 2 Quotation of variables

Variables	Percentage (%)
Facebook (F)	38.42%
YouTube (Y)	35.31%
Twitter (T)	26.27%

Twitter

The remaining 26.27% of social_media goes to Twitter. The company has 10% of the Twitter resources in programmatic. Forty-five percent in traffic and the remaining 45% in convert. This is achieved with the corresponding functions selected on each stream (T2p, T2t, and T2c) with the percentage on Twitter.

And about the remaining parameters:

- Twenty percent of marketplace resources are driven through the m2cm flow to content_marketing (cm).
- Fifty percent of personalized_ads (pa) resources are routed to the metrics factor via pa2m and 20% of video_marketing (vm) resources are routed to content_marketing (cm) via vm2cm.
- Ten percent of the resources collected in content_marketing (cm) is directed to the external youtubers factor and from there a negative percentage through the y2A flow to the competitive advantage A.
- Ten percent of content_marketing resources are routed through the cm2y stream to the external youtubers factor. At this point it has been defined the youtubers factor to negatively affect the product recognition by 5%. Twenty-five percent of Homepage_ads resources through Ha2m are directed to the metrics factor, and from there 60% are directed to the programmatic factor.
- From programmatic, 45% through p2CA is led to CA (Competitive Advantage). Sixty-five percent of traffic is led through t2m to the metrics factor. Thirty-five percent of convert resources are driven by competitive advantage5. All the resources gathered in CA and A feedback the model.

As it will be shown below, the system does not become viable, which is verified in the last stage of the process.

6 Modeling and Simulation

The system from the beginning of its operation and although the initial stocks that feed the positive route is at 100% of the initial percentage, it never manages to get positive for the A stock for 0–12tu. The consequence of this is the cancellation of the process since it is necessary both factors (CA, A) be positive.

The results from the operation of the model are shown in the following graph. The pores are on the vertical axis and the 12 months of the year are on the horizontal axis. Graphically shows the course of social_media, CA, A, and youtuber (Fig. 1).

The above configuration of the model will result in the following results (Fig. 2):

It is observed that although **CA** takes a positive value and equals **1384*social_media** (at t = 12tu) the **A** stock never takes a positive value and equals **−0.0142*social_media**. Consequently, the system will never manage to work under these circumstances.

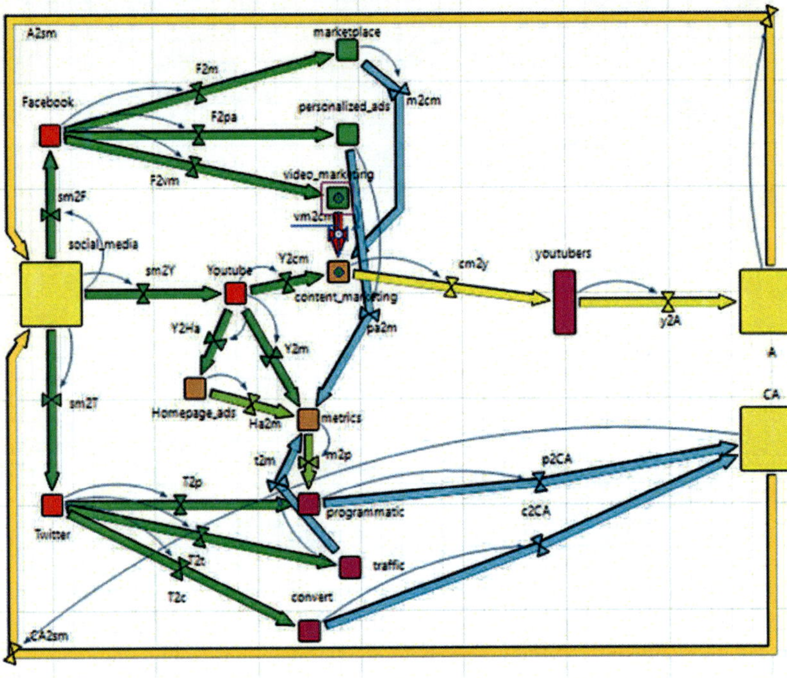

Fig. 1 Simulation model

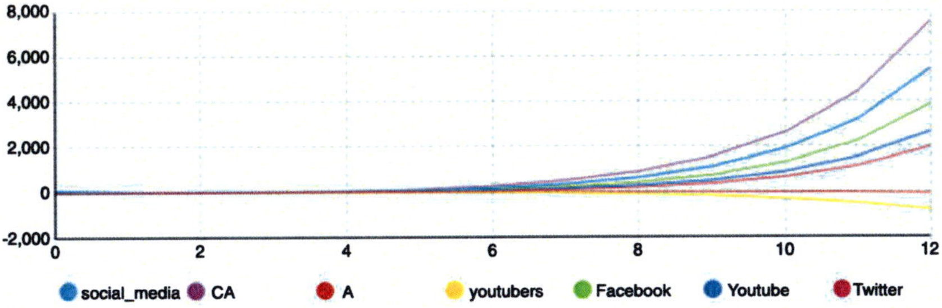

Fig. 2 Diagram during simulation phase

7 Optimization (Fig. 3)

The goal of the company is to gain a competitive advantage and recognition through social media. The company aims at the maximum possible competitive advantage and maximum recognition. This practically, in the model, means that after the initial capital that is invested and is 100% of the resources, the resources that feed back the

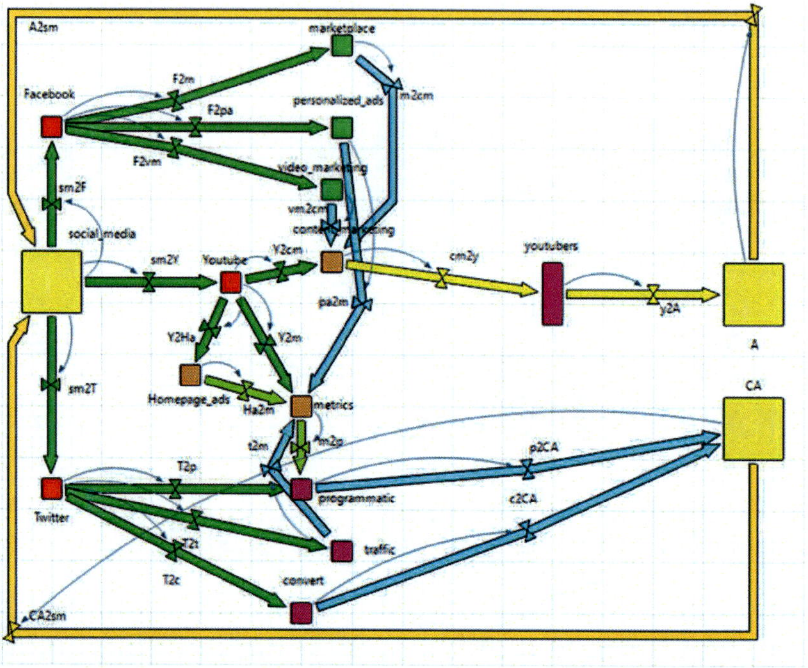

Fig. 3 Optimization model

Table 3 Quotation of variables in optimization

Variables	Percentage (%)
Facebook (F)	40.2%
YouTube (Y)	39.4%
Twitter (T)	20.4%

social media should lead it to its maximum possible price, as the bigger the sm the bigger it will be the CA–A. To achieve this goal modifications had to be made to the model values shown in the image below. Toward that direction, the energy percentages will vary. After many simulations the result is as the following table (Table 3; Fig. 4):

The maximum possible profit is achieved if the percentage of resources flowing from social_media to, Facebook, YouTube, and Twitter is differentiated. It is observed that after the fourth month in the process of optimization the key factors CA and A are increased. An important factor that positively affected the recognition is the youtubers factor that due to the positive impact 5% of the resources are directed to A through y2A.

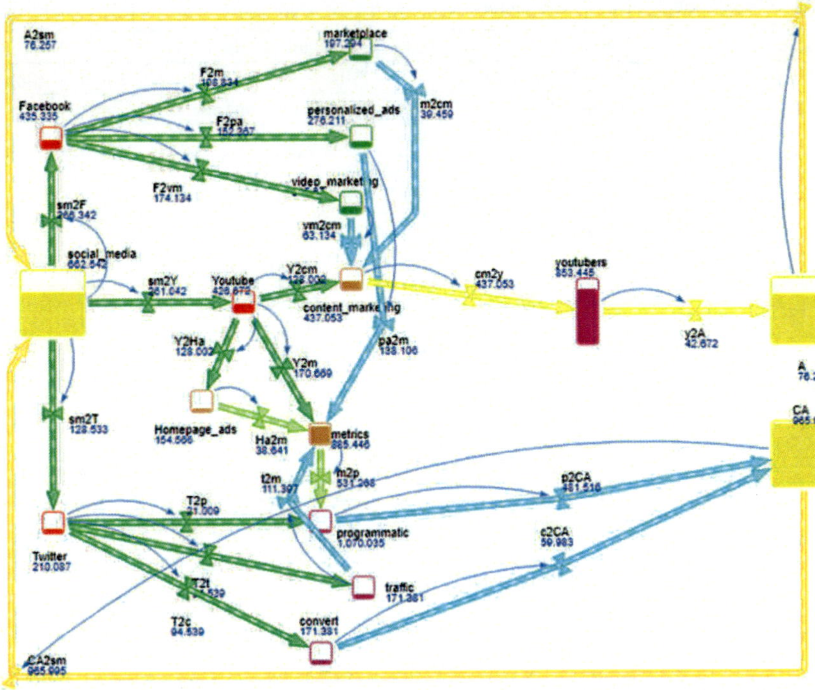

Fig. 4 Diagram during optimization phase (tu = 8)

Table 4 Explanation of flows and factors

Variables	Type	Explanation
F2m	Facebook*25/100	Flow from Facebook to marketplace
F2pa	Facebook*35/100	Flow from Facebook to personalized_ads
F2vm	Facebook*40/100	Flow from Facebook to video_marketing

More specifically:

(a) The following factors affect Facebook (Table 4):

(b) The following factors affect YouTube (Table 5):

(c) The following factors affect Twitter (Table 6):

(d) The following factors affect youtubers (Table 7):

(e) The following factors are from and to CA and A (Table 8):

It is observed that CA & A at **t = 0tu** begin from the initial value of 100% but then at the output, and at the same time, they both obtain a positive value as well. At

Table 5 Explanation of flows and factors

Variables	Type	Explanation
Y2cm	Youtube*30/100	Flow from YouTube to content_marketing
Y2m	Youtube*40/100	Flow from YouTube to metrics
Y2Ha	Youtube*30/100	Flow from YouTube to Homepage_ads

Table 6 Explanation of flows and factors

Variables	Type	Explanation
T2p	Twitter*10/100	Flow from Twitter to programmatic
T2t	Twitter*45/100	Flow from Twitter to traffic
T2c	Twitter*45/100	Flow from Twitter to convert

Table 7 Explanation of flows and factors

Variables	Type	Explanation
y2A	YouTubers*5/100	Flows to A from youtubers

Table 8 Explanation of flows and factors

Variables	Type	Explanation
A2sm	A	Flow from A to social_media
p2CA	Programmatic*45/100	Flow to CA from programmatic
c2CA	Convert*35/100	Flow to CA from convert
CA2sm	CA	Flow from CA to social_media

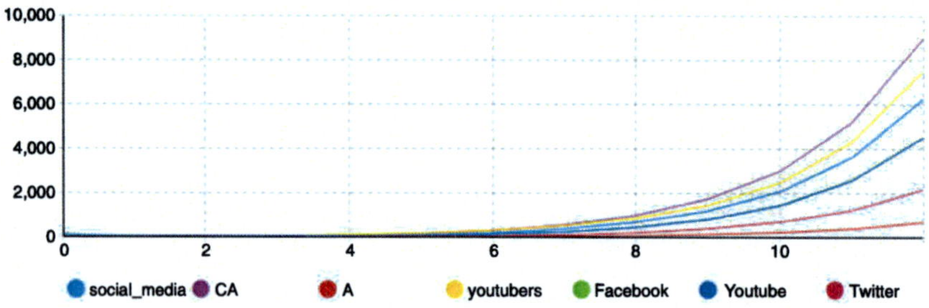

Fig. 5 Diagram during optimization phase

t = 12tu the **CA≈1.44*social_media & A ≈ 0.109*social_media** approximately. A result that, compared to modeling, shows a tremendous change (Fig. 5).

After the change in the distribution of resources, the model shows an increase in competitive advantage and recognition. The model is consequently fed back.

Therefore, the model after optimization is much more efficient.

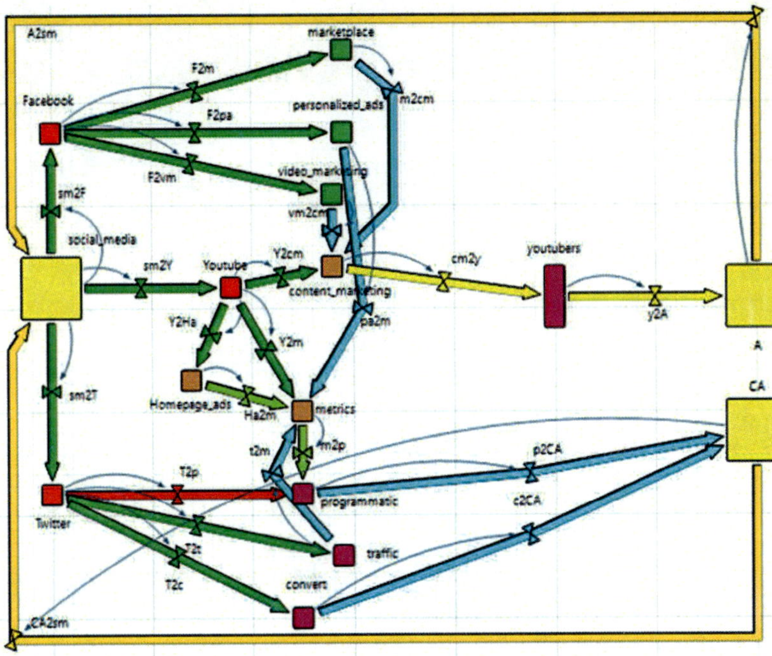

Fig. 6 Optimization model

Table 9 Explanation of flows and factors

Variables	Percentage (%)
Facebook (F)	37.14%
YouTube (Y)	28.79%
Twitter (T)	34.07%

(a) However, to achieve the examination of other scenarios, the viability of the system will be examined if it is considered that Twitter's programmatic returns to the previous state, targeting only a small portion of customers, with a lot of money. Based on this assumption, below the behavior of the system is listed (Fig. 6).

The parameter percentages that are applied to the above are as follows (Table 9):
The following diagram clarifies the behavior of the system (Fig. 7).
It is observed that at **t = 12tu** the **CA≈1.434*social_media** and **A** ≈ **0.0967* social_media** approximately, which means that the initial scope is succeeded.

(b) However, to achieve the examination of other scenarios, the viability of the system will be examined if it is considered that Twitter stops offering discount coupons from small businesses to users.

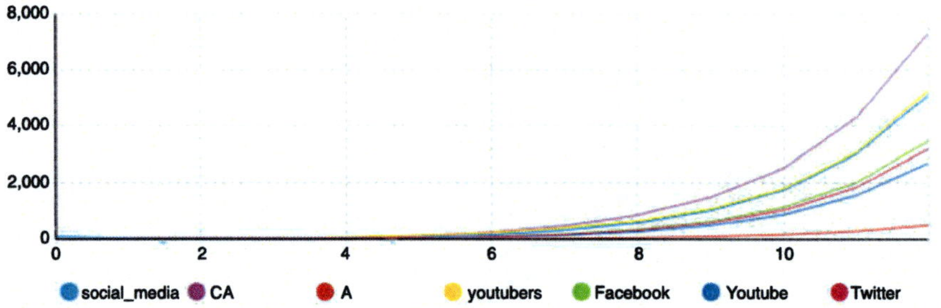

Fig. 7 Diagram during optimization phase

Table 10 Explanation of flows and factors

Variables	Percentage (%)
Facebook (F)	29.12%
YouTube (Y)	38.43%
Twitter (T)	32.45%

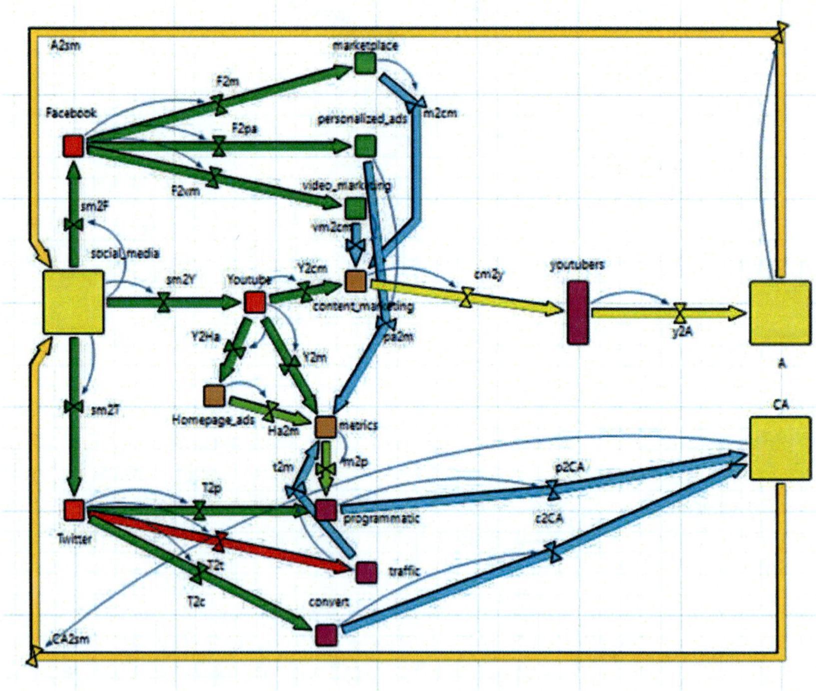

Fig. 8 Optimization model

Based on this assumption, below the behavior of the system is listed. The parameter percentages that are applied to the above are as follows (Table 10; Fig. 8):

The following diagram clarifies the behavior of the system (Fig. 9).

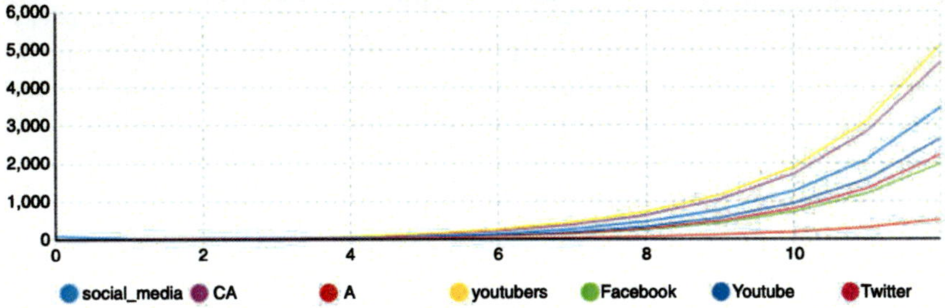

Fig. 9 Energy diagrams during optimization phase

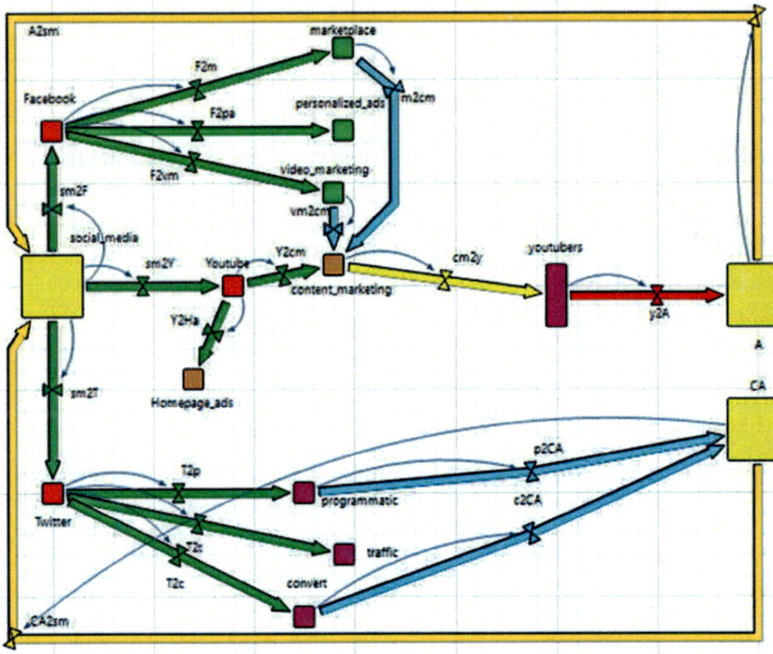

Fig. 10 Optimization model

It is observed that at **t = 12tu** the **CA≈1.34*social_media and A ≈ 0.1466*social_media** approximately, which means that the initial scope is succeeded.

(c) However, to achieve the examination of other scenarios, the viability of the system will be examined if it is considered that all mail servers broke down because of a cyber-attack and YouTube's owner resigned. Based on this assumption, below the behavior of the system is listed (Fig. 10).

Table 11 Explanation of flows and factors

Variables	Percentage (%)
Facebook (F)	31.82%
YouTube (Y)	29.01%
Twitter (T)	35.73%

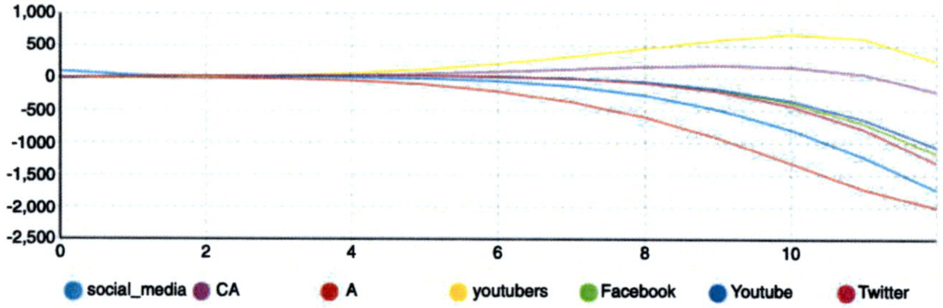

Fig. 11 Energy diagrams during optimization phase

The parameter percentages that are applied to the above are as follows (Table 11):

The following diagram clarifies the behavior of the system (Fig. 11).

It is observed that at **t = 12tu** the **CA≈0.1192*social_media and A ≈ 1.158*social_media** approximately but **youtubers = −0.1548*Green_Adv** which means that the initial scope is succeeded. Therefore, the model is considered to fail in the process.

8 Conclusion

This research is part of a research project examining and optimizing how useful and beneficial it is for the company to allocate all its resources to the social media sector as well as the impact it will have on consumers.

The businesses target is to have a return of at least the initial ROI (return of investment) resources. But to be as profitable as possible, the company's investment must be sought in an optimal version of the model with a different sharing of resources. After a study and several tests, the resources were distributed so that the company with the same financial investment had the maximum possible monetary gains. Optimization played a prominent role in this.

It is observed that social_media starts from the initial price of 100%, then decreases for a short time due to the investment and after the fourth month increases more especially after the optimization than before. So, it brings more money to the company. After the fourth month, there is an increase in CA and A's after optimization, where this is particularly important because they are the two main factors that will lead to the optimal increase of social_media. Therefore, the post-optimization model is more efficient.

In conclusion, social media is the modern way of transmitting experiences and as it is understood, no modern company can miss the place where the products and services they offer are discussed and judged. On the other hand, social media is a place for businesses where they can present their products and services, improve communication with their customers, exchange information and gain information about the competition.

As digital marketing evolves day by day, businesses are bound to adapt to change if they are to remain competitive [13, 14]. In fact, companies have no choice but to create a social media presence or not, but their concern is how well they will create it. Thus, the alignment of business strategies based on new technologies with the external environment may provide a sustainable competitive advantage to modern enterprises [15, 16].

Across this line of reasoning, emerging crises enforce a "place-based" typology of stakeholders [17], which fuel sustainable development especially in congested areas and tourist destinations [18–23] reenforcing human activity and wellbeing with several factors, such as digitalization, natural environment, and connectivity. Finally, except for digital advertising, future research on CSR and green marketing should be considered for the ethically sustainable use of social media [24–28].

References

1. Alalawan, A., Rana, N., Dwivedi, Y., Algharabat, R.: Social media in marketing: A review and analysis of the existing literature. Science Direct. **34**, 1177–1190 (2017)
2. Schwemmer, C., Ziewiecki, S.: Social Media Sellout: The Increasing Role of Product Promotion on YouTube, pp. 1–20 (2018)
3. Hoffman, L.D., και Fodor, M.: Can You Measure the ROI of Your Social MediaMarketing? pp. 41–49. MIT SLOAN Management Review (2010)
4. Johan Wigmo, J.: Social Media Marketing. Which Role Can Social Media Play as a Marketing Tool? Bachelor Thesis (2010)
5. Palmer, A., Koenig-Lewis, N.: An experiential, social network-based approach to direct marketing. Emerald Group Publishing Limited (2009)
6. Rosemary Thackeray, R., Neiger, B.L., Keller, H.: Integrating Social Media and Social Marketing: A Four-Step Process, pp. 165–168. Health Promotion Practice (2012)
7. Toudas, K., Kanellos, N.: Economic and accounting performance of Greek innovative firms through knowledge-based entrepreneurship. Journal of Accounting and Taxation. **1492**, 150–160 (2022). https://doi.org/10.5897/JAT2022.0530
8. Georgakopoulos, G., Toudas, K., Poutos, E., Kounadeas, T., Tsavalias, S.: Capital structure, corporate governance, equity ownership and their impact on firms' profitability and effectiveness in the energy sector. Energies. **15**(3625), 1–10 (2022). https://doi.org/10.3390/en15103625
9. Lydia, K.: Influencer marketing, what it is and how it is used by advertisers, pp. 1–90 (2020)
10. Athanasios, V.: Research on the use of YouTube in e-marketing, pp. 1–120 (2013)
11. Andrew, S.N., Eileen, F., Chen, Y.: How Does Brand-related User-generated Content Differ across Youtube, Facebook, and Twitter? pp. 102–113 (2012)
12. Trang P.: Tran: 'Personalized ads on Facebook: An effective marketing tool for online marketers', pp. 230–242 (2017)

13. Sakas, D.P., Reklitis, D.P., Trivellas, P., Vassilakis, C., Terzi, M.C.: The Effects of Logistics Websites' Technical Factors on the Optimization of Digital Marketing Strategies and Corporate Brand Name. Processes. **10**(5), 892 (2022)
14. Hatzithomas, L., Fotiadis, T.A., Coudounaris, D.N.: Standardization, adaptation, and personalization of international corporate social media communications. Psychol Mark. **33**(12), 1098–1105 (2016)
15. Reklitis, P., Trivellas, P.: Performance implications of aligning generic strategies with the business environment. International Journal of Management and Decision Making. **3**(3-4), 319–336 (2002)
16. Trivellas, P.: Investigating the impact of research and development strategy on firm performance. Key Engineering Materials. **495**, 306–309 (2012)
17. Trivellas, P., Anastasopoulou, A., Malindretos, G., Reklitis, P., Sakas, D.P.: A place based analysis of stakeholders' advancing sustainability in remote and isolated communities: The case of North Aegean Islands. IOP Conference Series: Earth and Environmental Science. **899**(1), 12044 (2021)
18. Trivellas, P., Kakkos, N., Vasiliadis, L., Belias, D.: Sustainability, Social Marketing and Host Attitudes About Tourism in the City of Larissa, pp. 89–95. Springer Proceedings in Business and Economics (2017)
19. Trivellas, P., Kakkos, N., Vasiliadis, L., Belias, D.: Residents' Perceptions Toward Cultural, Social and Economic Benefits and Costs of Tourism Industry: An Empirical Survey, pp. 171–181. Springer Proceedings in Business and Economics (2016)
20. Folinas, D., Fotiadis, T.: Marketing and Supply Chain Management: A Systemic Approach, 1st edn. Routledge (2017)
21. Fotiadis, T.: Strategic Marketing for High Technology Products: An Integrated Approach (Routledge Studies in Innovation, Organizations and Technology), 1st edn. Routledge (2018)
22. Fotiadis, T., Folinas, D., Vasileiou, K., Konstantoglou, A.: Marketing and the Customer Value Chain: Integrating Marketing and Supply Chain Management 1st Edition, Kindle Edition, 1st edn. Routledge (2022)
23. Katsikari, C., Hatzithomas, L., Fotiadis, T., Folinas, D.: Push and pull travel motivation: segmentation of the Greek market for social media marketing in tourism. Sustainability. **12**(11), 4770 (2020)
24. Reklitis, P., Trivellas, P., Mantzaris, I., Mantzari, E., Reklitis, D.: Employee perceptions of corporate social responsibility activities and work-related attitudes: The case of a greek management services organization. In: Gal, G., Akisik, O., Wooldridge, W. (eds.) Sustainability and Social Responsibility: Regulation and Re-porting, pp. 225–240. Springer Singapore, Singapore (2018). https://doi.org/10.1007/978-981-10-4502-8_10
25. Sakas, D.P., Kamperos, I.D.G., Reklitis, D.P., Giannakopoulos, N.T., Nasiopoulos, D.K., Terzi, M.C., Kanellos, N.: The effectiveness of centralized payment network advertisements on digital branding during the COVID-19 crisis. Sustainability. **14**, 3616 (2022). https://doi.org/10.3390/su14063616
26. Sakas, D.P., Reklitis, D.P.: Predictive model for estimating the impact of technical issues on consumers interaction in agri-logistics websites. In: Information and Communication Technologies for Agriculture—Theme IV: Actions, pp. 269–283. Springer International Publishing, Cham (2021). https://doi.org/10.1007/978-3-030-84156-0_14
27. Sakas, D.P., Giannakopoulos, N.T., Reklitis, D.P., Dasaklis, T.K.: The effects of cryptocurrency trading websites on airlines' advertisement campaigns. Journal of Theoretical and Applied Electronic Commerce Research. **16**, 3099–3119 (2021). https://doi.org/10.3390/jtaer16070169
28. Sakas, D.P., Reklitis, D.P.: The impact of organic traffic of crowdsourcing platforms on airlines' website traffic and user engagement. Sustainability. **13**, 8850 (2021). https://doi.org/10.3390/su13168850

Energy Saving of an RPAS (Remotely Piloted Aircraft System) Using Solar Panels, Modeling, and Optimization

Dimitrios A. Arvanitidis, Dimitrios K. Nasiopoulos, Dimitrios M. Mastrakoulis, and Panagiotis Reklitis

1 Introduction

Recently, the use of flying platforms such as unmanned aerial vehicles (UAVs) is rapidly growing for many applications in commercial, civilian, and military fields. The UAVs are very common with real-time monitoring, providing wireless coverage, remote sensing, search and rescue, delivery of goods, security and surveillance, precision agriculture, etc. UAVs have several key potential applications in wireless systems with their inherent attributes such as mobility, flexibility, and adaptive altitude. They can be equipped with panoptic cameras, sensors, etc., to be more likely to perform various tasks such as surveillance and military services [11], search and rescue [14], and remote sensing [4, 16].

There are two options available to increase the flight time. The first one is to increase the battery capacity, which in the conditions of current state-of-the-art

D. A. Arvanitidis (✉)
Department of Informatics and Telecommunications, University of Peloponnese, Tripoli, Greece
e-mail: darvanitidis@aua.gr

D. K. Nasiopoulos
Department of Agribusiness and Supply Chain Management, BICTEVAC LABORATORY Business Information and Communication Technologies in Value Chains laboratory, School of Applied Economics and Social Sciences, Agricultural University of Athens, Athens, Greece

D. M. Mastrakoulis
Department of Informatics and Telecommunications, University of Thessaly, Lamia, Greece
e-mail: dmastrakoulis@aua.gr

P. Reklitis
Department of Agribusiness and Supply Chain Management, Organizational Innovation and Management Systems, School of Applied Economics and Social Sciences, Agricultural University of Athens, Athens, Greece

D. P. Sakas et al. (eds.), *Computational and Strategic Business Modelling*, Springer Proceedings in Business and Economics,
https://doi.org/10.1007/978-3-031-41371-1_15

battery material technologies is a very limited option. Precisely, the battery might be too large for the drone or UAV to fly, or the material of the battery might be too expensive for the deployment to be feasible.

The second option is to charge the battery from an external source of energy intermittently. It can be either wired or wireless approaches.

Additionally, they play an important role in the telecommunications infrastructure supporting the Internet of Things (IoT) vision [9] and wireless sensor networks (WSNs). Commercial UAVs commonly use lithium and lithium-ion batteries that can maintain a flight time of about 20–40 min [1]. However, the energy limitations are a problem not only for UAVs but also for wireless sensor networks [8]. When UAVs do not use any alternative energy sources from the ambient environment, they may fail several tasks due to the limited operating time. Energy harvesting techniques are the process by which energy is derived from ambient sources such as solar energy, wind energy, and electromagnetic energy. They can be used to extend the lifetime of battery for low-power devices [2].

Moreover, staff members of high-tech companies with high levels of technical knowledge led to increased financial and economic performance, in comparison with companies, in the same industrial sector, employing less technically qualified staff [30]. A significant relation among major capital structure and corporate governance parameters and firm performance of 10 leading energy sector companies traded in the NYSE. Furthermore, pecking order and agency cost theories play an important role in the financing of these firms, while static trade and irrelevance theory find no support [31].

There is a lot of work being done on the use of PV (photovoltaic) panels in RPA, where they are used to increase flight duration and consequently recharge the battery [15]. What is most important to us is the fact that this helps significantly in increasing the flight duration of an RPA, given that the flights must cover a large area of land or forest.

2 Description of Hypothesis and Modeling Process (Table 1)

Table 1 Description of the forces applied

Solar energy	A key factor in the movement of a TRPA, as through it, any planned or not movement of the system takes place and through the solar flow, the resources go to solar energy.
Dynamic energy	An equally important factor of TRPA, which is applied to the system due to gravity but also the loss of energy, due to the movement of the UAV based on the altitude gradients and the configuration of the ground. Through the altitude flow the resources go to the surface.
Air resistance energy	The third factor of the model is the energy developed by the air resistance force, which is applied to the system in the opposite direction of motion. Through the airflow the resources go to air resistance.
Thermal energy	The fourth and final factor of TRPA is the energy applied to the system because of its activation, from the beginning of the movement. Through the heat flow the resources go to thermal loss.

3 Hypothesis

H1: *Finalenergy < 0/ percentages of energy factors vary.*
Each factor is assigned with some amount of energy from the initial TRPA. Nevertheless, the system is not vital.
H2: *Finalenergy > TRPA/40% energy consumption.*
The system is sustainable and with maximum energy savings. As a result, the UAV manages to save 40% of the energy it consumes.
H3: *Finalenergy>0 after T/2/*$E_{thermal}$*=0.*
Thermal energy turns to zero and as a result, the system achieves the continuation of the flight by managing to save energy in the second half of the journey.
H4: AltFinalenergy<0/33.4% functional panels.
Solar panels malfunctioned at 66.6% of the total surface and as a result, the alternative final energy is affected to such an extent that the UAV must land to recharge.

4 Methodology

The primary goal is the energy saving of an autonomous UAV, to which solar panels are attached to utilize solar energy. With the use of Anylogic PLE, the initial energy goes through the factors present in the model and it returns either in the form of energy saving and augmentation with the consequence of not requiring recharging, or in the form of its partial loss, in the possible application of the factor of different weather phenomena (rain, snow, night, and cloudy) and simultaneous for unbalanced factors malfunction a part of the solar panels surface. Time is referred to as time units (tu).

Initially, 100% of the energy is placed in TRPA in the context of faithful observance of the ECP (energy conservation principle) i.e., the sum of E_{DYN} (dynamic energy) and E_{KIN} (kinetic energy).

This practically means that after the initial energy that exists in the system and is 100% of the resources, the resources that supply the surface, air resistance and solar energy must lead it to its maximum possible value, which implies the ever-increasing value of the final energy amount, i.e., when the system tends to land. To land and with ECP in mind, it is reasonable that the kinetic energy will tend to zero, in contrast to the initial energy where the speed is constantly increasing for the system to achieve the purpose of the whole project and at the same time to exceed the sum of the other counter measures of forces applied.

Table 2 Energy types and supply quota in modeling

Type of energy	Percentage (%)
Dynamic energy	(–) 32
Air resistance energy	(–) 9
Solar energy	42
Thermal energy	(–) 17

5 Hypothesis Trials and Results

Initially, 100% of the resources of TRPA (initial stock) must be divided into the four factors which are solar energy, surface, air resistance, and thermal.

It is chosen to make a random distribution of energy in each stock. This is achieved by using a converter on the respective flow from TRPA to each factor. After researches, the following results are obtained regarding the behavior of the forces in the system. The green color in the following diagram indicates a positive to the system power and red indicates a negative one (Table 2).

5.1 *Solar Energy*

In this case, 42% of TRPA resources go to solar energy since the percentage of TRPA has been placed in solar flow. Then through the gain flow that determines the size of the energy received it ends up in the energy filter. When the resources reach the energy filter it must be ascertained whether the energy accumulated in the solar panels is greater than the sum of the other three energies under consideration, to make the model viable.

5.2 *Dynamic Energy*

The next 32% of the initial stock goes to the surface in the same way as previously explained. When the resources reach the altitude loss and by applying the given parameters, the magnitude of the reported energy applied to the system will emerge. In the same way these resources of the initial stock flow to the energy filter. When the resources reach the energy filter there is a condition that is satisfied, the resources flow to the total energy.

5.3 *Air Resistance Energy*

The next 9% of the initial stock goes to air resistance in the same way as previously explained. When the resources reach the airloss and by applying the parameters that are set, the size of the energy applied to the system will emerge. In the same way these resources of the initial stock flow into the energy filter. When the resources reach the energy filter the condition exists and if satisfied, the resources flow to total energy.

5.4 *Thermal Energy*

The last 17% of the initial stock goes to thermal in the same way as previously explained. When the resources reach the flow thermal loss and by applying the parameters that are set, the size of the energy applied to the system will emerge.

6 Modeling and Simulation

The system from the beginning of its operation and although the force that feeds the kinetic energy of the UAV is more than 40% of the initial energy, at no time for 0–15tu, it manages to get positive values.

The consequence of this is the cancellation of the flight due to a lack of energy to support the flight (Fig. 1).

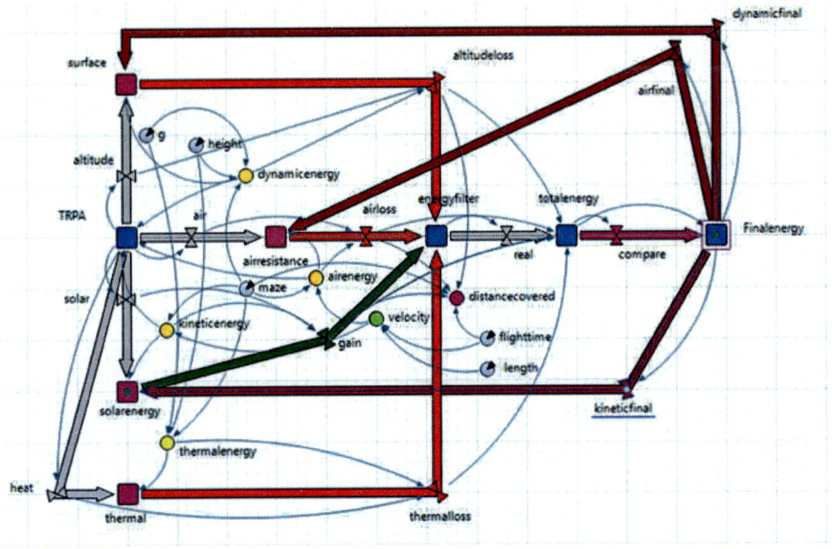

Fig. 1 Simulation model

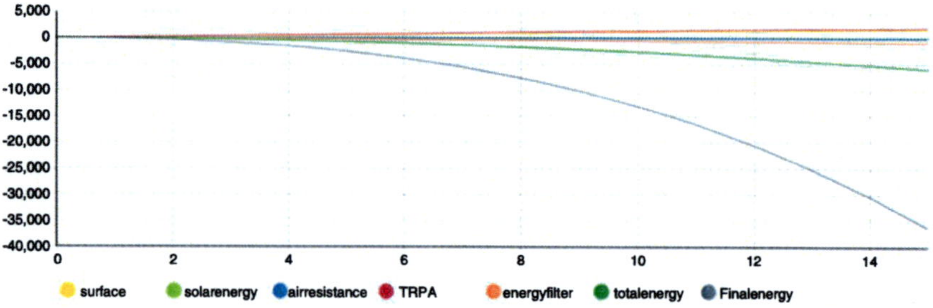

Fig. 2 Energy diagrams during the modeling phase

The above configuration of the model will result from the above quota (Fig. 2):

Subsequently, these resources of the initial stock flow into the energy filter. When the resources reach the energy filter the condition exists and if satisfied, the resources flow to the total energy.

It is observed that the system in modeling for time **t = 1.74tu** the **Finalenergy =--TRPA** and consequently the UAV will stop moving as it has lost all its energy. The simulation ran up to t = 15tu while the Finalenergy kept dropping. However, no point is detected (as shown in the diagram) where the Finalenergy curve enters the positive value axis.

7 Optimization

Energy savings can be achieved by consuming energy from the sun (solar panels) or by reducing the effect of some forces as well as few as possible stops for refueling or charging the battery with the corresponding benefit of not interrupting surveillance of the selected space.

This practically, in the system, means that after the initial energy that exists and is 100% of the resources, those that supply the surface, air resistance and solar energy must lead it to its maximum possible value, which implies the ever-increasing price of the final energy value, i.e., when the system tends to land.

With ECP in mind, it is reasonable that the kinetic energy will tend to be zero, in contrast to the initial energy where the speed is constantly increasing for the system to achieve the purpose of the whole project and at the same time to overcome the sum of the other counter measures of forces applied.

The maximum possible profit is achieved if the percentage of resources flowing from TRPA to solarenergy, surface, airresistance, and thermal is differentiated. Toward that direction, the energy percentages will vary. After many simulations, the result is as the following table (Table 3):

The aim of the system is to save energy and the smallest possible number of UAV stops to charge the battery. This means that in the system the final energy should be greater than the initial one in the short term. These percentages are placed on the converters bearing the name of each stream. More specifically:

Table 3 Energy types and supply quota in optimization

Type of energy	Percentage (%)
Dynamic energy	(–) 17.31
Air resistance energy	(–) 16.27
Solar energy	50.97
Thermal energy	(–) 15.45

Table 4 Explanation of flows and factors in the process of optimization in dynamic energy

Name	Type	Explanation
Maze	3.5	Weight in kg
G	9.81*3.6	Acceleration of gravity. The number 3.6 helps convert m/s to km/h
Height	1	Km
Dynamic energy	Maze*g*height	Dynamic energy
Altitude loss	–altitude*dynamicenergy*8/100	It is assumed that energy is reduced by 8%
Surface	Dynamicenergy	–
Altitude	TRPA*17.31/100	The flow to surface

Table 5 Explanation of flows and factors in the process of optimization in air resistance energy

Name	Type	Explanation
Maze	3.5	Weight in kg
Velocity	25	Velocity in km/h
Length	10	The surveillance area to be covered by the system in km
Airenergy	1/2*maze*velocity*velocity*40/100	Energy from air resistance
Airloss	–air*airenergy*2/100	It is assumed that energy is reduced by 2%
Airresistance	Airenergy	–
Air	TRPA*16.27/100	The flow to airresistance

(a) In the dynamic energy where $\mathbf{E_{\Delta YN}} = \mathbf{m*g*h}$, in the altitude loss the factor dynamic energy as well as the parameters maze, g and height are affected, consequently their configuration becomes as follows (Table 4):

(b) In the air resistance energy which is calculated with $E_A = 40\%*E_{KIN} = 40\%*1/2*m*v^2$, the airloss is affected by the airenergy factor as well as the parameters velocity, flight time, length, and maze, consequently their configuration becomes as follows (Table 5):

(c) In the kinetic energy where $E_{KIN} = 1/2*m*v^2$, the gain is affected by the solarenergy factor as well as the parameters velocity, flight time, length, and maze, consequently their configuration becomes as follows (Table 6):

Table 6 Explanation of flows and factors in the process of optimization in kinetic energy

Name	Type	Explanation
Maze	3.5	Weight in kg
Velocity	25	Velocity in km/h
Length	10	The surveillance area to be covered by the system in km
Kineticenergy	1/ 2*maze*velocity*velocity	Kinetic energy
Gain	Solar*kineticenergy*70/ 100	It is assumed that 70% of the energy is applied
Solarenergy	Kineticenergy	–
Solar	TRPA*50.97/100	The flow to solarenergy
Flighttime	4	Hours (h) or time units (tu)

Table 7 Explanation of flows and factors in the process of optimization in thermal energy

Name	Type	Explanation
Maze	3.5	Weight in kg
g	9.81*3.6	Acceleration of gravity. The number 3.6 helps convert m / s to km / h
Height	1	Km
Thermalenergy	Maze*g*height*40/ 100	Thermal energy
Thermalloss	- heat*thermalenergy*5/ 100	It is assumed that energy is reduced by 5%
Thermal	Dynamic energy	–
Heat	TRPA*15.45/100	The flow to thermal

(d) In the thermal energy where E_θ = m*g*h*40%, the gain is affected by the solarenergy factor as well as the parameters velocity, flighttime, length, and maze, consequently their configuration becomes as follows (Table 7; Fig. 3):

It is observed that TRPA at **t = 0tu** begins from the initial value of 100% but then at the output, i.e., at Finalenergy, it has a negative value, which means that the movement burdens the battery. At **t = 2.4tu,** it is obvious that the model is entering an optimization trajectory as Finalenergy > 0. However, at t = 15tu the **Finalenergy = 39,597*TRPA** approximately. A result that, compared to modeling, shows a tremendous change. Therefore, the model after optimization is much more efficient and it is easily understood that the cost savings are easily calculated from the sum Finalenergy-TRPA.

(a) However, to achieve the examination of other scenarios and the maximum possible energy in the system, it is assumed that the final energy of the system is redistributed to three of the four factors, not including the thermal energy, in percentages as follows (Table 8):

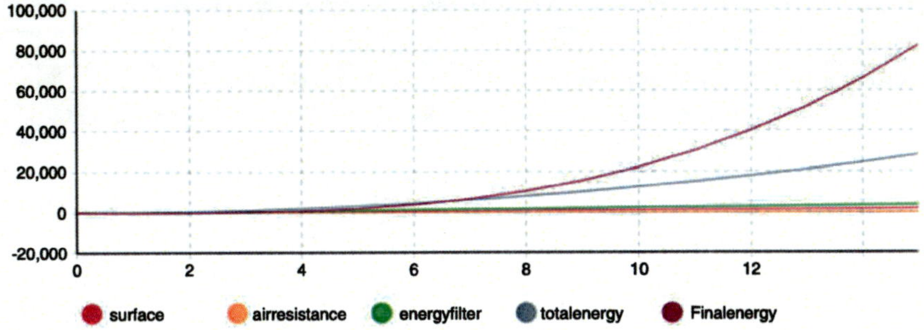

Fig. 3 Energy diagrams during the optimization phase

Table 8 Energy types and power supply quota in optimization (no thermal energy)

Type of energy	Percentage (%)
Dynamic energy	(−) 20.22
Air resistance energy	(−) 36.11
Solar energy	43.67

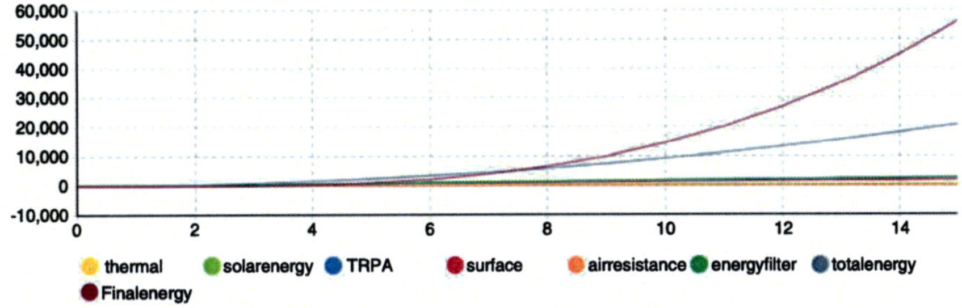

Fig. 4 Energy diagrams during optimization phase with thermal energy to zero

The following diagram clarifies the landscape in the field of energy flow in the system. More specifically, it is observed that TRPA at **t = 0tu** starts from the initial value of 100% but then at the output, i.e., at Finalenergy, it has a maximum negative value at **t = 2.18tu**, which means that the movement burdens the battery. At **t = 3.27tu** the system enters an optimization trajectory as Finalenergy > 0. Since the autonomy time of the UAV, is assumed to be 4-time units (**t = 4tu**), obviously, there is success to the energy savings plan since the system will not have to stop for charging (Fig. 4).

It is observed that at **t = 15tu** where Finalenergy equals the sum **totalenergy*284/100–TRPA** (where totalenergy equals the sum of all actions performed on the system), it manages to be exponentially greater than initial energy. More specifically the <u>**Finalenergy = 26,945*TRPA**</u> approximately. Therefore, the model after optimization is much more efficient and it is easily understood that the cost savings are easily calculated from the sum Finalenergy-TRPA. A result that,

Table 9 Explanation of flows and factors in the alternative optimization process

Name	Type	Explanation
Solardamage	−2/3*kineticenergy	Loss of solar energy
Badweather	−gain*90/100	Inextricably linked to the parameters of kinetic energy
Alternative	Totalenergy*10/ 100 + solardamage	–
AltFinalenergy	Alternative+totalenergy +badweather	Alternative final energy

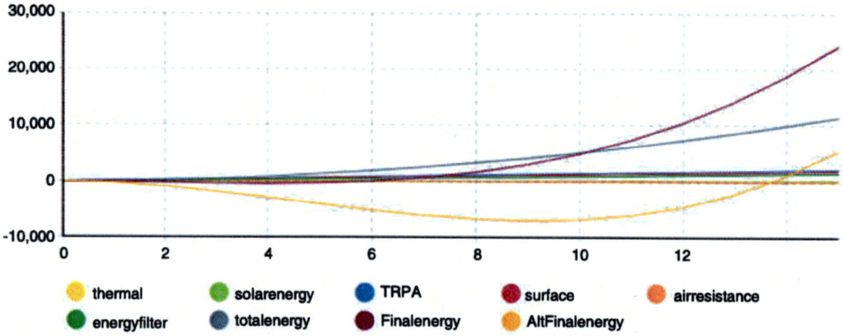

Fig. 5 Optimization model with AltFinalenergy included

compared to modeling, indicates a huge change. Therefore, the model after optimization is much more efficient and it is easily understood that the cost savings are easily calculated from the sum Finalenergy-TRPA.

(b) An alternative examination of the system flow will bring different behaviors and at the same time different results in the model. Therefore, it is assumed that conditions arise differently from before that significantly affect the system. One of them is climate change or the sun's absence and the second will be the loss (malfunction of a percentage of solar panels due to an unbalanced factor (e.g., hail)). In case the percentage of functional panels is set to 33.4%, the values of the parameters that apply will be configured as follows (Table 9; Figs. 5 and 6):

The following pictures clarify the landscape in the field of energy flow in the system. More specifically, it is observed that TRPA at **t = 0tu** starts from the initial value of 100% but then at the output, i.e., in Finalenergy it has a negative value, which means that the movement burdens the battery. At **t = 9.175tu** it is obvious that the model is approaching its maximum negative value where approximately AltFinalenergy = **−5559***TRPA.

Finally, at **t = 15tu** where AltFinalenergy is equal to the sum of **alternative + totalenergy + badweather**, has achieved to overcome the initial energy. Especially at **t = 13.71tu** the AltFinalenergy gets above zero. Therefore, the model

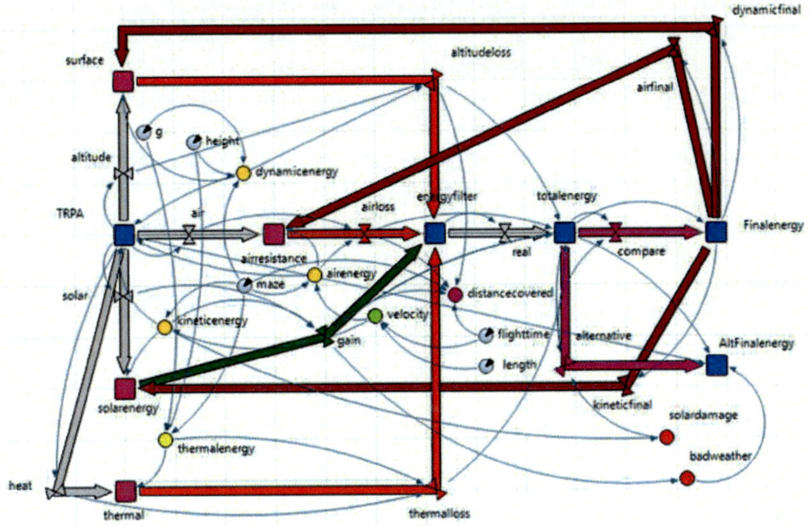

Fig. 6 Energy diagrams during the optimization phase with AltFinalenergy included

after optimization is still efficient and the saving value can be exported for **AltFinalenergy-TRPA**. Nevertheless, it is concluded that since it has been stated from the beginning as the autonomy time of the UAV's 4-time units (**t = 4tu**), obviously there isn't success to the energy savings plan since the system will have to stop for charging at least three (3) times on t = 4tu, t = 8tu and t = 12tu. Therefore, the system isn't considered to be functional.

8 Conclusion

This research is part of a research project examining and optimizing in the Anylogic PLE environment, how to save energy on an unmanned aircraft by exploiting test flight research data.

On the current model under consideration, the explanation given for the negative value of Finalenergy and AltFinalenergy is that every 4tu (as the battery life is predetermined) a landing will be required for recharging. Through this paper, it was accomplished to identify the necessity of additional energy sources that can be attached to an unmanned aerial vehicle.

The whole procedure concerns the movement of the aircraft with the addition or not of energy sources other than its built-in battery. The behavior of the aircraft was examined in conditions of the sun's radiation absence (night, solar eclipse, etc.) and in case of malfunction of the solar panels due to mechanical damage. Further research and application of additional alternative energy sources is required as well as the changes that the aircraft can make in its movement.

Of course, there are other alternative energy sources such as hydrogen cells or the addition of an additional battery, as well as many other ways to save energy by determining the aircraft's movement.

During the pandemic crisis and economic recession, the usage of unmanned aircraft as well as the exploitation of alternative energy sources may provide interesting solutions to the problem of routing in cities [17] and tourist locations. In this context, UAVs have attracted the interest of researchers in stakeholder analysis, and the sustainable supply chain domains in order to enrich human activity and well-being with connectivity and several factors, such as natural environment and critical infrastructure [18]. Indeed, the adoption of new technological alternatives such as unmanned aircraft may provide a viable solution for sustainable tourism and connectivity for the benefit of residents' quality of life [19–23]. Finally, future research in digital marketing, CSR, and market application could be considered in order to make the UAV projects viable [24–29].

References

1. Lee, B., Kwon, S., Park, P., Kim, K.: Active power management system for an unmanned aerial vehicle powered by solar cells, a fuel cell, and batteries. IEEE Transactions on Aerospace and Electronic Systems. **50**(4), 3167–3177 (2014)
2. Van Nguyen, C., Van Quyen, T., Le, A.M., Truong, L.H., Nguyen, M.T.: Advances in science. Technology and Engineering Systems Journal. **5**(1), 34–39 (2020) www.astesj.com
3. Dale, A.L., Kamran, M.: 'Efficiency Analysis for Long-Duration Electric MAVs', Infotech@Aerospace 26–29 September 2005. AIAA 2005-7090, Arlington (2005)
4. Hunt, E.R., Hively, W.D., Fujikawa, S., Linden, D., Daughtry, C.S., Mc-Carty, G.: Acquisition of nir-green-blue digital photographs from unmanned aircraft for crop monitoring. Remote Sensing. **2**(1), 290–305 (2010)
5. Giulio, A., Fabrizio, G.: Maximum range for battery-powered aircraft. Journal of Aircraft. **50** (2013)
6. Michel, H.: Eyes in the Sky: The Secret Rise of Gorgon Stare and How It Will Watch Us All. Arthur Houghton Mifflin Harcourt (2019) https://www.nature.com/articles/d41586-019-01792-5
7. Lance, W.T.: Range and endurance estimates for battery-powered. Journal of Aircraft. **48**(2), 2 (2011)
8. Nguyen, M., Nguyen, H., Masaracchia, A., Nguyen, C.: Stochastic-based power consumption analysis for data transmission in wireless sensor networks. EAI Endorsed Transactions on Industrial Networks and Intelligent Systems. **6**(19) (2019)
9. Motlagh, N.H., Taleb, T., Arouk, O.: Low-altitude unmanned aerial vehicles-based internet of things services: Comprehensive survey and future perspectives. IEEE Internet of Things Journal. **3**(6), 899–922 (2016)
10. Ohad, G., Aviv, R.: Optimizing electric propulsion systems for unmanned aerial vehicles. Journal Of Aircraft. **46**(4) (2009)
11. Lewis, P.: Cctv in the Sky: Police Plan to Use Military-Style Spy Drones, vol. 23, p. 1. The Guardian (2010)
12. Richard H. Van Atta, Project Leader and Michael J. Lippitz with Jasper C. Lupo Rob Mahoney Jack H. Nunn: 'Transformation and Transition: DARPA's Role in Fostering an Emerging Revolution in Military Affairs Volume 1', Overall Assessment/Institute for Defense Analyses, (2003)

13. Reed, S.G. Van Atta, R.H. Seymour, J.D.: Darpa technical accomplishments/An historical review of selected Darpa projects/Volume I, (1990)
14. Chandrasekharan, S., Gomez, K., Al-Hourani, A., Kandeepan, S., Rasheed, T., Goratti, L., Reynaud, L., Grace, D., Bucaille, I., Wirth, T., et al.: Designing and implementing future aerial communication network. IEEE Communications Magazine. **54**(5), 26–34 (2016)
15. Tan Chang, H.Y.: "Improving Electric Powered UAVs' Endurance by Incorporating Battery Dumping Concept", APISAT2014", 2014 Asia-Pacific International Symposium on Aerospace Technology, Northwestern Polytechnical University, Xi'an, Shaanxi, 710000 P.R.China/Available online at www.sciencedirect.com, (2014)
16. Niethammer, U., James, M., Rothmund, S., Travelletti, J., Joswig, M.: Uavbased remote sensing of the super-sauze landslide: Evaluation and results. Engineering Geology. **128**, 2–11 (2012)
17. Kakarountas, A., Spathoulas, G., Trivellas, P., Reklitis, P., Malindretos, G.: Survey on Routing Services for Smart Delivery in Urban Environments, vol. 9221790. SEEDA-CECNSM 2020 – 5th South-East Europe Design Automation, Computer Engineering, Computer Networks and Social Media Conference (2020)
18. Trivellas, P., Anastasopoulou, A., Malindretos, G., Reklitis, P., Sakas, D.P.: A place based analysis of stakeholders' advancing sustainability in remote and isolated communities: The case of North Aegean Islands. IOP Conference Series: Earth and Environmental Science. **899**(1), 12044 (2021)
19. Trivellas, P., Kakkos, N., Vasiliadis, L., Belias, D.: Sustainability, Social Marketing and Host Attitudes About Tourism in the City of Larissa. Springer Proceedings in Business and Economics, 89–95 (2017)
20. Trivellas, P., Kakkos, N., Vasiliadis, L., Belias, D.: Residents' Perceptions Toward Cultural, Social and Economic Benefits and Costs of Tourism Industry: An Empirical Survey, pp. 171–181. Springer Proceedings in Business and Economics (2016)
21. Folinas, D., Fotiadis, T.: Marketing and Supply Chain Management: A Systemic Approach, 1st edn. Routledge (2017)
22. Fotiadis, T.: Strategic Marketing for High Technology Products: An Integrated Approach (Routledge Studies in Innovation, Organizations and Technology), 1st edn. Routledge (2018)
23. Fotiadis, T., Folinas, D., Vasileiou, K., Konstantoglou, A.: Marketing and the Customer Value Chain: Integrating Marketing and Supply Chain Management 1st Edition, Kindle Edition, 1st edn. Routledge (2022)
24. Sakas, D.P., Reklitis, D.P.: The Impact of Organic Traffic of Crowdsourcing Platforms on Airlines' Website Traffic and User Engagement. Sustainability. **13**, 8850 (2021). https://doi.org/10.3390/su13168850
25. Sakas, D.P., Giannakopoulos, N.T., Reklitis, D.P., Dasaklis, T.K.: The Effects of Cryptocurrency Trading Websites on Airlines' Advertisement Campaigns. Journal of Theoretical and Applied Electronic Commerce Research. **16**, 3099–3119 (2021). https://doi.org/10.3390/jtaer16070169
26. Sakas, D.P., Reklitis, D.P.: Predictive model for estimating the impact of technical issues on consumers interaction in agri-logistics websites. In: Information and Communication Technologies for Agriculture—Theme IV: Actions, pp. 269–283. Springer International Publishing, Cham (2021). https://doi.org/10.1007/978-3-030-84156-0_14
27. Sakas, D.P., Reklitis, D.P., Trivellas, P., Vassilakis, C., Terzi, M.C.: The Effects of Logistics Websites' Technical Factors on the Optimization of Digital Marketing Strategies and Corporate Brand Name. Processes. **10**, 892 (2022). https://doi.org/10.3390/pr10050892
28. Sakas, D.P., Kamperos, I.D.G., Reklitis, D.P., Giannakopoulos, N.T., Nasiopoulos, D.K., Terzi, M.C., Kanellos, N.: The Effectiveness of Centralized Payment Network Advertisements on Digital Branding during the COVID-19 Crisis. Sustainability. **14**, 3616 (2022). https://doi.org/10.3390/su14063616

29. Reklitis, P., Trivellas, P., Mantzaris, I., Mantzari, E., Reklitis, D.: Employee Perceptions of Corporate Social Responsibility Activities and Work-Related Attitudes: The Case of a Greek Management Services Organization. In: Gal, G., Akisik, O., Wooldridge, W. (eds.) Sustainability and Social Responsibility: Regulation and Re-porting, pp. 225–240. Springer Singapore, Singapore (2018). https://doi.org/10.1007/978-981-10-4502-8_10
30. Toudas, K., Kanellos, N.: Economic and Accounting Performance of Greek innovative firms through Knowledge-based entrepreneurship. Journal of Accounting and Taxation. **1492**, 150–160 (2022). https://doi.org/10.5897/JAT2022.0530
31. Georgakopoulos, G., Toudas, K., Poutos, E., Kounadeas, T., Tsavalias, S.: Capital Structure, Corporate Governance, Equity Ownership and Their Impact on Firms' Profitability and Effectiveness in the Energy Sector. Energies. **15**(3625), 1–10 (2022). https://doi.org/10.3390/en15103625

Energy Saving of a Drone Using a Hydrogen Cell Battery, Modeling, and Optimization

Dimitrios A. Arvanitidis, Dimitrios K. Nasiopoulos, Dimitrios M. Mastrakoulis, and Panagiotis Reklitis

1 Introduction

The aim of the study is to determine when it is most appropriate, in terms of system mass, to use a battery or hybrid system (lithium battery and fuel cell battery) to power the drone. The model allows the combination of different types of fuel and battery systems. Model tuning data is obtained from commercial products. A hybrid system seems to be of interest for longer flights.

Today, most commercial drones are electric and battery-powered. Batteries have many disadvantages when it comes to using drones for some special applications. The main problem is the durability of the batteries. Nowadays, for commercial

D. A. Arvanitidis (✉)
Department of Informatics and Telecommunications, University of Peloponnese, Tripoli, Greece
e-mail: darvanitidis@aua.gr

D. K. Nasiopoulos
Department of Agribusiness and Supply Chain Management, BICTEVAC LABORATORY Business Information and Communication Technologies in Value Chains laboratory, School of Applied Economics and Social Sciences, Agricultural University of Athens, Athens, Greece

D. M. Mastrakoulis
Department of Informatics and Telecommunications, University of Thessaly, Lamia, Greece
e-mail: dmastrakoulis@aua.gr

P. Reklitis
Department of Agribusiness and Supply Chain Management, Organizational Innovation and Management Systems, School of Applied Economics and Social Sciences, Agricultural University of Athens, Athens, Greece

D. P. Sakas et al. (eds.), *Computational and Strategic Business Modelling*, Springer Proceedings in Business and Economics,
https://doi.org/10.1007/978-3-031-41371-1_16

drones, flight time is usually limited to 40 min [5]. In addition, the charge time of the batteries is still very long compared to the refill time for appliances such as internal combustion engines or fuel cells. One possible solution for extending the use of drones to more applications could be to develop a hybrid fuel system consisting of fuel cells and a battery. The fuel cells are powered by a hydrogen generator. Hydrogen is the lightest element in nature and has a high LHV of almost 120 MJ/kg, three times higher than gasoline [7]. Therefore, fuel cells could potentially have a high specific energy. Commercial aircraft have intermittent operating conditions and do not operate at full power throughout the flight. Therefore, the fuel cells could operate at constant power while the battery exceeds the power peaks in flight. Therefore, a hybrid system is of interest.

Moreover, staff members of high-tech companies with high levels of technical knowledge led to increased financial and economic performance, in comparison with companies, in the same industrial sector, employing less technically qualified staff [27]. A significant relation among major capital structure and corporate governance parameters and firm performance of 10 leading energy sector companies traded in the NYSE. Furthermore, pecking order and agency cost theories play an important role in the financing of these firms, while static trade and irrelevance theory find no support [28].

Another advantage of using fuel cells is that it takes a few minutes to refill the tank if hydrogen is used under pressure. In addition, at the end of the service life, most fuel cell components can be recycled or reused. One of the biggest disadvantages of fuel cells is the high cost [3]. The capacity of the model lies in the fact that if a value in the model is changed, the status and the results of the modeling change. Taking advantage of the capacity of the model, many different situations can be analyzed and arrive at process optimization. It is pointed out that in contrast to the above, the following model examines the motion of the drone with the simultaneous use of the lithium battery and the fuel cell battery, in this case, hydrogen.

2 Description of Hypothesis and Modeling Process (Tables 1 and 2)

For time t = 0 to T:

For time t = T/2 to T:

Table 1 Description of the forces applied (0-T/2)

Propeller energy	A key factor in the movement of this drone is the energy produced by the activation of the propellers, as it activates the movement of the system and through propflow ends up in the stock of Totalpropellerenergy.
Rechargeable battery energy	Another factor in the movement of the drone is the energy stored in an additional battery that has been placed in the system, as it charges the main battery and through rechflow ends up in the stock of Totalrechargeenergy.
Battery energy	Particularly, significant factor in the movement of the drone is the energy produced by the activation of the battery and through battflow ends up in the stock of Totalbatteryenergy.
Hydrogencell energy	Another factor for the movement of a drone is the energy stored in the hydrogen cells which through HCflow ends up in the stock of TotalHydroCellsenergy.

Table 2 Description of the forces applied (T/2-T)

Kinetic energy	A key factor in the movement of this drone is the kinetic energy, as through it any programmed or unplanned movement of the system takes place and through kinflow ends up to the stock of Totalkineticenergy.
Air resistance energy	The third factor of the drone is the energy developed by the air resistance force, which is applied to the system in the opposite direction of motion and through resflow ends up at the stock of Totalairresistance.
Thermal energy	Another factor of the drone is the energy applied to the system because of its activation, from the beginning of the movement and through thermflow ends up to the stock of Totalthermalenergy.
Dynamic energy	An equally important factor of the drone is the dynamic energy applied to the system due to gravity and energy loss due to the movement of the drone based on altitude gradients and ground configuration. Through dynflow ends up to the stock of Totaldynamicenergy.

3 Hypothesis

H1: $E_{fin} < 0$/*The system has only its battery as a power source.*

Only two powers affected the drone for the first half of the time.

H2: $E_{fin} > 0$/*Hydrogen cells and an extra battery are added to the system.*

Two extra powers were added to the system to contribute to the drone's energy saving.

H3: $E_{fin} < 0$/*The system has only battery as power source and for 0 to T/2 63% energy consumption of* E_{arx}.

The two powers that affect the system for the period 0 to T/2 are supplied with 63% of the initial energy and the next 37% goes to the energies affecting the drone for the period T/2 to T.

H4: $E_{fin} > 0$/*Hydrogen cells and an extra battery are added to the system and for 0 to T/2 63% energy consumption of* E_{arx}.

The powers that affect the system for the period 0 to T/2 are supplied with 63% of the initial energy and the next 37% goes to the energies affecting the drone for the period T/2 to T.

H5: *E_{fin} > 0/Hydrogen cells and an extra battery are added to the system and for 0 to T/2 37% energy consumption of E_{arx}.*

The powers that affect the system for the period 0 to T/2 are supplied with 37% of the initial energy and the next 63% goes to the energies affecting the drone for the period T/2 to T.

H6: *E_{fin} < 0/Extra Battery is being removed from the system.*

For the period 0 to T/2, the hydrogen cells have been removed from the system and so the percentage of that energy goes to the extra battery.

H7: *E_{fin} > 0/Hydrogen cells are being removed from the system.*

For the period 0 to T/2, the extra battery has been removed from the system and so the percentage of that energy goes to the hydrogen cells.

4 Methodology

The aim of the system is to save energy, which can be achieved by absorbing energy from the hydrogen cells or by adding a second battery and at the same time increasing the effect of some forces. The primary goal is to have as few stops as possible to refuel or charge the battery with the corresponding benefit of not interrupting the assigned mission. Practically the resources of the system that feedback the stocks, must lead it to its maximum possible value, which implies the ever-increasing value of the final value of energy, i.e., when the system tends to land. The purpose of the whole project is the non-interruption of the flight through the minimization of the system requirement for energy and at the same time the overcoming of the sum of the other opposite measures of forces that are applied. To achieve this goal, percentage changes had to be made. Time is referred to as time units (tu).

Initially, 100% of the energy is placed in E_{arx} in the context of faithful observance of the ECP (energy conservation principle) i.e., the sum of propeller and battery energy.

This practically means that after the initial energy that exists in the system and is 100% of the resources, the resources that supply the energy stocks must lead it to its maximum possible value, which implies the ever-increasing value of the final energy amount. It is therefore understandable that while the drone speed should reasonably be constantly increasing in order not to interrupt the flight, it should also be such, that the amount of energy (with a positive sign) feeds the flight to exceed the opposing forces.

Table 3 Energy types and supply quota in modeling

Type of energy	Percentage (%)
Propeller energy	(−) 14
Battery energy	36
Kinetic energy	22
Air resistance energy	(−) 8
Thermal energy	(−) 7
Dynamic energy	(−) 13

5 Hypothesis Trials and Results

Initially, 100% of the drone resources E_{arx} (initial stock) must be divided into the four factors which are Totalpropellerenergy, Totalbatteryenergy, Totalrechargeenergy, and TotalHydroCellsenergy.

It is initially selected not to apply the hydrogen cells as well as the additional battery with consequent distribution, to be done with as in the below table. This is achieved by using a converter on the respective flow from the drone to each factor.

After research, the following results are obtained regarding the behavior of the forces in the system. The green color in the following diagram indicates a positive to the system power and red indicates the negative ones (Table 3).

5.1 *Propeller Energy*

In this case, 14% of the drone 's resources go to Totalpropellerenergy since in the propflow flow the percentage has been placed on the drone. Then through the flow sum1 that determines the magnitude of the energy received it ends up in SUMENERGY1.

5.2 *Rechargeable Battery Energy*

Although no percentage of the original energy (0%) is available here, the flow would go to Totalrechargeenergy in the same way as previously explained. When the resources reached the rechflow, the given parameters would be applied, and the size of the corresponding action would be obtained.

5.3 Battery Energy

The next 36% of the initial factor goes to Totalbatteryenergy in the same way as above. When the resources reach sum3 and with the application of the given parameters, the magnitude of the reported energy applied to the system will emerge. In the same way these resources of the initial factor flow to SUMENERGY1.

5.4 Hydrogen Cell Battery Energy

Although no percentage of the original energy (0%) is available here either the flow would go to TotalHydroCellsenergy in the same way as above. When the resources reached the rechflow, the given parameters would be applied, and the size of the corresponding action would be obtained. Nevertheless, in SUMENERGY2, as mentioned above it can be observed the effect of four new forces in the middle of the flight's time ($E_{mid} = T/2$). For reasons of non-deviation from the initial position and considering the ECP, it is assumed that flowtotalsum1 is equal to ½*SUMENERGY1. The consequence of this is that the sum of the new actions is equal to the other 50% of E_{arx}.

It must be mentioned that when the resources reach SUMENERGY1 and after the four of the eight forces have affected the system, the sum of the remaining actions under consideration is added, to make the model viable. This statement is ascertained in the next stock of SUMENERGY2. All the amount of energy ends up in E_{fin}, where in case of a negative value or less than E_{arx}, it is reasonably concluded that there is not enough energy for the drone to swing or fly.

5.5 Kinetic Energy

In this case, 22% of the drone's resources go to Totalkineticenergy in the same way as above. When the resources reach sum5 and with the use of the given parameters, the magnitude of the reported energy applied to the system will emerge. In the same way, these resources flow to SUMENERGY2.

5.6 Air Resistance Energy

The next 8% of the initial value goes to Totalairresistance in the same way as above. When the resources reach sum6 and with the use of the given parameters, the magnitude of the reported energy applied to the system will emerge. In the same way, these resources flow to SUMENERGY2.

5.7 *Thermal Energy*

The next 7% of the initial value goes to Totalthermalenergy in the same way as previously explained. When the resources reach flowsum7 and by applying the parameters that are set, the size of the energy applied to the system will emerge. Subsequently, these resources flow into SUMENERGY2 as well.

5.8 *Dynamic Energy*

Finally, another 13% of the initial value is directed to the surface in the same way as above. When the resources reach sum8 and with the use of the given parameters, the magnitude of the reported energy applied to the system will emerge. In the same way, these resources flow to SUMENERGY2.

6 Modeling and Simulation

The system from the beginning of its operation and although the power that feeds the movement of the drone is at 35% of the initial energy, at no time for 0-15tu, it manages to get positive values. The consequence of this is the cancellation of the flight due to a lack of energy to support the flight (Fig. 1).

The above configuration of the model will result in the following results (Fig. 2):

These initial stock resources then flow into SUMENERGY1. When the resources reach SUMENERGY1 the condition exists and if satisfied, the resources flow to SUMENERGY2. It is observed that $E_{fin} = 3*E_{arx}$ at t = 0.65tu and equals to –0.06499*E_{arx} at the end of the mission (t = 15tu).

Consequently, the drone will stop moving as it has lost all its energy. The simulation ran up to t = 15tu while E_{fin} continued to rise in the negative axis. However, no point is detected (as shown in the diagram) where the E_{fin} curve enters the positive value axis.

7 Optimization

The aim of the system is to save energy, which can be achieved by absorbing energy from the hydrogen cells or by adding a second battery and at the same time increasing the effect of some forces. The main goal is to have as few stops as possible to refuel or charge the battery with the corresponding benefit of not interrupting the assigned mission. Practically for the system, it means that after the initial energy that exists and constitutes 100% of the resources, the resources that

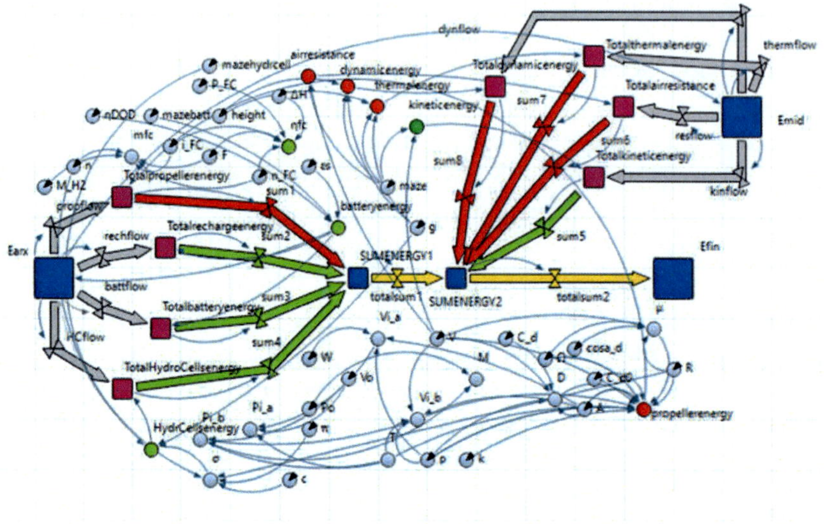

Fig. 1 Simulation model

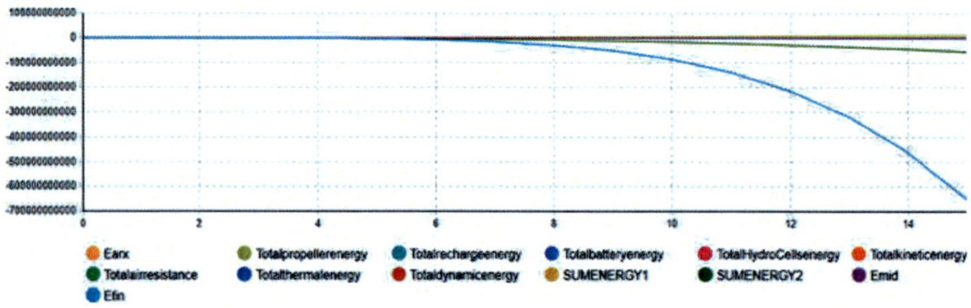

Fig. 2 Energy diagrams during the modeling phase

feedback the stocks must lead it to its maximum possible value, which implies the ever-increasing value of the final value of energy, i.e., when the system tends to land. The purpose of the whole project is the non-interruption of the flight through the minimization of the system requirement for energy and at the same time the overcoming of the sum of the other opposite measures of forces that are applied. With ECP in mind, it makes sense for the drone-driven force to tend to zero, as opposed to the principle of motion where the speed is constantly increasing for the system to achieve the purpose of the whole project and at the same time surpass the sum of the others. Countermeasures applied. The maximum possible profit is achieved if the percentage of resources flowing from the drone to the stocks is differentiated. Toward that direction, the energy percentages will vary. After many simulations, the result is as the following table (Table 4):

Table 4 Energy types and supply quota in optimization

Type of energy	Percentage (%)
Propeller energy	(−) 18.,21
Rechargeable battery energy	11.52
Battery energy	11.52
Hydrogen cell battery energy	8.75
Kinetic energy	21.77
Air resistance energy	(−) 8.53
Thermal energy	(−) 9.32
Dynamic energy	(−) 10.38

Table 5 Explanation of flows and factors in the process of optimization in propeller energy

Name	Type	Explanation
D	C_d*ρ*A*V*V/2	Traction (vertical) force
V	15	Speed (m/s)
k	1.10	Empirical factor
Vi_b	T/2*ρ*A*V	Inductive speed
ρ	0.8	Steadily
A	1.2	Propeller area
Ω	6	Angular speed of the propeller
R	29	Helix radius
C_d0	0.008	Vertical drag coefficient
μ	V*cosa_d/Ω*R	Advanced ratio
C_d	0.022	Vertical drag coefficient of the propeller of the drone center node
T	15	Rotor push
cosa_d	0	Coefficient of the drone angle of attack
Propellerenergy	D*V + k*T*Vi_b + (ρ*A*Ω *R*Ω*R*Ω*R*σ*C_d0* (1 + 4.6*μ*μ)/8)	Propeller energy
sum1	-Totalpropellerenergy	–
Propflow	E_{arx}*12.5/100	The flow to Totalpropellerenergy

The aim of the system is to save energy and the smallest possible number of drones stops to charge the battery. This means that in the system the final energy should be greater than the initial one in the short term. These percentages are placed on the converters bearing the name of each stream. More specifically:

(a) The propeller energy where Totalpropellerenergy = propellerenergy*propflow, is affected by the following factors (Table 5):

(b) The recharging energy (second battery) of the system, where, Totalrechargeenergy = batteryenergy*rechflow*20/100 is affected by the following factors (Table 6):

Table 6 Explanation of flows and factors in the process of optimization in recharge battery energy

Name	Type	Explanation
ε_S	163	Gravimetric efficiency index (E/m) amount of energy per unit mass
η_{DOD}	0.8	Battery discharge depth by affecting its life cycle
Mazebatt	0.78	Battery weight (kg)
Batteryenergy	ε_s*η_{DOD}*mazebatt*20%	Recharge energy
sum2	Totalrechargeenergy*20%	–
Rechflow	E_{arx}*12.5/100	The flow to Totalrechargeenergy

Table 7 Explanation of flows and factors in the process of optimization in battery energy

Name	Type	Explanation
ε_S	163	Gravimetric efficiency index (E/m) amount of energy per unit mass
η_{DOD}	0.8	Battery discharge depth affects by affecting its life cycle
Mazebatt	0.78	Battery weight (kg)
Batteryenergy	ε_s*η_{DOD}*mazebatt	Battery power
Sum3	Totalbatteryenergy	–
Battflow	E_{arx}*12.5/100	The flow to Totalbatteryenergy

Table 8 Explanation of flows and factors in the process of optimizing the hydrogen cell energy

Name	Type	Explanation
σ	n*R*c/π*R*R	Rotor rigidity
Mazehydrcell	1.4	Cell battery weight
n	2	Number of blades
π	314,159,265	Mathematical constant (π = P/δ)
R	29	Propeller radius
g	9.81	Acceleration of gravity in m/s^2
HydrCellsenergy	σ*mazehydrcell*g	Cell battery power
sum4	TotalHydroCellsenergy	–
HCflow	E_{arx}*12.5/100	The flow to TotalHydroCellsenergy

(c) In battery power (primarybattery) where Totalbatteryenergy = batteryenergy*battflow the following factors affect as follows (Table 7):

(d) The energy of hydrogen cells where TotalHydroCellsenergy = HydrCellsenergy*HCflow is influenced by the following factors as follows (Table 8):

(e) In kinetic energy where Totalkineticenergy = kineticenergy*kinflow, the following factors affect as follows (Table 9):

(f) The energy applied by the air resistance where Totalairresistance = airresistance*resflow*40/100 is affected by the following factors as follows (Table 10):

(g) The thermal factors that develop in the system where Totalthermalenergy = thermalenergy*thermflow*30/100, are affected by the following factors as follows (Table 11):

(h) The dynamic energy developed in the system where Totaldynamicenergy = dynamicenergy*dynflow is influenced by the following factors (Table 12):

Table 9 Explanation of flows and factors in the process of optimization in kinetic energy

Name	Type	Explanation
Maze (drone)	7	Drone weight (kg)
V	15	Speed in m/s
Kineticenergy	0.5*maze*V*V	Kinetic energy
sum5	Totalkineticenergy	–
Kinflow	E_{mid}*25/100	–
E_{mid}	E_{arx}/2	The energy in the middle of the system path

Table 10 Explanation of flows and factors in the process of optimization in air resistance energy

Name	Type	Explanation
Maze (drone)	7	Drone weight (kg)
V	15	Speed in m/s
Airresistance	0.5*maze*V*V	Air resistance energy
sum6	-Totalairresistance	–
Resflow	E_{mid}*25/100	–
E_{mid}	E_{arx}/2	The energy in the middle of the system path

Table 11 Explanation of flows and factors in the process of optimization in thermal energy

Name	Type	Explanation
Maze (drone)	7	Drone weight (kg)
g	9.81	Acceleration of gravity in m/s^2
Height	25	The height at which it flies
Thermalenergy	Maze*g*height	Thermal energy
sum7	–Totalthermalenergy	–
Thermflow	E_{mid}*25/100	–
E_{mid}	E_{arx}/2	The energy in the middle of the system path

Table 12 Explanation of flows and factors in the process of optimization in dynamic energy

Name	Type	Explanation
Maze (drone)	7	Drone weight (kg)
g	9.81	Acceleration of gravity in m/s^2
Height	25	The height at which it flies
Dynamicenergy	Maze*g*height	Dynamic energy
sum8	-Totaldynamicenergy	–
Dynflow	E_{mid}*25/100	–
E_{mid}	E_{arx}/2	The energy in the middle of the system path

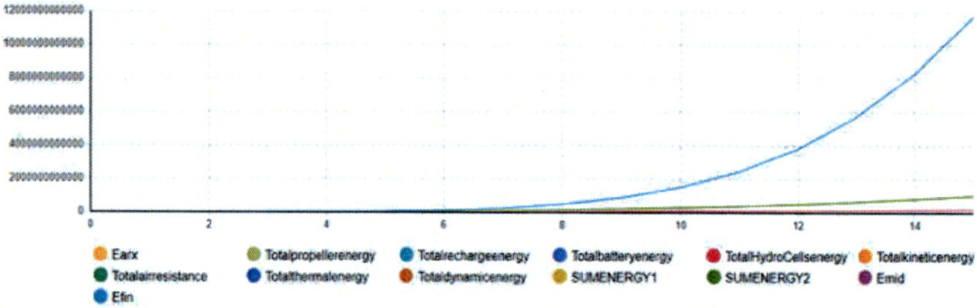

Fig. 3 Energy diagrams during optimization phase

It is observed that E_{arx} at **t = 0tu** begins from the initial value of 100% but then at the output, and at the same time at E_{fin}, it has a positive value as well. At t = 15tu the $\mathbf{E_{fin} \approx 7.333*10^{13}*E_{arx}}$ approximately. A result that, compared to modeling, shows a tremendous change (Fig. 3).

Therefore, the model after optimization is much more efficient and it is easily understood that the cost savings are easily calculated from the sum E_{fin}-E_{arx}. More specifically, it is observed that the drone at t = 0tu starts from the initial value of 100% and then at the output, i.e., at E_{fin} with geometric progression, the eventually produced energy increases. At t = 1.48tu the model is entering a maximum optimization trajectory as $E_{fin} > E_{arx}$. Finally, at t = 15tu the E_{fin} manages to be exponentially greater than the initial energy. Therefore, the model after optimization is much more efficient and it is easy to understand that the value of savings is easily calculated from the sum of E_{fin}-E_{arx}, thus achieving optimization.

(A) However, to achieve the examination of other scenarios and the maximum possible energy in the system, the viability of the system will be examined if:

  (a) Remove the extra battery and hydrogen cells.
  (b) It is considered that 63% of the total energy of the system is utilized in the first part of the traffic, in percentages as follows (Table 13):

Table 13 Energy types in optimization (no extra battery and hydro cells)

Type of energy	Percentage (%)
Propeller energy	(–) 41,18
Battery energy	21,82
Kinetic energy	12,49
Air resistance energy	(–) 7,32
Thermal energy	(–) 6,13
Dynamic energy	(–) 11,06

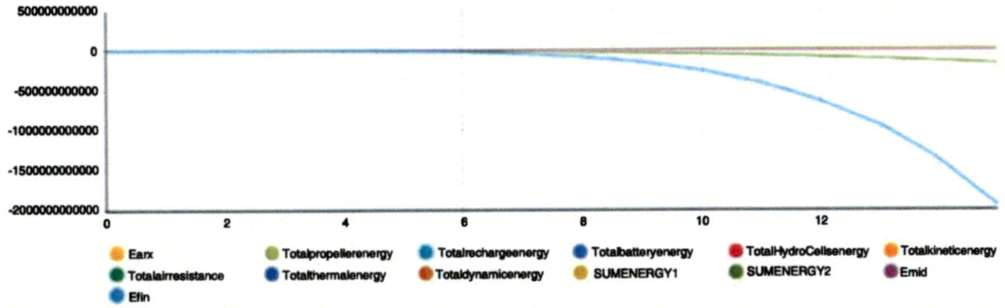

Fig. 4 Energy diagrams during optimization phase (no extra battery and hydro cells)

The following diagram clarifies the landscape in the field of energy flow in the system (Fig. 4).

It is observed that at **t = 0.5tu** the $\mathbf{E_{fin=}\text{-}2{*}E_{arx}}$ and at **t = 15tu** that $\mathbf{E_{fin} = -0{,}1196{*}10^{13}{*}E_{arx}}$ approximately, which means that the movement needs more energy than the battery can provide. At no point after 2.54tu, the drone manages to get a positive value. Therefore, the model after this simulation lacks efficiency and it is easily understood that there is no success to the energy savings plan since the system will have to stop charging, so the cost savings are easily calculated from the sum E_{fin}-E_{arx}. A result that, compared to modeling, indicates almost the same results (negative axis) with different values. Obviously, the model after optimization is the least efficient since the drone won't manage to perform correctly.

(B) An alternative examination of the system flow will bring different behavior and at the same time different results in the model. Therefore, here it is assumed that conditions arise differently from before that significantly affect the system.

   (a) Extra battery and hydrogen cells attach to the drone.
   (b) It is considered that 63% of the total energy of the system is utilized in the first part of the traffic, in percentages as follows (Table 14):

The following diagram clarifies the landscape in the field of energy flow in the system (Fig. 5).

It is observed that at **t = 4.5tu** the $\mathbf{E_{fin}}$ overcomes the value of SUMENERGY2 and at **t = 15tu** the $\mathbf{E_{fin} = 13{,}207{*}10^{13}{*}E_{arx}}$ approximately. That is almost twice the

Table 14 Energy types in optimization (no extra battery and hydro cells-63% in the first half)

Type of energy	Percentage (%)
Propeller energy	(−) 18.21
Rechargeable battery energy	14.76
Battery energy	14.76
Hydrogen cell battery energy	15.27
Kinetic energy	19.98
Air resistance energy	(−) 4.21
Thermal energy	(−) 7.73
Dynamic energy	(−) 5.08

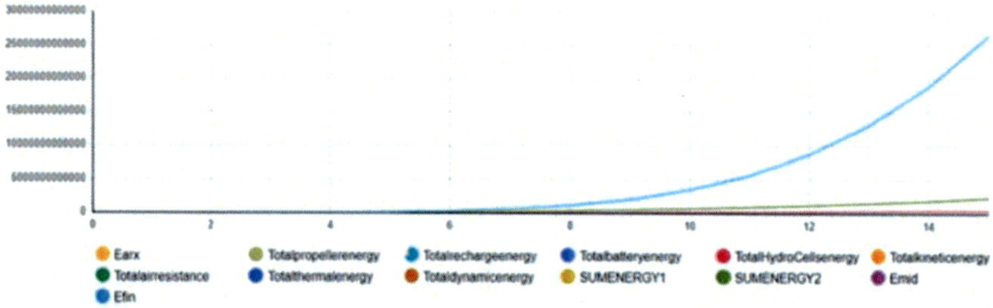

Fig. 5 Energy diagrams during the optimization phase (no extra battery and hydro cells-63% in the first half)

Table 15 Energy types in optimization (63% in the first half)

Type of energy	Percentage (%)
Propeller energy	(−)9.32
Rechargeable battery energy	7.81
Battery energy	7.81
Hydrogen cell battery energy	12.06
Kinetic energy	22.87
Air resistance energy	(−) 13.49
Thermal energy	(−) 12.82
Dynamic energy	(−) 13.82

previous result of the optimization process. Therefore, the model is considered to succeed in the energy saving process.

(C) A different approach to system behavior could be as follows:

(a) Add extra battery and hydrogen cells.
(b) It is considered that 37% of the total energy of the system is utilized in the first part of the traffic in percentages as follows (Table 15):

The following diagram clarifies the landscape in the field of energy flow in the system (Fig. 6).

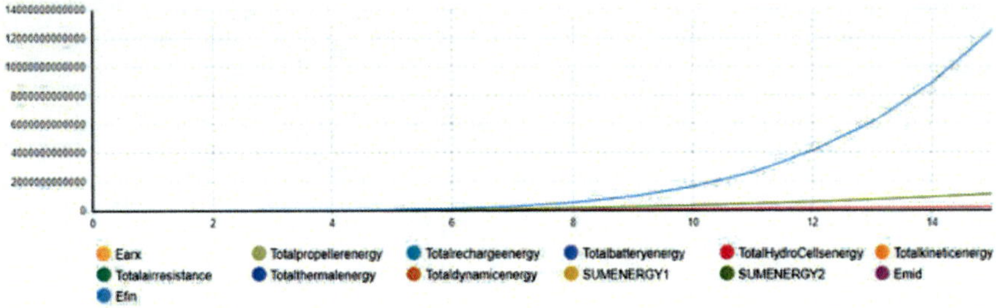

Fig. 6 Energy diagrams during the optimization phase (63% in the first half)

Table 16 Energy types in optimization (extra battery added and hydro cells off the system)

Type of energy	Percentage (%)
Propeller energy	(−) 18.21
Rechargeable battery energy	20.27
Battery energy	11.52
Hydrogen cell battery energy	0
Kinetic energy	21.77
Air resistance energy	(−) 8.53
Thermal energy	(−) 9.32
Dynamic energy	(−) 10.38

It is observed that at **t = 5.015tu** the $\mathbf{E_{fin}}$ equals the value of SUMENERGY2 and at **t = 15tu** the $\mathbf{E_{fin} = 10{,}543 * 10^{13} * E_{arx}}$ approximately. That is almost twice the previous result of the optimization process. Therefore, the model is considered to succeed in the energy saving process.

(D) Another approach to system behavior could be:

(a) Add extra battery and remove hydrogen cells.
(b) It is considered that 50% of the total energy of the system is utilized in the first part of the traffic and the rest to Emid, in percentages as follows (Table 16):

The following diagram clarifies the landscape in the field of energy flow in the system (Fig. 7).

More specifically, it is observed that the drone at t = 0tu starts from the initial value of 100% and then at the output, at E_{fin} with geometric progression, the eventually produced energy decreases. Therefore, the system can't be considered reliable. It is observed that at **t = 1tu** the $\mathbf{E_{fin=}\text{-}14 * E_{arx}}$ and at **t = 15tu** that $\mathbf{E_{fin} = -0{,}532 * 10^{11} * E_{arx}}$ approximately, which means that the movement needs more energy than the battery can provide. Obviously, at no point at all the drone manage to get a positive value.

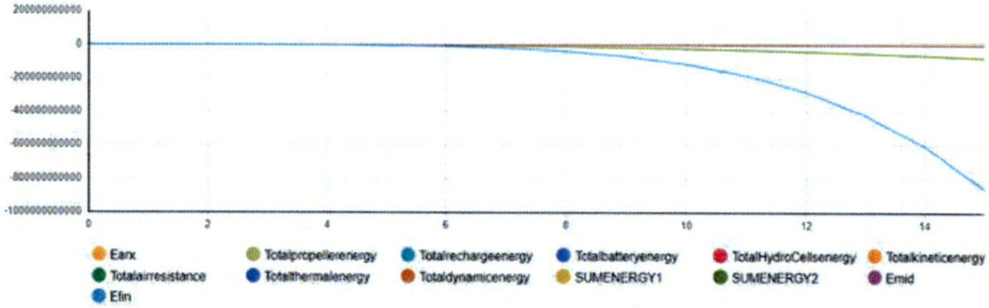

Fig. 7 Energy diagrams during the optimization phase (extra battery added and hydro cells off the system)

Table 17 Energy types in optimization (hydro cells added an extra battery off the system)

Type of energy	Percentage (%)
Propeller energy	(−) 18.21
Rechargeable battery energy	0
Battery energy	11.52
Hydrogen cell battery energy	20.27
Kinetic energy	21.77
Air resistance energy	(−) 8.53
Thermal energy	(−) 9.32
Dynamic energy	(−) 10.38

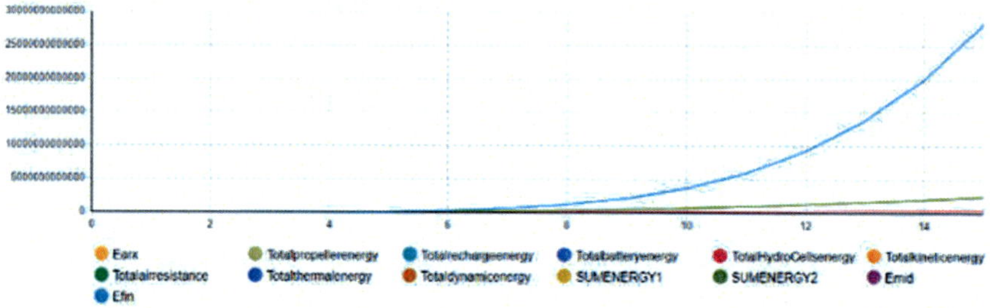

Fig. 8 Energy diagrams during optimization phase (hydro cells added an extra battery off the system)

(E) Another approach to system behavior could be as follows:

(a) Add hydrogen cells and remove the extra battery.
(b) It is considered that 50% of the total energy of the system is utilized in the first part of the traffic and the rest to E_{mid}, in percentages as follows (Table 17):

The following diagram clarifies the landscape in the field of energy flow in the system (Fig. 8).

It is observed that at **t = 4.49tu** the $\mathbf{E_{fin}}$ equals the value of SUMENERGY2 and at **t = 15tu** the $\mathbf{E_{fin} = 17{,}696*10^{13}*E_{arx}}$ approximately. That is almost by far the highest result of the optimization process and simultaneously the biggest amount of energy saved.

8 Conclusion

This research is part of a research project examining and optimizing how to save energy on a drone by exploiting test flight research data. The whole procedure concerns the movement of the aircraft with the addition of energy sources or not, such as an additional battery or hydrogen cells. Further research and application of additional alternative energy sources is required as well as the changes that the aircraft can make in its movement. In the present work, the behavior of the drone was examined if hydrogen cells and/or an additional battery were attached to it, as well as the variation of the quota received by the respective stock.

On the current model under consideration, the explanation given for the negative value of E_{fin} is that the drone will not be able to carry out any mission assigned to it unless an additional energy source is attached to the drone. In this drone, the hydrogen cells are a determinant factor and therefore the extra battery doesn't prove "strong" enough to keep it moving. Of course, there are other items that need to be considered such as the ability of the drone to store energy for the system to be able to use it to increase the distance traveled or increase the flight time.

Besides, the adoption of drones has been proven to be a promising alternative to the problem of connectivity and routing in cities [15] and tourist locations, supporting the preservation of tourist areas for the benefit of residents' well-being [16–19]. In line with this logic, stakeholders' analysis along the supply chain tapping on new technologies such as drones can offer a valuable alternative for sustainable development based on remoteness, connectivity, and cohesiveness during a crisis [20]. Future research in digital marketing and corporate social responsibility should be considered in order to raise funds for UAV projects [21–26].

References

1. Bradley, T.H., Moffitt, B.A., Mavris, D.N., Parekh, D.E.: Development and experimental characterization of a fuel cell powered aircraft. J. Power Sources. **171**(2), 793–e801 (2007)
2. Beibei, R., Shuzhi Sam, G., Chang, C., Cheng-Heng, F., Tong, H.L.: Modeling, control and coordination of helicopter systems. Springer (2012)
3. Camhi, J.: The drones report. Transp Logist Technol Rev. (2016)
4. Donateo, T., Ficarella, A., Spedicato, L., Arista, A., Ferraro, M.: A new approach to calculating endurance in electric flight and comparing fuel cells and batteries. Appl. Energy. **187**, 807e19 (2017)
5. Finnegan P.: World civil unmanned aerial systems market profile & forecast (2016)

6. Ustolin, F., Taccani, R.: 'Fuel cells for airborne usage: Energy storage Comparison', Universit_a degli studi di Trieste, Dipartimento di Ingegneria e Architettura, Trieste, Italy/ international journal of hydrogen energy (2018)
7. Huamin, Z., Xianfeng, L., Hongzhang, Z.: Li-S and Li-O2 batteries with high specific energy: Research and development. Springer (2017)
8. Jenkins, D., Vasigh, B.: 'The economic impact of unmanned aircraft systems integration in the United States', Julien P, Bergthorson JM. Sustainable Energy & Fuels Enabling the metal fuel economy: green recycling of metal fuels. R Soc Chem. **1**, 615–e625 (2017)
9. Apeland, J., Pavlou, D., Hemmingsen, T.: Suitability analysis of implementing a fuel cell on a multirotor drone. J. Aerosp. Technol. Manag. **12**. On-line version ISSN 2175-9146. São José dos Campos 2020 Epub Aug 10, 2020 (2020). https://doi.org/10.5028/jatm.v12.1172
10. Kim, K., Kim, T., Lee, K., Kwon, S.: Fuel cell system with sodium borohydride as hydrogen source for unmanned aerial vehicles. J. Power Sources. **196**(21), 9069e75 (2011)
11. Kim, T.: NaBH4 (sodium borohydride) hydrogen generator with a volume-exchange fuel tank for small unmanned aerial vehicles powered by a PEM (proton exchange membrane). Fuel Cell Energy. **69**, 721e7 (2014)
12. Rao, Z., Wang, S.: A review of power battery thermal energy management. Renew Sustain Energy. **15**(9), 4554e71 (2011)
13. Seddon, J.: Basic helicopter aerodynamics. BSP Professional Books (1990)
14. Yacoubi, M., Lemone, C., Bourguignon, A., Hendrick, P.: Study of the propulsion system of a VTOL MUAV using fuel cells. 9th National congress on theoretical and applied mechanics. p. 5 (2012)
15. Kakarountas, A., Spathoulas, G., Trivellas, P., Reklitis, P., Malindretos, G.: Survey on Routing Services for Smart Delivery in Urban Environments. SEEDA-CECNSM 2020 – 5th South-East Europe Design Automation, Computer Engineering, Computer Networks and Social Media Conference, 9221790 (2020)
16. Trivellas, P., Kakkos, N., Vasiliadis, L., Belias, D.: Sustainability, Social Marketing and Host Attitudes About Tourism in the City of Larissa, pp. 89–95. Springer Proceedings in Business and Economics (2017)
17. Trivellas, P., Kakkos, N., Vasiliadis, L., Belias, D.: Residents' Perceptions Toward Cultural, Social and Economic Benefits and Costs of Tourism Industry: An Empirical Survey, pp. 171–181. Springer Proceedings in Business and Economics (2016)
18. Fotiadis, T.: Strategic Marketing for High Technology Products: An Integrated Approach (Routledge Studies in Innovation, Organizations and Technology), 1st edn. Routledge (2018)
19. Fotiadis, T., Folinas, D., Vasileiou, K., Konstantoglou, A.: Marketing and the Customer Value Chain: Integrating Marketing and Supply Chain Management 1st Edition, Kindle Edition, 1st edn. Routledge (2022)
20. Trivellas, P., Anastasopoulou, A., Malindretos, G., Reklitis, P., Sakas, D.P.: A place based analysis of stakeholders' advancing sustainability in remote and isolated communities: The case of North Aegean Islands. IOP Conference Series: Earth and Environmental Science. **899**(1), 12044 (2021)
21. Reklitis, P., Trivellas, P., Mantzaris, I., Mantzari, E., Reklitis, D.: Employee perceptions of corporate social responsibility activities and work-related attitudes: The case of a Greek management services organization. In: Gal, G., Akisik, O., Wooldridge, W. (eds.) Sustainability and Social Responsibility: Regulation and Re-porting, pp. 225–240. Springer Singapore, Singapore (2018). https://doi.org/10.1007/978-981-10-4502-8_10
22. Sakas, D.P., Reklitis, D.P., Trivellas, P., Vassilakis, C., Terzi, M.C.: The Effects of Logistics Websites' Technical Factors on the Optimization of Digital Marketing Strategies and Corporate Brand Name. Processes. **10**, 892 (2022). https://doi.org/10.3390/pr10050892
23. Sakas, D.P., Giannakopoulos, N.T., Reklitis, D.P., Dasaklis, T.K.: The Effects of Cryptocurrency Trading Websites on Airlines' Advertisement Campaigns. J. Theor. Appl. Electron. Commer. Res. **16**, 3099–3119 (2021). https://doi.org/10.3390/jtaer16070169

24. Sakas, D.P., Kamperos, I.D.G., Reklitis, D.P., Giannakopoulos, N.T., Nasiopoulos, D.K., Terzi, M.C., Kanellos, N.: The Effectiveness of Centralized Payment Network Advertisements on Digital Branding during the COVID-19 Crisis. Sustainability. **14**, 3616 (2022). https://doi.org/10.3390/su14063616
25. Sakas, D.P., Reklitis, D.P.: Predictive model for estimating the impact of technical issues on consumers interaction in agri-logistics websites. In: Information and Communication Technologies for Agriculture—Theme IV: Actions, pp. 269–283. Springer International Publishing, Cham (2021). https://doi.org/10.1007/978-3-030-84156-0_14
26. Sakas, D.P., Reklitis, D.P.: The Impact of Organic Traffic of Crowdsourcing Platforms on Airlines' Website Traffic and User Engagement. Sustainability. **13**, 8850 (2021). https://doi.org/10.3390/su13168850
27. Toudas, K., Kanellos, N.: Economic and Accounting Performance of Greek innovative firms through Knowledge-based entrepreneurship. J Account Tax. **1492**, 150–160 (2022). https://doi.org/10.5897/JAT2022.0530
28. Georgakopoulos, G., Toudas, K., Poutos, E., Kounadeas, T., Tsavalias, S.: Capital Structure, Corporate Governance, Equity Ownership and Their Impact on Firms' Profitability and Effectiveness in the Energy Sector. Energies. **15**(3625), 1–10 (2022). https://doi.org/10.3390/en15103625

Energy Saving of a Drone in Order to Increase Flight Time and Distance Traveled, Modeling, and Optimization

Dimitrios A. Arvanitidis, Dimitrios K. Nasiopoulos, Dimitrios M. Mastrakoulis, and Panagiotis Reklitis

1 Introduction

The aim of the system under consideration is to extend the flight time and consequently the distance traveled through energy savings, which can be achieved by consuming energy from the solar panels attached to the drone.

This research corresponds to the pursuit of energy savings for an autonomous drone, to which solar panels are attached to achieve enough solar energy for its smooth movement. Then, if this stock is analyzed and passed through the factors present in the model, it is returned either in the form of saving and increasing energy with the consequence of not requiring recharging, or in the form of high energy consumption, due to the removal of additional cells that provide power other than the

D. A. Arvanitidis (✉)
Department of Informatics and Telecommunications, University of Peloponnese, Tripoli, Greece
e-mail: darvanitidis@aua.gr

D. K. Nasiopoulos
Department of Agribusiness and Supply Chain Management, BICTEVAC LABORATORY Business Information and Communication Technologies in Value Chains Laboratory, School of Applied Economics and Social Sciences, Agricultural University of Athens, Athens, Greece

D. M. Mastrakoulis
Department of Informatics and Telecommunications, University of Thessaly, Lamia, Greece
e-mail: dmastrakoulis@aua.gr

P. Reklitis
Department of Agribusiness and Supply Chain Management, Organizational Innovation and Management Systems, School of Applied Economics and Social Sciences, Agricultural University of Athens, Athens, Greece
e-mail: preklitis@aua.gr

D. P. Sakas et al. (eds.), *Computational and Strategic Business Modelling*, Springer Proceedings in Business and Economics,
https://doi.org/10.1007/978-3-031-41371-1_17

battery. The overall BESS control structure based on the modern third-order generator model is shown in the following Figs. 1 and 2 [1–3].

To simulate the SG excitation system and speed control system, the BESS inverters must have inertia and damping characteristics. A typical third-order SG model is based on the second-order rotor motion equation and the first-order transient electric equation [4–8].

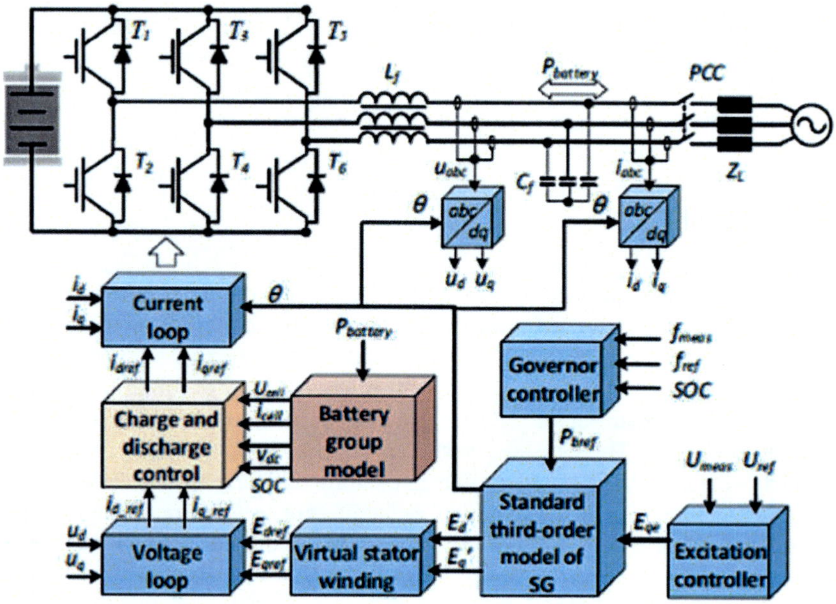

Fig. 1 Active support control strategy based on a third-class SG model

Fig. 2 Three-dimensional display for flexibility: amplitude, ramp rate, and duration

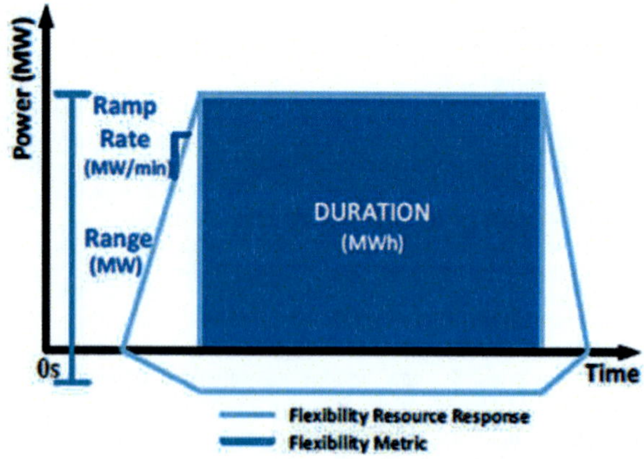

The flexibility of BESS can be analyzed in three dimensions [9]:

(a) In the absolute power output range (MW)
(b) At the power output change speed or ramp rate (MW/min)
(c) During energy levels (MWh)

$$\Delta P_{\text{battery}} = -(k_m{}^* \, \Delta f - D\Delta) \tag{1}$$

The BESS power unit regulation can then be expressed as a formula (2):

$$k_{\text{battery}} = -\Delta_{\text{battery}}/\Delta f = (K_{\text{m}} - D) \tag{2}$$

The difference factor of BESS can be expressed as a formula (3):

$$_{\text{battery}}\% = 100/k_{\text{battery}} = 100/(K_{\text{m}} - D) \tag{3}$$

BESS distributes a part of the power support according to its own difference factor when the power system load contributes to the output frequency modulation. The main purpose is a simple procedure for extending the flight time of a multicopter.

Many of the concepts presented here also apply to fixed-wing drones. The optimization process is a loop, so it is considered necessary for certain assumptions to be made. The rotation of the propellers creates a push and at the same time, allows the drone to take off and hover. In hover, the combined thrust of the propellers is equal to the total weight of the drone [3, 10, 11] (Fig. 3).

As a theoretical approach it could be assumed that the maximum thrust achieved by the propeller must be approximately twice the swing thrust. Efficiency is the ratio of output divided by input. The propellers convert mechanical energy into impulse [3].

$$\begin{aligned}\text{PropellerEfficiency} &= \text{Thrust}/\text{MechanicalPower} \\ &= \text{Thrust}/\text{Torque}^* \ \text{RotationSpeed}\end{aligned} \tag{4}$$

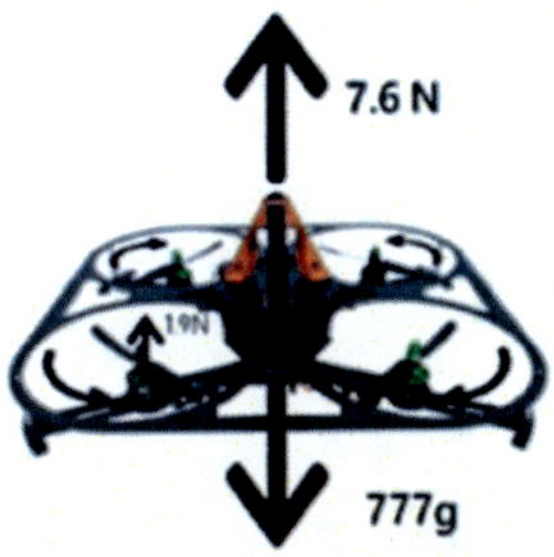

Fig. 3 Example of thrust and weight of a drone

Battery capacity can be expressed as the total flight time (FT) in hours, multiplied by the power produced (Power in Watt):

$$\text{FlightTime}^{*}\ \text{Power} = E_{\text{battery}} \tag{5}$$

The capacity of the battery (E_{battery}) is equal to the weight of the battery (W_{battery} in grams) multiplied by the energy density (sigmabattery in Wh/g):

$$E_{\text{battery}} = {}_{\text{battery}}{}^{*}\ W_{\text{battery}} \tag{6}$$

The total power (Power in Watt) is equal to the weight of the drone (W_{drone} (g) = W_{frame} (g) + W_{battery} (g)) divided by the performance of the propellers (in g/W):

$$\text{Power} = W_{\text{drone}}/\text{prop}_{\text{efficiency}} \tag{7}$$

Propeller performance is a function of the total weight of the drone with the number of propellers on the drone. Thus, by combining the equations, the flight time appears to be:

$$\text{FT} = {}_{\text{battery}}{}^{*}\ W_{\text{battery}}{}^{*}\ \text{prop}_{\text{efficiency}}/W_{\text{drone}} \tag{8}$$

Increasing the battery weight increases the ratio in the above equation but reduces the performance of the propeller. With the use of modeling is more efficient to simulate the factors and powers that apply to the system. This process draws useful conclusions before it is implemented. The Anylogic PLE program was used for the simulation. The capacity of the model lies in the fact that if a value in the model is differenced, the situation changes as well as the results of the modeling. Taking advantage of the capacity of the model, many different situations can be analyzed and arrive at process optimization. It is pointed out that in contrast to previous examinations, here the possibility will be examined through the following analysis to extend the flight time of the system, which reasonably implies the increase in the distance it can travel.

2 Description of Hypothesis and Modeling Process (Table 1)

Table 1 Description of the forces applied

Solar energy	A key factor in the movement of this drone is the energy received by the sun's rays, as it is very important in activating the movement of the system. Resources therefore flow from TRPA to Esolar stock through soperc.
Propeller energy	An additional key factor in the movement of this drone is the energy produced by the activation of the propellers, as it activates the movement of the system. Resources therefore flow from TRPA to Eprop stock through prperc.
Kinetic energy of the drone system	Another factor in the movement of this drone is the kinetic energy of the drone, i.e., "body" and battery as through it any programmed or unplanned movement of the system takes place. Resources are therefore created and flow from TRPA to Edrone stock through drperc.
Framework energy	A different factor compared to what has been analyzed so far is the framework energy. Resources therefore flow from TRPA to Eframe stock through frperc.
Battery energy	A particularly significant factor in the movement of the drone is the energy produced by the activation of the battery, as through it the system operates smoothly, and the main thing is the energy that pushes the drone in motion as it contains the unique force applied to the principle of motion $t = 0$. Resources therefore flow from TRPA to Ebattery stock through battperc.
Thermal energy	At the end of the movement of the system, some forces are applied which in the existing example will be interpreted as the denaturation into an energy that has a negative effect on the system. This is done by streaming resources to Ethermal stock through thperc.

3 Hypothesis

H1: $E_{\mathrm{fin}} < 0$/*The system has battery as a power source (no solar panels)*
Four powers affect the drone at the beginning of the process and one more later.

H2: $E_{\mathrm{fin}} > 0$/*Propellers and solar panels are added to the system*
Two extra powers were added to the system to contribute to the drone's energy saving.

H3: $E_{\mathrm{fin}} > 0$/*The system has no thermal losses*
The thermal losses don't affect the system so more energy is saved.

H4: $E_{\mathrm{fin}} < 0$/*The panels that are used to collect sun's energy are removed from the system*
The power from solar panels that boosted the system is no longer active.

H5: $E_{\mathrm{fin}} > 0$/*Solar panels are excluded from the system as well as the thermal losses*
Although the system loses a boost power it also gains the thermal loss exclusion.

H6: $E_{\mathrm{fin}} > 0$/*The battery is working properly for a 32%*
The battery isn't working at 100% properly something that means that the rest of the power must help so that the system remains vital.

H7: $E_{\mathrm{fin}} > 0$/*The battery is working properly for a 32% and number of propellers is differentiated*

The battery isn't working at 100% properly something that means that the rest of the power must help so that the system remains vital. Three out of four propellers are excluded from the system and so the system weight is changed.

4 Methodology

The aim of the system under consideration is to extend the flight time and consequently, the distance traveled through energy savings, which can be achieved by consuming energy from the solar panels attached to the drone. The pursuit is to have as few stops as possible to refuel or charge the battery, with the corresponding benefit of not interrupting the assigned mission. This practically means for the system, that after the initial energy that is provided and constitutes 100% of the resources, the resources that feed the stocks, must lead it to its maximum possible value, which implies the ever-increasing value of the final value of energy, that is, when the system tends to land. The purpose of the whole project is the non-interruption of the flight through the minimization of the system requirement for energy and at the same time the overcoming of the sum of the other opposite measures of forces that are applied. To achieve this goal, percentage changes had to be made. Time is referred to as time units (tu).

Initially, 100% of the energy is placed in TRPA in the context of faithful observance of the ECP (energy conservation principle), that is, the battery energy. This practically means that after the initial energy that exists in the system and is 100% of the resources, the resources that supply the energy stocks must lead it to its maximum possible value, which implies the ever-increasing value of the final energy amount. It is therefore understandable that while the drone speed should reasonably be constantly increasing in order not to interrupt the flight, it should also be such that the amount of energy (with a positive sign) feeds the flight to exceed the opposing forces.

5 Hypothesis Trials and Results

Initially, 100% of the drone resources must be divided into five factors which are $E_{BATTERY}$, E_{DRONE}, E_{PROP}, E_{FRAME}, and $E_{thermal}$. In the beginning, it is selected not to apply the solar cells with consequent distribution, to be done with as in Table 2. This is achieved by using a converter on the respective flow from the drone to each

Table 2 Energy types and supply quota in modeling

Type of energy	Percentage (%)
Propeller energy	29
Kinetic energy drone	13
Framework energy	(−) 20
Thermal energy	(−) 10
Battery energy	28

factor. This is achieved by using a converter on the respective flow from TRPA to each factor. Then around the middle of the system path and after all the actions have been finalized into the BESS (battery energy storage system), the performance of the system is examined among the removal of the solar panels. The bottom line is the effect of E_{thermal} is applied as mentioned above. After researches, the following results are obtained regarding the behavior of the forces in the system. The green color in the following diagram indicates a positive to the system power and red indicates a negative one (Table 2).

Propeller Energy
In this case, 29% of TRPA resources go to E_{PROP} through soperc flow. Then through the bess2 flow that determines the magnitude of the energy received it ends up in BESS.

Solar Energy
E_{SOLAR} will not be powered as it is considered that the solar panels have not yet been attached during the modeling.

Kinetic Energy of the Drone System
The next 13% of the initial factor TRPA goes to E_{DRONE} in the same way as mentioned above. When the resources reach bess3 and with the application of the given parameters the size of the mentioned energy that is applied in the system will emerge. The resources of TRPA end up in BESS.

Framework Energy
The next 20% of the initial factor goes to E_{FRAME}. When the resources reach bess4 and with the application of the given parameters the size of the reported energy applied to the system will emerge.

Thermal Energy
Various forces, thermal energy is generated, which here is considered to affect the system by a percentage of 10% of the initial value of thermal energy.

Battery Energy
The final 28% of the initial factor goes to E_{BATTERY} in the exact same way as above. When the resources reach bess5 and by applying the given parameters the size of the reported energy applied to the system will emerge. In the same way these resources of TRPA flow to BESS. When the system reached a specific point in the middle of the path (BESS), it was considered a test to examine the effect of solar energy on the system (alternative). Therefore, in a second route, the solar panels were removed and consequently the positive effect of this energy on the system. Nevertheless, the alternative system is deliberately not tested by thermal energy. As it will be shown below, the model does not become viable, which is verified in the next stage of Finalenergy. The total energy produced in the event of a negative value or less than the initial system energy (TRPA), reasonably concludes that the energy is not sufficient to continue the drone flight.

6 Modeling and Simulation

The system from the beginning of its operation and although the power that feeds the movement of the drone is at 41% of the initial energy, at no time for 0–15tu, it manages to get positive values. The consequence of this is the cancellation of the flight due to a lack of energy to support the flight (Fig. 4).

The above configuration of the model will result in the following results (Fig. 5):

These initial stock resources then flow into BESS. When the resources reach BESS, the condition exists and if satisfied, the resources flow to Finalenergy. It is observed that $\mathbf{E_{fin}}$ never takes a positive value and equals to $\mathbf{-2584*10^5*E_{arx}}$ at the end of the mission (t = 15tu).

Consequently, the drone will never manage to take off. The simulation ran up to t = 15tu while E_{fin} continued to rise in the negative axis. However, no point is

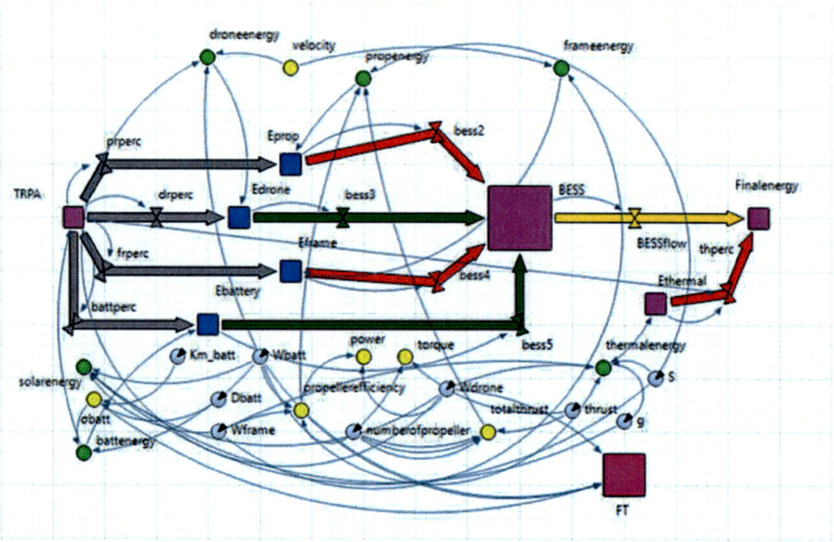

Fig. 4 Simulation model

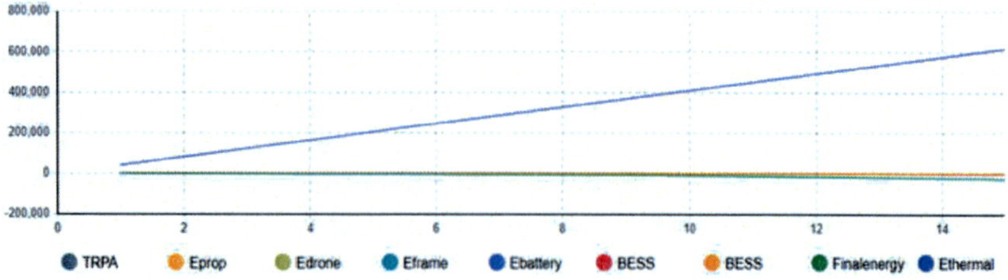

Fig. 5 Energy diagrams during the modeling phase

detected (as shown in the diagram) where the E_{fin} curve enters the positive value axis.

7 Optimization (Fig. 6)

The aim of the system is to save energy and at the same time extend the drone flight time, which can be achieved by absorbing energy from the solar cells. Practically for the system, it means that after the initial energy that exists and constitutes 100% of the resources, the ones that feedback the stocks, must lead it to its maximum possible value, which implies the ever-increasing value of the final value of energy, that is, when the system tends to land. The purpose of the whole project is the non-interruption of the flight through the minimization of the system requirement for energy and at the same time the overcoming of the sum of the other opposite measures of forces that are applied. With ECP in mind, it makes sense for the drone-driven force to tend to zero, as opposed to the principle of motion where the speed is constantly increasing for the system to achieve the purpose of the whole project and at the same time surpass the sum of the other countermeasures applied. The maximum possible profit is achieved if the percentage of resources flowing from the drone to the stocks is differentiated. Toward that direction, the energy percentages will vary. After many simulations, the result is as the following table (Table 3):

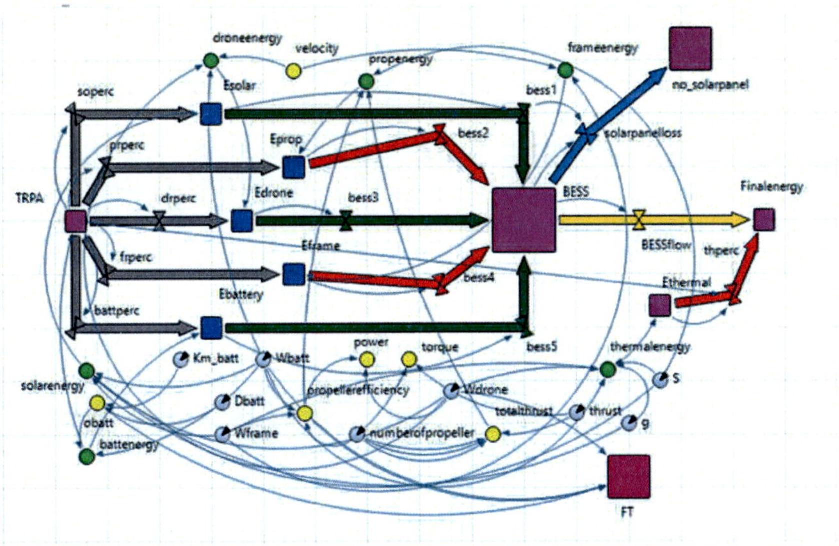

Fig. 6 Optimization model

Table 3 Energy types and supply quota in optimization

Type of energy	Percentage (%)
Solar energy	8.32
Propeller energy	(−) 15.31
Kinetic energy drone	28.34
Framework energy	(−) 24.88
Thermal energy	(−) 5.86
Battery energy	17.29

Table 4 Explanation of solar flows and factors

Name	Type	Explanation
W_{batt}	0.5	Battery weight
g	9.81	Acceleration of gravity in m/s^2
S	1000	The total distance examined
W_{frame}	0.2	Frame weight
W_{drone}	3.5	Drone weight
solarenergy	$(W_{batt} + W_{frame} + W_{drone})*g*S$	Solar energy
bess1	E_{solar}	The flow to BESS
soperc	–	The flow to the E_{solar}

Table 5 Explanation of propeller energy flows and factors

Name	Type	Explanation
Totalthrust	thrust*numberofpropeller	Pushing the system from all propellers (4)
Thrust	0.075	Push the system from a propeller
S	1000	The total distance examined
Numberofpropeller	4	Number of propellers
Propellerefficiency	(Wframe + Wbatt)/ numberofpropeller	Propeller efficiency
Wframe	0.2	Frame weight
Wbatt	0.5	Battery weight
Torque	numberofpropeller*thrust	Torque
Propenergy	totalthrust*S/propellerefficiency	Propeller energy
bess2	E_{prop}	The flow to BESS
Prperc	–	The flow to Eprop

Since the system's final energy should be greater than the initial one in the short term the above percentages are placed on the converters bearing the name of each stream. More specifically:

(a) In solar energy where E_{solar} = solarenergy*soperc, the following factors affect as follows (Table 4):

(b) In propeller energy where E_{prop} = propenergy*prperc, the following factors act as follows (Table 5):

Table 6 Explanation of drone kinetic energy flows and factor

Name	Type	Explanation
velocity	60	Speed in m/s
Wbatt	0.5	Battery weight
Wdrone	3.5	Drone weight
power	Wdrone/propellerefficiency	The power of propellers
droneenergy	0.5*(Wdrone + Wbatt)*velocity*velocity	Kinetic energy
Bess3	E_{drone}	The flow to BESS
drperc	–	The flow to E_{drone}

Table 7 Explanation of framework energy flows and factors

Name	Type	Explanation
velocity	60	Speed in m/s
Wframe	0.2	Frame weight
frameenergy	0.5*Wframe*velocity*velocity	Kinetic energy
Bess4	$-E_{frame}$	The flow to BESS
frperc	–	The flow to E_{frame}

Table 8 Explanation of battery energy flows and factors

Name	Type	Explanation
Wbatt	0.5	Battery weight
Σbatt	1/(Km_batt-Dbatt)	Energy density
Km_batt	97	Proportional battery power frequency ratio
Dbatt	20	Battery damping factor
Battenergy	σbatt*Wbatt	Battery power
Bess5	$-E_{battery}$	The flow to BESS
Battperc	–	The flow to $E_{battery}$

(c) The following factors affect the drone kinetic energy where E_{drone} = droneenergy*drperc (Table 6):

(d) The framework energy where, E_{frame} = frameenergy*frperc is influenced by the following factors as follows (Table 7):

(e) In the battery energy where, $E_{battery}$ = battenergy*battperc the following factors affect as follows (Table 8):

(f) In thermal energy where, $E_{thermal}$ = thermalenergy*<u>**x**</u>/100 (<u>**x**</u> depends on the scenario) the following factors affect as follows (Table 9):

From the previous diagrams, the differences in the flow of time are obvious. **No_solarpanel** is analyzed as follows (Table 10):

Table 9 Explanation of thermal energy flows and factors

Name	Type	Explanation
Wbatt	0.5	Battery weight
Wdrone	3.5	Drone weight
Wframe	0.2	Frame weight
G	9.81	Acceleration of gravity in m/s^2
S	1000	The total distance examined
thermalenergy	(Wdrone + Wbatt + Wframe)*g*S	Thermal energy
BESSflow	BESS	The flow from BESS to Finalenergy
Finalenergy	thperc + BESSflow	Final action
Thperc	$-E_{thermal}$	The flow to Finalenergy
BESS	bess2 + bess3 + bess1 + bess5 + bess4	The sum of the actions that affect the system

Table 10 Explanation of no_solarpanel flows and factors

Name	Type – value
no_solarpanel	Solarpanelloss
solarpanelloss	BESS – bess1
BESS	bess2 + bess3 + bess1 + bess5 + bess4
bess1	E_{solar}

Table 11 Explanation of flows and factors in the "ideal" optimization process

Name	Type	Explanation
FT	σbattery*Wbattery*propefficiency/ Wdrone	Flight time
σbatt	1/(Km_batt-Dbatt)	Difference rate
Km_batt	97	Proportional power frequency factor
Dbatt	20	Depreciation rate
Wbatt	0.5	Battery weight
Wframe	0.2	Frame weight
propellerefficiency	(Wframe + Wbatt)/numberofpropeller	Propeller efficiency
numberofpropeller	4	Number of propellers

In the process of modeling as well as in optimization, the value is positive. Nevertheless, as the aim here is to increase the flight time, the **FT** factors are analyzed as follows (Table 11):

It is observed that TRPA at **t = 0tu** begins from the initial value of 100% but then at the output, and at the same time at E_{fin}, it has a positive value as well. At $t = 15$tu the $\mathbf{E_{fin} \approx 5403*10^8*E_{arx}}$ approximately. A result that, compared to modeling, shows a tremendous change (Fig. 7).

Therefore, the model after optimization is much more efficient and it is easily understood that the cost savings are easily calculated from the sum E_{fin}-TRPA. More specifically, it is observed that the drone at $t = 0$tu starts from the initial value of

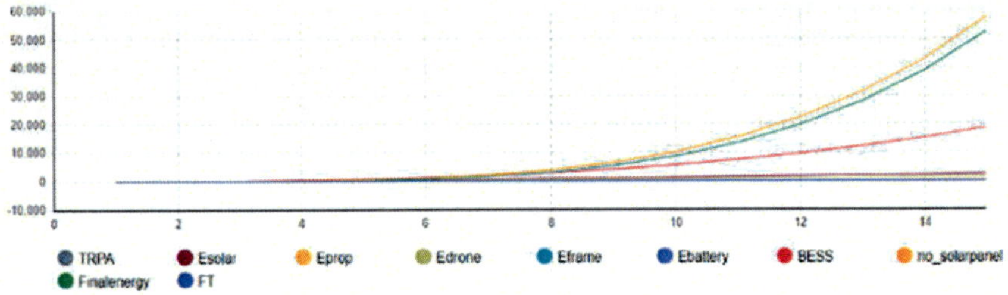

Fig. 7 Energy diagrams during the optimization phase

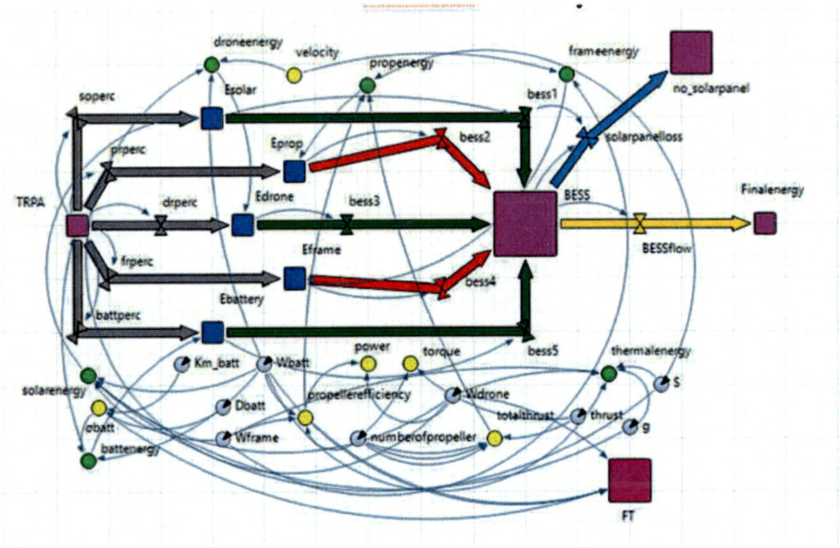

Fig. 8 Simulation model (no thermal losses)

100% and then at the output, that is, at E_{fin} with geometric progression, the eventually produced energy increases. At t = 3.78tu the model is entering a maximum optimization trajectory as E_{fin} > TRPA. Finally, at t = 15tu the E_{fin} manages to be exponentially greater than the initial energy. Therefore, the model after optimization is much more efficient and it is easy to understand that the value of savings is easily calculated from the sum of E_{fin}-TRPA, thus achieving optimization.

A. However, to achieve the examination of other scenarios and the maximum possible energy in the system, the viability of the system will be examined if it is considered that there are no thermal effects to the system. Based on this, below the behavior of the system is listed (Fig. 8).

The energy percentages that are applied to the above are (Table 12):

The following diagram clarifies the landscape in the field of energy flow in the system (Fig. 9).

Table 12 Energy types in optimization (no thermal losses)

Type of energy	Percentage (%)
Solar energy	7.51
Propeller energy	(−) 18.47
Kinetic energy drone	24.89
Framework energy	(−) 29.15
Battery energy	19.98

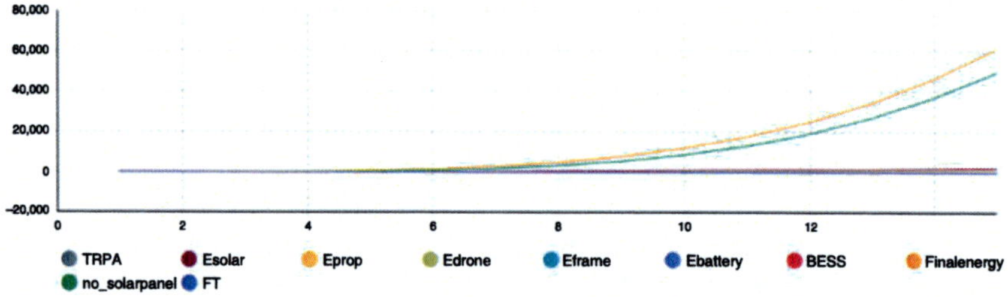

Fig. 9 Energy diagrams during the optimization phase (no thermal losses)

Table 13 Energy types in optimization (no solar panels)

Type of energy	Percentage (%)
Propeller energy	(−) 14.07
Kinetic energy drone	23.48
Framework energy	(−) 21.13
Thermal energy	(−) 3.02
Battery energy	38.30

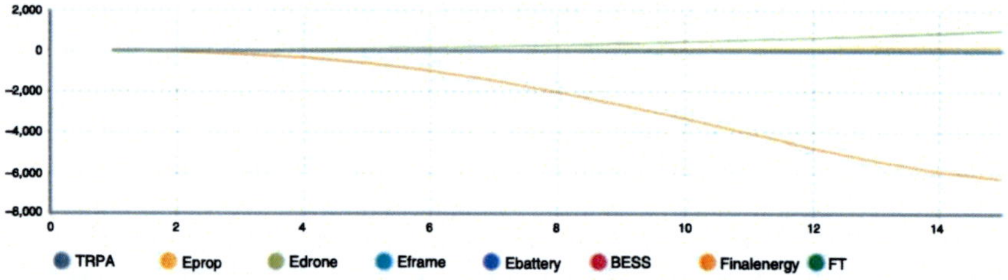

Fig. 10 Energy diagrams during the optimization phase (no solar panels)

It is observed that at $\mathbf{t = 15tu}$ the $\mathbf{E_{fin} \approx 6302*10^8*E_{arx}}$ approximately, which means that the energy saving plan is succeeded. Obviously, the model after optimization is very efficient since the drone manages to perform correctly.

B. An alternative examination of the system flow will bring different behavior and at the same time different results in the model. Therefore, here it is assumed that the system receives no effect from the sun (no solar panels) (Table 13).

The following diagram clarifies the landscape in the field of energy flow in the system (Fig. 10).

It is observed that at **t = 15tu** the $\mathbf{E_{fin}} \approx \mathbf{-0{,}6461*10^8*E_{arx}}$ approximately. That is almost twice the previous result of the optimization process. Therefore, the model is considered to fail in the energy saving process.

C. A different approach to system behavior could be as follows:

(a) Removal of solar panels and thermal power.
(b) Thermal power and therefore the system will develop as follows (Fig. 11):

The energy percentages that are applied to the above are (Table 14):

The following diagram clarifies the landscape in the field of energy flow in the system (Fig. 12).

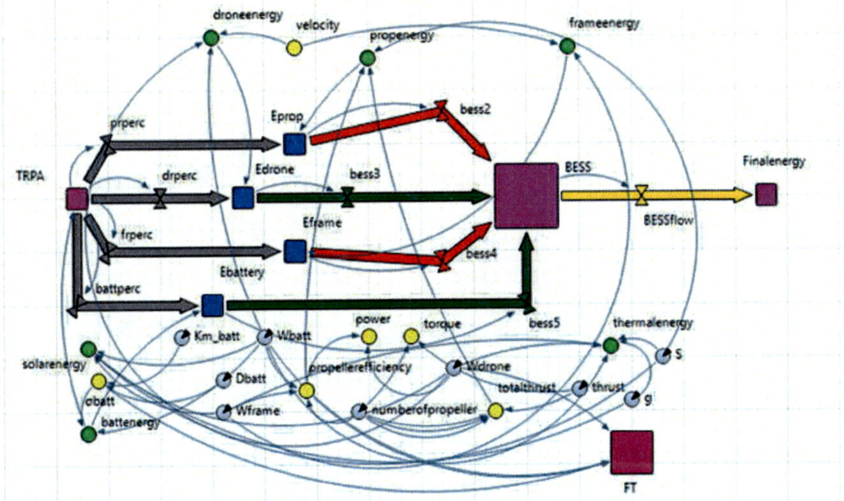

Fig. 11 Simulation model (no solar cells and no thermal losses)

Table 14 Energy types in optimization (no solar cells and no thermal losses)

Type of energy	Percentage (%)
Propeller energy	(−) 21.81
Kinetic energy drone	22.89
Framework energy	(−) 26.09
Battery energy	29.21

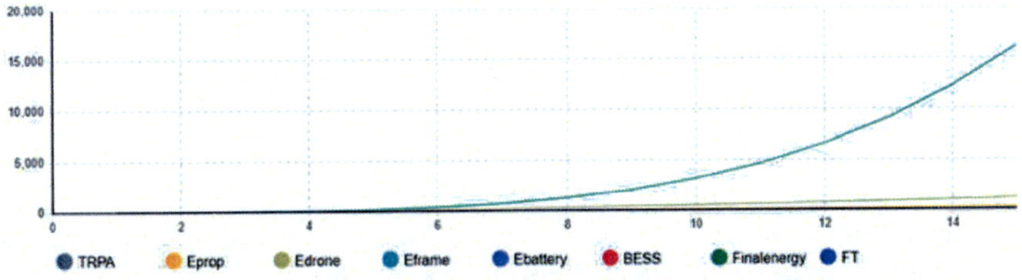

Fig. 12 Energy diagrams during the optimization phase (no solar cells and no thermal losses)

Table 15 Energy types in optimization (battery is 68% less efficient)

Type of energy	Percentage (%)
Solar energy	11.17
Propeller energy	(–) 19.42
Kinetic energy drone	27.19
Framework energy	(–) 19.93
Thermal energy	(–) 9.43
Battery energy	12.86

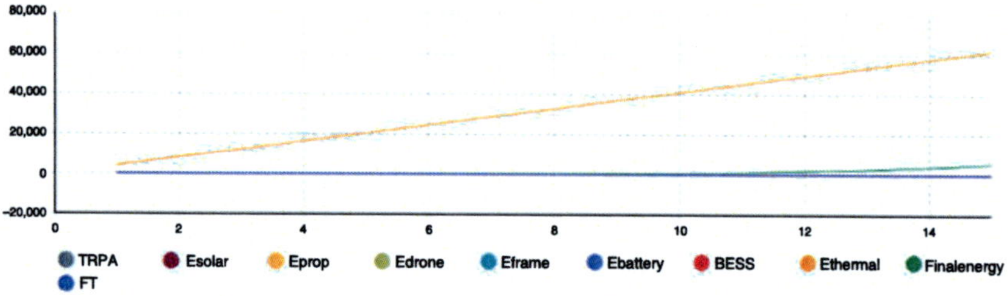

Fig. 13 Energy diagrams during the optimization phase (battery is 68% less efficient)

Table 16 Energy types in optimization (battery is 68% less efficient)

Type of energy	Percentage (%)
Solar energy	11.17
Propeller energy	(–) 19.42
Kinetic energy drone	27.19
Framework energy	(–) 19.93
Thermal energy	(–) 9.43
Battery Energy	12.86

It is observed that at **t = 15tu** the $\mathbf{E_{fin}} \approx \mathbf{1665*10^5*E_{arx}}$ approximately. Therefore, the model is considered to succeed in the energy saving process.

D. Another approach to system behavior could be that 68% of battery malfunctions. In accordance with that the energy percentages will develop as follows (Table 15):

The following diagram clarifies the landscape in the field of energy flow in the system (Fig. 13).

More specifically, it is observed that the drone at $t = 0$tu starts from the initial value of 100% and then at the output, at E_{fin} with geometric progression, the eventually produced energy decreases. Therefore, the system can't be considered reliable. It is observed that at **t = 3.88tu** the $\mathbf{E_{fin}} \approx \mathbf{-4567*10^6*E_{arx}}$ and at **t = 15tu** that $\mathbf{E_{fin}} \approx \mathbf{5{,}76*10^5*E_{arx}}$ approximately, which means that the system manages to complete its mission although the battery is almost at 30% of its efficiency.

E. Finally, it is assumed that the drone is left with one propeller instead of four and because of that the total weight is decreased by 1.05 kg (about 350 g/propeller). To understand the significance of that action here it is assumed that the system has a malfunctioned battery as in H6 (Tables 16 and 17):

Table 17 New values after modification to the system

Name	New Value
Numberofpropeller	1
Wdrone	2.45

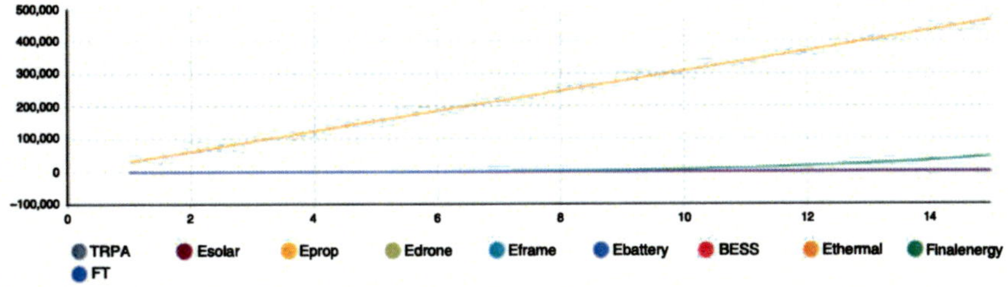

Fig. 14 Energy diagrams during the optimization phase (battery 32% working, propellers excluded from system)

The following diagram clarifies the landscape in the field of energy flow in the system (Fig. 14).

More specifically, it is observed that the drone at $t = 0$tu starts from the initial value of 100% and then at the output, at E_{fin} with geometric progression, the eventually produced energy decreases. Therefore, the system can't be considered reliable. It is observed that at **t = 4.86tu** the $\mathbf{E_{fin}}$ gets >0 and about $\mathbf{8.28{*}E_{arx}}$ and at **t = 15tu** that $\mathbf{E_{fin}} \approx \mathbf{4585{*}10^5{*}E_{arx}}$ approximately, which means that the system manages to complete its mission although the battery is almost at 30% of its efficiency. In contrast to all the above hypotheses and because of the modifications that took place, the FT gained in this one is 0.028tu totally instead of 0.005tu.

8 Conclusion

This research is part of a research project examining and optimizing how to save energy on a drone by exploiting test flight research data. The whole procedure concerns the movement of the aircraft with the addition of solar cells or not and at the same time the presence or exclusion of the thermal losses that can be applied by nature to the drone. Further research and application of additional alternative energy sources are required as well as a way to reduce in significant level the thermal loss.

In this work, the behavior of the drone was examined if solar cells are attached to it, as well as the importance of the battery existence for the current drone and finally the ability of solar panels to increase the energy of the system to such an extent that it significantly extends flight time and simultaneously the distance traveled. On the current model under consideration, the explanation given for the negative value of E_{fin} is that the drone will not be able to carry out any mission assigned to it unless an additional energy source is attached to the drone. In this drone, the solar cells are not a determinant factor to increase the flight time of the drone. Considering the dependent variables of flight time, it is understood that by decreasing the weight of the drone and/or decreasing the number of propellers, the desired result can be achieved. Of course, there are other items that need to be considered such as the ability of the drone to store energy for the system to be able to use it to increase the distance traveled or increase the flight time.

Regarding relevant managerial and policy implications, for example, the complexity of routing in cities [12] as well as in tourist locations may be resolved by the adoption of drones [13]. In particular, residents at tourist destinations may consider new technologies such as UAVs, as an interesting alternative policy in order to leverage sustainable supply chain, economic benefits, and human well-being, preserving at the same time the natural environment and community cohesiveness [12, 14, 15]. Finally, future research in CSR and digital advertising could be beneficial for UAV companies in order to raise funds for their projects [16–21].

References

1. Bergen, A.R.: Power Systems Analysis. Pearson Education India, Delhi (2009)
2. Hatziargyriou, N., et al.: Microgrids. IEEE Power Energy Mag. **5**(4), 78–94 (2007)
3. Ge, Y., Sun, T., Liu, C., Wang, C., Shan, J., Wang, X.: Paper ID: 941/Third-order synchronous machine model based active support control of BESS and its contribution analysis for primary frequency response. In: International Conference on Applied Energy, Västerås, Sweden (2019)
4. Katiraei, F., Mohammad, R.I., Peter, W.L.: Micro-grid autonomous operation during and subsequent to islanding process. IEEE Trans. Power Delivery. **20**(1), 248–257 (2005)
5. Kim, Y.S., Kim, E.S., Moon, S.I.: Frequency and voltage control strategy of standalone microgrids with high penetration of intermittent renewable generation systems. IEEE Trans. Power Syst. **31**(1), 718–728 (2016)
6. Micallef, A., et al.: Secondary control for reactive power sharing and voltage amplitude restoration in droop-controlled islanded microgrids. In: 2012 3rd IEEE International Symposium on Power Electronics for Distributed Generation Systems (PEDG). IEEE (2012)
7. Olivares, D.E., et al.: Trends in microgrid control. IEEE Trans. Smart Grid. **5**(4), 1905–1919 (2014)
8. Serban, I., Marinescu, C.: Control strategy of three-phase battery energy storage systems for frequency support in microgrids and with uninterrupted supply of local loads [J]. IEEE Trans. Power Electron. **29**(9), 5010–5020 (2014)
9. Han, H., et al.: Review of power sharing control strategies for islanding operation of AC microgrids. IEEE Trans. Smart Grid. **7**(1), 200–215 (2016)
10. Thresher, R., Michael, R., Paul, V.: To capture the wind. IEEE Power Energy Mag. **5**(6), 34–46 (2007)
11. Yingjie, T., Muttaqi, K.M., Paul, C.P., et al.: Enhanced frequency regulation using multilevel energy storage in remote area power supply systems [J]. IEEE Trans. Power Syst. (2018). https://doi.org/10.1109/TPWRS.2018.2867190
12. Fotiadis, T.: Strategic Marketing for High Technology Products: An Integrated Approach Routledge Studies in Innovation, Organizations and Technology, 1st edn. Routledge, London (2018)
13. Kakarountas, A., Spathoulas, G., Trivellas, P., Reklitis, P., Malindretos, G.: Survey on routing services for smart delivery in urban environments. In: SEEDA-CECNSM 2020 – 5th South-

East Europe Design Automation, Computer Engineering, Computer Networks and Social Media Conference, 9221790 (2020)
14. Folinas, D., Fotiadis, T.: Marketing and Supply Chain Management: A Systemic Approach, 1st edn. Routledge, Abingdon (2017)
15. Fotiadis, T., Folinas, D., Vasileiou, K., Konstantoglou, A.: Marketing and the Customer Value Chain: Integrating Marketing and Supply Chain Management, 1st Edition, Kindle edn. Routledge, Abingdon (2022)
16. Sakas, D.P., Reklitis, D.P.: The impact of organic traffic of crowdsourcing platforms on airlines' website traffic and user engagement. Sustainability. **13**, 8850 (2021). https://doi.org/10.3390/su13168850
17. Sakas, D.P., Kamperos, I.D.G., Reklitis, D.P., Giannakopoulos, N.T., Nasiopoulos, D.K., Terzi, M.C., Kanellos, N.: The effectiveness of centralized payment network advertisements on digital branding during the COVID-19 crisis. Sustainability. **14**, 3616 (2022). https://doi.org/10.3390/su14063616
18. Sakas, D.P., Giannakopoulos, N.T., Reklitis, D.P., Dasaklis, T.K.: The effects of cryptocurrency trading websites on airlines' advertisement campaigns. J. Theor. Appl. Electron. Commer. Res. **16**, 3099–3119 (2021). https://doi.org/10.3390/jtaer16070169
19. Reklitis, P., Trivellas, P., Mantzaris, I., Mantzari, E., Reklitis, D.: Employee perceptions of corporate social responsibility activities and work-related attitudes: the case of a Greek management services organization. In: Gal, G., Akisik, O., Wooldridge, W. (eds.) Sustainability and Social Responsibility: Regulation and Reporting, pp. 225–240. Springer Singapore, Singapore (2018). https://doi.org/10.1007/978-981-10-4502-8_10
20. Sakas, D.P., Reklitis, D.P., Trivellas, P., Vassilakis, C., Terzi, M.C.: The effects of logistics websites' technical factors on the optimization of digital marketing strategies and corporate brand name. Processes. **10**, 892 (2022). https://doi.org/10.3390/pr10050892
21. Sakas, D.P., Reklitis, D.P.: Predictive model for estimating the impact of technical issues on consumers interaction in agri-logistics websites. In: Information and Communication Technologies for Agriculture—Theme IV: Actions, pp. 269–283. Springer International Publishing, Cham (2021). https://doi.org/10.1007/978-3-030-84156-0_14

Assessment of the Berkeley Solar Drone's Ability to Save Energy and Assist in the Development of Fifth Generation Mobile Networks with IoT Capability, Modeling and Optimization

Dimitrios A. Arvanitidis, Dimitrios K. Nasiopoulos, Dimitrios M. Mastrakoulis, and Panagiotis Trivellas

1 Introduction

The aim of the system under consideration is to extend the flight time and, consequently, the distance traveled through energy savings, which can be achieved by consuming energy from the solar panels attached to the drone. This research corresponds to the pursuit of energy savings for an autonomous drone, to which solar panels are attached to achieve enough solar energy for its smooth movement. Then, if this stock is analyzed and passed through the factors present in the model, it is returned either in the form of saving and increasing energy with the consequence of not requiring recharging, or in the form of high energy consumption, due to the removal of additional cells that provide power other than the battery.

D. A. Arvanitidis (✉)
Department of Informatics and Telecommunications, University of Peloponnese, Tripoli, Greece
e-mail: darvanitidis@aua.gr

D. K. Nasiopoulos
Department of Agribusiness and Supply Chain Management, BICTEVAC LABORATORY Business Information and Communication Technologies in Value Chains Laboratory, School of Applied Economics and Social Sciences, Agricultural University of Athens, Athens, Greece

D. M. Mastrakoulis
Department of Informatics and Telecommunications, University of Thessaly, Lamia, Greece
e-mail: dmastrakoulis@aua.gr

P. Trivellas
Department of Agribusiness and Supply Chain Management, Organizational Innovation and Management Systems, School of Applied Economics and Social Sciences, Agricultural University of Athens, Athens, Greece

D. P. Sakas et al. (eds.), *Computational and Strategic Business Modelling*, Springer Proceedings in Business and Economics,
https://doi.org/10.1007/978-3-031-41371-1_18

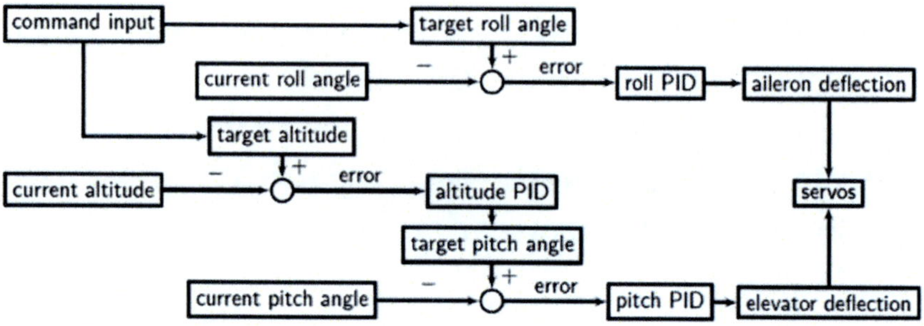

Fig. 1 Autopilot control loop subset

Nevertheless, if it becomes possible to develop a method for autonomous and cheap periodic energy recovery, a drone can be built for efficient continuous flight, limited not only by fuel but also by the life of the material. The main advantage of a multi-day flight platform is its ability to be expanded by additional electronic means [1]. Fast-growing platforms can act as ad-hoc communication hubs, invaluable for emergency response functions. To simplify the problem, the aircraft will use a program to maintain the aircraft interconnection with the pilots. A practical case for automation here is an on-board autopilot operating on commands. A damping mechanism to reduce Dutch roll oscillations falls into this role. The next outer loop consists of an altitude controller and a direction and speed controller, in terms of rolling, deflection, and acceleration position (throttle), to be able to keep these parameters constant as required for the coordination of turns [2] (Fig. 1).

Andre Noth's doctoral dissertation, "Design of Solar Powered Airplanes for Continuous Flight," sheds light on the relationship between various aspects of aircraft design [3–5]. The following equations have been adapted from his methodology. The interdependence between variables complicates the mathematical models used to predict the physical properties of the aircraft.

Equation (1) adds the masses that for any set of parameters, remain constant and do not depend on other values:

$$m_1 = m_{av} + m_{pl} \tag{1}$$

Equation (2) derives the mass of the frame (air) from the constant of the structural mass:

$$m_2 = kaf\, r^{x2}\, b^{x1} \tag{2}$$

Equation (3) determines the mass of the solar panels:

$$m_3 = (k_{sc} + k_{enc})A_{sc} \tag{3}$$

Equation (4) derives the mass of the Maximum PowerPoint Tracker (MPPT):

$$m_4 = k_{mppt}{}^{*}I_{max}{}^{*}\eta_{sc}{}^{*}\eta_{cbr}{}^{*}\eta_{mppt}{}^{*}A_{sc} \tag{4}$$

Equation (5) shows that the mass depends on the T_{night} overnight duration and the efficiency with which the k_{bat} batteries can be discharged:

$$m_5 = T_{night}{}^{*}P_{Etot}/d_{chrg}{}^{*}k_{bat} \tag{5}$$

Equation (6) derives the propulsion of the total mass:

$$m_6 = k_{prop}{}^{*}P_{Eflight} \tag{6}$$

To aggregate the size of the total mass, the individual masses are combined:

$$m_{tot} = \sum_{i=1}^{6} mi \tag{7}$$

To pump the mechanical power required for the flight, it is considered the total mass of the aircraft and its design. The drag coefficient has two components:

- The CDafl drag coefficient
- The drag coefficient

These are added to reach the CD. Lack of FAA clearance limits the aircraft to a flight ceiling much lower than the target altitude.

$$P_{mech} = m_{tot}{}^{2/3*}(1/b) * C_D C_L{}^{2/3*}\mathrm{sqrt}\left(2rg^3/\rho\right) \tag{8}$$

The necessary electrical power for the flight is calculated by the following type:

$$P_{Eflight} = P_{mech}/\eta_{ctrl}{}^{*}\eta_{mot}{}^{*}\eta_{grb}{}^{*}\eta_{plr} \tag{9}$$

To calculate the total electrical power used:

$$P_{Etot} = P_{Eflight} + 1^{*}\left(P_{av} + P_{pld}\right)/\eta\mathrm{bec} \tag{10}$$

Finally, the necessary energy production of solar cells:

$$A_{sc} = P_{Etot}{}^{*}\left(\pi/2^{*}\eta_{sc}{}^{*}\eta_{cbr}{}^{*}\eta_{mppt}{}^{*}\eta_{wthr}\right) * \left(1 + \left(T_{night}/T_{day}\right) * 1/\eta_{chrg}{}^{*}\eta_{dchrg}\right) * 1I_{max} \tag{11}$$

C_L	0.8	Airfoil lift coefficient
$C_{D_{afl}}$	0.013	Airfoil drag coefficient
$C_{D_{par}}$	0.006	Parasitic drag coefficient
I_{max}	950 W/m^2	Maximum irradiance
k_{bat}	726545.455 J/kg	Energy density of lithium-polymer batteries
k_{sc}	0.32 kg/m^2	Mass density of solar cells
k_{enc}	0.26 kg/m^2	Mass density of encapsulation
k_{mppt}	0.00042 kg/W	Mass to power ratio of MPPT
k_{prop}	0.008 kg/W	Mass to power ratio of propulsion group
k_{af}	0.044852191 kg/m^3	Structural mass constant
m_{av}	0.2 kg	Mass of avionics
η_{bec}	0.65	Efficiency of step-down converter
η_{sc}	0.169	Efficiency of sollar cells
η_{cbr}	0.9	Efficiency of the curved solar panels
η_{chrg}	0.95	Efficiency of of battery charge
η_{ctrl}	0.95	Efficiency of motor controller
η_{dchrg}	0.95	Efficiency of battery discharge
η_{grb}	0.97	Efficiency of gearbox
η_{mot}	0.85	Efficiency of motor
η_{mppt}	0.97	Efficiency of MPPT
η_{plr}	0.85	Efficiency of propeller
N_{wthr}	0.7	Irradiance margin factor
m_{pld}	0.5 kg	Payload mass
r	10.0	Aspect ratio
P_{av}	8.5 W	Power of avionics
x_1	3.1	Airframe mass wingspan coefficient
x_2	-0.25	Airframe mass aspect ratio coefficient
P_{pld}	0.5 W	Payload power consumption
p	0.8 kg/m^3	Air density (2500 m altitude)
T_{day}	34200 s	Day duration
T_{night}	52200 s	Night duration
b	3 m	Wingspan

Fig. 2 The input values used for the calculations

Figure 2 contains the input values used for the calculations. After using a computer program to calculate a fixed point for equations and initial inputs, the model generates the following values for the level properties [2]:

Unmanned aerial vehicles (UAVs) are gaining the attention of the research community and the telecommunications industry with the rapid development of the Internet of Things (IoTs). For data collection and dissemination, IoT-based technologies are provided [6, 7]. The estimated flight time depends on three parameters [4, 8–11]:

(a) The available battery capacity mounted on the UAV
(b) The discharge limit
(c) The average power of the amplifier

And consequently:

$$AT = C_B{}^* D_L / AA_D \tag{12}$$

The discharge limit is calculated at 80%, while the AAD is calculated using the following Eq. (13):

$$AA_D = c^* P_W / \Delta V \tag{13}$$

Accordingly, Eq. (14) is formulated as follows:

$$AT = C_B{}^* D_L / (AA_D + E_U) \tag{14}$$

Scenarios for the Development of Knowledge About Energy

The specific use factor depends on the user's location, density as well as the UAV coverage radius [12]. User density is usually divided into three categories:

(a) Sparsely populated areas
(b) Residential area (city)
(c) Densely populated areas [13]

It is observed that the existence of a larger number of cells (more UAVs) requires transmission power. When there are fewer cell numbers, a high transmission power (≥53.9 W) is required. This result is incorporated mathematically using the following equation [13–15]:

$$E_U = P_T / \Delta V \tag{15}$$

Scenario 1 (a): height increasing

To calculate UAV (AT) time without any clear usage, these steps are followed:

- Calculate the average draw Amp using Eq. (2)
- Calculate UAV time using Eq. (1)

Scenario 1 (b): No clear use (UAV is hovering)

UAV flight altitude not only affects base station coverage and user density but also results in higher power usage. Here, examining link quality can also be an important point focusing on the high mobility environment of a drone in the network [6–9, 16–20]. Increasing the altitude by 15 m leads to an additional power consumption of 0.5 A [21–23]. Equation (2) to cover the Amps for the height factor (AHF) is configured as in Eq. 16.

$$AA_D = C^* ((P_W / V) + A_{HF}) \tag{16}$$

The following are the steps for calculating UAV time (AT) without any specific use but considering the altitude factor.

2 Description of Hypothesis and Modeling Process

Scenario 1 (Table 1)

Scenario 2 (Table 2)

Table 1 Description of the forces applied

Kinetic energy	A key factor in the movement of this drone is the kinetic energy of the drone, as through it any programmed or unplanned movement of the system takes place. Resources are therefore created and flow from BSD to IoT stock through flow9.
Air resistance energy	Another factor is the energy produced by the air resistance. Resources therefore flow from BSD to the IoT stock through flow11. Then the stock of FLIGHT "breaks" into two pieces. In Scenario 1 = scenario1flight, the air resistance is applied.
Dynamic energy	The last factor is the dynamic energy presented in the system. Resources therefore flow from the BSD to the IoT stock through flow10. Then the stock of FLIGHT "breaks" into two pieces. In Scenario 2 = scenario2hover, the dynamic energy is applied.

Table 2 Description of the forces applied

Solar energy	A key factor in the movement of this drone is the energy received by the sun's rays, as it is very important in activating the movement of the system. The energy percentages of BSD go to E_kin since in flow8 the percentage is placed on BSD. It then flows to the IoT via flow1, which determines the amount of energy received.
Dynamic energy	An additional key factor in the movement of this drone is the energy produced by the activation of the propellers, as it activates the movement of the system. The initial factor energy percentages go to E_dyn in the same way as above through flow10, with the significant difference that it goes directly without the intervention of FLIGHT and IoT. Respectively, when the resources reach flow6 and by applying the given parameters, the magnitude of the reported energy applied to the system will emerge. In the same way these resources of the initial factor flow to FINALENERGY2.
Kinetic Energy	Another factor in the movement of this drone is the kinetic energy of the drone, i.e., "body" and battery as through it any programmed or unplanned movement of the system takes place. The energy percentages of the initial factor go to E_air in the same way as mentioned above through flow11, with the significant difference that it goes directly, without the intervention of FLIGHT and IoT. Respectively, when the resources reach flow8 and by applying the given parameters, the magnitude of the reported energy applied to the system will emerge. In the same way these resources of the initial factor flow to FINALENERGY1.

3 Hypothesis

H1: $E_{fin} < 0$/*The system has battery as a power source (no solar panels)*
The system is fully functional.

H2: $E_{fin} > 0$/*The solar panels are fully functional and there are no losses*
No losses affect the system's functionality.

H3: $E_{fin} > 0$/*Solar panels are 30% functional*
Two extra powers were added to the system to contribute to the drone's energy-saving

H4: $E_{fin} < 0$/*The system lacks the presence of solar panels and has 50% less losses*
The power from solar panels that boosted the system is no longer active

H5: $E_{fin} > 0$/*The drone can no longer move but only hover*
Although the system loses a boost power it also gains the thermal loss exclusion

4 Methodology

The aim of the system under consideration is to extend the flight time and consequently, the distance traveled through energy savings, which can be achieved by consuming energy from the solar panels attached to the drone. The pursuit is to have as few stops as possible to refuel or charge the battery, with the corresponding benefit of not interrupting the assigned mission. This practically means for the system, that after the initial energy that is provided and constitutes 100% of the resources, the resources that feed the stocks, must lead it to its maximum possible value, which implies the ever-increasing value of the final value. of energy, i.e., when the system tends to land. The purpose of the whole project is the non-interruption of the flight through the minimization of the system requirement for energy and at the same time the overcoming of the sum of the other opposite measures of forces that are applied. To achieve this, percentage changes had to be made. Time is referred to as time units (tu). Initially, 100% of the energy is placed in BSD in the context of faithful observance of the ECP (Energy Conservation Principle). This practically means that after the initial energy that exists in the system and is 100% of the resources, the resources that supply the energy stocks must lead it to its maximum possible value, which implies the ever-increasing value of the final energy amount. It is therefore understandable that while the drone speed should reasonably be constantly increasing in order not to interrupt the flight, it should also be such that the amount of energy (with a positive sign) feeds the flight to exceed the opposing forces.

5 Hypothesis Trials and Results

It should first be clarified that the Drone starts its movement with the aim of moving to an area within the city that does not have access to the network. It will perform a movement consisting of two (2) distinct stages. The first concerns the movement of the system in space until it reaches the required area and when it returns from it. The second concerns the drone hover.

In this case, several stocks will be analyzed and several factors that exist in the model will be influenced. In more detail, from BSD it will continue to FLIGHT, then to IoT where from there it will be divided into *scenario1flight* and *scenario2hover*. Finally, after all the factors and parameters are affected, the final energy is divided into FINALENERGY1 and FINALENERGY2, respectively.

Initially, 100% of the energy is placed in the BSD in the context of faithful observance of the ADE. (Energy Conservation Principle) that will affect the system during the take-off process at time $t = 0$, i.e., P_Etot (Total system power). During the movement of the system, other forces will be affected because of the production of energies E_kin (kinetic energy), E_dyn (dynamic energy), E_air (airresistance energy).

After researches, the following results are obtained regarding the behavior of the forces in the system. The green color in the following diagram indicates a positive to the system power and red indicates a negative ones (Table 3).

Kinetic Energy
In this case, 39% of BSD resources go to E_kin through flow12. E_kin is applied there, where through flow7 it ends up in FLIGHT.

Air Resistance Energy
The next 19% of the initial factor BSD goes to Eair in the same way as mentioned above. Respectively, when the resources reach flow11 and with the application of the given parameters the size of the mentioned energy that is applied in the system will emerge. Then the stock of FLIGHT "breaks" into two pieces. In Scenario 1 = scenario1flight the air resistance is applied. The resources of BSD end up into scenario1flight.

Dynamic Energy
The final 42% of the initial factor goes to E_{dyn} in the exact same way as above. Respectively, when the resources reach flow13 and by applying the given parameters the size of the reported energy applied to the system will emerge. In the same way these resources of BSD flow to FLIGHT. Then the stock of FLIGHT "breaks" into two pieces. In Scenario 2 = scenario2hover the dynamic energy is applied.

Table 3 Energy types and supply quota in modeling

Type of energy	Percentage (%)
Kinetic energy	39
Air resistance energy	(−) 19
Dynamic energy	(−) 42

6 Modeling and Simulation

The system from the beginning of its operation, at no time for 0–14tu (on the 14.2tu the energy is granted with a positive value), manages to get positive values. The consequence of this is the cancellation of the flight due to a lack of energy to support the flight (Fig. 3).

The above configuration of the model will result in the following results (Fig. 4):

These initial stock resources then flow into FLIGHT. When the resources reach FLIGHT, the condition exists and if satisfied, the resources flow to IOT to be divided further into FINALENERGY 1&2. It is observed that **FINALENERGY2** can't achieve a positive value at no time until 14.2tu and equals **22,4189*BSD**. Considering that the battery's drone lasts for approximately four-time units, the drone will never manage to complete its mission.

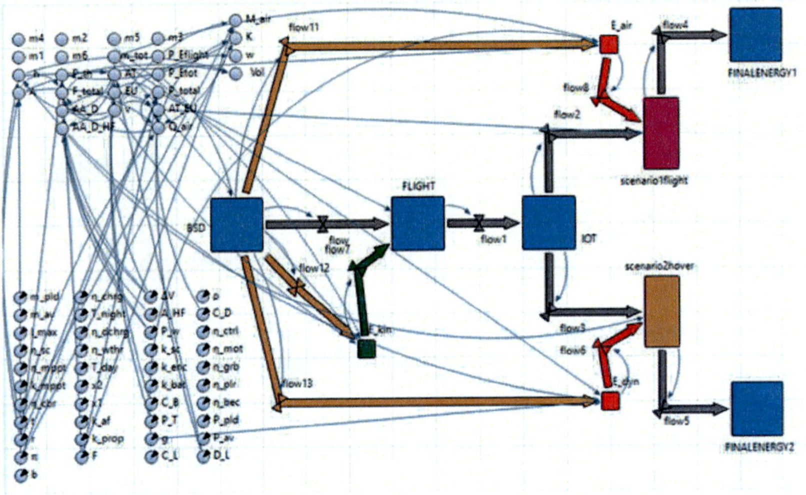

Fig. 3 Simulation model

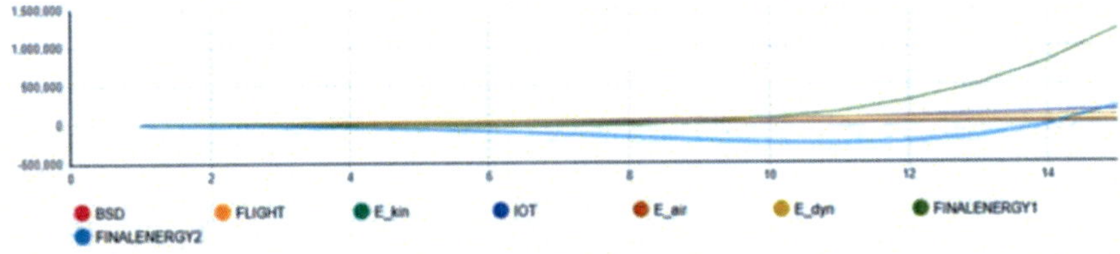

Fig. 4 Energy diagrams during the modeling phase

7 Optimization (Fig. 5)

The aim of the system is to save energy and at the same time extend the drone flight time, which can be achieved by absorbing energy from the solar cells. Practically for the system, it means that after the initial energy that exists and constitutes 100% of the resources, the ones that feedback the stocks must lead it to its maximum possible value, which implies the ever-increasing value of the final value of energy, i.e., when the system tends to land. The purpose of the whole project is the non-interruption of the flight through the minimization of the system requirement for energy and at the same time the overcoming of the sum of the other opposite measures of forces that are applied. With ECP in mind, it makes sense for the drone-driven force to tend to zero, as opposed to the principle of motion where the speed is constantly increasing for the system to achieve the purpose of the whole project and at the same time surpass the sum of the other countermeasures applied. The maximum possible profit is achieved if the percentage of resources flowing from the drone to the stocks is differentiated. Toward that direction, the energy percentages will vary. After many simulations, the result is as in Table 4:

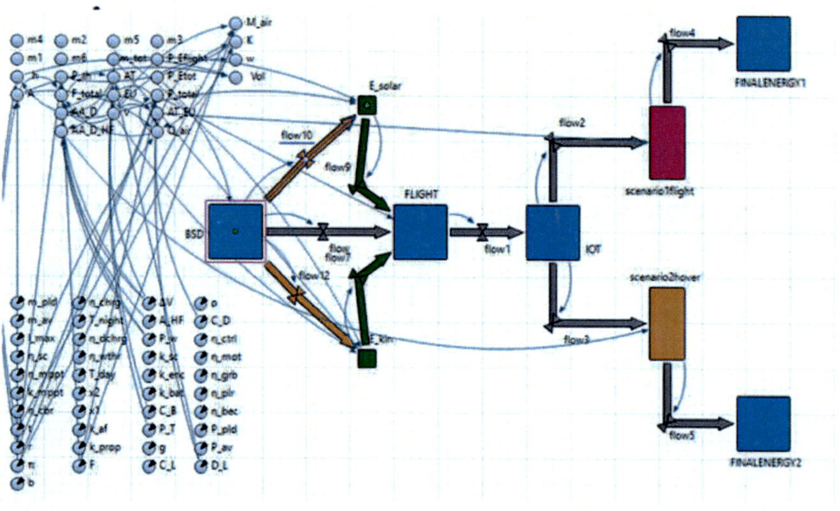

Fig. 5 Optimization model

Table 4 Energy types and supply quota in optimization

Type of energy	Percentage (%)
Kinetic energy	66.33
Air resistance energy	(−) 8.41
Dynamic energy	(−) 5.51
Solar energy	19.75

Since the system's final energy should be greater than the initial one in the short term the above percentages are placed on the converters bearing the name of each stream. More specifically:

(a) In solar energy where Esolar = m_tot*v*v*x/200, the following factors affect as follows (Table 5):

(b) The dynamic energy where E_dyn = m_tot*g*h, is influenced by the following factors (Table 6):

(c) The kinetic energy of the drone where E_kin = m_tot*v*v/2, is affected by the following factors (Table 7):

Table 5 Explanation of solar flows and factors

Name	Type	Explanation
m1	m_av + m_pld	Weight m1
m2	k_af*sqrt(r*sqrt(x1)*sqrt(x2))	Weight m2
m3	(k_sc + k_enc)*Asc	Weight m3
m4	k_mppt*I_max*η_sc*η_cbr*η_mppt*Asc	Weight m4
m5	T_night*P_Etot/η_dchrg*k_bat	Weight m5
m6	k_prop*P_Eflight	Weight m6
m_tot	m1 + m2 + m3 + m4 + m5 + m6	Weight m_tot
v	sqrt(F/2*Q_air*A)	Βάρος drone
flow10	BSD*x/100	Flow to E_solar
flow9	E_solar	Flow to FLIGHT
E_solar	m_tot*v*v*x/200	Solar energy

Table 6 Explanation of flows and factors of dynamic energy

Name	Type	Explanation
m1	m_av + m_pld	Weight m1
m2	k_af*sqrt(r*sqrt(x1)*sqrt(x2))	Weight m2
m3	(k_sc + k_enc)*Asc	Weight m3
m4	k_mppt*I_max*η_sc*η_cbr*η_mppt*Asc	Weight m4
m5	T_night*P_Etot/η_dchrg*k_bat	Weight m5
m6	k_prop*P_Eflight	Weight m6
m_tot	m1 + m2 + m3 + m4 + m5 + m6	Weight m_tot
g	9.81	Acceleration of gravity in m/s^2
flow6	−E_dyn*x/100	Flow to scenario2hover
flow13	BSD*x/100	Flow to E_dyn
h	v*t	The height at which the drone hovers
t	15	The time it takes to test the model
E_dyn	m_tot*g*h	Dynamic energy

Table 7 Explanation of kinetic energy flows and factors

Name	Type	Explanation
m1	m_av + m_pld	Weight m1
m2	k_af*sqrt(r*sqrt(x1)*sqrt(x2))	Weight m2
m3	(k_sc + k_enc)*Asc	Weight m3
m4	k_mppt*I_max*η_sc*η_cbr*η_mppt*Asc	Weight m4
m5	T_night*P_Etot/η_dchrg*k_bat	Weight m5
m6	k_prop*P_Eflight	Weight m6
m_tot	m1 + m2 + m3 + m4 + m5 + m6	Weight m_tot
V	sqrt(F/2*Q_air*A)	Drone speed
flow12	BSD*x/100	Flow to E_kin
flow7	E_kin	Flow to FLIGHT
E_kin	m_tot*v*v/2	Kinetic energy

Table 8 Explanation of flows and factors of air resistance energy

Name	Type	Explanation
m1	m_av + m_pld	Weight m1
m2	k_af*sqrt(r*sqrt(x1)*sqrt(x2))	Weight m2
m3	(k_sc + k_enc)*Asc	Weight m3
m4	k_mppt*I_max*η_sc*η_cbr*η_mppt*Asc	Weight m4
m5	T_night*P_Etot/η_dchrg*k_bat	Weight m5
m6	k_prop*P_Eflight	Weight m6
m_tot	m1 + m2 + m3 + m4 + m5 + m6	Weight m_tot
g	9.81	Acceleration of gravity in m/s^2
flow8	−E_air	Flow to scenario1flight
flow11	BSD*20/100	Flow to E_air
h	v*t	The height at which the drone hovers
t	15	The time it takes to test the model
E_air	m_tot*g*h*x/100	Air resistance energy

Table 9 Explanation of flows and factors of air resistance energy

Name	Type
BSD	P_Etot
Flow	BSD
FLIGHT	P_total + flow9 + flow7 + flow
flow1	FLIGHT
IOT	flow1

(d) In the air resistance energy where E_air = m_tot*g*h*20/100, the following factors affect as follows (Table 8):

Common elements of the two scenarios (Table 9):

Scenario 1 (Table 10)

Scenario 2 (Table 11)

Parameter tables (Tables 12 and 13):

It is observed that **BSD** at **t = 0tu** begins from the initial value of 100% but then at the output, and at the same time at **IOT**, it has a positive value as well. At t = 15tu the **FINALENERGY1** ≈ **6589*10^3*BSD** and

Table 10 Explanation of flows and factors of air resistance energy

Name	Type
flow2	IOT
scenario1flight	AT_E_U + flow2 + flow8
flow4	scenario1flight
FINALENERGY1	flow4

Table 11 Explanation of Flows and Factors of Air Resistance Energy

Name	Type
flow3	IOT
scenario2hover	AA_D_HF + flow3 + flow6
flow5	scenario2hover
FINALENERGY2	flow5

Table 12 Explanation of other variables

Name	Type
A	π*r*r
P_th	K*sqrt(F*F*F)/r
F_total	2*Q_air*A*v*v
AA_D	m_tot*P_w/ΔV
AA_D_HF	m_tot*((P_w/ΔV) + A_HF)
AT	C_B*D_L/AA_D
EU	P_T/ΔV
P_E_{flight}	$P_{mech}/\eta_{ctrl}*\eta_{mot}*\eta_{grb}*\eta_{plr}$
P_E_{tot}	$P\_E_{flight} + (P_{av} + P_{pld})/\eta_{bec}$
P_total	sqrt(F*F*F/2*Q_air*π*r*r)
AT_E_U	C_D*D_L/(AA_D + EU)
Q_air	M_air/A*v*t
M_air	F*t/2*v
K	sqrt(1/2*π*Q_air)
W	2*v
Vol	A*h
m_pld	0.5
m_av	0.2
I_max	950

Table 13 Explanation of other variables

Name	Type	Name	Type
η_sc	0.169	A_HF	15
η_mppt	0.97	P_w	170
k_mppt	0.00042	k_sc	0.32
η_cbr	0.9	k_enc	0.26
T	15	k_bat	726545.455
R	10	C_B	8.8
Π	3.14	P_T	6.088
B	3	C_L	0.8
η_chrg	0.95	P	0.8
T_night	52200	C_D	0.013
η_dchrg	0.95	η_ctrl	0.95
η_wthr	0.7	η_mot	0.85
T_day	34200	η_grb	0.97
x2	−0.25	η_plr	0.85
x1	3.1	η_bec	0.65
k_af	0.044852191	P_pld	0.5
k_prop	0.008	P_av	8.5
F	0.95	D_L	0.80
ΔV	36		

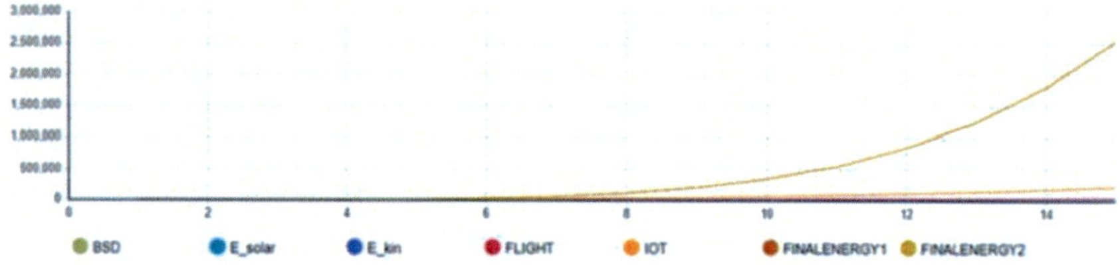

Fig. 6 Energy diagrams during the optimization phase

FINALENERGY2 ≈ 6608*10^3*BSD approximately. A result that, compared to modeling, shows a tremendous change (Fig. 6).

Therefore, the model after optimization is much more efficient and it is easily understood that the cost savings are easily calculated from the sum FINALENERGY1 + FINALENERGY2-BSD. More specifically, it is observed that the drone at t = 0tu starts from the initial value of 100% and then at the output, i.e., at IOT with geometric progression, the eventually produced energy increases. Therefore, the model after optimization is much more efficient and it is easy to understand that the value of savings is easily calculated from the sum of IOT – BSD, thus achieving optimization.

A. However, to achieve the examination of other scenarios and the maximum possible energy in the system, the viability of the system will be examined if it is considered that the solar panels retain only 30% functionality. Based on this, below the behavior of the system is listed (Fig. 7).

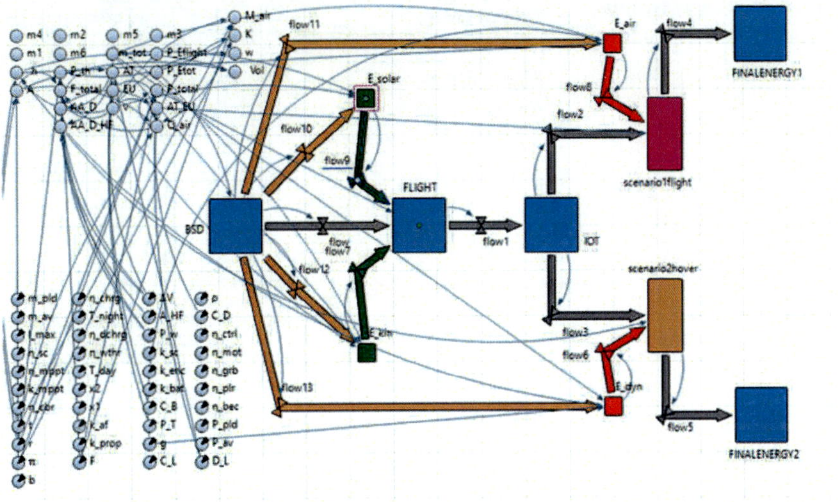

Fig. 7 Simulation model (30% functionality)

Table 14 Energy types in optimization (30% functionality)

Type of energy	Percentage (%)
Solar energy	19.47
Dynamic Energy	(−) 23.79
Kinetic energy drone	31.52
Air resistance energy	(−) 25.22

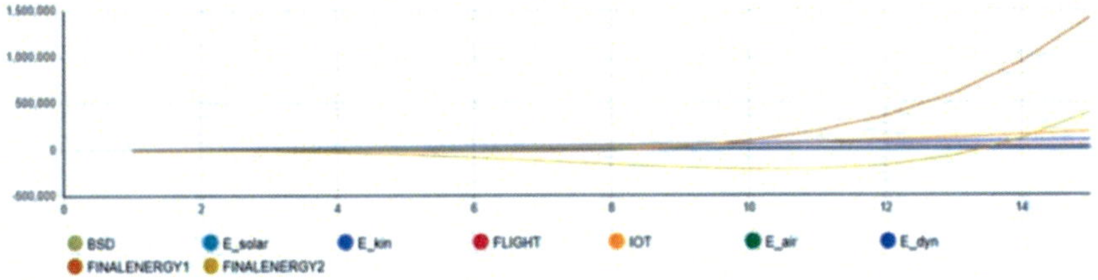

Fig. 8 Energy diagrams during optimization phase (30% functionality)

The energy percentages that are applied to the above are (Table 14):

The following diagram clarifies the landscape in the field of energy flow in the system (Fig. 8).

It is observed that at **t = 15tu** the **FINALENERGY1 ≈ 3661*10^3*BSD** and **FINALENERGY2 ≈ 1006*10^3*BSD** approximately, which means that energy saving plan has not succeeded since the FINALENERGY2 turns positive at $t = 14$tu.

B. An alternative examination of the system flow will bring different behaviors and at the same time different results in the model. Therefore, here it is assumed that the system receives no effect from the sun (no solar panels) and at the same time the other losses of the system are 50% smaller (Table 15).

Table 15 Energy types in optimization (no solar panels)

Type of energy	Percentage (%)
Dynamic energy	(−) 38.14
Kinetic energy drone	37.65
Air resistance energy	(−) 24.21

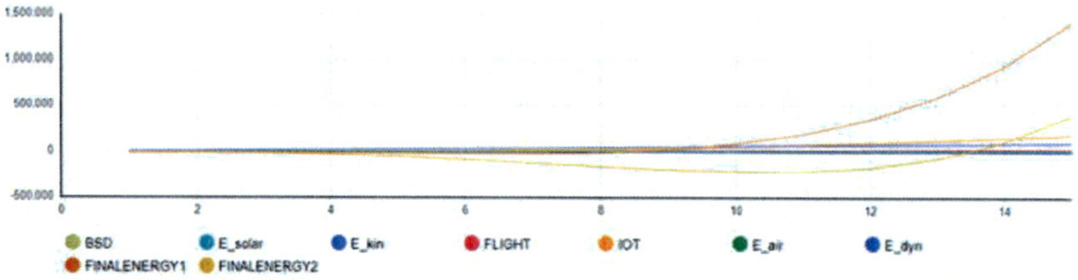

Fig. 9 Energy diagrams during optimization phase (no solar panels)

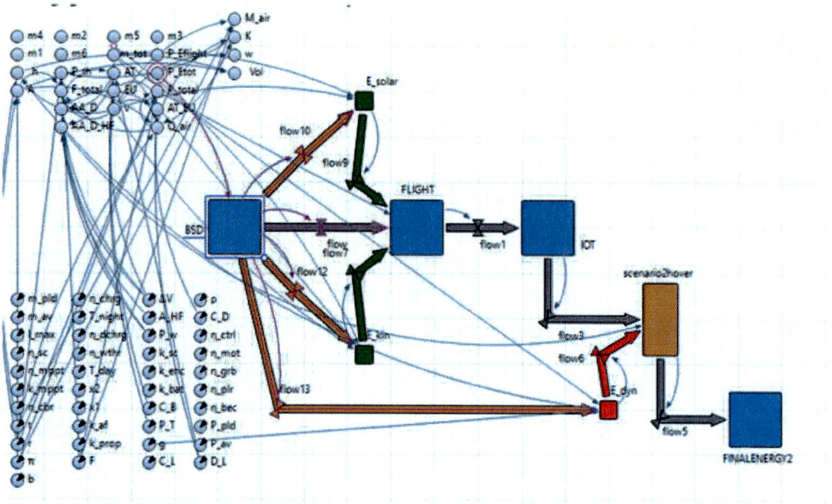

Fig. 10 Simulation model (hover)

The following diagram clarifies the landscape in the field of energy flow in the system (Fig. 9).

It is observed that at **t = 15tu** the **FINALENERGY1 ≈ 3170*10^3*BSD** and **FINALENERGY2 ≈ 5159*10^2*BSD** approximately. That is almost the same as the previous result of the optimization process. Therefore, the model is considered to fail in the energy-saving process.

C. A different approach to system behavior could be the assumption that somehow the drone reaches the desired area to assist in the development of fifth-generation mobile networks with IoT. So now the hovering process is examined only (Fig. 10):

Table 16 Energy types in optimization (hover)

Type of energy	Percentage (%)
Solar energy	27.98
Dynamic energy	(−) 34.19
Kinetic energy drone	37.83

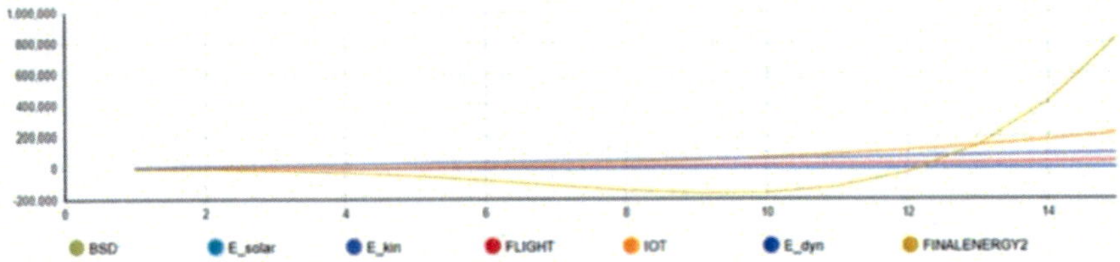

Fig. 11 Energy diagrams during the optimization phase (hover)

The energy percentages that are applied to the above are (Table 16):

The following diagram clarifies the landscape in the field of energy flow in the system (Fig. 11).

It is observed that at **t = 15tu** the $\mathbf{E_{fin}} \approx$ **2151*10^3*BSD** approximately. Nevertheless, the energy turns positive at t = 12tu. Therefore, the model is considered not to succeed in the energy-saving process.

8 Conclusion

This research is part of a research project examining and optimizing how to save energy on a drone by exploiting test flight research data. The whole procedure concerns the movement of the aircraft with the addition of solar cells or not and at the same time the presence or exclusion of the thermal losses that can be applied by nature to the drone. Further research and application of additional alternative energy sources are required as well as a way to reduce in significant level the thermal loss. In the present work, the behavior of the drone was examined if solar cells are attached to it, as well as the importance of the battery existence for the current drone and finally the ability of solar panels to increase the energy of the system to such an extent that it significantly extends flight time and the time that the drone can perform hovering over an area in order to assist in the development of fifth-generation mobile networks.

On the current model under consideration, the explanation given for the negative value of FINALENERGY1 is that the drone will not be able to carry out any mission assigned to it unless an additional energy source is attached to it. The explanation given for the negative value of FINALENERGY2 is that the drone will never manage to hover over the assigned area and so the whole project must be abandoned.

Of course, there are other items that need to be considered such as the ability of the drone to store energy in one or more internal or external batteries for the system to be able to use it to minimize the drone's vulnerabilities that may come up by using only one battery system.

Several managerial implications may be proposed, such as the adoption of drones in order to resolve routing issues in cities [24] and tourist locations. Following this rationale, researchers have highlighted the crucial role of these new technologies (drones) as a key pillar of sustainable development bridging human activity and well-being with connectivity and several factors, such as the natural environment and critical infrastructure [25–28]. Additionally, future research in digital advertising and CSR is crucial for the expansion and viability of UAV projects [29–34].

References

1. Sawalmeh, Othman, N.S., Shakhatreh, H.: Efficient deployment of multi-UAVs in massively crowded events. Sensors (Basel, Switzerland). **18**, 3640 (2018)
2. Maharbiz, M., Hargreaves, Z., Kim, B., Nguyen, Q., Albert, O.: Berkeley Solar Drone Light Basin. Laboratory University of California, Berkeley (2011)
3. Noth, A.'e.: Design of Solar Powered Airplanes for Continuous Flight. ETH, Zürich (2008)
4. Carl Engel and Adam Woodworth.: "Solar Powered UAV Project", (2006). http://www.rcgroups.com/forums/showthread.php?t=572000
5. Hall, D.W., et al.: A Preliminary Study of Solar Powered Aircraft and Associated Power Trains. Tech. rep. 3699 (1983)
6. Kumar, K., Kumar, S., Kaiwartya, O., Kashyap, P.K., Lloret, J., Song, H.: Drone assisted flying ad-hoc networks: mobility and service oriented modeling using neuro-fuzzy. Ad Hoc Netw. **106**, 102242 (2020)
7. Kumar, K., Kumar, S., Kaiwartya, O., Sikandar, A., Kharel, R., Mauri, J.L.: Internet of unmanned aerial vehicles: QoS provisioning in aerial ad-hoc networks. Sensors. **20**, 3160 (2020)
8. Mukherjee, A., Dey, N., Kausar, N., Ashour, A.S., Taiar, R., Hassanien, A.E.: A disaster management specific mobility model for flying ad-hoc network. In: Emergency and Disaster Management: Concepts, Methodologies, Tools, and Applications. IGI Global, Hershey (2019)
9. Perabathini, B., Tummuri, K., Agrawal, A., Varma, V.S.: Efficient 3D placement of UAVs with QoS assurance in ad hoc wireless networks. In: 2019 28th International Conference on Computer Communication and Networks (ICCCN). IEEE, Valencia (2019)
10. Murray, C.J.: The Quest for Perpetual Flight. Design News (2005)
11. Lubritto, C., Petraglia, A., Vetromile, C., Curcuruto, S., Logorelli, M., Marsico, G., D'Onofrio, A.: Energy and environmental aspects of mobile communication systems. Energy. **36**, 1109 (2011)
12. Fotouhi, A., Qiang, H., Ding, M., Hassan, M., Giordano, L.G., Garcia-Rodriguez, A., Yuan, J.: Survey on UAV cellular communications: practical aspects, standardization advancements, regulation, and security challenges. IEEE Commun. Surv. Tutorials. **21**(4), 3417–3442 (2019)
13. Newcome, L.R.: Unmanned Aviation. A Brief History of Unmanned Aerial Vehicles. American Institute of Aeronautics and Astronautics, Reston (2004)
14. Wang, L., Hu, B., Chen, S.: Energy efficient placement of a drone base station for minimum required transmit power. IEEE Wireless Commun. Lett. (2018). https://doi.org/10.1109/LWC.2018.2808957

15. Asif-Ur-Rahman, M., Afsana, F., Mahmud, M., Kaiser, M.S., Ahmed, M.R., Kaiwartya, O., James-Taylor, A.: Toward a heterogeneous mist, fog, and cloud-based framework for the internet of healthcare things. IEEE Internet Things J. **6**, 4049 (2019)
16. Ullah, F., Abdullah, A.H., Kaiwartya, O., Kumar, S., Arshad, M.M.: Medium access control (MAC) for wireless body area network (WBAN): superframe structure, multiple access technique, taxonomy, and challenges. HCIS. **7**, 1 (2017)
17. Qureshi, K.N., Abdullah, A.H., Kaiwartya, O., Ullah, F., Iqbal, S., Altameem, A.: Weighted link quality and forward progress coupled with modified RTS/CTS for beaconless packet forwarding protocol (B-PFP) in VANETs. Telecommun. Syst.: Modell. Anal. Des. Manag. **75**, 145 (2016)
18. Qureshi, K.N., Abdullah, A.H., Kaiwartya, O., Iqbal, S., Butt, R.A., Bashir, F.: A dynamic congestion control scheme for safety applications in vehicular ad hoc networks. Comput. Electr. Eng. **72**, 774 (2018)
19. Marchese, M., Moheddine, A., Patrone, F.: IoT and UAV integration in 5G hybrid terrestrial-satellite networks. Sensors (Basel, Switzerland). **19**, 3704 (2019)
20. Majeed, S., et al.: Unmanned aerial vehicles optimal airtime estimation for energy aware deployment in IoT-enabled fifth generation cellular networks. EURASIP J. Wirel. Commun. Netw. **2020**, 1 (2020)
21. Lu, J., Wan, S., Chen, X., Fan, P.: Energy-efficient 3D UAV-BS placement versus mobile users' density and circuit power. In: 2017 IEEE Globecom Workshops (GC Wkshps). IEEE, Singapore (2017)
22. Greg Goebel.: "Modern US Endurance UAVs", (2010) . http://www.vectorsite.net/twuav_13.html
23. Ahmed, M.N., Abdullah, A.H., Chizari, H., Kaiwartya, O.: F3TM: Flooding Factor based Trust Management Framework for secure data transmission in MANETs. J. King Saud Univ. Comput. Inf. Sci. **29**, 269 (2017)
24. Kakarountas, A., Spathoulas, G., Trivellas, P., Reklitis, P., Malindretos, G.: Survey on routing services for smart delivery in urban environments. In: SEEDA-CECNSM 2020 – 5th South-East Europe Design Automation, Computer Engineering, Computer Networks and Social Media Conference, p. 9221790. IEEE, Piscataway (2020)
25. Trivellas, P., Anastasopoulou, A., Malindretos, G., Reklitis, P., Sakas, D.P.: A place based analysis of stakeholders' advancing sustainability in remote and isolated communities: the case of North Aegean Islands. IOP Conf. Ser.: Earth Environ. Sci. **899**(1), 012044 (2021)
26. Trivellas, P., Kakkos, N., Vasiliadis, L., Belias, D.: Sustainability, social marketing and host attitudes about tourism in the city of Larissa. In: Springer Proceedings in Business and Economics, pp. 89–95. Springer, Cham (2017)
27. Trivellas, P., Kakkos, N., Vasiliadis, L., Belias, D.: Residents' perceptions toward cultural, social and economic benefits and costs of tourism industry: an empirical survey. In: Springer Proceedings in Business and Economics, pp. 171–181. Springer, Cham (2016)
28. Folinas, D., Fotiadis, T.: Marketing and Supply Chain Management: A Systemic Approach, 1st edn. Routledge, Abingdon (2017)
29. Sakas, D.P., Reklitis, D.P., Trivellas, P., Vassilakis, C., Terzi, M.C.: The effects of logistics websites' technical factors on the optimization of digital marketing strategies and corporate brand name. Processes. **10**, 892 (2022). https://doi.org/10.3390/pr10050892
30. Reklitis, P., Trivellas, P., Mantzaris, I., Mantzari, E., Reklitis, D.: Employee perceptions of corporate social responsibility activities and work-related attitudes: the case of a Greek management services organization. In: Gal, G., Akisik, O., Wooldridge, W. (eds.) Sustainability and Social Responsibility: Regulation and Re-porting, pp. 225–240. Springer Singapore, Singapore (2018). https://doi.org/10.1007/978-981-10-4502-8_10
31. Sakas, D.P., Kamperos, I.D.G., Reklitis, D.P., Giannakopoulos, N.T., Nasiopoulos, D.K., Terzi, M.C., Kanellos, N.: The effectiveness of centralized payment network advertisements on digital branding during the COVID-19 crisis. Sustainability. **14**, 3616 (2022). https://doi.org/10.3390/su14063616

32. Sakas, D.P., Giannakopoulos, N.T., Reklitis, D.P., Dasaklis, T.K.: The effects of cryptocurrency trading websites on airlines' advertisement campaigns. J. Theor. Appl. Electron. Commer. Res. **16**, 3099–3119 (2021). https://doi.org/10.3390/jtaer16070169
33. Sakas, D.P., Reklitis, D.P.: Predictive model for estimating the impact of technical issues on consumers interaction in agri-logistics websites. In: Information and Communication Technologies for Agriculture—Theme IV: Actions, pp. 269–283. Springer International Publishing, Cham (2021). https://doi.org/10.1007/978-3-030-84156-0_14
34. Sakas, D.P., Reklitis, D.P.: The impact of organic traffic of crowdsourcing platforms on airlines' website traffic and user engagement. Sustainability. **13**, 8850 (2021). https://doi.org/10.3390/su13168850

Modeling and Simulation of Examining the Performance of a Drone for Long-Distance Endurance

Dimitrios M. Mastrakoulis, Dimitrios K. Nasiopoulos, Dimitrios A. Arvanitidis, Elli Ch. Zara, and Alexandros Plagiannakos

1 Introduction

The aim of the system under consideration is to extend the flight time and, consequently, the distance traveled through energy savings, which can be achieved by consuming energy from the solar panels attached to the drone.

This research corresponds to the pursuit of energy savings of an autonomous drone, to which extra batteries are attached to achieve enough energy for its smooth movement [1]. Then, if this stock is analyzed and passed through the factors present in the model, it is returned either in the form of saving and increasing energy with the consequence of not requiring recharging, or in the form of high energy consumption, due to the removal of additional batteries that provide power other than the main battery. There are many other aircraft parameters that determine the endurance of any aircraft, however, for simplicity, the endurance calculation will be estimated as shown in the equation below [2]:

D. M. Mastrakoulis
Department of Informatics and Telecommunications, University of Thessaly, Lamia, Greece
e-mail: dmastrakoulis@aua.gr

D. K. Nasiopoulos
Department of Agribusiness and Supply Chain Management, BICTEVAC LABORATORY Business Information and Communication Technologies in Value Chains Laboratory, School of Applied Economics and Social Sciences, Agricultural University of Athens, Athens, Greece

D. A. Arvanitidis (✉) · E. C. Zara · A. Plagiannakos
Department of Informatics and Telecommunications, University of Peloponnese, Tripolis, Greece
e-mail: darvanitidis@aua.gr

D. P. Sakas et al. (eds.), *Computational and Strategic Business Modelling*, Springer Proceedings in Business and Economics,
https://doi.org/10.1007/978-3-031-41371-1_19

$$\text{Endurance} = \text{Battery Capacity}/\text{Current} \quad (1)$$

The durability of the aircraft will depend to a large extent on:

- Aircraft size
- Weight
- Lifting weight

A large fixed-wing aircraft with a large wingspan would be much more robust than a small quadrotor. Durability also plays an important role in increasing the range of the aircraft.

How far an unmanned aerial vehicle can travel is called range. The range of the aircraft depends on [1, 3–9]:

- The amount of electricity produced by the aircraft
- Endurance
- Flight speed
- Aerodynamic performance

Using mathematical relations, the range of a drone can be easily determined. The range calculation for both fixed wing drones and quadrotors can be calculated using the following equation.

The kV value is the number of revolutions per minute that the motor will rotate when 1 V is applied to the motor. The "pitch" value is the pitch (in inches) of the drone propeller. Durability is the time in hours that an aircraft can stay in the air.

$$\text{Range} = \text{KV}^{*}\text{V}^{*}60^{*}\text{Pitch}^{*}\text{Endurance}/12^{*}5260 \quad (2)$$

This equation will allow a general estimate of the total drone range. The following parameters are required to calculate the range accurately:

- Length of the wing
- Weight
- The lift coefficient of the spoiler used in the aircraft

Any entity concerned with aircraft safety should be particularly concerned about the endurance and range capabilities of different types of drone categories.

For further explanation, an example of using a large aircraft with a fixed wing will be presented. With this example, the demonstration will include a set that deals with drone safety, the range of a possible attack, and the speed at which the attack can occur.

To further adjust the scenario, a few parameters are required for the large, fixed wing, so that the strength and range can be simulated. Below are the parameters of "drone for example" [1, 9–15].

- Battery capacity (Ah) = 20 Ah
- Current (amps) during the flight = 10 A
- Voltage = 14.8 V

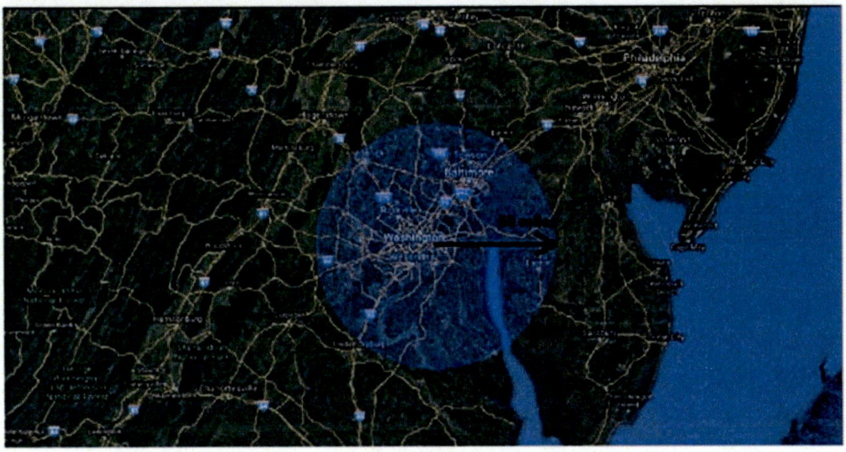

Fig. 1 Drone approach area of interest in Washington, DC

- kV = 1000
- Pitch = 7″

Using these parameters, strength, range, and speed can be determined. These values were calculated using a more accurate method of determining speed, strength, and range.

- Endurance = 2 hours
- Range = 100 miles
- Speed = 50 mph

Somehow, one entity that could deal with drone safety is a soccer field within a big city. For this example, the FedEx field of action in Washington, DC, is selected. However, this example can be applied to any event or location (Fig. 1).

As can be seen from the picture, this drone may have come from an extremely large area near the point where the event takes place. For the drone to approach the area of interest from the outer ridge of the aircraft range and retreat to its original position will take only 2 hours.

This example illustrates why a system designed to detect potential threats is the key to any entity focusing on protecting itself from the effects of hostile drones.

2 Description of Hypothesis and Modeling Process (Table 1)

Table 1 Description of the forces applied

Analysis of forces at the beginning of motion:	
Battery energy	A key factor in the movement of this drone is the energy stored in the battery that is attached to this system. Resources therefore flow from Initialenergy to Batteryenergy stock through battflow. Then the parameters that have been set for the harmonious movement and operation of the drone are applied.
Kinetic energy	An additional key factor in the movement of this drone is the kinetic energy of the drone, i.e., "body" and battery as through it any programmed or unplanned movement of the system takes place. Resources are therefore created and flow from Initialenergy to Kineticenergy stock through kinflow. Then the parameters that have been set for the harmonious movement and operation of the drone are applied.
Wing battery	A key factor in the movement of this drone is the energy produced by the movement of the wings, as it activates the movement of the system. Resources therefore flow from Initialenergy to Propenergy stock through propflow. Then the parameters that have been set for the harmonious movement and operation of the drone are applied.
Analysis of the added forces at the end of the motion:	
Air resistance energy	At the end of the system movement, a force is applied which in the existing model will be interpreted as the denaturation into an energy that negatively affects the system. This is done by flowing resources from Airresistance stock through airflow to Finalenergy. Then the parameters that have been set are applied and play an important role in the movement of the system and its operation.
Back up wing energy	Another key factor in the movement of this drone is the energy produced by the movement of the wings, as it activates the movement of the system. Resources therefore flow from Initialenergy to Propenergy stock through propflow. Then the parameters that have been set for the harmonious movement and operation of the drone are applied.

3 Hypothesis

H1: *Finalenergy* > 0, *maxrange* < 0*/The system has kinetic energy*

Three powers affect the drone at the beginning of the process and one more (negative) later

H2: *Finalenergy* > 0, *maxrange* > 0*/Battery that carries the same amount of energy as the drone's wings both energies are added to the system*

One extra power added to the system to contribute to the drone's energy saving.

H3: *Finalenergy* > 0, *maxrange* > 0*/The system has no resistance from the environment (air)*

The environmental losses don't affect the system so more energy is saved.

H4: *Finalenergy* > 0, *maxrange* > 0*/Both batteries collecting wing energy are damaged*

The energy collected from the wing batteries that boosted the system is now excluded.

H5: *Finalenergy < 0, maxrange < 0/Main battery is excluded from the system and environment performs full resist to the system*

The system doesn't manage to keep working with the use of both wing batteries.

4 Methodology

The aim of the system under consideration is to extend the flight time and consequently, the distance traveled through energy savings, which can be achieved by consuming energy from the solar panels attached to the drone. The pursuit is to have as few stops as possible to refuel or charge the battery, with the corresponding benefit of not interrupting the assigned mission. This practically means for the system, that after the initial energy that is provided and constitutes 100% of the resources, the resources that feed the stocks, must lead it to its maximum possible value, which implies the ever-increasing value of the final value of energy, i.e., when the system tends to land. The purpose of the whole project is the non-interruption of the flight through the minimization of the system requirement for energy and at the same time the overcoming of the sum of the other opposite measures of forces that are applied. To achieve this goal, percentage changes had to be made. Time is referred to as time units (tu).

Initially, 100% of the energy is placed in Initialenergy in the context of faithful observance of the ECP (Energy Conservation Principle). This practically means that after the initial energy that exists in the system and is 100% of the resources, the resources that supply the energy stocks must lead it to its maximum possible value, which implies the ever-increasing value of the final energy amount. It is therefore understandable that while the drone speed should reasonably be constantly increasing in order not to interrupt the flight, it should also be such that the amount of energy (with a positive sign) feeds the flight to exceed the opposing forces.

The first element being used is Initialenergy. Initially, this chapter seeks to examine and analyze how it will make possible the autonomy of the drone and consequently increase the flight time and the total distance it will travel. Then, if this stock is analyzed and goes through the factors that exist in the model, it will end up in Finalenergy, which is, as will be seen below, a key "ingredient" for calculating the maxrange. It is pointed out that the examined operating time of the system is from $t = 0$ to $t = 15$.

Initially, 100% of the energy that will affect the system during the take-off process at time $t = 0$ is placed in Initialenergy, and more specifically, the battery energy (Batteryenergy). During the movement of the system, other forces will act because of the production of energies Kineticenergy (kinetic energy), Wingenergy (energy produced by the propellers), Batteryenergy (battery energy) in the intermediate motion of the BackupWingenergy (energy produced by the backup propellers), and at the end of the motion the Airresistance (energy produced by the air resistance).

5 Hypothesis Trials and Results

Initially, 100% of the drone's resources must be shared among the three factors which are Batteryenergy, Kineticenergy, and Wingenergy as shown in Fig. 2 below. Initially, the distribution of energy is chosen to be as shown in Table 2. This is achieved by using a converter on the respective flow from the Initialenergy to each factor. Then and around the middle of the system path all the actions are gathered in Systemenergy. Finally, the effect of Airresistance is applied as mentioned above.

After researches, the following results are obtained regarding the behavior of the forces in the system. The green color in the following diagram indicates a positive to the system power and red indicates a negative ones.

Battery Power
In this case, 30% of the initial factor goes to Batteryenergy. When the resources reach sum1 and by applying the given parameters the size of the reported energy applied to the system will be obtained. In the same way these resources of the initial factor flow to Systemenergy.

Kinetic Energy
The next 8% of the initial factor goes to Kineticenergy in the same way as above. Respectively, when the resources reach sum2 and with the application of the given parameters, the magnitude of the mentioned energy that is applied in the system will emerge. In the same way these resources of the initial factor flow to Sumenergy.

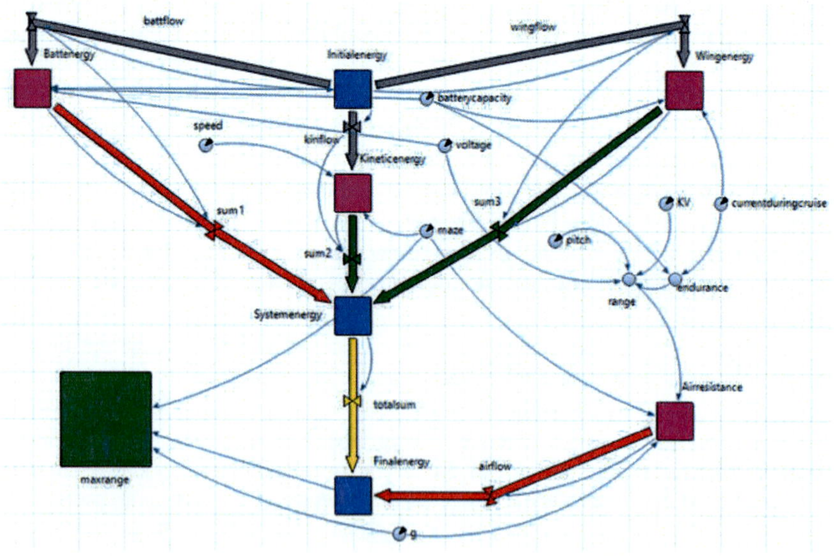

Fig. 2 Simulation model

Table 2 Energy types and supply quota in modeling

Types of energy	Percentage (%)
Kinetic energy	8
Battery energy	(–) 30
Wing energy	62

Wing Energy

The next 62% of Initialenergy's resources go to Wingenergy after the percentage has been placed on Initialenergy in the propflow stream. Then through the flow sum3 that determines the size of the energy received, it ends up in Sumenergy.

As it will be shown below, the model does not become viable, which is verified in the next stage of Finalenergy. The total energy produced in the event of a negative value or less than the initial system energy (TRPA), reasonably concludes that the energy is not sufficient to continue the drone flight.

6 Modelling and Simulation

The system from the beginning of its operation and although the power that feeds the movement of the drone is at 70% of the initial energy, it never manages to get positive for 0–15tu. Nevertheless, the maxrange (the parameter that defines the distance it has been traveled) never until 15tu gets over zero (>0). The consequence of this is the cancellation of the flight due to a lack of energy to support the flight.

The above configuration of the model will result to the following results (Fig. 3):

These initial stock resources flow into Systemenergy. When the resources reach Systemenergy the condition exists and if satisfied, the resources flow to Finalenergy. It is observed that although **Finalenergy** takes a positive value and equals to **1643*10^8*Initialenergy** at $t = 14.45$tu the maxrange never takes a positive value and equals $-5{,}12*10^{19}$.

Consequently, the drone will never manage to take off. The simulation ran up to $t = 15$tu and at no point before 14.45tu is detected (as shown in the diagram), where the Finalenergy curve enters the positive value axis. It is considered that the battery lasts for 12tu. After that point, the drone won't be able to perform a flight.

7 Optimization (Fig. 4)

The aim of the system is to save energy and at the same time extend the drone flight time, which can be achieved by absorbing energy from the wing batteries. Practically for the system, it means that after the initial energy that exists and constitutes 100% of the resources, the ones that feedback the stocks must lead it to its maximum possible value, which implies the ever-increasing value of the final value of energy,

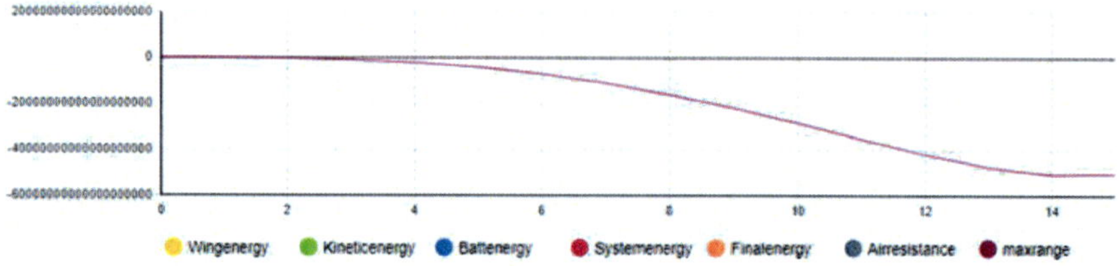

Fig. 3 Energy diagrams during modeling phase

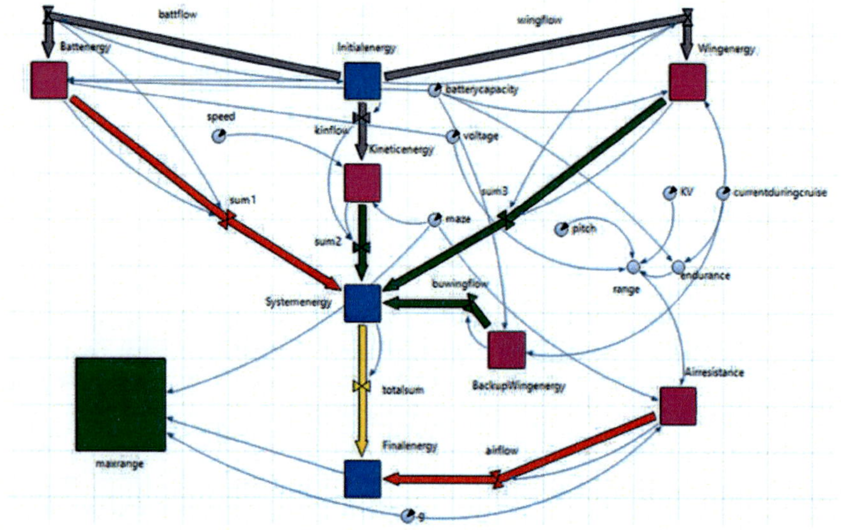

Fig. 4 Optimization model

i.e., when the system tends to land. The purpose of the whole project is the non-interruption of the flight through the minimization of the system requirement for energy and at the same time the overcoming of the sum of the other opposite measures of forces that are applied. With ECP in mind, it makes sense for the drone-driven force to tend to zero, as opposed to the principle of motion where the speed is constantly increasing for the system to achieve the purpose of the whole project and at the same time surpass the sum of the other countermeasures applied. The maximum possible profit is achieved if the percentage of resources flowing from the drone to the stocks is differentiated.

Toward that direction, the energy percentages will vary. After many simulations, the result is as the following table (Table 3):

After a study, the maximum possible profit is achieved if the percentage of resources flowing from the drone to the stocks is differentiated. Everyone will receive different percentages of the energy initially given to the system under

Table 3 Energy types and supply quota in optimization

Types of energy	Percentage (%)
Wing energy	30.67
Battery energy	(–) 10.28
Kinetic energy	59.05

Table 4 Explanation of battery energy Flows and factors

Name	Type – price	Explanation
voltage	14.8	Voltage
batterycapacity	20	Battery capacity
battflow	Initialenergy*10/100	Battery power flow – percentage of the original
sum1	–Battenergy*battflow	Flow to Sumenergy
Systemenergy	buwingflow + sum2 + sum1 + sum3	The sum of all energies

Table 5 Explanation of drone kinetic flows and factors

Name	Type – price	Explanation
Speed	50*1609.34	50 miles per hour (1609.34 converts it to km/h)
maze	3.5	Drone weight
Kineticenergy	0.5*maze*speed*speed	The kinetic energy
kinflow	Initialenergy*30/100	The flow of kinetic energy – percentage on the original
sum2	Kineticenergy* kinflow	The flow to Sumenergy
Systemenergy	buwingflow + sum2 + sum1 + sum3	The sum of all energies

consideration. These percentages are placed on the converters that bear the name of each stream. More specifically:

(a) The following factors affect battery power where Battenergy = batterycapacity*voltage/2 (Table 4):

(b) The kinetic energy of the drone where Kineticenergy = 0.5*maze*speed*speed, is influenced by the following factors (Table 5):

(c) The wing factors where Wingenergy = batterycapacity*currentduringcruise, are affected by the following factors (Table 6):

(d) In the backup wing energy where BackupWingenergy = batterycapacity * currentduringcruise, the following factors affect as follows (Table 7):

(e) The energy produced by the air resistance where Airresistance = maze*g*range, is influenced by the following factors as follows (Table 8):

Table 6 Explanation of flows and factors of wing energy

Name	Type – price	Explanation
batterycapacity	20	Battery capacity
Currentduringcruise	10	The value of the current during the movement of the system
Wingenergy	batterycapacity*currentduringcruise	Wing energy
sum3	Wingenergy*wingflow	The flow to Sumenergy
propflow	Initialenergy*60/100	Wing energy flow – percentage on the original
Systemenergy	buwingflow + sum2 + sum1 + sum3	The sum of all energies

Table 7 Explanation of flows and factors of wing energy

Name	Type – price	Explanation
batterycapacity	20	Battery capacity
Currentduringcruise	10	The value of the current during the movement of the system
BackupWingenergy	batterycapacity*currentduringcruise	Spare wing energy
buwingflow	BackupWingenergy	Wing energy flow – percentage on the original
Systemenergy	buwingflow + sum2 + sum1 + sum3	The sum of all energies

Table 8 Explanation of flows and factors of air resistance energy

Name	Type – price	Explanation
range	KV*voltage*60*pitch*endurance/12*5260	The range of strength
Pitch	7	The extent of the wings
KV	1000	Constantly equal to
endurance	batterycapacity/currentduringcruise	System battery life
voltage	14.8	Voltage
currentduringcruise	10	The value of the current during the movement of the system
g	9.81*3600	Acceleration to speed in km/h^2
airflow	–Airresistance*3/100	Air resistance flow

From the previous diagrams, the differences in the flow of time are obvious.

It is observed that Initialenergy at **t = 0tu** begins from the initial value of 100% but then at the output, and at the same time at Finalenergy, it has a positive value as well. At t = 15tu, the **Finalenergy≈9922*10^{11}* Initialenergy,** approximately, a result that, compared to modeling, shows a tremendous change (Fig. 5).

Therefore, the model after optimization is much more efficient and it is easily understood that the cost savings are easily calculated from the sum Finalenergy – Initialenergy. More specifically, it is observed that the drone at t = 0tu starts from the

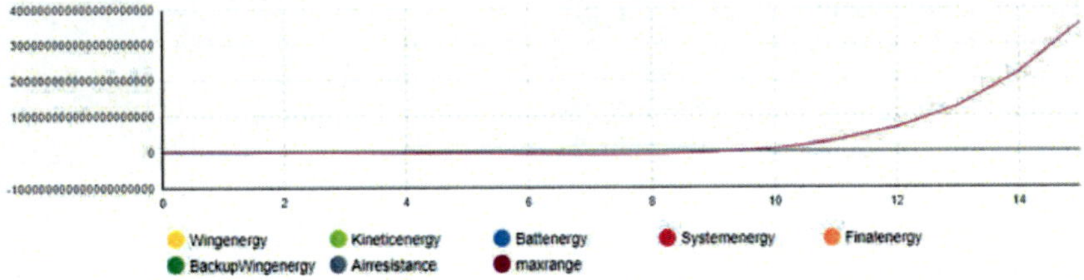

Fig. 5 Energy diagrams during the optimization phase

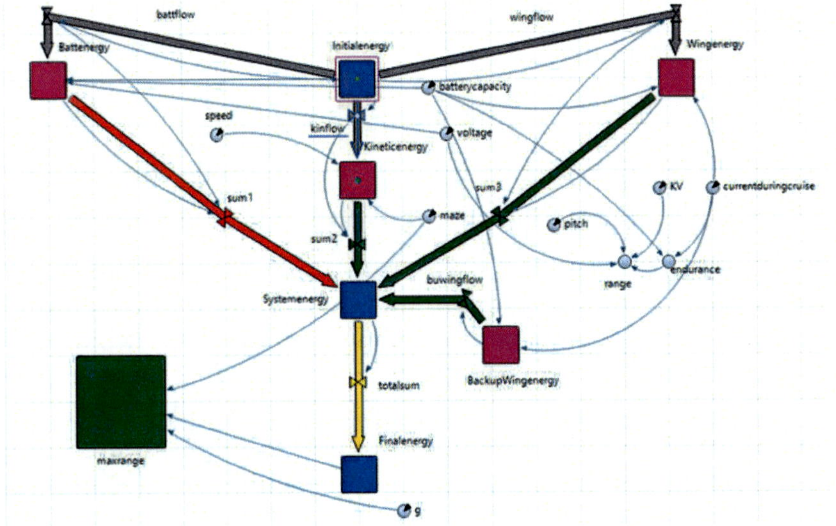

Fig. 6 Simulation model (no airresistance)

initial value of 100% and then at the output, i.e., at Finalenergy with geometric progression, the eventually produced energy increases. At $t = 9.35$tu the model is entering a maximum optimization trajectory as Finalenergy > Initialenergy and maxrange > 0. Finally, at $t = 15$tu the Finalenergy manages to be exponentially greater than the initial energy. Therefore, the model after optimization is much more efficient and it is easy to understand that the value of savings is easily calculated from the sum of Finalenergy – Initialenergy, thus achieving optimization.

A. However, to achieve the examination of other scenarios and the maximum possible energy in the system, the viability of the system will be examined if it is considered that there is no resistance effect from the environment to the system. Based on this, below the behavior of the system is listed (Fig. 6).

Table 9 Energy types in optimization (no airresistance)

Types of energy	Percentage (%)
Wing energy	23.49
Battery energy	(−) 37.82
Kinetic energy	38.69

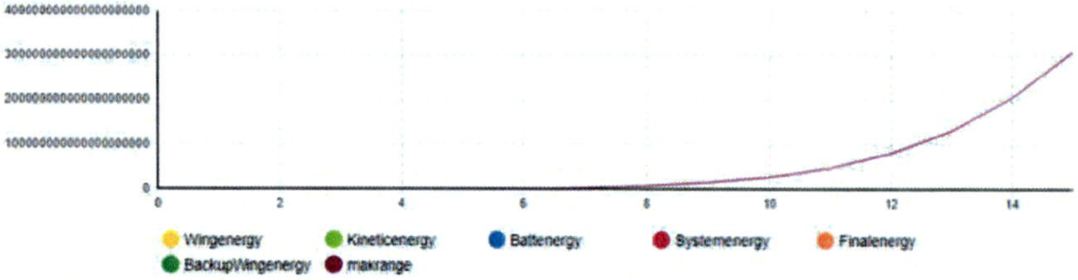

Fig. 7 Energy diagrams during optimization phase (no airresistance)

The energy percentages that are applied to the above are (Table 9):

The following diagram clarifies the landscape in the field of energy flow in the system (Fig. 7).

It is observed that at **t** = **15tu** the **Finalenergy≈7393*10^11^*Initialenergy** approximately and maxrange gets above zero, which means that the energy saving plan has succeeded. Obviously, the model after optimization is very efficient since the drone manages to perform correctly.

B. An alternative examination of the system flow will bring different behavior and at the same time different results in the model. Therefore, here it is assumed that the system receives no effect from the energy stored in the system from the wings (both main and backup) (Fig. 8 and Table 10).

The following diagram clarifies the landscape in the field of energy flow in the system (Fig. 9).

It is observed that at **t** = **11.63tu** both the **Finalenergy** and **maxrange** are **>0** and that at **t** = **15tu** the **Finalenergy** ≈ **4,5*10^11^*Initialenergy** approximately. That is almost twice the previous result of the optimization process. Therefore, the model is considered to succeed in the energy-saving process.

C. A different approach to system behavior could be:

(a) Removal of the system's main battery
(b) Airresistance to maximum and, therefore, the system will develop as follows (Fig. 10):

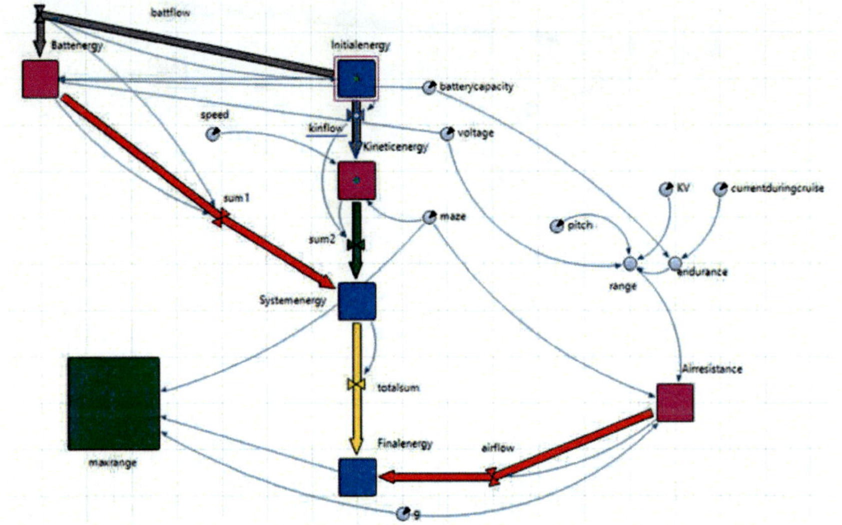

Fig. 8 Simulation model (no wingenergy)

Table 10 Energy types in optimization (no wingenergy)

Types of Energy	Percentage (%)
Battery energy	(–) 69.31
Kinetic energy	30.69

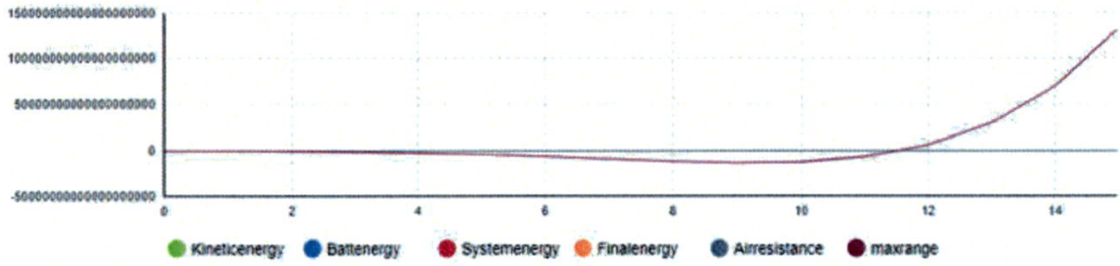

Fig. 9 Energy diagrams during optimization phase (no wingenergy)

The energy percentages that are applied to the above are (Table 11):

The following diagram clarifies the landscape in the field of energy flow in the system (Fig. 11).

It is observed that at **t = 15tu**, the **Finalenergy** ≈ **2638*10**^{12}***Initialenergy,** approximately, and **maxrange < 0**. Therefore, the model is considered to fail in the energy-saving process.

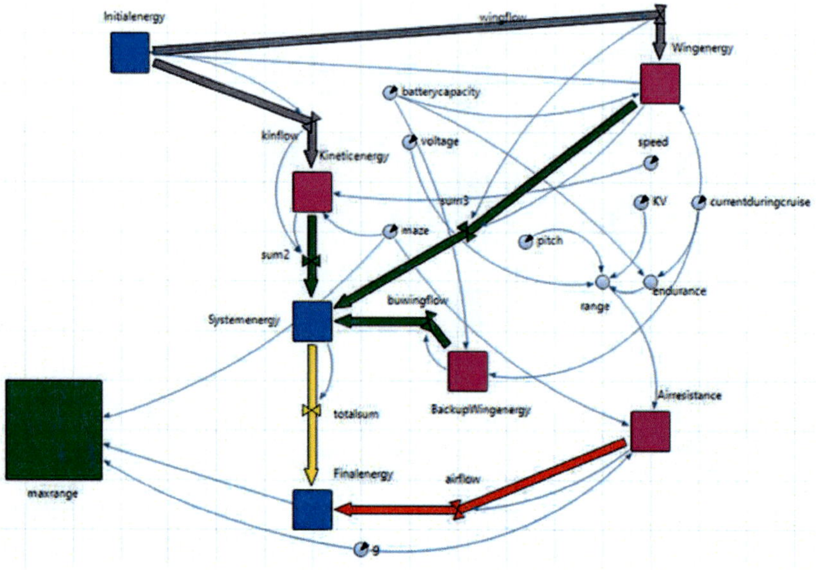

Fig. 10 Simulation model (no main battery and airresistance maximum)

Table 11 Energy types in optimization (no main battery and airresistance maximum)

Types of Energy	Percentage (%)
Wing energy	61.87
Kinetic energy	38.13

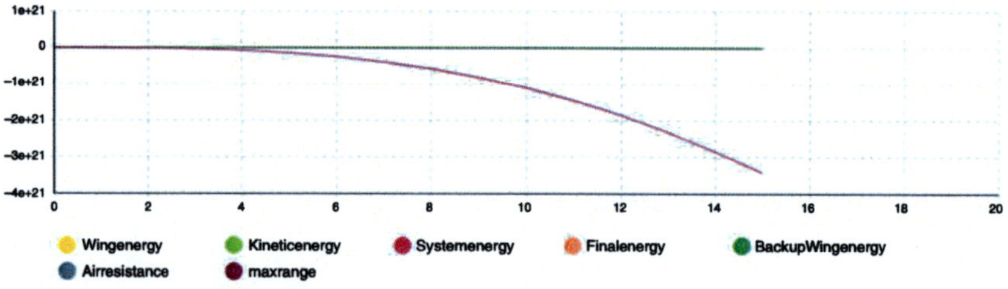

Fig. 11 Energy diagrams during optimization phase (no main battery and airresistance maximum)

8 Conclusion

This research is part of a research project examining and optimizing how to save energy on a drone by exploiting test flight research data. The whole procedure concerns the movement of the aircraft with the addition of two batteries supplied from the wings or not and at the same time the presence or exclusion of the environmental effect that can be applied by nature to the drone. Further research and application of additional alternative energy sources is required as well as a way to reduce in significant level the environmental effect. In the present work, the behavior of the drone was examined if wing batteries are attached to it, as well as the importance of the battery existence for the current drone and finally the ability of the environmental effect to decrease the energy of the system to such an extent that it significantly doesn't allow the continuity of the flight. Therefore, it also examines the maximum range that the drone will manage to fly if the system's final energy is above zero.

On the current model under consideration, the explanation given for the negative value of Finalenergy is that the drone will not be able to carry out any mission assigned to it unless an additional energy source is attached to the drone or there are no environmental effects. In this drone, the energy from the wings is not a determinant factor to increase the flight time of the drone. Considering the dependent variables of flight time, it is understood that by decreasing the maze or the dimensions of the drone to succeed in less air resistance, the desired result can be achieved.

Referring to managerial implications, the usage of drones for the resolution of supply chain problems and complexities such as routing in cities [16] and tourist destinations has been suggested.

Considering policy implications involving all actors of the supply chain ecosystem, drones may offer an alternative technological solution for establishing connectivity and remoteness in isolated communities or tourist destinations, during pandemic and economic crises, assuring human activity and residents' well-being as well as connectivity and several other factors, such as natural environment preservation and cohesiveness [17–19]. Finally, future research in corporate social responsibility should be considered with the assistance of green digital marketing in order to ensure the viability of UAV companies [20–25].

References

1. Jodeh, N., et al.: Development of Small Unmanned Aerial Vehicle Research Platform: Modeling and Simulating with Flight Test Validation. American Institute of Aeronautics and Astronautics (2012)
2. Selig, M.S.: Real-Time Flight Simulation of Highly Maneuverable Unmanned Aerial Vehicles. University of Illinois at Urbana–Champaign (2014)
3. Liu, D.-N., et al.: Flight Modeling and Simulation for Dynamic Soaring with Small Unmanned Air Vehicles. College of Aerospace Science and Engineering, National University of Defense Technology (2017)
4. Gundlach, J.: Designing Unmanned Aircraft Systems. American Institute of Aeronautics and Astronautics, Reston (2012)
5. Hung, J.Y., et al.: On parallel hybrid-electric propulsion system for unmanned aerial vehicles. Prog. Aerosp. Sci. **51**, 1 (2012)

6. Biomo, J.-D.M.M., et al.: An enhanced Gauss-Markov mobility model for simulations of unmanned aerial ad hoc networks. In: 2014 7th IFIP Wireless and Mobile Networking Conference (WMNC), Vilamoura, Portugal (2014)
7. Ro, K., Kamman, J.W.: Modeling and Simulation of Hose-Paradrogue Aerial Refueling Systems. Western Michigan University (2012)
8. Doherty, P., et al.: The WITAS Unmanned Aerial Vehicle Project. Linköping University, Sweden (2000)
9. Drews, P.L.J., et al.: Hybrid Unmanned Aerial Underwater Vehicle: Modeling and Simulation. IEEE, Chicago (2014)
10. Wu, H., et al.: Modeling and simulation of dynamic ant colony's labor division for task allocation of UAV swarm. Elsevier Phys. A: Stat. Mech. Appl. **491**, 127 (2018)
11. Bajić, M., et al.: Modeling and Simulation of Very High Spatial Resolution UXOs and Landmines in a Hyperspectral Scene for UAV Survey. Department of IT and Computer Sciences, Zagreb University of Applied Sciences, Zagreb (2021)
12. Tang, Q., et al.: Modeling and simulation of A2G channel based on UAV array. In: 2020 IEEE 6th International Conference on Computer and Communications (ICCC), Chengdu, China (2020)
13. Garcia, R., Barnes, L.: Multi-UAV Simulator Utilizing X-Plane. Springer Science (2009)
14. Jia, R., et al.: 3D geometry-based UAV-MIMO channel modeling and simulation. China Commun. **15**(12), 64–74 (2018)
15. Spyridis, Y., et al.: Modelling and Simulation of a New Cooperative Algorithm for UAV Swarm Coordination in Mobile RF Target Tracking. Department of Electronic and Electrical Engineering, The University of Sheffield, Sheffield (2021)
16. Kakarountas, A., Spathoulas, G., Trivellas, P., Reklitis, P., Malindretos, G.: Survey on routing services for smart delivery in urban environments. In: SEEDA-CECNSM 2020 – 5th South-East Europe Design Automation, Computer Engineering, Computer Networks and Social Media Conference, 9221790 (2020)
17. Trivellas, P., Anastasopoulou, A., Malindretos, G., Reklitis, P., Sakas, D.P.: A place based analysis of stakeholders' advancing sustainability in remote and isolated communities: the case of North Aegean Islands. IOP Conf. Ser.: Earth Environ. Sci. **899**(1), 012044 (2021)
18. Trivellas, P., Kakkos, N., Vasiliadis, L., Belias, D.: Sustainability, social marketing and host attitudes about tourism in the city of Larissa. In: Springer Proceedings in Business and Economics, pp. 89–95. Springer, Cham (2017)
19. Trivellas, P., Kakkos, N., Vasiliadis, L., Belias, D.: Residents' perceptions toward cultural, social and economic benefits and costs of tourism industry: an empirical survey. In: Springer Proceedings in Business and Economics, pp. 171–181. Springer, Cham (2016)
20. Reklitis, P., Trivellas, P., Mantzaris, I., Mantzari, E., Reklitis, D.: Employee perceptions of corporate social responsibility activities and work-related attitudes: the case of a Greek management services organization. In: Gal, G., Akisik, O., Wooldridge, W. (eds.) Sustainability and Social Responsibility: Regulation and Re-porting, pp. 225–240. Springer Singapore, Singapore (2018). https://doi.org/10.1007/978-981-10-4502-8_10
21. Sakas, D.P., Kamperos, I.D.G., Reklitis, D.P., Giannakopoulos, N.T., Nasiopoulos, D.K., Terzi, M.C., Kanellos, N.: The effectiveness of centralized payment network advertisements on digital branding during the COVID-19 crisis. Sustainability. **14**, 3616 (2022). https://doi.org/10.3390/su14063616
22. Sakas, D.P., Reklitis, D.P., Trivellas, P., Vassilakis, C., Terzi, M.C.: The effects of logistics websites' technical factors on the optimization of digital marketing strategies and corporate brand name. Processes. **10**, 892 (2022). https://doi.org/10.3390/pr10050892
23. Sakas, D.P., Reklitis, D.P.: Predictive model for estimating the impact of technical issues on consumers interaction in agri-logistics websites. In: Information and Communication Technologies for Agriculture—Theme IV: Actions, pp. 269–283. Springer International Publishing, Cham (2021). https://doi.org/10.1007/978-3-030-84156-0_14

24. Sakas, D.P., Giannakopoulos, N.T., Reklitis, D.P., Dasaklis, T.K.: The effects of cryptocurrency trading websites on airlines' advertisement campaigns. J. Theor. Appl. Electron. Commer. Res. **16**, 3099–3119 (2021). https://doi.org/10.3390/jtaer16070169
25. Sakas, D.P., Reklitis, D.P.: The impact of organic traffic of crowdsourcing platforms on airlines' website traffic and user engagement. Sustainability. **13**, 8850 (2021). https://doi.org/10.3390/su13168850

Part III
Business Transformation

Konstadinos Kutsikos

Description: Driving business growth in a complicated and uncertain economic environment has been the goal for all businesses across the globe for the past few years. The ever-evolving role and needs of customers can only add to this turmoil. For example, how do we design and execute one-to-one marketing activities in a world rocked by social media, where "one" is really the influencer for an unknown group of people? Although the overarching goal is about exploring how internal and external marketing activities may influence (or be influenced by) business transformation activities, we welcome contributions from academics and practitioners that explore trends and developments in other fields that affect business transformation activities. Such fields include (but are not limited to): Business Strategy (e.g. real options strategic management), Organizational Development (e.g. networked organizations), Leadership, Innovation Management, Change Management, Information Management (e.g. enterprise interoperability).

K. Kutsikos
Business School, University of the Aegean, Mytilene, Greece

Distance Education and Digital Disruptions in the COVID-19 Era

Georgios Grigoriadis

1 Introduction

The term "Conventional Education" is widely considered education in the classroom, where the teacher and the learner are in the same place at the same time. Nowadays, the interest focuses on the perception, the attitude, and the view of the distance learning as an "open," continuous and accessible process, which corresponds to the learning needs of all types and is formed for all people. Since the mid-1990s, Information Computing Technologies have given additional impetus to distance learning, contributing significantly to its spread, giving simultaneously a new dynamic to this field. The techno-centric approaches, which initially prevailed, led finally to create educational environments which lack critical thinking. Based on the above, the following research questions arose and in this project, an attempt is made to find out the answers. The first question concerns the transition of traditional education structure into distance education in combination with the use of new technologies. In particular, the subject transition to a contemporary way of education is considered to be important due to the existing pandemic period. Secondly, innovative technological achievements could not leave the field of education without a decisive impact. Accordingly, the next question emerged about the reformation of the education process and how individuals can adopt the discussed new model. In recent years, the focus of research has been on the pedagogical dimension of Information Technologies in distance learning. Creative integration of Information Technologies in a wider social context and the wide use of technology within pedagogical terms create the right conditions under which collaborative environments could be developed which would encourage exploratory and critical thinking

G. Grigoriadis (✉)
Hellenic Open University, Patras, Greece
e-mail: std134217@ac.eap.gr

D. P. Sakas et al. (eds.), *Computational and Strategic Business Modelling*,
Springer Proceedings in Business and Economics,
https://doi.org/10.1007/978-3-031-41371-1_20

[1, 2]. *Recently, due to the situation formed by the coronavirus pandemic, a successful change took place in the field of education which replaced the communication and material posting tools that are used in the learning process. distance learning, both synchronous and asynchronous, revealed new ways to approach knowledge subjects and establish communication between teachers and students.*

2 Education Models

The traditional form of education based on the physical presence of the students and the teachers has changed considerably and has shifted to another state of education. According to Papalampropoulos P. [3], there are various theories of education available in international literature. The education models could be separated into two basic categories. We can support that traditional education is a way of learning that happens "in person" and in real time (between teachers and students). As a result, all didactic interactions took place necessarily in the same place and at the same time, with workshops, lectures, seminars, manuals, and supervision by trainers/teachers [4, 5].

The second type of educational model is distance learning. The term distance learning, in general, is used to describe a teaching process during which one or more learners are remotely located from their teacher/trainer. In such a situation, using technology is necessary to ensure that basic communication can successfully be established among people involved in the educational process and of course the learners' access to the teaching material [6]. At the same time, communication should aim at establishing an agreeable learning environment for the student, without affecting academic progress. Distance learning revealed new ways to approach knowledge subjects and establish a communication form between teachers and students. To some extent, the education community reached closer to the digital world and a new dimension to the various types of learning was given [3].

In the context of rapid developments that take place in the twenty-first century, traditional teaching has failed to fully correspond to new social realities. Hence, technological progress determines the future of education, identifying distance learning as an up-to-date suggestion, which takes advantage of the positive aspects of traditional teaching, also taking advantage of its own features. Compared to traditional teaching, distance learning requires more organization, preparation, and cooperation among all stakeholders involved in its implementation. This sense of collaboration is really important, since it strengthens bonds in the teaching team, and generates a stronger sense of closeness, constructive interaction, and cooperation among both teachers and students [7].

3 New Technologies

The majority of educational structures worldwide, it is attempting to develop new learning methods that are based on the use of digital technologies. The rapid growth of the Internet and its specific applications, combined with the introduction of a multitude of new Information and Communication Technologies (ICT), has sparked a revolution in the form, function, and possibilities of educational activities, currently introducing and establishing the terms "distance learning" and "e-learning" [3]. Furthermore, this growth is creating new data in Open and Distance Education significantly altering the environment of learning, communication and interaction with the introduction of Open Educational Resources-OERs, Massive Open Online Courses (MOOCs), Learning Analytics-LA and Educational Data Mining-EDM mainly in Higher Education. As a result, the new trends in education have been influenced by new trends in technology. Digital learning is an opportunity with great potential to develop and improve teaching quality [8]. It also results in changing the traditional teaching models. However, since the 2010s, it has been acknowledged that the digital revolution has affected distance learning rather more slowly, compared to other fields. Progress in distance learning, as a result of the digital revolution, was slow and gradual, and therefore, breakthrough change did not take place directly [9]. The introduction of new technologies to the education sector, combined with their spectacular development, has changed the landscape, when it comes to how knowledge is transmitted in distance learning [10]. The use of "Emerging Technologies" should produce a radical change in learning processes.

4 Interaction of Distance Learning and the Teaching Process

In distance education, learning is based on the interaction between learners and their teachers and between learners at each other, going beyond traditional methods of interaction, since emphasis is placed on communication through ICT (e.g., email, radio, television, videotapes, computers, and teleconferencing). Another key characteristic of teaching through distance learning is that the orientation of educational methods is changed, since a certain part of responsibility concerning the teaching material, bears down on the teacher regardless of its form. The tools of synchronized and unsynchronized communication are used to help each student/learner to be educated through an online learning environment. Thus, learning becomes more "independent", as it is managed more by the learner now, compared to the past.

There seems to be general confusion about the options that refer to ways that innovation, in general, and digital innovation, in specific, can be promoted in a distance education environment. It is also required to adopt and accept the idea that the use of digital technology allows gradually the implementation of better methods to achieve new learning and teaching objectives. In short, to accomplish innovation in learning and teaching, a multi-strategy approach is required, which will include systematic educational innovations, such as ways to address new approaches that will result in the development of a future practice to facilitate the work of those involved in the education process [11].

The new teaching model emerges as an opportunity to improve education in the twenty-first century. Along with direct response to address the problem, it is also required to develop sound planning, which will result from utilizing existing human resources in the field of distance education, combined with a methodical approach based on pedagogical requirements and strategic partnerships driven by a vision of enhanced efficiency and quality of the standards applied to education provided [12, 13]. The purpose is not to provide teaching on its own, but rather to facilitate the students' learning process. The point of this effort made by teachers is to show learners how to approach learning [14].

5 Conclusion

In the context of rapid developments that take place in the twenty-first century, traditional teaching has failed to fully correspond to new social realities. Hence, technological evolution determines the future of education, identifying distance learning as an up-to-date suggestion, which takes advantage of the positive aspects of traditional teaching, while adding new features. It has been observed that the new teaching model can mitigate the differences between formal, nonformal, and informal learning to benefit learners, indicating possible practices that may be applied, in order to harness the potential of informal learning applied to learners. Moreover, it attempts to promote methods, through which open and lifelong learning can be boosted by modern digital networks. In terms of transforming the educational framework in a more human-centered direction, it was shown in research that a change in the characteristics that determine the human identity, i.e., the adoption of a humanitarian mentality, will lead to the desired situation, where the educational framework will be characterized by philosophical and humanitarian principles. Another way to interlink technology with learning and strengthen their relationship is through online lessons, with the simultaneous development of web courses and other interactive material that can boost communication among learners. In this way, a virtual knowledge environment with the use of distance learning is established, in the form of a knowledge management structure, which strengthens interaction among users [15, 16]. In addition, virtual classrooms can yield benefits through the interaction that takes place during teleconferencing.

Other conclusions that can be drawn refer to the possibility that is offered by digital tools towards the implementation of more effective and interesting teaching for students/learners with disabilities or special learning needs, ensuring higher motivation and autonomy in learning, even in an emergency, such as the period of the coronavirus pandemic. By connecting various types of teaching applications, the function of distance learning has all the required characteristics that are used in the context of a traditional education system. This is educational modernization, since it offers maximum autonomy to learners, which can be implemented without the direct supervision of teachers, in a learning environment that guides and supports the process.

References

1. Jonker, T., Sikkema, K., Konstantelou, A., Kutsikos, K., Vignati, A., Zurlo, F.: Creativity and ICT to drive new entrepreneurship education. Cumulus Work. Pap. **33**, 97–104 (2017)
2. Kutsikos, K.: Leadership & disruptive innovation: developing the next generation of entrepreneurs and corporate innovators. In: Remenyi, D. (ed.) 5th Teaching Innovation and Entrepreneurship Excellence Awards, pp. 36–49. ACPI, Reading (2019)
3. Papalamprakopoulos, P.: Teaching in Distance Learning. The Role of Teacher. Academia, Patras (2020)
4. Alaxiotis, N., Kalavrouziotis, I., Lionarakis, A., Orfanoudakis, T., Stauropoulos, H.: Open, conventional, and dual-mode universities: current state in Europe and worldwide. In: 10th International Conference in Open & Distance Learning, Patras (2019)
5. Dimitrios, N.K.: Open source web applications. How it spread through the internet and their contribution to education. Adv. Inf. Process. Manag. **1**(1), 82–84 (2011)
6. Drosatos, D.: Innovation in the Distance Learning Education Through the Use of ICT. International Conference in Open & Distance Learning, Athens (2019)
7. Armakolas, S., Panagiotopoulos, C.: Distance learning through videoconferencing: the effects of technological factors. Athens: Open Educ. – J. Open Distance Educ. Educ. Technol. **16**(1), 22–43 (2020)
8. Allen, I., Seaman, J.: Conflicted: Faculty and Online Education. Babson Survey Research Group (2012)
9. Sofos, A., Kostas, A., Paraschou, B.: Distance Learning & Technology. Association of Greek Academic Libraries, Athens (2015)
10. Neira, E., Salinas, J., de Benito, B.: Emerging technologies (ETs) in education: a systematic review of the literature publishes between 2006 and 2016. Manizales: Int. J. Emerg. Technol. Learn. **12**, 128 (2017)
11. Canellopoulos, A., Koutsoumpa, M.: The Research on Videoconferencing in Distance Education A Bibliographical Review. Open Education, Athens (2019)
12. Kutsikos, K., Konstantopoulos, N., Sakas, D., Verginadis, Y.: Developing and managing digital service ecosystems: a service science viewpoint. J. Syst. Inf. Technol. **16**(3), 233–248 (2014)
13. Anastasiadis, P.: Distance learning in the era of COVID-19, focusing on the Greek Education, Creta: Open Distance Educ. Educ. Technol. 16(2), 20–48 (2020)
14. Stringkas, K., Tsimpiris, A.: Teacher Training with Distance e-Learning. Open Education, Patras (2019)
15. Dimitrios, N.K.: Modeling the development of the online conference's services. Libr. Rev. **65**(3), 160–184 (2016)
16. Dimitrios, N.K.: Free software – open source software. A powerful tool for developing creativity in the hands of the student. Adv. Inf. Process. Manag. **1**(1), 78–81 (2011)

Motivations of Business Ecosystem Participants for Value Co-creation: An Overview

Margaret Mastropetrou, Konstadinos Kutsikos, and George Bithas

1 Introduction

The purpose of this review is to discuss the motivations of a business ecosystem's participants and the drivers to join a network and a value co-creation process. This information will be analyzed and combined or intertwined, in order to overcome the knowledge frontier and lead to the exploration of new ideas, such as the connection of stakeholders' motives to their assigned roles in the ecosystem. The importance of this research effort lies in the fact that although the concepts of stakeholders' motivations and their roles have been explored separately, it is the first time that they are analyzed together in a business ecosystem level. After gaining an in-depth insight on the existing studies, concerning the research question, the purpose is to draw a general conclusion, which will in turn lead to identifying more specific gaps and proposing guidelines for future research.

M. Mastropetrou (✉) · G. Bithas
Growth Transformation & Value Engineering (W.A.V.E.) Lab, Business School, University of the Aegean, Chios, Greece
e-mail: mastropetrou@aegean.gr

K. Kutsikos
Growth Transformation & Value Engineering (W.A.V.E.) Lab, Business School, University of the Aegean, Chios, Greece

Business Administration Program, School of Social Sciences, Hellenic Open University, Patras, Greece

D. P. Sakas et al. (eds.), *Computational and Strategic Business Modelling*,
Springer Proceedings in Business and Economics,
https://doi.org/10.1007/978-3-031-41371-1_21

2 Background

Value is not usually created in a single organization, but rather co-created in complex networks [1], such as business ecosystems. This collaboration contributes to the complementary services for a greater goal [3, 17, 23, 26]. Value co-creation or "the joint, collaborative, concurrent, peer like process of producing new value" [12, 27] is considered to be co-created by the interaction of multiple stakeholders, rather than two single entities [19, 24, 25]. Frow P. et al. [11] suggest that exchange happens because no single actor has all the resources to function effectively in isolation and, in order to co-create value, needs to take part in resource integration practices. This may happen even when the priorities, goals, and values of the participants appear to conflict instead of align. This interaction in the context of a complex network can be expressed as a "stakeholder ecosystem" [13], in which a stakeholder "is any person or group that is affected by how an organization pursues its own goals" [10].

Provided that business ecosystems foster value co-creation by the stakeholders, exploring the reasons why stakeholders join an ecosystem is necessary. In other words, the identification of stakeholders' motives is important, in order for the focal firm to set up effective resource integration practices, which can consequently lead to worthwhile value co-creation. Ultimately, all stakeholders may be viewed as resource integrators that collectively co-create value [21, 22, 29].

In order to understand the motivations of stakeholders, the role of each stakeholder in the ecosystem has to be explored as well as the way each motivation leads to the adoption of a specific role. The reason for this connection is because roles are fundamental for the viability of the business ecosystem [5].

According to Iansiti M. and Levien R. [15], the most important member of an ecosystem is the "keystone" or ecosystem leader, which is supported by the "niche players" or "complementors" that form the majority of the ecosystem and generate the greatest part of innovations and created value. Based on the basic roles, Bithas G. et al. [5] have proposed a model where roles are described in terms of three sets of attributes (role elements): service innovation, collaboration and resilience.

In order to model and analyze the relationships among the stakeholders in a business ecosystem, which includes human organization and other forms of social structures, the i* framework is proposed. I* is an agent-oriented goal modeling framework. Goal modeling aims to determine what various actors want and how (and whether) those wants will be achieved. I* stands for distributed intentionality [35] building on the premise that actors don't merely interact with each other through actions or information flows but relate to each other at an intentional level.

3 Methodology

The methodology followed is the systematic literature review (SLR). This method was chosen because it is the most transparent, inclusive, and explanatory type of study [9] and because a systematic mapping study has been already conducted, providing a roadmap for this current study. The steps for the SLR include scoping, planning, conducting the search, and reporting the results.

Firstly, it was clarified that although the concepts of ecosystems and value co-creation were previously studied, the perspective was different and the examination of those concepts in combination with stakeholders' motives has not taken place. Thus, a gap in the literature is identified. Then, after identifying the purpose of the review, the research question that emerged, using the CIMO logic [9], was as follows: What is the relationship between a business ecosystem role and a new participant's motivation? Before moving on, a review protocol was developed in order to enhance the quality of the review [20], find possible mistakes [6] and increase the reliability and repeatability of the review [6]. The inclusion criteria mainly concern the inclusion of studies that provide scientific research findings and reviews or reports, concerning the topic of business ecosystems and value co-creation, not in general, but in relation with stakeholders' motivations, intentions, goals, and roles. The next step is the search of scientific databases, using keywords, synonyms, timeframes, and backward and forward searching. The study selection includes the review of title and abstract of each paper and the matching of the inclusion-exclusion criteria. This process can be repeated several times, and some modifications are made. After that, the quality assessment takes place, for which the JBI checklists were used in order to evaluate the selected studies. Finally, the data extraction is a crucial step, where after obtaining the list of papers from the literature search, information is taken from each paper to serve as the raw material for the synthesis step [33]. This information, along with the assessment from the previous step, were all presented in a table.

4 Results

In order to answer the review question, what motivations drive stakeholders to join a business ecosystem, an exploration of several studies has taken place. Stakeholders see several opportunities in joining a business ecosystem. Since no single firm can provide the value offerings alone, ecosystem players need to have strong collaboration and relationship with community members, groups, and organizations to form a win-win situation for themselves and for end users. After all, ecosystems provide opportunities to participate in diversified fields and explore a huge market. It is found that in business companies, the motivation of people is, at least, made up of three components: extrinsic factors (such as salary, incentives, and job recognition), intrinsic factors (such as job satisfaction and learning), and also prosocial factors related to satisfying the needs of the main stakeholders of an organization [2, 31, 32]. A more analytical description of these motivations and aspirations follows:

1. Reputation enhancement motives: The ecosystem's positive image is expected to enhance each stakeholder's reputation [34]. The attractiveness of the orchestrator and the available alternatives affect the degree of attachment of the complementor [16].
2. Experimentation motives: In order to develop new products/services within the ecosystem, and beyond it. Actors expect to hybridize solutions and tools from other actors and transfer them into their specific ecosystem [34].
3. Relationship motives: In order to develop new partnerships and collaborations. Co-creation allows stakeholders to connect with each other [34].
4. Communication: A multicentered flow communication enables stakeholders to co-create a shared storytelling and build a shared identity of the ecosystem [34].
5. Mapping practices: Mapping practices identify and organize micro-specialized competences into complex resource integration, meaning that the value created by each stakeholder's activity might be used and produce serendipitous results by other actors in the ecosystem [34].
6. Smoother decision-making: Multi-stakeholder co-creation is enabled by formalized shared processes, managed by dedicated teams, where decision-making is smoothed by the designated presence of a decision maker [34]. This is critical to maintain harmony among ecosystem players and to ensure a healthy cooperative ecosystem [18].
7. New skills: Multi-stakeholder co-creation depends on and builds individual characteristics: creativity, flexibility, and negotiation are requisite skills to build shared value [30, 34].
8. Standardization: Ecosystem players have their own proprietary solutions. They try to influence and aspire their solution to evolve as de facto standards. They aspire to influence toward standardization, which could go in their favor to capture bigger market share [18].
9. Social motives: Get products to market with a larger goal of bringing better quality of life to the society [18].
10. Diversification: Business goal is to exploit market potential by developing vertical solutions in various industrial sectors [18].

In order to ensure the sustainable development of the ecosystem, along with the efficient value co-creation process, the connection between the participants' motives and the roles they take up has to be explored. The role of each complementor in the ecosystem is important as it is a sign of the engagement and the intensity of value co-creation [5]. Thus, it would be very interesting to explore how each motive may lead to certain roles. According to Bithas G. et al. [5], the roles are defined according to the axes of their model, which include value co-creation, resources capitalization, and the intensity of value co-creation. Various roles are assigned according to the different combinations of positions along the axes.

For the modeling of the relationships among the ecosystem participants, the use of i* framework has been explored. The literature review has provided enough evidence that i* is the best framework for the purposes of the review question. Firstly, it represents the intentional goals of each actor and how the goals can be achieved [35], as well as the interactions between them. Furthermore, it is used to obtain an understanding about organizational relationships and the rationales behind them, thus emphasizing social aspects by providing a graphical depiction of system actors, intentions, dependencies, and responsibilities [14]. Therefore, researchers consider i* as a modeling approach that contributes in analyzing and re/designing organizations [7]. It also offers a more open and strategic conception of agents than other goal-oriented requirements frameworks [8]. Finally, it helps participants and stakeholders understand the implications of an organization design need not capture the actual problem-solving knowledge [36].

5 Conclusions and Future Research

The literature has investigated participants' intentions mostly from the strategic and organizational perspective. Identifying stakeholder motives is important for the focal organization in setting up effective resource integration practices, which can then lead to worthwhile value co-creation. For this reason, establishing the connection between stakeholders' motivations and their assigned roles is essential for value co-creation practices, as well as for the viability of the ecosystem. This concept is not studied in the academic literature nor under a managerial perspective. Thus, it provides material for further investigation. Based on the model of Bithas G. et al. [5], the relation between the ecosystem participants' motivations and their assigned roles in the network will be explored in the future study. Another point for future consideration is the adaptation of i* modeling framework into an actual ecosystem, aiming to present the intentions and relations of stakeholders in a real-life example, which in turn will definitely open more paths to explore [4, 28].

References

1. Adner, R.: Ecosystem as structure: an actionable construct for strategy. J. Manag. **43**(1), 39–58 (2017)
2. Bastons, M., Mas, M., Rey, C.: Pro-stakeholders motivation: uncovering a new source of motivation for business companies. J. Manag. Organ. **23**(5), 621–632 (2017)
3. Betz, C., Jung, R.: Value creation in business ecosystems – a design theory for a reference model. In: AMCIS 2021 Proceedings 4 (2021)
4. Bithas, G., Sakas, D.P., Kutsikos, K.: Business transformation through service science: a path for business continuity. In: Strategic Innovative Marketing, pp. 159–164. Springer, Cham (2017)
5. Bithas, G., Kutsikos, K., Warr, A., Sakas, D.: Managing transformation within service systems networks: a system viability approach. Syst. Res. Behav. Sci. **35**, 469–484 (2018)

6. Brereton, P., Kitchenham, B.A., Budgen, D., Turner, M., Khalil, M.: Lessons from applying the systematic literature review process within the software engineering domain. J. Syst. Softw. **80**, 571–583 (2017)
7. Chung, L., Nixon, B., Mylopoulos, J.: Non-Functional Requirements in Software Engineering. Kluwer Academic Publishers, London (2000)
8. Dardenne, A., van Lamsweerde, S., Fickas, S.: Goal directed requirements acquisition. Sci. Comput. Program. **20**(3), 50 (1993)
9. Denyer, D., Tranfield, D.: Producing a systematic review. In: Buchanan, D., Bryman, A. (eds.) The Sage Handbook of Organizational Research Methods, pp. 671–689. Sage Publications, London (2009)
10. Freeman, R.E.: Strategic Management: A Stakeholder Approach. Pitman, Boston (1984)
11. Frow, P., McColl-Kennedy, J.R., Hilton, T., Davidson, A., Payne, A., Brozovic, D.: Value propositions: a service ecosystems perspective. Mark. Theory. **14**(3), 327–351 (2014)
12. Galvagno, M., Dalli, D.: Theory of value co-creation: a systematic literature review. Manag. Serv. Qual. **24**, 643–683 (2014)
13. Gyrd-Jones, R.I., Kornum, N.: Managing the co-created brand: value and cultural complementarity in online and offline multi-stakeholder ecosystems. J. Bus. Res. **66**(9), 1484–1493 (2013)
14. Horkoff, J., Yu, E., Ghose, A.: Interactive goal model analysis applied-systematic pro-cedures versus ad hoc analysis. In: van Bommel, P., et al. (eds.) PoEM 2010, LNBIP 68, pp. 130–144. Springer, Berlin, Heidelberg (2010)
15. Iansiti, M., Levien, R.: The Keystone Advantage. What the New Dynamics of Business Ecosystems Mean for Strategy, Innovation, and Sustainability. Harvard Business School Press, Boston (2004)
16. Jacobides, M.G.: In the ecosystem economy, what's your strategy? The five questions you need to answer. Available at: https://hbr.org/2019/09/in-the-ecosystem-economy-whats-your-strategy (2019)
17. Jacobides, M.G., Cennamo, C., Gawer, A.: Towards a theory of ecosystems. Strateg. Manag. J. **39**(8), 2255–2276 (2018)
18. Kar, S., Chakravorty, B., Sinha, S., Gupta, M.P.: Analysis of stakeholders within IoT ecosystem. In: Kar, A.K., et al. (eds.) Digital India, Advances in Theory and Practice of Emerging Markets. Springer International Publishing AG, part of Springer Nature (2018)
19. Ketonen-Oksi, S., Valkokari, A.: Innovation ecosystems as structures for value co-creation. Technol. Innov. Manag. Rev. **9**(2), 24–34 (2019)
20. Kitchenham, B., Charters, S.: Guidelines for Performing Systematic Literature Reviews in Software Engineering. EBSE Technical Report, Software Engineering Group, School of Computer Science and Mathematics, Keele University, Department of Computer Science, University of Durham (2007)
21. Kutsikos, K., Kontos, G.: A systems-based complexity management framework for collaborative e-government services. Int. J. Appl. Syst. Stud. **4**(1–2), 1–16 (2011)
22. Kutsikos, K., Kontos, G.: Enabling business interoperability: a service co-creation viewpoint. In: Enterprise Interoperability V, pp. 191–201. Springer, Berlin (2012)
23. Kutsikos, K., Kontos, G.: Business opportunity identification through dynamic in-formation management in collaborative business ventures. Procedia Soc. Behav. Sci. **73**, 616–624 (2013)
24. Kutsikos, K., Mentzas, G.: A service portfolio model for value creation in net-worked enterprise systems. In: Cezon, M., Wolfsthal, Y. (eds.) Towards a Service-Based Internet Lecture Notes in Computer Science, pp. 101–109. Springer, Berlin (2011)
25. Kutsikos, K., Mentzas, G.: Managing value creation in knowledge-intensive business service systems. In: Kantola, J., Karwowski, W. (eds.) Knowledge Service Engineering Handbook, pp. 123–138. CRC Press, USA (Taylor & Francis Group), Boca Raton (2012)
26. Kutsikos, K., Sakas, D.: A framework for enabling service configuration decisions: the case of IT outsourcing providers. Procedia Soc. Behav. Sci. **148**, 604–610 (2014)
27. Kutsikos, K., Konstantopoulos, N., Sakas, D., Verginadis, Y.: Developing and managing digital service ecosystems: a service science viewpoint. J. Syst. Inf. Technol. **16**(3), 233–248 (2014)

28. Mastropetrou, M., Bithas, G., Kutsikos, K.: Digital transformation in the luxury industry – a systematic mapping study. In: Proceedings of the 12th EuroMed Academy of Business conference, pp. 731–746. Euromed Press, Nicosia (2019)
29. Merz, M.A., He, Y., Vargo, S.L.: The evolving brand logic: a service dominant logic perspective. J. Acad. Mark. Sci. **37**(3), 328–344 (2009)
30. Nasiopoulos, D.K., et al.: Free software – open source software. A powerful tool for developing creativity in the hands of the student. Adv. Inf. Process. Manag., 78–81 (2011)
31. Nasiopoulos, D.K., et al.: Simulation of generation of new ideas for new product development and IT services. AIP Conf. Proc. **1644**, 60–68 (2015)
32. Nasiopoulos, D.K., et al.: Modeling of market segmentation for new IT product development. AIP Conf. Proc. **1644**, 51–59 (2015)
33. Okoli, C.: A guide to conducting a standalone systematic literature review. Commun. Assoc. Inf. Syst. **37**, 43 (2015)
34. Pera, R., Occhiocupo, N., Clarke, J.: Motives and resources for value co-creation in a multi-stakeholder ecosystem: a managerial perspective. J. Bus. Res. **69**, 4033–4041 (2016)
35. Yu, E.: Towards modelling and reasoning support for early-phase requirements engineering. In: Proceedings of the IEEE International Symposium on Requirements Engineering (RE 97), pp. 226–235. IEEE CS Press (1997)
36. Yu, E., Mylopoulos, J.: An actor-dependency model of organizational work: with application to business process reengineering. In: Proceedings of the Conference on Organizational Computing Systems, pp. 258–268. ACM, Milpitas (1993)

The Emergence of Business Ecosystems in the Healthcare Industry

Dimitra Giannaraki and Konstadinos Kutsikos

1 Introduction

The healthcare sector is one of the largest, most complicated, and dynamic in the global economy, consisting of diverse entities with a wide range of activities. The latter are nowadays affected by globalization, technological disruption, and fierce competition. Sociodemographic, regulatory, and economic challenges are further stressing healthcare providers to deliver healthcare at higher levels of affordability, accessibility, and quality.

Key questions then emerge: Can a single entity achieve the above by itself? At what cost? And how sustainable would that be? Organizations in other industries facing similar challenges have been exploring collaborative solutions. James F. Moore [1, pp. 75–83] suggested that a firm is embedded in a business environment, named business ecosystem [2, 3], and needs to co-evolve with other firms through developing beneficial relationships of collaboration and competition with other organizations, just as symbiotic activities occur in natural ecosystems [1]. The ecosystem perspective in the healthcare industry could help industry players focus on business model innovation, based on coexistence, service co-development, and

D. Giannaraki (✉)
Business Administration Program, School of Social Sciences, Hellenic Open University, Patras, Greece

Quality Assurance Department, Pharmathen International S.A., Rodopi, Greece
e-mail: std139877@ac.eap.gr

K. Kutsikos
Business Administration Program, School of Social Sciences, Hellenic Open University, Patras, Greece

Growth Transformation & Value Engineering (W.A.V.E.) Lab, Business School, University of the Aegean, Chios, Greece

D. P. Sakas et al. (eds.), *Computational and Strategic Business Modelling*,
Springer Proceedings in Business and Economics,
https://doi.org/10.1007/978-3-031-41371-1_22

value co-creation. Indeed, they could join forces toward achieving a common goal: to deliver better care to more people at an affordable cost.

Based on these observations, this article highlights how the business ecosystem concept can be applied in the healthcare industry. It intends to identify and emphasize the drivers and trends that may force healthcare organizations to reassess their strategies, engage in collaborative business model innovation, and implement healthcare business ecosystemic models.

2 Methodology

The research methodology that was followed is the design science research (DSR) methodology, due to its ability to identify and explore practical field problems [4]. In the context of this article, the first two DSR steps are briefly presented: (a) the clarification of field problem to be addressed and (b) the literature-based exploratory research.

Activities of the first DSR step were triggered by the exploration of the present status of healthcare service systems to identify their complexities. Literature-based research was then based on a systematic mapping study due to its ability to provide a broad review of a research area and identify the primary studies in a specific field [4–7]. This was then followed by the definition of research questions, the circumstantial research based on selected search strings, the filtering of findings according to inclusion and exclusion criteria, results extraction, and their qualitative assessment and classification.

Overall, two research questions were formed: (a) What are the main elements that nowadays indicate a high level of unsustainability of traditional healthcare systems? (b) How are healthcare models reshaped in the context of the business ecosystems' revolution?

3 Drivers for Disruption

The application of the aforementioned research methodology quickly uncovered an important trend: The global healthcare industry is experiencing tremendous uncertainty and rapid change, which is inevitably leading to transformation. Its future is now shifting to well-being and customer-centric models, while healthcare organizations are expected to restructure their strategies to overcome the major inefficiencies of the industry [8].

The first driver for such changes is the level of healthcare expenditures, which is a function of a wide range of demographic, social, and economic factors, heavily depending on internal financing and the organizational structure of each healthcare system. Nowadays, a steady increase of global healthcare expenditures is observed, attributed to primary healthcare for the rehabilitative and preventive management of illnesses and diseases [8–10]. On one hand, chronic diseases have registered a tremendous increase, denoting the necessity for intensive curative and rehabilitative healthcare that is usually of high complexity and cost [11, 12]. On the other hand, over the last 50 years, there is a remarkable increase in life expectancy, which is expected to be further increased in the future [13]. This trend is linked to the rapid growth of aging population, who require a greater part of healthcare expenditures. Considering, as well, the rife contingency of the aging population to suffer from chronic diseases, it is inevitable to experience more pressure on healthcare systems.

The second driver is the lack of patients' empowerment and engagement in healthcare issues [14]. In a society characterized by massive consumerism and information deluge, customers' awareness is enhanced, resulting in higher demands and expectations of goods and services. In the healthcare industry, patients that have greater awareness of health issues anticipate being treated like customers and receive improved value, in terms of treatment relative to costs [12, 15]. They expect convenience, improved affordability, accessibility, and control over their decisions on medical treatments, including personalized solutions [13, 16].

These drivers create a breeding ground for significant transformation [12, 17, 18]. Healthcare organizations must reassess their strategies in order not only to reduce costs but also to extent their offerings to more specialized, digital, friendly-in-use, affordable and personalized services. At the same time, they realize that development of such services requires extensive resources and capabilities, making it increasingly difficult for a single organization to manage them alone [14]. Cost-sharing and dynamic resources' exchange among entities are suggested to overcome these complications and enable healthcare providers to create and deliver value to patients [19, 20].

Hence, collaborative approaches are the key solution to overhaul problems, reduce risks and costs, and develop new opportunities and perspectives [21–23]. On the one side of the spectrum, collaboration enables healthcare organizations to spend less on investments that are outside their core competency, such as new technologies or marketing, and concentrate their resources in areas that have the specialization to compete [24, 25]. On the other side of the spectrum, collaboration enables disruptors from other industries to contribute new capabilities and resources, leading to new healthcare services and potentially new business models [26–28].

4 Healthcare Business Ecosystems

The imperative of collaboration-driven innovation that has emerged led to the adoption of the business ecosystems approach in the healthcare industry [29]. A healthcare business ecosystem (HBE) could be defined as a coordinated network of interdependent and interacting enterprises (within and/or outside the healthcare industry) that share and combine their resources to produce a set of healthcare capabilities of value across all HBE stakeholders, such as patients, providers, payers, pharmaceuticals, etc. HBE's objective is not only to deliver enhanced and efficient experiences for patients and stakeholders, such as individualized and aggregated services to patients, or increased providers' productivity, but also to address solutions to the healthcare industry's inefficiencies [22, 30].

To achieve that, a HBE is not a stable network with prefixed partners and roles. It is an evolving and constantly changing environment that aggregates and orchestrates its stakeholders according to available resources, value propositions, and the demands of the market [17]. Various HBE models have emerged to incorporate the preventive and customer-centered healthcare delivery [13, 14, 31]. The rest of this section is focused on two ecosystemic models.

4.1 Medical "6 Ps" Healthcare Ecosystem

A "6 Ps" HBE, whose stakeholders are entities that look for medical solutions, consists mainly of medical stakeholders, organized around the first "P", i.e., the patient. The distinct role and features of each stakeholder are:

- Patients linked to a HBE are characterized by improved engagement, access to personalized healthcare and preventive assistance.
- Providers are all the professionals and organizations that provide care. They can now work together in multidisciplinary teams, forming a vast network of medical information about treatments, medications, and therapies. Each such stakeholder can then offer more accurate, personalized, and efficient care.
- Payers, the organizations that cover healthcare costs, have now a central, interacting role with providers-patients by focusing their services on costs transparency and reduction, to make health accessible to everyone.
- Pharmaceuticals and pharmacies that relate to patients and providers form the core of services and products delivery. By integrating and sharing all information that is generated within the HBE, their activities are improved in terms of offering more accurate and efficient medication study, process, and delivery.
- Policy makers, that regulate healthcare operations, are called to play a supportive but fundamental role in a HBE establishment: facilitate the collaboration among multiple entities to consider the marketer patients' needs and adapt their policies respectively [32].

It is evidenced that a "6 Ps" medical-based HBE intends to offer effective, accessible, and affordable care by organizing and managing its medical stakeholders around patients' needs.

4.2 Holistic Healthcare Ecosystem

A 360° HBE model is suggested by McKinsey & Company (2020), and it could be characterized as a holistic one, with a wide range of features at operational and strategic level [30]:

- Home care, near-home care, and self-care are critical advances that the holistic HBE aims to establish on a large scale. Telehealth and virtual care are going to be delivered mainly through patients' engagement by remote- and self-monitoring practices. Traditional care will also be delivered in different but more affordable and convenient ways near the patient's location.
- The modalities of traditional healthcare will be connected with consumers in a HBE by improving healthcare teams' coordination, support in scheduling, accessibility, and quality.
- Daily life activities form a key part of this HBE model, focusing on public actions, behaviors, and habits through tracking them. As a result, a significant pool of health and wellness data can be collected. Tracking has a great role in patients' actions and behaviors that in turn affect health state and well-being. It also identifies a broad framework for health status and preventive actions.
- Financial support is a fundamental part of HBE, as every stakeholder operation or patient activity has to be supported by economically efficient solutions.
- Last but not least, social care is also in scope since patients' health is indirectly related to some community elements (e.g., transportation services) and social determinants or needs (e.g., faith institutions and family support) that have to be considered in order to be served as well.

All these features are organized around the patient or healthy people, who define contextually their needs and drive the emergence of tailored services within the HBE, either for treatment or for nontraditional healthcare provision [30, 33].

5 Discussion

Following these strategic developments in the healthcare industry, many industry players may rethink their strategy and move away from traditional business models toward the creation of business ecosystems in order to deliver increased value. Our research identified the following managerial guidelines for a healthcare organization interested in engaging in HBE development:

1. Define the scope of the ecosystem. The first step is to define the healthcare ecosystem scope for the co-creation of healthcare services and/or products. This is directly linked to the target customer groups of the HBE, combined with the target stakeholder organizations for its formation.
2. Value proposition. The second step is to assure that value will be clearly delivered to all HBE stakeholders (including target customers). It is mandatory to focus on

patients and healthy individuals and create tangible value propositions in the form of healthcare services, as well as an effective omnichannel approach to improve their experience throughout their care journey. In terms of HBE stakeholders, they need to be committed to collaboration so as to coordinate their value co-creation activities.

3. Technology integration. Collaboration between HBE stakeholders can be facilitated through innovative digital technologies. The medical data generation, storage, efficient management, and their sharing among the HBE stakeholders are usually enabled through digital platforms. Hence, the selection of the most appropriate digital infrastructure that supports the healthcare ecosystem's operations and management practices is a mandatory step.
4. Governance. The aforementioned activities require a masterful orchestration of governance practices within the healthcare ecosystem. Depending on the HBE model, stakeholders shall be organized with distinct roles and rules of partnership. A committed leader acts as the core driver to build and organize the ecosystem. Fair business agreements need to be clearly defined to manage the varying levels of cooperation. An important aspect of governance application is the continuous identification and invitation of new stakeholders in the HBE to ensure growth and success.

6 Conclusions

In a constantly evolving and changing business environment, the dynamic and complex healthcare industry is confronted with many inefficiencies, leading the sector to a transformation future. It is evidenced that healthcare ecosystems may act as a catalyst for healthcare organizations to balance costs of care and the empowerment and engagement of patients in their own health conditions. In this article, two fundamental healthcare ecosystemic models were identified: (a) a medical one, which accrues and connects stakeholders from the healthcare industry in order to offer improved solutions for patients, and (b) a holistic one, which is focused not only on healthcare but also on well-being of customers, where ecosystem stakeholders from various industries offer holistic care and wellness services.

References

1. Moore, J.F.: Predators and prey: a new ecology of competition. Harv. Bus. Rev. **71**(3), 75–83 (1993)
2. Jacobides, M.G., Cennamo, C., Gawer, A.: Towards a theory of ecosystems. Strateg. Manag. J. **39**(8), 2255–2276 (2018)
3. Kutsikos, K., Konstantopoulos, N., Sakas, D., Verginadis, Y.: Developing and managing digital service ecosystems: a service science viewpoint. J. Syst. Inf. Technol. **16**(3), 233–248 (2014)

4. Petersen, K., Feldt, R., Mujtaba, S., Mattsson, M.: Systematic mapping studies in software engineering. In: 12th International Conference on Evaluation and Assessment in Software Engineering (EASE 2008), vol. 12, pp. 1–10. University of Bari, Bari (2008)
5. Aken, V.J.E., Romme, A.G.L.: Reinventing the future: adding design science to the repertoire of organization and management studies. Org. Manag. J. **6**(1), 5–12 (2009)
6. Kitchenham, B.A., Charters, S.: Guidelines for performing systematic literature reviews in software engineering. Technical Report (EBSE 2007-001). Keele University and Durham University Joint Report (2007)
7. Petersen, K., Vakkalanka, S., Kuzniarz, L.: Guidelines for conducting systematic mapping studies in software engineering: an update. Inf. Softw. Technol. J. **64**, 1–18 (2015)
8. World Health Organization: Global Spending on Health 2020: Weathering the Storm. World Health Organization, Geneva (2020) https://www.who.int/publications/i/item/9789240017788
9. World Health Organization: Primary Health Care. World Health Organization, Geneva (2021) https://www.who.int/news-room/fact-sheets/detail/primary-health-care
10. OECD/European Union: Health at a Glance: Europe 2020: State of Health in the EU Cycle. OECD Publishing, Paris (2020). https://doi.org/10.1787/82129230-en
11. Roser, M., Ortiz-Ospina, E., Ritchie, H.: Life expectancy. https://ourworldindata.org/life-expectancy (2019)
12. Carroll, N., Kennedy, C., Richardson, I.: Challenges towards a Connected Community Healthcare Ecosystem (CCHE) for managing long-term conditions. J. Int. Soc. Gerontechnol. **14**, 64–77 (2016)
13. Cordina, J., Gilbert, G., Griffin, N., Kumar, R.: Next-Generation Member Engagement During the Care Journey. McKinsey & Company. https://www.mckinsey.com/industries/healthcare-systems-and-services/our-insights/next-generation-member-engagement-during-the-care-journey (2019)
14. Palumbo, R.: Contextualizing co-production of health care: a systematic literature review. Int. J. Public Sect. Manag. **29**(1), 72–90 (2016)
15. Porter, M.E., Lee, T.H.: The strategy that will fix health care. Harv. Bus. Rev. **91**(12), 24 (2013)
16. Schulz, F., Kalla, S., Heinke, M., Förthmann, M.: Race is on: digital platforms for medical technology. Accenture. https://www.accenture.com/us-en/insights/life-sciences/digital-platforms-medical-technology (2018)
17. Frow, P., McColl-Kennedy, J.R., Payne, A.: Co-creation practices: their role in shaping a health care ecosystem. Ind. Mark. Manag. **56**, 24–39 (2016)
18. Bithas, G., Kutsikos, K., Sakas, D., Konstantopoulos, N.: Business transformation through service science: the road ahead. Procedia Soc. Behav. Sci. **175**, 439–446 (2015)
19. Bithas, G., Kutsikos, K., Warr, A., Sakas, D.: Managing transformation within service systems networks: a system viability approach. Syst. Res. Behav. Sci. **35**(4), 469–484 (2018)
20. Bithas, G., Sakas, D.P., Kutsikos, K.: Business transformation through service science: a path for business continuity. In: Kavoura, A., Sakas, S., Tomaras, P. (eds.) Strategic Innovative Marketing. Springer Proceedings in Business and Economics, pp. 159–164. Springer, Cham (2017)
21. Harrison, M.: A 5-point model for value-based health care. Harvard Business Review. https://hbr.org/2019/10/a-5-point-model-for-value-based-health-care?ab=at_art_art_1x1&registration=success (2019)
22. Ciasullo, M.V., Cosimato, S., Palumbo, R., Storlazzi, A.: Value co-creation in the health service ecosystems: the enabling role of institutional arrangements. Int. Bus. Res. **10**(12), 222–238 (2017)
23. Kutsikos, K., Kontos, G.: Enabling business interoperability: a service co-creation viewpoint. In: Poler, R., Doumeingts, G., Katzy, B., Chalmeta, R. (eds.) Enterprise Interoperability V, vol. 5, pp. 191–201. Springer, Berlin (2012)
24. Nasiopoulos, D.K., et al.: Simulation of generation of new ideas for new product development and IT services. AIP Conf. Proc. **1644**, 60–68 (2015)

25. Nasiopoulos, D.K., et al.: Modeling of market segmentation for new IT product development. AIP Conf. Proc. **1644**, 51–59 (2015)
26. Judah, R., Radin, J., Leste, T., Gisby, S., D'Amico, K., Israel, A., Mahoney, N.: New roads to the health innovation ecosystems of tomorrow. Deloitte. https://www2.deloitte.com/us/en/pages/life-sciences-and-healthcare/articles/innovation-ecosystems-in-health-care (2020)
27. Kutsikos, K., Mentzas, G.: Managing value creation in knowledge-intensive business service systems. In: Kantola, J., Karwowski, W. (eds.) Knowledge Service Engineering Handbook, pp. 123–138. CRC Press (Taylor & Francis Group), Boca Raton (2012)
28. Kutsikos, K., Mentzas, G.: A service portfolio model for value creation in networked enterprise systems. In: Cezon, M., Wolfsthal, Y. (eds.) Towards a Service-Based Internet Lecture Notes in Computer Science, pp. 101–109. Springer, Berlin (2011)
29. Moore, J.F.: The rise of a new corporate form. Wash. Q. **21**(1), 167–181 (1998)
30. Singhal, S., Kayyali, B., Levin, R., Greenberg, Z.: The Next Wave of Healthcare Innovation: The Evolution of Ecosystems. McKinsey & Company. https://www.mckinsey.com/industries/healthcare-systems-and-services/our-insights/the-next-wave-of-healthcare-innovation-the-evolution-of-ecosystems (2020)
31. Batra, N., Judah, R., Betts, D., Thomas, S.: Forces of change. The future of health. https://www2.deloitte.com/za/en/pages/life-sciences-and-healthcare/articles/the-future-of-health.html (2019)
32. HCL Technologies. Empowering the new healthcare ecosystem. https://www.hcltech.com/ (2011)
33. Singhal, S., Carlton, S.: The Era of Exponential Improvement in Healthcare? McKinsey & Company. https://www.mckinsey.com/industries/healthcare-systems-and-services/our-insights/the-era-of-exponential-improvement-in-healthcare (2019)

The Current State of Artificial Intelligence

Djamilya M. Maksubova, Navrat M. Umargadzhieva, Bariyat Sh. Dadaeva, and Patimat G. Aripova

1 Introduction

Artificial intelligence (AI)—what is it? The pursuit of AI hinges on replicating the behavior and complexity of humankind. The development of computer technology has led to the emergence of many studies aiming to learn, recognize, and synthesize human speech; to create vision systems that can recognize people's faces; to teach cars how to drive without drivers; etc. Systems exhibiting levels of human behavior are called AI systems. The study of these systems is an independent scientific discipline that itself incorporates other scientific disciplines.

According to one definition, artificial intelligence is a science and technology that includes a set of tools that allow computers to answer questions and draw expert conclusions on the basis of their accumulated knowledge—that is, knowledge that was not put into them by their developers. Other scientific disciplines incorporated into AI include computer science, informational technology, neuroscience, and other technologies created from it.

AI systems can be divided into two classes: strong (or general) AI and weak (or applied) AI. First, strong, or universal, AI is comparable to human intelligence: This type of AI can learn how to mimic people, and not inferior on level development most of people, as in many ways even surpasses them. There are many more strict definitions but for understanding given definitions enough.

All other AI systems are subsumed under weak AI, insofar as they can do only one thing, such as searching for inquiries on the Internet, diagnosing a specific disease, etc. Such AI increase efficiency and productivity. AI systems can already perform many tasks better than humans can, and in the near future, these systems

D. M. Maksubova · N. M. Umargadzhieva · B. S. Dadaeva · P. G. Aripova (✉)
Federal State Budgetary Educational Institution of Higher Professional Education, Dagestan State University, Republic of Dagestan, Makhachkala, Russia

D. P. Sakas et al. (eds.), *Computational and Strategic Business Modelling*, Springer Proceedings in Business and Economics,
https://doi.org/10.1007/978-3-031-41371-1_23

will be increasingly improved. The process of working on projects with weak AI requires significant preparation. Machines can separately carry out tasks such as searching for information, recognizing speech and people, mimicking natural language, drawing logical conclusion, and others by using powerful machine-learning tools, but they are developing rapidly, pushing AI technologies forward every year and progressing toward the creation of strong AI. In connection with weak AI, mention should be made of the so-called AI effect. As soon as AI has achieved a previously unthinkable result on a task, it ceases to be considered an AI task—i.e., it is devalued. This effect reflects Larry Tesler's theorem, where AI is whatever hasn't been done yet.

Scientists working on AI have attempted to solve the issue of how to represent knowledge. Their solutions have made it possible to create so-called expert systems (ESs, systems that help in decision-making by using their knowledge bases); the methods of self-learning for machines have become crucial, especially in attempts to replicate the work of the human nervous system. The latter resulted in the creation of artificial neural networks (ANNs). Thus, at the heart of all research on AI lies the idea of modeling the processes of human thinking with a computer. After all, the science of AI is a cognitive science; i.e., it studies the acquisition (collection, accumulation, and perception) of knowledge. Once AI is comparable to the human intellect, it will have an unlimited range of applications and will change our existence.

In January 2016, the founder of the World Economic Forum, Davos Claus Schwab, named artificial intelligence one of the main driving forces of the fourth industrial revolution: "This Fourth Industrial Revolution is coming at us, as a tsunami; its basic driving forces are achievements in the field of artificial intelligence, robotics, nanotechnology, the Internet of Things, and other sciences" [10].

There has been a qualitative transition from the computing era to the cognitive era (in terms futurologists and the Second Machine Age), at which point a new type of computer quickly learned how to deal with structured, unstructured, and fuzzy-structured data and thus started replacing the work of people, for a large number of cognitive tasks [2].

2 The Three Waves of Artificial Intelligence

AI has been in use for over half a century, during which it has sometimes piqued the interest of researchers and has sometimes almost completely disappeared.

The first works on AI date back to the 1950s. They started with attempts to create the following three types of programs:

1. A program for playing chess was started in 1954 by analysts A. Newell, J. Shaw, and G. Simon. A. Turing and K. Shannon volunteered to help them, as well as a group of Dutch psychologists. In 1957, the chess program (NSS) was completed.

The work was based on heuristics, which regulates decisions in the absence of theoretical grounds [9].

2. Machine translation programs that translate one natural language to another started to be developed in the USSR in 1954–1957, first translating English and Chinese on a BESM-2 computer at the Institute of Precise Mechanic and Computing Equipment (ITMiVT), under the direction of L.N. Queen. In 1954, computers developed by the IBM Corporation, under the leadership of Professor Leon Dostert, were able to translate 60 Russians phrases into English by using dictionaries, consisting of 250 total pairs of words and on the basis of six grammar rules. The results seemed promising, but the problem turned out to be much more difficult than expected. The goal of the translation was to teach computers not only rules but also exceptions, and the technical limitations of computers at that time obviated their ability to complete this task. Nevertheless, attempts to create such systems set powerful impetuses to develop mathematical linguistics.
3. The third program of the 1950s, namely one for automatic proof theorems, underwent such massive development that, in 1960, the universal solver tasks (GPS) was invented, which could automatically prove theorems from planimetry and solve algebraic problems, among other tasks [9].

In 1963, John McCarthy marked the first wave of AI by creating the first language for programming AI tasks: the list-processing (LISP) language. The advent of this language opened up functional programming. Note that these first high-level languages were procedural. The second wave of AI, which began in the late 1960s, was associated with the advent of logic programming (language prologue, 1971) and the boom in so-called expert systems (ESs). These were also the beginnings of AI, where ESs specialized in managing knowledge, interviewing experts in subject areas, manually compiling knowledge bases (BZ), drawing logical conclusions—all of which were important elements in self-learning. In addition, there were problems with experts who did not share their knowledge or who stopped doing so as soon as they realized that the introduction of ESs would lower their professional status because novice specialists could use ESs to achieve high-level results.

The creation of ESs generated great interest in the problem of representing knowledge in computer systems. As a result, semantic networks, frame systems, production systems (systems based on rules), and combinations of these started to appear over time. Meanwhile, work on machine translation systems for texts continued and made significant progress. The development of such systems became a particularly urgent task during the Cold War, when the United States gathered a large number of materials in Russian, but there were not enough human resources for their rapid translation. "In the 1990s, project companies IBM Candide was involved ten-year-old experience translations of the transcripts of the meetings of the Canadian Parliament, published in French and English—about three million sentences. Since these were official documents, their translations have been made with meeting extremely high standards. By the standards of that time the amount of data was huge. This technology, known as 'statistical machine translation,' has transformed the task of water into one big mathematical problem" [8]. However, improvements to the quality of machine translation stalled.

The second wave also includes the creation of advanced programs for games such as checkers and chess. In 1974, the first World Chess Championship featuring chess programs spelled victory for the Soviet chess program Kaissa (M. Donskoy, A. Arlazarov, A. Bitman, and A. Uskov). The success of Kaissa turned out to be a worldwide sensation because everyone predicted the victory of the US program. According to M.V. Donskoy, although Kaissa played better than earlier programs did, programs capable of defeating grandmasters were still far. In general, they were founded on formal logic, which is applicable to formalizable tasks such as logic games.

The current resurgence of interest in AI marks the third wave and differs from the previous ones in both amplitude and coverage. Solving AI problems now requires technical facilities, widespread wireless networks, the Internet, and highly advanced work in this area. The third wave started with the famous victory of the US chess program Deep Blue (2.5:3.5) over the then world chess champion, Garry Kasparov, in the six-game match.

At the beginning of the third wave, the so-called genetic programming allowed for the imitation of biological processes to solve certain tasks, such as searching for optimal solutions. Since then, the focus of AI has been subject to the vicissitudes of changing directions and the attention paid to specific programs. An intelligent software agent is an active software system or software and a hardware system (e.g., robots) that autonomously performs the function (or set of functions) for which it was created. Somet of the main problems of these agents include agents' communication, the development of languages for such communication, the coordination of agent behavior, the distribution of roles in agent coalitions, and the collective behavior of agents [3].

The current stage has been characterized by the very rapid development of ANN technologies—networks that simulate the work of biological neurons in living beings. The simplest ANN consists of three layers of artificial neurons: an input layer, hidden layers, and an output layer. The input layer receives signals from the outside world; in the hidden layers, which may consist of one to five layers, these signals are processed and transferred to the output layer; and the output layer forms the results.

Others systems have been developed thanks to new technologies. The first wave of AI development saw machine translations that have since achieved high accuracy, ESs, etc. In 2006, a Google Translate system was developed on the basis of using Big Data. The trillion words used by Google Translate in 2006 consisted of scattered snippets of Internet content. It became a "training set" for AI to calculate and learn which English words precede and follow other English words. New words that appear in natural languages can be added to the system, and obsolete ones can be removed. "Her re-water is more precise, although very far from perfection: by mid 2012. it has covered more than 60 languages and is now even capable of receiving voice input on the 14 languages for instant translation" [8].

One of the questions for modern AI is whether this third wave will end up as another disappointment while interest in AI declines or whether it will result in concerted efforts by scientific communities around the world to achieve progress in specific subject areas. According to the author, the stakes for AI are now very high and its development will only increase.

3 The History of AI's Achievements

Numerous scientific works on AI have been devoted to computer vision. This direction of AI is associated with the development deep learning (more on that below). For the first time, computers were capable of performing some visual classification tasks better than people can. For example, the accuracy of IBM Watson at determining the optimal treatment for lung cancer was 90%, outperforming the accuracy of oncologists' diagnoses by 40%.

Machine learning (also called statistical learning) is an important concept in AI. The basis of this technology was laid by Arthur Samuel in 1959, when he proposed working on teaching computers without specifically using programmed algorithms. In the simplest sense, programming is learning, but for AI to improve, it must carry out certain tasks and exercises itself.

Machine education is a framework that uses a database of training examples the computer or neural network can configurate (learning), after which it recognizes and classifies new incoming data according to the configuration of the training. Algorithms and methods teach computers to make decisions on the basis of available data. Adding training examples improves their results on recognition tasks, and in this way, self-learning takes place. Using this technology on a large database of photos, a computer taught itself to recognize faces, and it did so this with higher accuracy than a human could reach.

The real breakthrough in machine learning occurred in early 2016, when the Google AlphaGo program was able to beat Go champion Lee Sedol. This game is among the most intellectually challenging games in the world, even more difficult than chess (the board is composed of 19 × 19 cells, and there are many more positions than the corresponding number in chess), in which winning requires sorting through all the possible moves. Observing others' victories allowed AlphaGo to engage in deep machine training (deep learning, or DL), which is now the current trend in developing AI. The term *deep learning* is applied to artificial neural networks (ANNs) with more than one hidden layer, so formally, *deep* also indicates a more complex multilayer neural network architecture (see p. 134). What is unique to deep learning is that the machine itself finds signs (key traits that it uses to separate one class of objects from another) and structures them hierarchically: from the simplest to the most complex. Deep learning encompasses various technologies and works on large datasets. The program Alpha Go at first analyzed 29.4 million moves in 160,000 games between professional players, and then two copies of the program started playing against each other, adding new batches to the training set.

Having played millions of games, the program learned to evaluate the most advantageous positions on the board [5].

Deep-learning technology is now an integral part of research in the field of speech and image recognition, for creating control systems for uncrewed vehicles, diagnosing diseases, solving other complex problems, and carrying out other complex tasks. The development of deep-learning technology started with the distribution of deep learning implemented by IBM in the summer of 2017. Training via DL significantly reduces training time. What current AI cannot do and what separates it from total AI is the inability to use previously acquired skills to learn new ones; AI cannot learn new skills on the basis of previously acquired ones. It lacks the ability to generalize from its accumulated knowledge so as to use it in different contexts.

The widespread adoption of smartphones has spawned the likewise widespread usage of speech assistants, which incorporate elements of AI. The well-known applications of AI include Google (Google), Siri (Apple) Cortana (Microsoft), Echo (Amazon), and Alice (Yandex), among others, which are already used by millions of people. Data applications are also available on tablets and personal computers. By training on more and more datasets from these users, over time, these programs will become more intelligent and irreplaceable.

The next important projects for developing AI require detection structures that replicate those of the human brain. Such projects are very expensive, and their adoption has therefore been limited to a few countries and large corporations. This reverse design, which suggests that the subtleties of the human brain and how it functions must first be isolated before constructing hardware and software. Eventually, scientists hope to create a computer with human-level AI. Several important projects (with investments in the billions of dollars) are currently aiming to achieve this result. Simulating the human brain is part of a dedicated international project, namely the Human Brain Project (HBP), which includes a team from the Swiss Federal Polytechnic School in Lausanne, under the guidance of Professor Henry Markram (Henry Markram), and more than a hundred scientific groups from around the world. The project aims to synthesize all the knowledge we have on the brain into a single complete model brain housed inside a supercomputer. This project is expected to be completed by the end of 2023. Another project, the Brain Activity Map Project (2013), was designed to ensure that in 10 years, US scientists would be able to record and map the activity of each neuron in a human brain.

About a dozen less-capacious but no-less-important projects are currently studying the human brain. In one of them, Blue Brain, which is based in Switzerland, scientists are studying how neuronal ensembles work. The Synapse Project, funded by the US Defense Advanced Research Projects Agency (DARPA) and IBM, has created physical copies of the human brain in the form of special microcircuits with artificial neurons. This AI application is called neuromorphic electronics [1]. China is also developing similar projects. However, researchers' efforts have been hampered by insufficiently exhaustive brain modeling and the current technological limitations on contemporary supercomputers: This task would requires tens of exaflops, but our current supercomputers have achieved benchmark scores only one-tenth of that.

Among the other global projects that are using AI methods and system analytics is Human Genome Project, where AI is planned to be used to determine the DNA sequences of all life on the planet. For example, Amazonia Third Way Initiative is developing a project titled Earth Bank of Codes, which aims to collect the genetic sequence of the natural world [6].

4 State-Sponsored Studies on AI in Russia

Russia entered the third wave of AI with a significant delay. AI research is funded by the Russian Foundation for Basic Research (RFBR) and the Russian Science Foundation (RSF), but because of a lack of coordination, funding for research and development is generally inadequate, and the results are incongruent. The current best AI developments in Russia are military robots.

Currently in Russia, AI projects are focusing on speech recognition, facial recognition, image recognition, and autonomous vehicles, among others. Large Russian companies are already using AI for their businesses. Machine-learning company MTS predicts that customers will take advantage of AI services to save themselves time, which will at the same time increase sales [2].

Sberbank, together with MIPT, is developing a project to use AI technology in a call center. The objective of this neurointelligence project, called iPavlov, is to develop algorithms using deep learning to create conversational AI. This technology should reduce the burden on employees and raise the quality of service for clients [3]. The Yandex company develops AI applications for production optimization in industry. In addition, the company introduced the CatBoost machine-learning method, the successor method of Matrixnet, which is used in the delivery of almost all of Yandex's services [3]. ABBYY uses AI technology in decision-making. FlexiCapture, for example, is a versatile platform using AI to process information. It determines the type of document, extracts data from it, and sends those data to information systems, such as CRM, accounting systems, and procurement management systems [3]. Finally, AI applications are being developed for robots and uncrewed aerial vehicles (UAVs), such as drones.

Universities independently conducts scientific research on AI, and large-scale research projects on AI featuring university collaborations are nearly to completely absent. The greatest concern here is teaching and learning cognitive sciences. Many of these disciplines simply do not have qualified teachers. Training programs lag far behind the needs of the labor market.

In general, there are about a dozen strong teams in Russia working on AI.

The Federal Research Center in Russia, "Informatics and Control" (FIC PS RAS), has developed several applications for AI [2]:

1. Exactus Expert is a system for evaluating the quality of scientific publications on semantic research and analysis
2. Exactus Patent is used for patent information for semantic research and analysis

3. Exactus Like is used for detecting similar texts and calculating the degree of semantic proximity
4. Text Appliance conducts system analyses on unstructured information.

The main problem is that Russia has so few research fundamentals to develop products. There is a gap between the science, which survived the "dashing '90s," and the implementation of its developments as finished products [7].

5 Main Directions for Future Work in AI

Works in the field of AI is now underway in many countries and is being deployed broadly. Such works can be grouped under the following directions:

1. In deep learning, efforts have been aimed at significantly reducing the training time of the ANN and reducing the volume of teaching examples. An ideal ANN should learn in real time.
2. Robots should synthesize their answers on the basis of the body of knowledge that was loaded into them, according to the context and sequence of words. Robots should learn to correlate incoming information with their knowledge base and apply it to other topics [8].
3. Neuromorphic microcircuits and computers based on them, such as microchips, have already been released by corporations such as IBM and Intel. They ca contain up to 4096 artificial neurons and up to 256 million synapses, so this direction is very promising.
4. Work continues on mapping the human brain and modeling how it works.
5. Speech interfaces require systems and services that can recognize and clearly understand speech. They must also be able to discern the contexts of speech because context is an essential part of natural language. Studies have shifted to developing systems capable of interacting with humans through dialogue, not just reacting to default requests. Broadcast search systems and a number of popular voice query applications, especially those on smartphones, have already become commonplace, and the volume of such queries will only increase.
6. The development of brain–computer interfaces started rapidly, has since slowed down, but will continue.
7. Researchers are studying the group behavior of robots and the interactions between robots and humans in the course of performing operations. Their efforts have focused on determining effective distribution tasks that incorporate both humans and machines. This direction is important for both military robots, service robots, and robots employed in production.
8. AI is being implemented in the Internet of things (IoT), Industry 4.0, and other parts of our digital world to process the vast numbers of data generated by such systems, because without AI, completing such tasks would be too onerous.
9. The navigation systems for autonomous vehicles use AI in cities. This is one of the most promising directions from a market viewpoint. This technology should

be available on uncrewed vehicles and other autonomous vehicles starting in 2022 and be widespread by 2025.

10. Researchers should consider developing a new class of systems, such as a "human intelligence on demand" service, robots as a service (RaaS), etc.
11. Hyperpersonalization services could keep users constantly connected to the network; then on the basis of their past data and by using behavior modeling, applications using AI could make suggestions for which services they should offer and which decisions should be made under various circumstances.
12. Future works need to solve AI's catastrophic forgetting. When an AI system is confronted with something that it has never been trained on, it will constantly repeat the same mistakes. Attempts to retrain current systems have led to a phenomenon called catastrophic forgetting, where AIs forget information and knowledge from previous tasks while they are learning new ones, because its study of the new subject matter violates its prior knowledge [7].
13. A regulatory framework for using robots still needs to be developed. "Robots as subjects of law have not yet been legally formalized in any country in the world, since for this it is necessary to officially recognize that a robot is not just a thing but a creation, endowed, albeit artificial, but with intelligence, as well as with its own will. Therefore, there are still no laws in the world that direct or prohibit robots directly." Legislators face the difficult task of balancing interests in technical progress, security for societies and states, and people's individual needs. In 2008 in South Korea, introduced a law promoting the development and spread of smart robots. In February 2017, the European Parliament adopted the Civil Law Rules on Robotics, which established ethical principles that aim to protect humankind from robots [4].
14. AI still requires further development in understanding context, reasoning, precedents, and more.
15. According to a number of experts, current developments in the field AI lack the hardware and software to properly combine the virtual and real worlds, though it is being attempted in robotics, such as with tactile hand-surface robots.

6 Conclusion

Over the past decades, the possibility or impossibility of modeling human thinking has remains one of the major philosophical problems in the field of AI because that modeling could allow AI to gain consciousness. The question is no longer entirely hypothetical and obviously remains crucial for predicting the future of the technology, not only Russia but also other parts of the planet. Half a century of the history of AI development has shown that such predictions are unreliable. Although AI can train on neural networks, they can do so only in the ways that they have already mastered. Other scientists are studying such problems by using modeling and simulations [8, 9].

References

1. Belov, S., Katkalo, V.: Lack of artificial intelligence. Access mode: https://www.vedomosti.ru/opinion/articles/2017/03/21/681987-defitsit-iskusstvennogo-intellectual. Last accessed 10 Feb 2022
2. Bessmertny, I.A.: Artificial Intelligence, p. 132. St. Petersburg State University ITMO, St. Petersburg (2010)
3. Brink, X., Richards, D., Feverolf, M.: Machine Education, p. 336. Peter, St. Petersburg (2017)
4. Vasilyeva, D.: Trends in the development of artificial intelligence (2018)
5. Artificial intelligence (AI) as a key factor in the digitalization of the global economy. Mode access: https://www.crn.ru/news/detail.php?ID=117544. Last accessed 10 Feb 2022
6. Omarov, O.A., Omarova, N.O., Omarova, P.K.: Formation of waves in explosive processes on a cathode with and without an external magnetic field, 2018. J. Phys. Conf. Ser. **1094**, 012028 (2018). https://doi.org/10.1088/1742-6596/1094/1/012028
7. Omarov, O.A., Omarova, N.O., Omarova, P.K.: The construction of a pulsed-periodic thermonuclear reactor using high pressure gases in strong magnetic fields 2018. J. Phys. Conf. Ser. **1094**, 012021 (2018). https://doi.org/10.1088/1742-6596/1094/1/012021
8. Dimitrios, N.K., Sakas, D.P., Vlachos, D.S.: The role of information systems in creating strategic leadership model. Procedia Soc. Behav. Sci. **73**, 285–293 (2013)
9. Nasiopoulos, D.K.: Logistics companies reengineering. Change of leadership and human resources planning through management information systems. In: Business Intelligence and Modelling, vol. 6, pp. 207–217. Springer, Cham (2021)
10. World Economic Forum, Davos Claus Schwab, January 2016

Evaluating an Impact of Digital Transformation

Ilya Vasilyevich Solntsev and Tatiana Vyacheslavovna Bart

1 Introduction

Despite the presence of a large number of studies on the digital economy [1]; Institute for Emerging Market Research of the SKOLKOVO Business School [2, 6, 8], most of them only indirectly affect the effects generated by digital transformation.

Litvintseva and Karelin [3] studied the impact of digital factors on the well-being of the population in the constituent entities of the Russian Federation, measured through gross regional product per capita. The digital component of the quality of life of the population was determined on the basis of the Russian digital index. As a result, it was revealed that the positive impact of digitalization is observed in economically developed and rich regions of Russia, while in less developed regions, this impact may be accompanied by digital risks.

In the work of Eskindarov et al. [10], the risks and benefits of digital transformation in Russia are analyzed.

Wang et al. [34] have shown that for Chinese companies, having a digital transformation strategy has a positive impact on short-term and long-term financial results.

Important aspects of digital transformation are: increasing customer awareness [23], influence on purchasing decisions [18], and increasing the ability to evaluate products and services on the Internet.

I. V. Solntsev · T. V. Bart (✉)
Marketing & Sports Business Department, Financial University Under the Government of the Russian Federation, Moscow, Russian Federation
e-mail: tbart@mail.ru

D. P. Sakas et al. (eds.), *Computational and Strategic Business Modelling*,
Springer Proceedings in Business and Economics,
https://doi.org/10.1007/978-3-031-41371-1_24

Digital products are more user-friendly [37] and available through various channels 24/7 [12, 35].

Organizations, in turn, better understand their customers [24] and adjust their internal processes, resources, and costs accordingly [11, 16, 28, 36] and develop new types of relationships with them [36].

Tolboom [33] has analyzed the impact of digital transformation across nine dimensions of the Canvas business model and found that almost all of them contain at least one element that is expected to change moderately or even more. The greatest impact can be made on the value proposition of organizations, customer segments, relationships with them, and the resources used. The author assumes that due to digital transformation, the range of products and services, their availability and usability will increase; productivity will increase, whereas prices will slightly decrease. There will be increased awareness of products and services in the market, and it will become easier to evaluate them. In addition, customers will have more shopping opportunities, and the level of service will increase.

Solntsev and Petrenko [9], considering approaches to the formation of a digital transformation strategy, note that the assessment of the relevant effects can be carried out at the state level (GVA, jobs, import substitution, and social sphere) and at the level of an individual company, taking into account industry specifics. The authors also classify emerging effects into tangible and intangible ones and provide a list of indicators that characterize the effectiveness of digital transformation.

At the same time, even in those works where the effects of digital transformation are considered, approaches to their assessment are not proposed. The purpose of this study is to eliminate this methodological gap and form increments of scientific knowledge through the solution of the following tasks:

1. Analyze and classify the effects generated by digital transformation.
2. Suggest approaches to assessing these effects at the level of the state and companies.

2 Effects of Digital Transformation

For the most detailed analysis of the effects of digital transformation, see the OECD report "Measuring the Digital Transformation. A Roadmap for the Future" [29]:

1. Growth in labor productivity – the volume of output (value added) produced per unit of labor input (the number of employees or, if data permits, the number of hours worked). The cost for each industry is calculated relative to the economy as a whole (that is, GDP per person employed in each country). Interestingly, on average across the OECD, the level of labor productivity in the "information industries" is about 65% higher than in the rest.

2. Cross-industry effects and gross value added. The development of information technologies ensures the growth of production volumes not only in the information industries but also allows the development of related sectors; one example is the glass of a smartphone screen. The global demand for goods and services in the information industries through international trade and investment can stimulate the activity of many other leading domestic industries. Thus, high-tech products in the present-day market are the result of numerous stages of production in many countries. Combining the value added generated by the information industries with the value added of other domestic industries, embodied in the global demand for information sector products, represents the first step towards defining an "extended information footprint".
3. New jobs. Although digitalization often results in the replacement of human labor, the development of new products and the need for the resources needed creates a demand for human capital. For example, according to the OECD, for every 10 additional jobs created in the European Union between 2011 and 2017, four were in IT-intensive occupations. According to the study Lyashok et al. [5], new technologies lead to increased employment and reduced working hours. In addition, digitalization has a positive impact on wages [4].
4. Well-being. The OECD identifies 11 parameters of well-being and calculates a special Better Life Index.[1] The impact of digital transformation on these dimensions is ambiguous – there are both risks and opportunities in areas such as work–life balance, social connections, governance, and civic engagement. New technologies allow workers to be more mobile. Often there is no need to be physically at the workplace all the time. Remote work allows you to manage your time more flexibly and can contribute to increased job satisfaction and work–life balance. At the same time, the employer's expectations of constant communication and an increase in the number of working hours outside of working hours are increasing. It should also be noted that access to and use of teleworking tools is biased towards highly skilled workers and that the burden of reconciling work and family life through teleworking can often fall on women more than men [19].

   Work is closely related to self-esteem and the well-being of people. In particular, the Internet can be useful when looking for a new job. The Internet also provides a new arena for people to engage in civic and political debate. This aspect of digital transformation is sometimes seen as a risk as online political participation is thought to exacerbate ideological divisions. However, recent research has found only limited evidence that political polarization may be related to online media use [21].

   Satisfaction is typically measured through face-to-face surveys. Ideally, indicators of the welfare impact of digital transformation should reflect not only people's use of digital technologies but also how much this use makes them more or less satisfied with their lives.

[1] https://www.oecdbetterlifeindex.org/ru/#/11111111111

5. Success in education. At the state level, this group of indicators primarily assesses the impact of digital technologies (primarily the Internet) on learning outcomes at school. For example, the Program for International Student Assessment (PISA) is conducted every 3 years for 15-year-old students nearing the end of their compulsory education. PISA assesses how well they can apply what they learned in school to real-life situations. Across the OECD, 90% of schoolchildren enjoy using digital devices, and 61% reported forgetting time when using them. About 55% of students in OECD countries report feeling unwell when not connected to the Internet. In countries such as France, Greece, Portugal, and Sweden, this ratio reaches about 80% compared to about 40% in Estonia and Slovenia.

   At the corporate level, educational programs and advanced training of employees can also become one of the objects of evaluation in the framework of digital transformation projects. According to the OECD, employees of companies with a high digital intensity are more likely to participate in training than others. In addition, workers performing non-standard tasks or tasks that require intensive use of information technology usually have relatively higher skills. Company-based learning helps motivate and reward employees and align their competencies with the needs of the firm. Training can also help reduce income inequality and equip low-skilled workers with the skills they need to manage digital transformation. However, evidence suggests that much of the training is aimed at upskilling middle- and high-skilled workers rather than low-skilled workers.
6. Health. Digital technologies have a multidirectional impact on human health. For example, online resources can help people better understand and manage their health. The demand and popularity of online booking are influenced by a wide range of factors, including an aging population that increases the need for healthcare, the skills people have, and the extent to which online booking offers better services compared to other channels. In addition, in 2018, half of all people aged 16–74 in the OECD accessed health information online.

   Information technology is aimed at optimizing working time and providing new ways of organizing work. However, the Internet and other digital tools have dramatically increased the flow of information managed by employees, which has a direct impact on perceived stress levels. Research has shown new forms of information flows in a wide range of areas such as investment analysis, management decision-making, pricing, and many others. These information flows pass through a range of digital media such as email, intranets and messaging systems [22]. The resulting information overload is associated with technostress, "a form of stress associated with people's attempts to cope with ever-evolving information technology and changing physical, social, and cognitive responses" [13, 17, 30]. Workplace information overload reduces job satisfaction and results in poorer health [27, 30], while perceived email overload is associated with burnout and reduced work engagement [31].
7. New business opportunities. Government support can help people build digital businesses.

8. Social integration. Digital technologies have revolutionized many everyday activities, from communicating with others to shopping and banking, searching for information, and entertainment content. Social networks allow people to maintain existing social relationships and build new ones. At the same time, data on the impact of social networks on real social connections and mental health is ambiguous. Some scholars warn that social networking may crowd out face-to-face communication and reduce the quality of face-to-face contact [32]. However, there is plenty of evidence to support the idea that social media does increase social capital [20]. However, it is likely that not all sectors of society benefit equally from social media. For example, people with physical disabilities (including the elderly) may benefit more from online networks but are less likely to have the skills needed to access and use them.

3 Risks of Digital Transformation

Digital transformation not only creates a number of opportunities but also comes with various risks that can affect the lives and well-being of people. Thus, a constant connection to the Internet poses a potential risk to free time and the mental health of workers. The time spent on email outside of work and the employer's expectation that employees are available at all times have been shown to significantly reduce people's satisfaction with their work–life balance [15]. Workers who perform computer-intensive jobs are more likely to experience anxiety outside the workplace than those whose work does not require intensive use of information technology. Connected devices could also allow employers to track behavior and performance in new ways.

Another negative effect is disinformation and cyberbullying, the distribution of which has become more accessible with the use of digital technologies. While the former is not necessarily illegal and largely depends on people's ability to think critically, exposure to cyberbullying can lead to serious mental health problems such as anxiety, depression, and even self-harm [25].

An important aspect is information security. Digitization of information and network connectivity creates new challenges for protecting sensitive data and network communications. At the corporate level, it is important to approve a formal information security policy. National digital security strategies describe how countries prepare for and respond to attacks on their digital networks. They can be considered an important aspect of national preparedness in terms of managing digital security risks.

Finally, individuals are also increasingly providing (sometimes unknowingly) personal information to service providers and online platforms. The digitization of information and improved network connectivity create new challenges for protecting personal data, while attacks occur on a regular basis. In addition, people's privacy can be affected by the extraction of additional information that can be obtained by analyzing existing data to identify patterns and correlations, many of which are not necessarily considered personal data.

Among the negative manifestations of digitalization, it is also necessary to take into account environmental aspects. The production and use of information products are associated with the formation of "greenhouse gases" such as carbon dioxide (CO_2). The amount of CO_2 produced by the information industry in relation to the amount of output produced varies greatly between countries. Data from the United Nations Environmental and Economic Accounting shows that in most European countries, less than 5 tons of CO_2 is produced for every million dollars of information industry output. This figure largely depends on the volume of production of information products and the extent to which each country relies on fossil fuels for electricity generation. The products of digital industries also pose an environmental problem at the end of their life cycle, leading to an increase in e-waste. Improper and unsafe handling and disposal, including by open burning, pose significant risks to the environment and human health, and hinder the achievement of sustainable development goals. The proliferation of digital technologies and device types also leads to the generation of electronic waste – E-waste [14].

E-waste also refers to all items of electrical and electronic equipment that have been discarded. This includes cooling and freezing equipment, screens and monitors, lamps, as well as small appliances and telecommunications equipment – mobile phones, personal computers and printers. In OECD countries, there were 17 kg of e-waste per capita in 2016, equivalent to 41 kg per $100,000 of GDP. E-waste per capita ranges from nearly 30 kg per person in Norway to 8 kg in Turkey and less in countries such as China and India, broadly in line with digitalization in these countries.

However, it should be noted that in some cases technology can replace other polluting activities. For example, the use of telecommuting technologies can help reduce commuting emissions. The environmental problems posed by e-waste, which often contains hazardous and environmentally harmful substances, can be mitigated through recycling and reuse.

Eskindarov et al. [10] additionally highlight the following digital transformation risks:

1. Dependence on the Internet and foreign technological solutions
2. Reduction of jobs with low and medium skill levels
3. Digital inequality at the level of countries, regions, companies, and citizens, which allows IT companies to influence decision-making by economic entities, shape public opinion, and influence political processes
4. Reducing the possibilities of control by the state, including the implementation of protectionist measures to protect national producers
5. Decrease in the income of Russian companies, caused by a shift in consumer demand towards foreign products and, as a result, a reduction in tax payments
6. Formation of a new shadow sector of the economy (Internet fraud, DarkNet, and cyber-attacks) and the need to establish responsibility for "digital" crimes

4 Assessing the Effects of Digital Transformation

For the state, the result of assessing the socio-economic effects of the implementation of digital transformation projects can be a monetary assessment of the growth in gross domestic product, calculated on the basis of gross value added and tax payments. At the same time, the increase in gross value added should be discounted taking into account the project implementation period. The average yield of long-term federal loan bonds with a maturity of 10 years can be used as a discount rate.

The increase in gross value added is formed due to two components:

1. The IT industry that ensures the release of Russian products:
2. Related industries that ensure the production of the necessary Russian products (cross-industry effects):

By analogy with gross value added, tax and other payments to the budget system of the Russian Federation are divided into payments by IT companies and their counterparties.

The main advantage of this approach can be considered as a monetary assessment of the effects generated by digital transformation. At the same time, it is good practice to analyze "natural" indicators that take into account industry specifics. On the example of oil and gas companies, such metrics include: growth in reserves, production, depth of processing; reduction of repair time, downtime, loss of hydrocarbons; productivity growth (refineries and individual fields); energy saving; and number of automated filling stations.

Companies with state participation can also disclose the volume of increases in tax and dividend payments to the budget, but in this case, it will be quite difficult to allocate the part attributable to digital transformation projects.

Also, to assess the effects of digital transformation, a cost approach to management can be used, including such an indicator as economic value added:

$$EVA = (ROIC - WACC) \times CE \quad (3)$$

where:

ROIC – return on invested capital;
WACC – cost of invested capital;
CE – capital invested in DH.

Value will only be created if the return on digital investment exceeds the cost of capital. In the case under consideration, the volume of investments in digital transformation projects with a division into various sources of capital and their cost can be used as invested capital. Thus, the main advantage of this approach can be considered a net increment in the value of the company, realized through digital transformation. At the same time, the calculation of profitability will be difficult, because digital projects are often embedded in existing business processes and may not generate funds on their own. In this situation, it will be necessary to calculate the "digital" increment in operating profit after tax (NOPAT), which will be the basis for calculating ROIC.

Many digital transformation projects are aimed at solving social problems and improving the comfort level of citizens. To assess non-material effects associated primarily with the growth of satisfaction of citizens and employees (which is the result of the introduction of digital products), surveys built according to the "willingness to pay" method can be used, which allows to estimate the amount that the respondent is ready to give for a specific product/service/experience/convenience.

The main advantages of this technique include the simplicity and accuracy of the results. At the same time, it is necessary to note the complexity of forming a sample, which should take into account the individual characteristics of the respondents: gender, age, and income level. It is also extremely important to correctly formulate questions, accurately describing the subject of research and the conditions in which the choice will be made.

5 Conclusion

As part of this study, the effects generated by digital transformation projects were analyzed. The results showed that these effects are formed not only at the company (or project) level but also affect a whole block of government issues in the economy and social sphere. Despite the presence of a large database of works devoted to the digital economy, they only indirectly affect the problem of assessing the generated effects and do not contain methodological proposals or a criteria platform.

The author proposes the following approaches to assessing the effects of digital transformation:

1. At the state level – through an assessment of the growth in gross domestic product, calculated on the basis of the growth in gross value added and tax payments.
2. At the corporate level – through the calculation of economic value added.
3. At the level of assessing non-material effects, including the growth in well-being and satisfaction – through the use of the "willingness to pay" methodology.

Other scientists with similar cases are studying similar problems, using modeling and simulation [38, 39].

Further research in this area may be aimed at a more detailed study of methods for assessing each of the considered effects, as well as at justification of a special digital index by analogy with the OECD digital government concept.

References

1. World Bank: Competition in the Digital Age: Strategic Challenges for the Russian Federation (2018). https://www.vsemirnyjbank.org/ru/country/russia/publication/competing-in-digital-age. Date accessed: 11 Jul 2021
2. Institute for Emerging Markets Research of the SKOLKOVO Business School (IEMS): The digital life of Russian million-plus cities. Model. Dynamics. Examples. (2016) https://iems.skolkovo.ru/downloads/documents/SKOLKOVO_IEMS/Research_Reports/SKOLKOVO_IEMS_Research_2016-11-30_ru.pdf. Date accessed: 11 Jul 2021
3. Litvintseva, G.P., Karelin, I.N.: Effects of digital transformation of the economy and the quality of life of the population in Russia. Terra Econ. **18**(3), 53–71 (2020). https://doi.org/10.18522/2073-6606-2020-18-3-53-71
4. Lukyanova, A.L.: Digitization and the gender pay gap. Economic Policy. T. **16**(2), 88–117 (2021). https://doi.org/10.18288/1994-5124-2021-2-88-117
5. Lyashok, V.Y., Maleva, T.M., Lopatina, M.V.: The impact of new technologies on the labor market: past lessons and new challenges. Economic Policy T. **15**(4), 62–87 (2020). https://doi.org/10.18288/1994-5124-2020-4-62-87
6. NRU HSE: Digital transformation of industries: starting conditions and priorities: Ts75 reports. to XXII Apr. int. scientific. conf. on the problems of economic and social development, Moscow, 13–30 April. 2021 / G. I. Abdrakhmanova, K. B. Bykhovsky, N. N. Veselitskaya, K. O. Vishnevsky, L. M. Gokhberg and others; hands. ed. count P. B. Rudnik; scientific. ed. L. M. Gokhberg, P. B. Rudnik, K. O. Vishnevsky, T. S. Zinina; Nat. issled. University "Higher School of Economics". – M.: Izd. House of the Higher School of Economics, 2021. – 239, [1] p. – ISBN 978-5-7598-2510-4 (in the region). – ISBN 978-5-7598-2270-7 (e-book).
7. RANHIGS: Digital Transformation Strategy: Write to Execute / S83 ed. E. G. Potapova, P. M. Poteeva, M. S. Shklyaruk. – M.:
8. RANEPA: (2021) 184 p. http://strategy.cdto.ranepa.ru/Strategy_CDTO_RANEPA.pdf. Date accessed: 12 Jul 2021
9. Solntsev, I.V., Petrenko, E.S.: Strategy of digital transformation in industry: structure and target indicators. Issues of Innovative Economics. **11**(2), 681–702 (2021). https://doi.org/10.18334/vinec.11.2.112287
10. Eskindarov, M.A., Maslennikov, V.V., Maslennikov, O.V.: Risks and chances of the digital economy in Russia. Finance: Theory Pract. **23**(5), 6–17 (2019). https://doi.org/10.26794/2587-5671-2019-23-5-6-17
11. Agarwal, R., Dhar, V.: Editorial—big data, data science, and analytics: the opportunity and challenge for IS research. Inf. Syst. Res. **25**(3), 443–448 (2014)
12. Andal-Ancion, A., Cartwright, P.A., Yip, G.S.: The digital transformation of traditional businesses. MIT Sloan Manag. Rev. **44**(4), 34–41 (2003)
13. Arnetz, B.B., Wiholm, C.: Technological stress: psychophysiological symptoms in modern offices. J. Psychosom. Res. **43**, 35–42 (1997). https://doi.org/10.1016/S0022-3999(97)00083-4
14. Baldé, C.P., Forti, V., Gray, V., Kuehr, R., Stegmann, P.: The Global E-waste Monitor. United Nations University (UNU), International Telecommunication Union (ITU) & International Solid Waste Association (ISWA), Bonn/Geneva/Vienna (2017)
15. Belkin, L.Y., Becker, W.J., Conroy, S.A.: Exhausted, but unable to disconnect: after-hours email, work–family balance and identification. Proceedings. **2016** (2016). https://doi.org/10.5465/ambpp.2016.10353abstract
16. Bharadwaj, A., El Sawy, O., Pavlou, P., Venkatraman, N.: Digital business strategy: toward a next generation of insights. MIS Q. **37**(2), 471–482 (2013)
17. Brod, C.: Technostress: The Human Cost of the Computer Revolution. Addison-Wesley Publishing Company, Reading, USA (1984)
18. Clemons, E.K.: How information changes consumer behavior and how consumer behavior determines corporate strategy. J. Manag. Inf. Syst. **25**(2), 13–40 (2008)

19. Dettling, L.J.: Broadband in the labor market: the impact of residential high-speed internet on married women's labor force participation. ILR Rev. **70**(2), 451–482 (2016). https://doi.org/10.1177/0019793916644721
20. Dienlin, T., Masur, P., Trepte, S.: Reinforcement or displacement? The reciprocity of FtF, IM, and SNS communication and their effects on loneliness and life satisfaction. J. Comput.-Mediat. Commun. **22**, 71–87 (2017). https://doi.org/10.1111/jcc4.12183
21. Elizabeth Dubois & Grant Blank: The echo chamber is overstated: the moderating effect of political interest and diverse media. Inf. Commun. Soc. **21**(5), 729–745 (2018). https://doi.org/10.1080/1369118X.2018.1428656
22. Eppler, M.J., Mengis, J.: The concept of information overload: a review of literature from organization science, accounting, marketing, mis, and related disciplines. Inf. Soc. **20**(5), 325–344 (2004). https://doi.org/10.1080/01972240490507974
23. Hennig-Thurau, T., Malthouse, E.C., Friege, C., Gensler, S., Lobschat, L., Rangaswamy, A., Skiera, B.: The impact of new media on customer relationships. J. Serv. Res. **13**(3), 311–330 (2010)
24. Granados, N., Gupta, A.: Transparency strategy: competing with information in a digital world. MIS Q. **37**(2), 637–642 (2013)
25. Lindert, J.: Cyber-bullying and it its impact on mental health. Eur. J. Pub. Health. **27**(suppl_3) (2017). https://doi.org/10.1093/eurpub/ckx187.581
26. Javier López González & Marie-Agnes Jouanjean, 2017. Digital Trade: Developing a Framework for Analysis OECD Trade Policy Papers 205, OECD Publishing
27. Misra, S., Stokols, D.: Psychological and health outcomes of perceived information overload. Environ. Behaiv. **44**(6), 737–759 (2012). https://doi.org/10.1177/0013916511404408
28. Markus, M.L., Loebbecke, C.: Commoditized digital processes and business community platforms: new opportunities and challenges for digital business strategies. MIS Q. **37**(2), 649–654 (2013)
29. OECD: Measuring the Digital Transformation. A Roadmap for the Future. 260 pages (2019). https://doi.org/10.1787/9789264311992-en
30. Ragu-Nathan, T.S., Tarafdar, M., Ragu-Nathan, B.S., Tu, Q.: The consequences of technostress for end users in organizations: conceptual development and empirical validation. Inf. Syst. Res. **19**(4), 417–433 (2008). https://doi.org/10.1287/isre.1070.0165
31. Reinke, K., Chamorro-Premuzic, T.: When email use gets out of control: understanding the relationship between personality and email overload and their impact on burnout and work engagement. Comput. Hum. Behav. **36**, 502–509 (2014)
32. Rotondi, V., Stanca, L., Tomasuolo, M.: Connecting alone: smartphone use, quality of social interactions and well-being. J. Econ. Psychol. **63**, 17–26 (2017). https://doi.org/10.1016/j.joep.2017.09.001
33. Tolboom, I.H.: The Impact of Digital Transformation. A Survey Based Research to Explore the Effects of Digital Transformation on Organizations. Delft University of Technology, Faculty of Technology, Policy and Management (2016) https://repository.tudelft.nl/islandora/object/uuid:d1d6f874-abc1-4977-8d4e-4b98d3db8265/datastream/OBJ1/download (дата обращения: 11.07.2021)
34. Wang, H., Feng, J., Zhang, H., Li, X.: The effect of digital transformation strategy on performance: the moderating role of cognitive conflict. Int. J. Confl. Manag. **31**(3), 441–462 (2020). https://doi.org/10.1108/IJCMA-09-2019-0166
35. Weill, P., Woerner, S.: Optimizing your digital business model. MIT Sloan Manag. Rev. **54**(3), 71–78 (2013)
36. Westerman, G., Calmejane, C., Bonnet, D.: Digital Transformation: A Roadmap for Billion-Dollar Organizations. MIT Center for Digital Business, 1–68 (2011)

37. Yoo, Y.: Computing in everyday life: a call for research on experiential computing. MIS Q. **34**(2), 213–231 (2010)
38. Sakas, D., Vlachos, D., Nasiopoulos, D.: Modelling strategic management for the development of competitive advantage, based on technology. J. Syst. Inf. Technol. (2014)
39. Trivella, L., Dimitrios, N.K.: Knowledge management strategy within the higher education. The case of Greece. Procedia-Soc. Behav. Sci. **175**, 488–495 (2015)

Part IV
10th Symposium on Management Challenges (IANOS): Crisis Kills or Links?

Panagiotis Trivellas

This symposium attempts to cast light on concurrent advances on the fields of Organisational Behaviour, HRM, Innovation and Knowledge Management, Strategic Leadership, Sustainability, SCM, Cultural Heritage and Higher Education. Diverse and modern perspectives on the development and evolution of the relative disciplines on regional and national level are discussed and debated.

However, does the current crisis impose threats to survival (***Knowledge Innovation Leadership Learning Strategy: KILLS***?) or yield opportunities for changes and synergies to growth (***Leadership Innovation Networking Knowledge Strategy: LINKS***?)? Are the same fundamental cornerstones such as leadership, strategy, knowledge and innovation produce both excellence and perish?

In the light of the current recession, this symposium aims to track different views and discuss and publish research on the challenges in the field of management which influences societies, cultures, networks, organizations, teams and individuals. Furthermore, this session puts special emphasis on the investigation of those business processes fostering innovation and facilitating management transitions from dominant structures to more evolutionary, developmental paradigms.

P. Trivellas
Agricultural University of Athens, Athens, Greece

Towards the Development of Context-Sensitive Agrifood Entrepreneurial Skills Frameworks

Anna Anastasopoulou, Haris Marentakis, and Panagiotis Trivellas

1 Introduction

In agrifood ecosystems in Europe that are constantly changing and experiencing major disruptions (e.g., the ongoing COVID-19 pandemic or the war in Ukraine) much depends on finding creative ways to adjust to new circumstances, demonstrating resourcefulness, flexibility, and readiness for new learning. New learning that targets the upskilling and the reskilling of the labor force in the agrifood sector, as well as the identification of people with specialized skills, are two key factors in creating dynamic and resilient agrifood ecosystems.

Between 2012 and 2015, the European Commission funded a pilot project which included a survey of 2205 young farmers under 40, focus group discussions, and interviews with key stakeholders, to provide a comprehensive assessment of their specific needs and inform EU policy-making [42]. This project revealed that the knowledge needs of the interviewed young farmers are focused on entrepreneurial skills related to marketing, finance, networking, and communication followed by technological and managerial skills. The regulatory framework for EU rural development 2014–2020 already highlighted that knowledge transfer and diffusion should be adapted to the needs of rural actors, who require an appropriate level of training. Young farmers constitute an important group as they can contribute the most to fostering sustainability, innovation, resilience, and resource efficiency, which are necessary to achieve the SDGs and EU objectives. Besides, the declining number of young farmers in the agricultural sector [8] holds back generational renewal and raises concerns about the knowledge capital in this sector. According to the new

---

A. Anastasopoulou (✉) · H. Marentakis · P. Trivellas
Department of Agribusiness & Supply Chain Management, ORIMAS Lab,
Agricultural University of Athens, Thiva, Greece
e-mail: ptrivel@aua.gr

D. P. Sakas et al. (eds.), *Computational and Strategic Business Modelling*,
Springer Proceedings in Business and Economics,
https://doi.org/10.1007/978-3-031-41371-1_25

Common Agricultural Policy of the EU, agriculture is required to achieve higher levels of production of safe and quality food, while preserving the natural environment [16]. The evolution of agriculture, as well as the challenges faced by farm businesses in multiple crisis contexts, require an appropriate level of training, cultivating not only technological and managerial but also entrepreneurial skills. In line with this logic, farm entrepreneurship should enable socially and environmentally responsible farming, promoting resilient food systems with properly functioning supply chains, as well as other strategic choices, such as those related to the succession and diversification of farms.

Innovations, automation, rapid and often unpredictable changes in production and consumption, and global and regional natural and social phenomena, such as climate change, biodiversity loss, and population aging, are just some of the factors that may shape what agrifood businesses and employees need to know and be able to do. Moreover, there is a great degree of uncertainty over what new agrifood industries will emerge and the skills that will be in high demand in the future. Therefore, there is a need to develop a better understanding of the relationship between skills and agrifood labor market needs, such as trends for skills supply and demand and skill mismatches, and adopt solutions aiming to bridge gaps between the skills available in the labor market, the demands for current and future skills, and the education and training opportunities, and provisions.

A solution to bridge such gaps is to design instruments that facilitate communication and the development of shared understandings regarding occupations, skills, and qualifications. This is exactly what so-called skills frameworks are aimed for. They attempt to standardize skills taxonomies and ontologies, thus offering a common skills language for policy-makers, administrators, individuals, education and training providers, and the business and workplace world, to help the sharing of information on sectors, careers, jobs, and skills required for occupations and job roles.

Given that the knowledge needs of particularly young farmers are primarily focused on entrepreneurial skills, this study aims to set the basis for a discussion and empirical validation of a framework for entrepreneurial skills in agrifood ecosystems across Europe. Toward this aim, the following steps have been taken:

- Review of existing skills frameworks
- Mapping of entrepreneurial concepts with existing skills frameworks
- Preliminary literature review on entrepreneurial skills in agrifood ecosystems
- Mapping of entrepreneurship-related concepts discussed in the reviewed agrifood literature to identify gaps and needs for further research

2 Identification and Mapping of Entrepreneurial Skills in Three Comprehensive Skills Frameworks: The Cases of O*Net, SCT, and ESCO

In this chapter three of the most developed comprehensive frameworks of capacities and skills (O*Net, SCT, and ESCO) are reviewed to identify and map "entrepreneurship-related" concepts to corresponding capacities, skills, and/or abilities in these taxonomies.

In the USA, the Occupational Information Network (O*Net) framework [25, 26] is aimed to specify the character of occupations (via job-oriented descriptors) and people (via worker-oriented descriptors). In the context of O*Net, skills are defined as "... a person's level of proficiency or competency to perform a task" ([27], p. 464). Furthermore, abilities are defined as "... relatively enduring basic capacities for performing a wide range of different tasks" (ibid., p. 457). The O*Net framework adopts a skills hierarchy that includes a total of 35 basic and cross-functional skills. It further identifies 52 abilities organized hierarchically into four ability types (cognitive, physical, psychomotor, and sensory), six work values, defined as aspects of work that are important to a person's satisfaction, and seven work styles, which are personal characteristics that can affect how well someone performs a job.

In Canada, the "Skills and Competencies Taxonomy" (SCT) is meant to address the needs of the world of work to improve the comparability and application of skills concepts throughout sectors and occupations. It defines skills as "developed capacities that an individual must have to be effective in a job, role, function, task, or duty" [17]. Personal abilities and attributes are defined as the "inherent and developed aptitudes that facilitate the acquisition of knowledge and skills to perform at work" (ibid.). Finally, competencies are defined as "the combined utilization of personal abilities and attributes, skills and knowledge to effectively perform a job, role, function, task, or duty." At the level of skills, the taxonomy distinguishes 47 total skills organized into five skill types (foundational, analytical, technical resource management, and interpersonal). Furthermore, the SCT taxonomy includes 85 abilities organized into four ability types and personal attributes.

The "European Skills, Competences, Qualifications and Occupations" (ESCO) represents the European response to address the need to link skills to occupations. Currently, the ESCO skills database offers access to an impressive list of 13,890 "skill concepts" [25]. The ESCO skills pillar applies a wide definition of skills which includes skills, competencies, and knowledge by skill type. In ESCO skill is defined as the ability to apply knowledge and use know-how to complete tasks and solve problems. Competence is the proven ability to use knowledge, skills, personal, social, and/or methodological abilities, in work or study situations, and professional and personal development. Finally, knowledge means the outcome of the assimilation of information through learning. Knowledge is the body of facts, principles, theories, and practices that are related to a field of work or study. The ESCO framework, unlike O*Net and SCT, does not adopt a strictly hierarchical structure to organize competencies/skills, apart from the case of transversal skills and competencies.

In the following paragraphs, the focus is on the place of entrepreneurship as skill, ability, or personal attribute and knowledge in the three frameworks under review. In the ESCO framework, under the "life" category of transversal skills and competencies is included the "applying entrepreneurial and financial skills and competencies" concept which is defined as follows: "Manage own and others finances and resources. Demonstrate perseverance, openness to opportunity and risks, the ability to mobilize resources and willingness to learn from experience" [11]. This skill concept is hierarchically related to two narrower concepts. The first is "show entrepreneurial spirit," which is described as follows: "Develop, organize and manage an own business venture, identifying and pursuing opportunities and mobilizing resources, keeping in mind a profitability perspective. Demonstrate a proactive attitude and determination to achieve success in business" [12]. The second skill concept is "manage financial and material resources," which is defined as "...undertake effective financial planning, using credit, savings, investments, and pensions to achieve short and long term goals, using financial advice and guidance services with a critical mindset, comparing deals and offers when acquiring products or services, and actively selecting appropriate insurance products" [13]. Further search in the ESCO database produced two more results. The first was about entrepreneurship as a "knowledge concept" in the knowledge sub-domain "management and administration," which refers to "the development, organization, and management of an own business venture" [14]. The second result is "social entrepreneurship," which is another "knowledge concept" defined as "the process of creating, managing and scaling-up a venture in order to address social challenges" [15].

The SCT framework uses the term "Entrepreneurial Mindset," defined as "the personal characteristic of identifying and seizing opportunities and moving from idea to achievement" [18]. Entrepreneurship is considered as a personal quality, "... most often developed from social context, background, and individual experiences, which can make a difference to work performance" (ibid). Finally, the O*Net framework does not include entrepreneurship among its list of 35 skills and 42 personal abilities and attributes. Both the O*Net and SCT frameworks use the term "enterprising" as a personal interest for work environments and outcomes. In O*Net, "enterprising occupations" are defined as those which "... frequently involve starting up and carrying out projects. These occupations can involve leading people and making many decisions. Sometimes they require risk taking and often deal with business" [24]. In SCT "enterprising occupations" "... are characterized by the dominance of activities that entail the interaction with others to attain organizational or self-interest goals. They frequently involve starting up and carrying out projects, influencing, leading, or mobilizing people and making decisions and sometimes require risk taking and dealing with business. They foster enterprising competencies and achievements. Workers in those occupations are audacious, popular, self-confident, sociable, and possess leadership and speaking abilities..." [19].

Overall, all three skills frameworks do not offer a comprehensive taxonomy of entrepreneurial skills. However, they do offer the ground to identify potential competencies, skills, abilities, and knowledge that may contribute to the development of an entrepreneurial skills framework. Focusing on the aforementioned definitions adopted by the ESCO, O*Net, and SCT frameworks, we identified 15 separate "entrepreneurship-related" concepts (e.g., perseverance, openness to opportunity and risk-taking, management of resources, and identifying and seizing opportunities). This is indicative of the multidimensionality of the concept of "entrepreneurship" as competence or skill. The list of "entrepreneurship-related" concepts is presented in the first column of Table 1. In the next three columns of Table 1, the competencies, skills, abilities, and knowledge concepts of the three aforementioned frameworks that best correspond to each of the 15 "entrepreneurship-related" concepts in the list are mapped.

Moreover, the list of 15 entrepreneurship-related concepts that were identified in our analysis corresponds almost perfectly to 14 out of 15 competencies (apart from "Ethical and sustainable thinking") included in the "EntreComp: the European Entrepreneurship Competence Framework," which was developed by the European Commission as a reference framework "... to explain what is meant by an entrepreneurial mindset" ([9], p. 3). This finding provides a strong indication that the list of 15 "entrepreneurship-related" concepts and the corresponding competencies, skills, abilities, and knowledge of the ESCO, O*Net, and SCT taxonomies can offer a solid basis to deepen our inquiry into entrepreneurial skills as these are specifically identified and discussed in the academic literature in the agri-food sector.

3 A Preliminary Review of the Literature on Entrepreneurial Skills in Agrifood Ecosystems

In this section, we perform a preliminary semi-systematic literature review of empirical research papers and project reports on the agrifood sector, which defines, analyzes, and discusses competencies, skills, abilities, and knowledge that relate directly or indirectly to entrepreneurship. A semi-systematic literature review was adopted because entrepreneurship skills in agrifood have been conceptualized differently and studied within diverse disciplines, something which hinders a full systematic review process [32]. The initial set of papers and resources has been retrieved through Google Scholar, using combinations of keywords: "farming business" and "entrepreneurship skills." Search results were narrowed to contain the keywords "skills assessment/evaluation" and "skills development." A final list of 21 papers for review was constructed after keeping only those that reported empirical research findings.

In the Greek context, research has empirically explored the linkage between enhanced performance and managerial skills profiles, which include several entrepreneurial competencies such as leadership, creativity, and innovation [34, 35]. Furthermore, Trivellas et al. [36] mapped the skills and competencies profiles of

Table 1 Mapping "entrepreneurship-related" concepts to the corresponding competencies, skills, abilities, and knowledge of the ESCO, O*Net, and SCT taxonomies

	Entrepreneurship-related concepts	ESCO (EU)	O*NET (USA)	SCT (Canada)
1	Perseverance	Show determination	Persistence	Perseverance
2	Proactivity	Taking a proactive approach	Initiative	Initiative
3	Openness to opportunity and risk-taking	Identify new business opportunities Demonstrate willingness to take risks		Tolerance of ambiguity Risk-taking Self-confidence Risk management
4	Determination to achieve success	Show determination	Achievement Effort	Achievement Competitiveness
5	Willingness to learn from experience	Demonstrate the willingness to learn	Active learning	Active learning Continuous learning
6	Leading/influencing/mobilizing people	Lead others	Leadership	Leadership
7	Management of resources	Manage resources	Resource management	Resource management
8	Financial planning	Create a financial plan to optimize financial performance	Economics and accounting	Finance
9	Critical mindset	Think critically	Critical thinking	Critical thinking
10	Comparing deals and offers	Working with numbers and measures	Mathematical reasoning	Mathematical reasoning
11	Identifying and seizing opportunities	Show entrepreneurial spirit	Enterprising	Entrepreneurial mindset
12	Moving from idea to achievement/starting up and carrying out projects	Thinking creatively and innovatively Show initiative Work independently	Fluency of ideas Initiative Independence Innovation	Fluency of ideas creativity Initiative Independence Innovation
13	Decision-making	Make decisions	Judgment and decision making	Decision making
14	Interaction with others	Communicating Build trust	Active listening social skills	Interpersonal skills
15	Speaking abilities	Using more than one language	Oral comprehension Oral expression	Active listening Oral comprehension Oral expression

managers from more than 150 firms in the Greek agrifood sector and revealed the implications for organizational and supply chain effectiveness of marketing and logistics skills that included skills related to entrepreneurship, such as strategic and communication skills.

Sorensen et al.'s [33] semi-systematic literature review on agriculture sustainability identified, amongst others, entrepreneurship-related skill categories, including systemic perspective, lifelong learning, knowledge integration, network building, and technology knowledge. De Wolf and Schoorlemmer [4], in the context of the ESoF project [31], synthesized the results of research conducted in six European countries into five farmers' skills categories needed for success: (1) professional, (2) management, (3) opportunity, (4) strategic and (5) co-operation/networking, and highlighted that the last three groups are related to entrepreneurial skills. Extending this rationale, Vesala [38] argued that these complex, higher-level skills may be conceptualized as meta-level entrepreneurial skills. Dias et al. [5] argued that developing entrepreneurial skills and attitudes in the European agricultural sector sets the ground for success not only in diversified activities but also in core business activities. Pyysiäinen et al. [29] pointed out that different mixes of entrepreneurial skills are required for conventional farming and business diversification in the Finish context. Building on Chen et al. [3] findings, they conceptualize two subgroups of skills: (a) an essential specific category encompassing marketing, management, and financial control, and (b) a general entrepreneurial attitude (meta-level) category consisting of innovation and risk-taking. Zainalbidin et al. [41] identified as important for entrepreneurial work culture in the farming community in Malaysia the following factors: innovativeness, risk-taking, profit orientation, visionary, managerial skills, and self-confidence.

Despite the importance of entrepreneurial skills for business success in the agrifood sector, several studies indicate that these are not adequately developed. Dias et al. [6], in an extensive literature review on agricultural entrepreneurship from 2013 to 2017, concluded that farmers have lower entrepreneurial skills than entrepreneurs in other sectors, there are fewer female entrepreneurs in this sector, and younger farmers have fewer opportunities to become entrepreneurs. Phelan and Sharpley [28] argued that farmers in the North West of England lack many of the identified critical business and entrepreneurial skills and competencies for succeeding in farm tourism, such as emotional coping, critical evaluation, networking, self-awareness, environmental scanning, business concept, goal setting, and negotiation. Xaba [40] discussed major gaps in strategic planning, marketing, production, ICT, and financial management skills, which are essential for the entrepreneurial development of South African black farmers. Wiafe and Yona [39] elaborated on quantitative survey data from 200 smallholder farmers from central Malawi to find that only a small portion of farmers received proper and adequate training in the development of business management and entrepreneurial skills. Agbolosoo and Anaman [1], exploring the required entrepreneurial skills of small potato farmers in a region of Kenya, concluded that most farmers possess medium-level abilities that can be enhanced through training. McLennan's [21] study on young farmers in Australia discussed the barriers to achieve entering or expanding agricultural business and the business skills required to overcome those barriers.

As it is implied above, context is of paramount importance when considering gaps in entrepreneurship-related skills. Results from a survey on required skills in the agrarian sector in Armenia [20] emphasized future trends in competitiveness, technology, and the circular food economy. Khachatryan [20] identified several entrepreneurial skills, such as the capacity to learn, and the capacity to adapt to new situations, among the most important ones according to local employers in the agrarian sector. McElwee [22] reviewed research in six European countries and concluded that the heterogeneity of farmers in terms of farm size, geography, location, etc. should be taken into account, as well as the degree and the mix of existing and required entrepreneurial skills. The proposed skills-set includes generic business and management skills, marketing, financial, business planning, communication, and collaboration. Based on an empirical study of 3000 farms in the UK, McElwee and Bosworth [23] discussed different farmers' diversification approaches considering the degree of entrepreneurship activities, associated skills, and farmer profiles, as well as factors such as the farm size, ownership, employment, business organization, age, and educational background. Similarly, through an extended survey on young farmers' required skills in 28 EU member countries, Zondag et al. [42] concluded that several factors should be taken into account for the design of skills development programs (i.e., specific needs, geographical area, business sector, level of education, and farm situation), while very often farmers seem not to be aware of the technological and entrepreneurial skills needed.

The entrepreneurship-related concepts identified in the reviewed papers and reports presented above were mapped to the corresponding competencies, skills, abilities, and knowledge of the "Skills and Competencies Taxonomy" (SCT). This is anticipated to provide us with insights regarding the relative emphasis (and gaps, areas for further research) that is placed in the agrifood literature on different categories of entrepreneurship competencies/skills, abilities, and knowledge as these were identified in Table 1.

As is shown in Fig. 1, in the agrifood literature reviewed, the majority of references on entrepreneurship-related skills have to do with "resource management" ($n = 56$) (e.g., management skills, organizational management, HRM, and business planning, are only some of the most used terms). The categories "from ideas to achievement" ($n = 25$) (e.g., vision, creativity, adaptability, and innovation), "financial planning" ($n = 21$), (e.g., budget, pricing, and cost management), "opportunities and risks" ($n = 17$) (e.g., dealing with uncertainty, risk-taking, and challenge acceptance), and "communication" ($n = 17$) are following. On the other hand, the entrepreneurship-related concepts of "mathematical reasoning," "perseverance," "proactivity," "critical mindset," and "new learning," although important for entrepreneurship are much less discussed in the agrifood literature reviewed.

4 Discussion

Because of their potential impact, frameworks for entrepreneurial skills in agrifood ecosystems have to be backed by valid, up-to-date, and comprehensive scientific evidence. Furthermore, such evidence as well as the methods through which data were collected and analyzed should be open to public scrutiny. This helps to ensure

Fig. 1 Mapping "entrepreneurship-related" concepts identified in the literature review to the SCT taxonomy

that entrepreneurial skills in agrifood ecosystems meet high-quality standards. Moreover, it is necessary to engage in independent, rigorous, and comprehensive scientific research on skills and their links to specific job tasks, and occupations in the agrifood sector. Indicative of the above is the finding of our analysis of the relevant literature that "mathematical reasoning," "proactivity," "perseverance," "critical mindset," and "new learning" are competencies/skills that need to be further explored empirically to help promote entrepreneurship in agrifood ecosystems.

One source of data for the development of an entrepreneurial skills framework for the agrifood eco-systems is experts. Experts, academics, and researchers with an in-depth knowledge of the agrifood sector can play a crucial role in the identification of entrepreneurial skills types and individual skills, the development of skill descriptions, the creation of skill hierarchies and ontologies, and the rating of entrepreneurial skills in terms of importance and skill level for specific types of businesses in the sector (e.g., [2]). On the other hand, a wide variety of practitioners in the field, from large food supply-chain managers to small farmer-owners, are likely to be the most valid sources of information on the day-to-day tasks that are required to start-up and successfully run a business in an agrifood ecosystem (e.g., [26]). They can also offer invaluable insights regarding the relative importance and the level of mastery of entrepreneurial skills required in the process of setting up a business in an agrifood ecosystem.

As previous studies have demonstrated that there is a gap in skills perceptions between employees and employers (e.g., Tsirkas et al. [37]), it is of utmost importance to engage and draw upon the experiences and perspectives of different actors in the field. Job posting information and information on job seekers and employers in the agrifood ecosystem published online are sources that can also offer invaluable insights. Big data mining and analytics that cluster and map skills to occupations are a recent addition to the methods adopted to assist the construction of skills/occupations frameworks [7, 30].

An interesting finding of our analysis is that resource management skills are the most studied among the 15 entrepreneurship-related concepts identified in Table 1. However, resource management represents only a subset of the skills needed to promote an entrepreneurial spirit in the agrifood sector. A much broader spectrum of skills is needed to spark entrepreneurial behaviors and endeavors. It is thus of great importance to discuss the conceptual/theoretical underpinnings, the methods, and the data on the basis of which such frameworks are developed, validated, and updated. The development of an entrepreneurship skills framework that responds to the needs of a particular agrifood ecosystem has to build on synergies between stakeholders and take into consideration all the different stakeholder voices in the field, not only as implementers but also as active participants in the conceptualization, design, and update of such a framework. This approach calls for a wider change in paradigm that moves away from top-down generic taxonomies and "global" skills descriptions to place-based, bottom-up, and evidence-based understandings of the multiplicity of perspectives, life experiences, and local wisdom on what kind of skill combinations and skills' mastery are needed to set-up and successfully run a business within a specific agrifood ecosystem.

Acknowledgments We acknowledge support of this work by the project "SMART AGRICULTURE AND CIRCULAR BIO-ECONOMY – SmartBIC." (MIS MIS5047106) which is implemented under the Action "Reinforcement of the Research and Innovation Infrastructure," funded by the Operational Programme "Competitiveness, Entrepreneurship and Innovation" (NSRF 2014-2020) and co-financed by Greece and the European Union (European Regional Development Fund).

References

1. Agbolosoo, J.A., Anaman, R.: Assessment of entrepreneurial behavior skills among small farmers: an exploratory study. SEISENSE J. Manage. **4**(3), 17–33 (2021)
2. Burgoyne, T.C., Reeder, M.C., Allen, M.T.: O* NET® Analyst Ratings of Occupational Skills: Analysis Cycle 22 Results. Final Report (No. 091). Human Resources Research Organization (2021)
3. Chen, C.C., Greene, P.G., Crick, A.: Does entrepreneurial self-efficacy distinguish entrepreneurs from managers? J. Bus. Ventur. **13**(4), 295–316 (1998)
4. de Wolf, P., Schoorlemmer, H.: Exploring the significance of entrepreneurship in agriculture. In: Rudmann, C. (ed.) Entrepreneurial Skills and their Role in Enhancing the Relative Independence of Farmers, pp. 27–34. Research Institute of Organic Agriculture, FiBL, Frick, Switzerland (2008)
5. Dias, C.S., Rodrigues, R.G., Ferreira, J.J.: Agricultural entrepreneurship: Going back to the basics. J. Rural. Stud. **70**, 125–138 (2019a)

6. Dias, C.S., Rodrigues, R.G., Ferreira, J.J.: What's new in the research on agricultural entrepreneurship? J. Rural. Stud. **65**, 99–115 (2019b)
7. Djumalieva, J., Sleeman, C.: An open and data-driven taxonomy of skills extracted from online job adverts. In: Larsen, C., Rand, S., Schmid, A., Dean, A. (eds.) Developing Skills in a Changing World of Work, pp. 425–454. Rainer Hampp Verlag (2018)
8. European Commission: Overview of CAP Reform 2014-2020. Agricultural Policyperspectives Brief (No. 5). DG Agriculture and Rural Development, Unit for Agricultural Policy Analysis and Perspectives (2013)
9. European Commission: EntreComp: the European Entrepreneurship Competence Framework. Publications Office of the European Union, Luxembourg (2018) https://ec.europa.eu/social/main.jsp?catId=738&pubId=8201. Last accessed 10 Jun 2016
10. European Commission: Skills & Competencies. (2022). https://esco.ec.europa.eu/en/classification/skill_main. Last accessed 01 Jun 2022
11. European Commission.: Skills & Competencies. Applying entrepreneurial and financial skills and competencies. (n.d.-a). http://data.europa.eu/esco/skill/T6.5. Last accessed 01 Jun 2022
12. European Commission: Skills & Competencies. Show entrepreneurial spirit. (n.d.-b). http://data.europa.eu/esco/skill/bdcf429c-5ccf-4c3d-bb61-4c987573a35e. Last accessed 01 Jun 2022
13. European Commission: Skills & Competencies. Manage financial and material resources. (n.d.-c). http://data.europa.eu/esco/skill/1ddc021d-d818-4731-904f-842d7491f419. Last accessed 01 Jun 2022
14. European Commission: Skills & Competencies. Entrepreneur. (n.d.-d) http://data.europa.eu/esco/skill/658605f2-1c95-49f0-bd98-0af7b15ad0b0. Last accessed 01 Jun 2022
15. European Commission: Skills & Competencies. Social Entrepreneurship. (n.d.-e). http://data.europa.eu/esco/skill/9f72c931-f94d-4faa-83e4-8674c399f5d2. Last accessed 2022/06/01
16. European Council and Council of the EU: Common agricultural policy 2023–2027. (2022). https://www.consilium.europa.eu/en/policies/cap-introduction/cap-future-2020-common-agricultural-policy-2023-2027/. Last accessed 01 Jun 2022
17. Government of Canada: Skills and Competencies Taxonomy. (2021a). https://noc.esdc.gc.ca/SkillsTaxonomy/SkillsTaxonomyWelcome. Last accessed 01 Jun 2022
18. Government of Canada: Personal Abilities and Attributes. (2021b). https://noc.esdc.gc.ca/SkillsTaxonomy/PersonalAbilitiesAttributes. Last accessed 01 Jun 2022
19. Government of Canada:. Interests. (2021c). https://noc.esdc.gc.ca/SkillsTaxonomy/Interests. Last accessed 01 Jun 2022
20. Khachatryan, L.: Survey: Identification of Skills Needed in the Agrarian Sector and Motivation of Youth to Pursue a Career in the Agricultural Sector in Armenia, Project Report. EU4Youth – SAY YES Skills for Jobs (2019)
21. Maclennan, A.: DPI Young Farmers Survey Report v.1.2. Young Farmers Business Program. NSW Department of Primary Industries (2017)
22. McElwee, G.: Farmers as entrepreneurs: developing competitive skills. J. Dev. Entrep. **11**(03), 187–206 (2006)
23. McElwee, G., Bosworth, G.: Exploring the strategic skills of farmers across a typology of farm diversification approaches. J. Farm Manag. **13**(12), 819–838 (2010)
24. O*Net Online: Interests (2022). https://www.onetonline.org/find/descriptor/browse/1.B.1. Last accessed 01 Jun 2022
25. O*Net Resource Center:. About O*NET (2022a). https://www.onetcenter.org/overview.html. Last accessed 01 Jun 2011
26. O*Net Resource Center: O*NET® Questionnaires (2022b). https://www.onetcenter.org/questionnaires.html. Last accessed 01 Jun 2022
27. Peterson, N., Mumford, M., Borman, W., Jeanneret, P., Fleishman, E., Levin, K., Campion, M., Mayfield, M., Morgeson, F., Pearlman, K., Gowing, M., Lancaster, A., Silver, M., Dye, D.: Understanding work using the occupational information network (O*Net): implications for practice and research. Pers. Psychol. **54**(2), 451–492 (2001)

28. Phelan, C., Sharpley, R.: Exploring entrepreneurial skills and competencies in farm tourism. Local Econ. **27**(2), 103–118 (2012)
29. Pyysiäinen, J., Anderson, A., McElwee, G., Vesala, K.: Developing the entrepreneurial skills of farmers: some myths explored. Int. J. Entrep. Behav. Res. **12**(1), 21–39 (2006)
30. Rentzsch, R., Staneva, M.: Skills-Matching and Skills Intelligence through curated and data-driven ontologies. In: Schumacher, C. (ed.) Proceedings of the DELFI Workshops, pp. 46–58. Gesellschaft für Informatik e.V.z, Bonn (2020, September)
31. Rudmann, C. (ed.): Entrepreneurial Skills and their Role in Enhancing the Relative Independence of Farmers. Research Institute of Organic Agriculture, FiBL, Frick, Switzerland (2008)
32. Snyder, H.: Literature review as a research methodology: An overview and guidelines. J. Bus. Res. **104**, 333–339 (2019)
33. Sørensen, L.B., Germundsson, L.B., Hansen, S.R., Rojas, C., Kristensen, N.H.: What skills do agricultural professionals need in the transition towards a sustainable agriculture? A qualitative literature review. Sustainability. **13**(24), 13556 (2021)
34. Trivellas, P., Drimoussis, C.: Investigating leadership styles, behavioural and managerial competency profiles of successful project managers in Greece. Procedia Soc. Behav. Sci. **73**, 692–700 (2013)
35. Trivellas, P., Reklitis, P.: Leadership competencies profiles and managerial effectiveness in Greece. Procedia Econ. Financ. **9**, 380–390 (2014)
36. Trivellas, P., Reklitis, P., Marinagi, C., Tsoulfas, G.T.: Examining gaps in business and logistics skills and their performance implications in the agrifood supply Chain in Greece. In: Kavoura, A., Kefallonitis, E., Giovanis, A. (eds.) Strategic Innovative Marketing and Tourism. Springer Proceedings in Business and Economics, pp. 199–205. Springer, Cham (2019)
37. Tsirkas, K., Chytiri, A.P., Bouranta, N.: The gap in soft skills perceptions: a dyadic analysis. Educ. Train. **62**(4), 357–377 (2020)
38. Vesala, K.: A Theoretical and methodological approach to the study of the assessment and development of entrepreneurial skills in the farm context. In: Vesala, K.M., Pyysiäinen, J. (eds.) Understanding Entrepreneurial Skills in the farm context, pp. 23–54. Research Institute of Organic Agriculture, Frick, Switzerland (2008)
39. Wiafe, T., Yona, L.: Farm business management skills: a missing link for smallholder farmers: a case of Malingude, Malawi. Eur. J. Bus. Innov. Res. **4**(3), 53–73 (2016)
40. Xaba, G.G.: Identifying Entrepreneurial Skills Required by South African Black Farmers: An Entrepreneurial Model (Doctoral dissertation). University of the Witwatersrand, Faculty of Commerce, Law and Management, Johannesburg (2014)
41. Zainalabidin, M., Golnaz, R., Ira, A., Amin, M.A., Ezhar, T.: Work culture and developing agri-entrepreneurial skills among farmers. Am. J. Econ. Bus. Adm. **3**(3), 490–497 (2011)
42. Zondag, M.-J., Koppert, S., de Lauwere, C., Sloot, P., Pauer, A.: Needs of Young Farmers. Report I of the Pilot Project: Exchange Programmes for Young Farmers. European Commission, Directorate-General for Agriculture and Rural Development (2015) https://data.europa.eu/doi/10.2762/13075

Assessment of the Legal Framework of Greece in Terms of the Applicability of Green Innovation in the Supply Chain Sector

Nikos Kanellos and Christos Kalantzis

1 Introduction

In Greece, over the last decade, there has been a growing interest in the fields of innovation, especially in the field of environment and the redesign of resource management policies [1]. The combined activity in these areas can be described as "green innovation", where the term "green" describes a direction toward the benefit of the natural environment and the term "innovation" mainly refers to the application of research [2]. The interest in this direction is indirectly justified by the current environmental issues and directly justified by the available institutional, legislative, and financial incentives, as formed in a National and European framework for dealing with the former.

Indeed, issues of wide influence are able and do shape various trends in human activities. In coordinated terms, however, human behaviors can be oriented through the legal framework that governs them. All human activities are expected to be in harmony with the legal relationships that describe the variables of the system to which they belong. This system can be of any nature if it is governed by specific legislation. An example is an administrative system, such as a country. In the current study, that administrative system is the country of Greece.

So, at the national level, there are laws and regulations governing the field of innovation and the field of the environment. However, there is considerable exploratory interest in whether this broader framework can be applied effectively in specific areas. Areas with a large footprint in the environment and a significant presence in human activities. Such an area of activity is that of the supply chain.

N. Kanellos (✉) · C. Kalantzis
Department of Agribusiness and Supply Chain Management, School of Applied Economics and Social Sciences, Agricultural University of Athens, Athens, Greece
e-mail: nikos.kanellos2@aua.gr

D. P. Sakas et al. (eds.), *Computational and Strategic Business Modelling*, Springer Proceedings in Business and Economics,
https://doi.org/10.1007/978-3-031-41371-1_26

A supply chain is behind every economic asset and its management. So, taking into consideration the scope of its activities, the reduction of its environmental footprint is a major step for green growth. Specifically, supply chain management involves the integrated approach to business operations and strategies that involve the flow of materials and information from the inbound and outbound ends of a business [3]. The companies involved in these processes are the same companies that are under pressure and incentives, from overlying systems, to reduce their environmental footprint, because of climate change and the need for this issue to be addressed directly by the private sector too.

In the bibliography, the above is defined as a green supply chain. The green supply chain is a process that aims to minimize the ecological impact of a product or service. It covers all phases of a product's life, from the extraction of raw materials, the design of products, and their distribution to their final use by consumers, and their disposal at the end of their life cycle [4]. Although the green supply chain made its appearance in research, it has yet to appear effectively in action, through an integrated implementation.

Green supply chain innovation is critical for attaining operational and financial excellence, rather than merely financial performance [5]. Beyond a shared topic of multiple interpretations put up by academics via concurrent techniques, green supply chain innovation has been interpreted by businesses and literature [6, 7]. Firms may be driven to undertake green operations more effectively in order to generate higher performance [8], as well as to integrate with customers and suppliers [9]. This integration may also be seen as a critical operational capability from the perspective of green supply chain innovation [10].

Therefore, research requires a complete understanding of the relationship between green supply chain innovation and green innovation. Moreover, green supply chain innovation gives a chance for enterprises to learn about green supply chain innovation in an acceptable manner. Earlier green supply chain innovation studies have ignored the potential connection between different types of green supply chain innovation [10, 11], while others have overlooked the impact of technical advances and operational competencies on green supply chain innovation and corporate success [12].

2 Purpose of this Paper

Starting from the interest that has been noted in the field of green innovation and its application in supply chain systems, the purpose of this chapter is to evaluate the legal framework of Greece in terms of its ability to support the combined development of the above sectors. As the object of this study concerns the current legislative provisions, qualitative, and posterior evaluations will be carried out. The main objective of this chapter is to assess the legal framework in terms of strengths and weaknesses that are maintained in the current situation, which may facilitate or complicate, respectively, the implementation of practical innovation in supply chain processes. At the same time, it is important to be evaluated for its prospects and potential failures that may occur in future scenarios. A suitable tool that meets the above guidelines and is suitable for cases of instructions and regulation is the SWOT (strengths, weaknesses, opportunities, and threats) analysis.

3 Methodology

The SWOT analysis, according to the bibliography, concerns the analysis of the internal and external environment of an organization, to identify internal advantages that can be used to take advantage of opportunities and avoid external threats, while limiting its weaknesses [13]. The method's usefulness applies to cases beyond organizations and businesses, such as plans and programs.

Starting with the internal environment of the legal framework of interest, to evaluate its strengths and weaknesses, the main goal and its objectives will be identified first. Next, regarding the external environment, the factors that can play the role of opportunity or threat will be identified too. At the same time, the major institutional framework related to supply chain, innovation, and environmental issues will be mapped, as it will be used as an evaluation background. Finally, the subjects, the described actions, and the requested results of the targets will be used as criteria in the SWOT analysis.

3.1 Current Situation of the Legal Network

The existing legal framework needs to be identified as a reference point for the targets that will be determined, as it may contain objectives and directions, but also as an object of evaluation. For this reason, the main points of the legal framework of the country are listed below. Specifically, the regulations and directions that need a general application in time and place were collected. That being said, it does not include special provisions that do not apply to the whole country and that may change in the short term.

The Constitution of Greece

When approaching the legal framework for any purpose, it is necessary to include the constitution of Greece. The purpose of the constitution is, among other things, to indicate how the power will be exercised and to set guidelines so that the rest of the regulations and institutional provisions concern the application of the articles it contains. The following are the articles related to the current study.

- **Article 16**: This article establishes that art, science, research, and teaching are free and that their development and promotion are the responsibility of the state [14].
- **Article 22**: This article establishes that work is a right protected by the State, which must ensure adequate employment conditions [14]. Those conditions are thoroughly described.
- **Article 24**: The aim of this article is the protection of the environment and specifically states, among other things, that the protection of the natural and cultural environment is an obligation of the State and a right that everyone has [14].
- **Article 101**: This article mainly consolidates the administrative function of the state and creates the level at which the legislation will be applied (Constitution of Greece 2013).
- **Article 106**: In this article, the state is appointed as a planner and a coordinator of the country's economic activity in order, among other things, to protect the general interest and to bring economic growth in all sectors [14].

Articles 16, 22, and 24 have been selected as they relate directly to the subject of this study, while articles 101 and 106 have been selected as they form an important part of the framework defining how the institutional and legal framework is implemented.

Law 4302/2014

Law 4302 of 2014, as published by the Greek Official Gazette 225/A'/2014, targets directly the regulation of supply chain issues [14]. Specifically, issues of all material are included with some of the basic articles that represent the structure of the legislation being listed below.

- Part A′
  - Chapter A′ – General provisions

    Article 1 – Definitions
    Article 2 – Scope
    Article 3 – Regime for Exercising Supply Chain Activities
    Article 5 – Green Supply Chain
    Article 6 – Supply Chain Promotion Actions
    Article 7 – Council Establishment for Supply Chain Development and Competitiveness

  - Chapter B′ – Storage and Distribution Centers

    Article 8 – Installation and Operation of Storage and Distribution Centers
    Article 10 – Traffic Connections

- Part B′ – National Corporate/Commercial Parks of Supply Chain and other Provisions

Article 14 – National Corporate/Commercial Park of Supply Chain

It is noted that as is the case with most legislation, the above law has been amended several times. For example, the Article on the green supply chain was added by Par. 3 of Article 48 of Law 4442/2016.

Law 4310/2014

The object of Law 4310 of 2014, as published by the Greek Official Gazette 258/A'/2014, is the promotion of research, technological development, and innovation [15]. Specifically, it lays the foundations for the involvement of stakeholders around these objects, describes their development incentives of a monetary or rewarding nature, and discusses other provisions of interest.

It is important that this law aims to implement Article 16 of the Constitution and that in 2016 it was amended by Law 4386, without of course differentiating its main body and its initial goals. The following are the indicative chapters that are included.

- Chapter A′ – Objectives Purposes Meaning of Terms
- Chapter B′ – Planning and Implementation Instruments
- Chapter C′ – Enhancing Technological Development and Innovation Research
- Chapter D′ – Promoting Excellence in Research, Technological Development, and Innovation Researcher Mobility
- Chapter E′ – Nuclear Energy Management, Technology, and Radiation Protection Greek Atomic Energy Committee (GAEC)
- Chapter F′ – Transitional Authorizing Delegated Provisions Supervised Organizations

Law 4864/2021

This legislation includes regulations for strategic investments and improvement of the funding environment through the acceleration of processes, creates a framework for tech companies, and deals with other urgent provisions for development [16]. It was published by the Greek Official Gazette 237/A'/2021 as a response to the rapid growth of start-ups and the tendency to apply research to find solutions to contemporary issues.

Parts of significance in the content of the legislation are the relevant provisions for the development law, which are described in particular with Law 4887 of 2022, and the third part that legislatively frames the spin-off companies, i.e., the companies that implement in business a research result [17].

Special Framework for Spatial Planning and Sustainable Development of Industry

The present concerns spatial planning and is published by the Greek Official Gazette 151/AAP/2009. Its purpose is to transform the spatial structure of the national sector of industry in the direction of sustainable development, which includes three dimensions: environment, society, and economy [18]. This includes, but is not limited to, the following goals:

- Green to protect the environment and promote sustainability
- Social for equality and cohesion
- Economic nature for prosperity and development of the entrepreneurship sector

Innovation, the environment, and the supply chain are often mentioned within the above national plan, sometimes as objectives and sometimes as parts of systems and processes.

3.2 *Target and Sub-targets*

For the sake of the present study, the target and sub-targets of the legal framework are identified to be used later as evaluation guidelines. The target is derived directly from the purpose of the study and is specifically the implementation of green innovation in the supply chain. To achieve the above goal, individual sub-targets are identified, as formulated below.

Sub-target 1 – Creation of Incentive Framework for Innovation and Green Entrepreneurship

Encouraging incentives of a monetary and rewarding nature in green and innovative actions and initiatives. These include the promotion of green Corporate Social Responsibility (CSR) and funding programs.

Sub-target 2 – Connection of Research and Business Sectors

A necessary condition for innovation is the business application of research results. In addition to cultivating a climate, several issues that may arise during the transfer of the idea to the market, such as intellectual property or funding, need to be addressed.

Sub-target 3 – Modernization of Supply Chain Operations and Facilitation of Communications Between Individual Flows

Digital transformation and optimal coordination between logistics process flows can result in resource savings. In fact, for a company, it is very important economically to effectively coordinate them, especially the main logistics functions that absorb the basic costs [19]. It also includes the most efficient way of storage, transport, and distribution of products.

Sub-target 4 – Use of Smart Technologies

Use of new technologies, such as IoT, to optimize processes and maximize their performance and reduce environmental footprint. Examples are smart retail and smart logistics [20].

Sub-target 5 – Insurance of Flexibility and Adaptability Against Modern Issues

The supply chain, as the process of delivering to the consumer the product that meets one or some of its needs, must respond to potential crises that are occurring or may occur in the future. An example is the transport sector that, as is expected, will be affected by climate change [21].

Sub-target 6 – Usability of the Legal Framework

An important factor is the usability of the legal framework, how easily one navigates it and locates what he is looking for, and whether the communication with the competent implementing authorities is contributory.

3.3 The External Factors

To evaluate the legal framework of Greece in terms of the applicability of green innovation in the supply chain, it is necessary to identify the influential factors of the external environment. The external environment, in this case, includes the factors outside the legal framework of Greece that influence it and, at the same time, can't be controlled by it. Taking into consideration the number of processes and entities the supply chain includes; the variety of influential factors keeps increasing. Supply chains facilitate the production of everything, from computers and cars to lifesaving medicines and food, and support world trade in goods that are worth almost $20 trillion. End products have up to many thousands of parts, sourced from diverse geographies around the world [22]. To lay the groundwork for the SWOT analysis, main and modern factors were identified, as they follow.

The Economy

Economic trends and issues have a great influence on the activity of all sectors, including supply chain or innovation. The situation or the objective of any economic system, such as a country or the whole world, affects individual economic activities, like the supply chain or the funding of innovative initiatives. Numerous factors can intervene with the selected sectors for this study. For example, the inflation of an economic system brings a general increase in the prices of goods and services, affecting the supply chain directly. One more factor that affects the supply chain is unemployment, as an important factor in the demand for products. Moreover, in the case of economic growth of a system, funding directly green innovation is much more approachable financially.

Climate Change and Environment

One of the most challenging issues the whole world needs to address is climate change. Climate change already made its appearance through many events and there is a need for integrated adaptations, as it is expected to further intensify in the future and affect every part of our lives. The supply chain is expected to undergo a lot of difficulties too. For example, drought has already impacted supply chains across the world. In Europe specifically, drought has had a significant impact on the flow of goods through waterways by reducing the transport flows [22]. Of course, climate change has become a keystone for the growth of green innovation, while the planet's resources determine greatly major supply chain functions, like the distribution of products, or variables, like the prices of products.

Geopolitics

Any main event in geopolitics that may arise can influence significantly the supply chain, as the global market and its prices are changing. A recent example is the major changes in the supply chain regarding products that come from Russia, whose geopolitical behavior brought difficulties for Europe to commerce with.

Global Health

Events of health interest, on a global scale, have a significant impact on the supply chain, just like the outburst of the COVID-19 pandemic. In this case, the majority of industries had to change and adapt to the emerging issues and needs, while extreme social changes were taking place. The pandemic shifted the industrial activity so that some sectors flourished and some others declined, with a great impact on the supply chain. For example, because of the pandemic crisis, some shortages emerged, resulting in supply chain activity holding up and abrupt price increases.

3.4 SWOT Analysis

To qualitatively evaluate the legal framework, the SWOT analysis method will be used, the criteria of which are the strengths, weaknesses, threats, and opportunities of the subject under consideration. The advantages arise from the positive effects that the legal framework is going to have on the targets, while respectively, the weaknesses are reflected as the negative effects. In terms of opportunities, the current situation will be studied in terms of its advantages and the possible perspectives will be determined. At the same time, the potential threats arise mainly from the development that the weaknesses may touch off or from external factors of the spatial, institutional and economic system, called Greece. Figure 1, presented below, shows the summarized SWOT analysis.

Strengths

The current legal framework provides for several incentives to implement innovation and green ideas. Specifically, the Green Fund and the 2022 Development Fund are

Strengths Adequate financial incentives Connection of research and entrepreneurship Promotion of digital transformation	*Weaknesses* Complicated and not easy to use Does not take sufficient measures for the underdeveloped technological infrastructure Insufficient emphasis on supply
Opportunities Growing development of the innovation sector Economic benefits for the private sector Business management optimization	*Threats* Limited flexibility and adaptability Inability to implement interventions Modern crises

Fig. 1 SWOT analysis

some major tools for mobilizing green actions, which could be implemented, among other subjects, in supply chain issues. At the same time, with the creation of a specific legal framework for spin-off companies, the connection of research with the market was significantly supported, while the digital transformation promoted in spatial directions is a tool for the modernization of the supply chain.

Weaknesses

A major weakness, which characterizes the entire Greek legal framework, is its complexity which makes it inaccessible and not easy to use. This is due to the many parts and ad hoc amendments, which result in scattered information. At the same time, the country's technological infrastructure, such as that of the Internet, is not sufficiently developed and often makes it difficult to use smart technologies. Therefore, the legal framework has not been able to adequately coordinate the implementation of competent interventions. The legal framework for the implementation of innovative ideas in the supply chain is insufficiently developed, as there are mainly partial reports without specific directions in this area.

Opportunities

The existing legal framework largely manages to cover the range of issues around green innovation and the supply chain. It is an opportunity for the research sector and the market to take advantage of it and develop a joint approach. Minimizing the resources required in supply chain processes could also mean minimizing financial costs, while through digital transformation, the optimization of procedures and streamlining of communication could lead to new and more efficient business management methods.

Threats

The legal framework around the axes of work is newly formed and quite complex because of the various amendments it has received. This may limit its ability to support the supply chain and green innovation industry on current issues and threats, such as climate change, that is, to reduce its flexibility and adaptability. Furthermore, the provisions contain several proposals and directions without any clarification for their implementation, which may limit the practical development of the sectors under this study.

4 Discussion

From the completion of the SWOT analysis, some qualitative conclusions and even more topics for discussion emerge. The country has been governed in recent years by a climate and a desire for green growth, and importance has been given, among other things, to resolving contemporary environmental issues. In this context, the interest in green innovation in research and entrepreneurship has increased, while current issues such as climate change and Russia's geopolitical behavior in 2022 have raised issues to be resolved particularly in the supply chain. The legal framework trying to respond has been significantly shaped, with a certain extent of success. It has achieved significant support for new ventures in the field of green innovation and in connecting the academic world with the business while providing directions for digital transformation.

However, the legal framework as a whole remains complex, and regarding areas of supply chain and green innovation, ad hoc enactments and amendments have been made that have not yet been fully adapted to address the issues completely. While the supply chain has acquired its framework, it does not have sufficient directions for green innovation. Great interest is presented in the development of the supply chain sector and how it will be formed through the existing and potential legal interventions. Innovation in Greece has developed a significant momentum that may bring many changes in various industries. One of these sectors is the supply chain and specifically its management. It is possible that the administrative functions around it will be formed and the costs in its partial processes will change to a considerable extent.

At the same time, the various modern crises that occur are expected to affect the entire supply chain in a variety of ways. The readiness and adaptability of the existing legal framework to ensure its sustainable operation will play a decisive role, elements during the SWOT Analysis didn't appear. Moreover, not only interesting but also alarming is the set of interventions that will arise from the legal framework. In particular, whether the discussion concerns legislation or guidelines, proposing specific interventions to achieve the goals and their successful implementation are crucial elements for any possible development scenario, of any sector.

Ultimately, the institutional framework has met current needs and supports the implementation of green innovation in areas such as the supply chain. He has taken care of logistics issues and motivated green and innovative applications. However, it has been formulated for specific or immediate problems or needs and has not yet been sufficiently tested in the implementation, while it also has weaknesses that are found in the entire institutional framework of the country. Therefore, the legal framework can support the implementation of green innovation in the supply chain, but its weaknesses need to be corrected to be able to respond to potential problems in the internal environment of the supply chain, but also the external.

5 Future Work

Future research could be extended beyond the supply chain sector, toward more innovative sectors like knowledge-based firms [23, 24], cryptocurrency organizations [25–28], centralized payment services and platforms [29] and the wider scope of logistics operations and customer analysis [30–33]. Other scientists with similar cases are studying similar problems, using modeling and simulation [34–36].

References

1. GSEE Labor Institute: Green Economy, Social Cohesion and Employment. INEGSEE (2011)
2. Law no. 4310, Article 2: Hellenic Government Gazette. (2014)
3. Stevens, G.C.: Integrating the supply chain. Int. J. Phys. Distrib. Mater. Manag. (1989)
4. Fortes, J.: Green Supply Chain Management: A Literature. Otago Management Graduate Review (2009)
5. Kong, T., Feng, T., Huo, B.: Green supply chain integration and financial performance: a social contagion and information sharing perspective. Bus. Strateg. Environ. **30**, 2255–2270 (2021). https://doi.org/10.1002/bse.2745
6. Setyadi, A.: Does green supply chain integration contribute towards sustainable performance? Uncertain Supply Chain Manag. **7**, 121–132 (2019). https://doi.org/10.5267/j.uscm.2018.10.012
7. Kong, T., Feng, T., Huang, Y., Cai, J.: How to convert green supply chain integration efforts into green innovation: a perspective of knowledge-based view. Sustain. Dev. **28**, 1106–1121 (2020). https://doi.org/10.1002/sd.2062
8. Hu, R., Shahzad, F., Abbas, A., Liu, X.: Decoupling the influence of eco-sustainability motivations in the adoption of the green industrial IoT and the impact of advanced manufacturing technologies. J. Clean. Prod. **339**, 130708 (2022). https://doi.org/10.1016/j.jclepro.2022.130708
9. Zhu, Q., Geng, Y., Fujita, T., Hashimoto, S.: Green supply chain management in leading manufacturers: case studies in Japanese large companies. Manag. Res. Rev. **33**, 380–392 (2010). https://doi.org/10.1108/01409171011030471
10. Lo, S.M., Zhang, S., Wang, Z., Zhao, X.: The impact of relationship quality and supplier development on green supply chain integration: a mediation and moderation analysis. J. Clean. Prod. **202**, 524–535 (2018). https://doi.org/10.1016/j.jclepro.2018.08.175
11. Wong, C.Y., Wong, C.W.Y., Boon-itt, S.: Effects of green supply chain integration and green innovation on environmental and cost performance. Int. J. Prod. Res. **58**, 4589–4609 (2020). https://doi.org/10.1080/00207543.2020.1756510
12. Al-Ghwayeen, W.S., Abdallah, A.B.: Green supply chain management and export performance: the mediating role of environmental performance. J. Manuf. Technol. Manag. **29**, 1233–1252 (2018). https://doi.org/10.1108/JMTM-03-2018-0079
13. Panagiotou, G.: Bringing SWOT into focus 2003. Bus. Strateg. Rev. (2003)
14. Law no. 4302: Constitution of Greece. Hellenic Government Gazette. (2014)
15. Law no. 4310: Constitution of Greece. Hellenic Government Gazette. (2014)
16. Law no. 4864: Constitution of Greece. Hellenic Government Gazette. (2021)
17. Law no. 4887: Constitution of Greece. Hellenic Government Gazette. (2022)
18. Hellenic Government Gazette: Approval of a Special Framework for Spatial Planning and Sustainable Development for industry and its strategic environmental impact study. (2009)
19. Salavou, E., Kyriakidou, O.: Management Corporate Functions. Rosili (2010)
20. Kalantzis, C.: Possible IoT applications in smart cities. Envinow.gr. (2019)

21. Kalantzis, C.: Climate change: its impact on transport systems. Envinow.gr. (2021)
22. McKinsey Global Institute: Economic conditions outlooks, 2020. (2020, 10 May 2022). https://www.mckinsey.com/business-functions/strategy-and-corporate-finance/our-insights/economic-conditions-outlooks-2020
23. Kanellos, N.S.: A conceptual framework for analyzing knowledge-based entrepreneurship. In: 4th Symposium on Business and Management and Dynamic Simulation Models Supporting Management Strategies (2011) http://hdl.handle.net/10797/13771
24. Sakas, D.P., Nasiopoulos, D.K., Reklitis, P.: Modeling and simulation of the strategic use of the internet forum aiming at business success of high-technology companies. In: Sakas, D., Nasiopoulos, D. (eds.) Strategic Innovative Marketing. IC-SIM 2017. Springer Proceedings in Business and Economics. Springer, Cham (2019). https://doi.org/10.1007/978-3-030-16099-9_22
25. Sakas, D.P., Giannakopoulos, N.T., Kanellos, N., Tryfonopoulos, C.: Digital marketing enhancement of cryptocurrency websites through customer innovative data process. PRO. **10**(5), 960 (2022a). https://doi.org/10.3390/pr10050960
26. Sakas, D.P., Giannakopoulos, N.T., Kanellos, N., Migkos, S.P.: Innovative cryptocurrency trade websites' marketing strategy refinement, via digital behavior. IEEE Access. **10**, 63163–63176 (2022b). https://doi.org/10.1109/ACCESS.2022.3182396
27. Sakas, D.P., Giannakopoulos, N.T., Reklitis, D.P., Dasaklis, T.K.: The effects of cryptocurrency trading websites on airlines' advertisement campaigns. J. Theor. Appl. Electron. Commer. Res. ***16***(*7*), 3099–3119 (2021). https://doi.org/10.3390/jtaer16070169
28. Sakas, D.P., Kamperos, I.D.G., Reklitis, D.P., Giannakopoulos, N.T., Nasiopoulos, D.K., Terzi, M.C., Kanellos, N.: The effectiveness of centralized payment network advertisements on digital branding during the COVID-19 crisis. Sustain. For. ***14***(*6*), 3616 (2022c). https://doi.org/10.3390/su14063616
29. Sakas, D.P., Giannakopoulos, N.T.: Harvesting crowdsourcing platforms' traffic in favour of air forwarders' brand name and sustainability. Sustainability. **13**(15), 8222 (2021a). https://doi.org/10.3390/su13158222
30. Sakas, D.P., Giannakopoulos, N.T.: Big Data Contribution in Desktop and Mobile Devices Comparison, Regarding Airlines' Digital Brand Name Effect. Big Data and Cognitive Computing. **5**(4), 48 (2021b). https://doi.org/10.3390/bdcc5040048
31. Sakas, D.P., Reklitis, D.P.: Predictive model for estimating the impact of technical issues on consumers interaction in agri-logistics websites. In: Bochtis, D.D., Pearson, S., Lampridi, M., Marinoudi, V., Pardalos, P.M. (eds.) Information and Communication Technologies for Agriculture—Theme IV: Actions. Springer Optimization and Its Applications 185. Springer, Cham (2021). https://doi.org/10.1007/978-3-030-84156-0_14
32. Terzi, M.C., Sakas, D.P., Seimenis, I.: The contribution of the Scientific Committee in the Development of Conferences. Procedia Soc. Behav. Sci. **73**, 373–382 (2013). https://doi.org/10.1016/j.sbspro.2013.02.064
33. Nasiopoulos, D.K., Sakas, D.P., Vlachos, D.S.: Modeling the scientific dimension of academic conferences. Procedia Soc. Behav. Sci. **147**, 576–585 (2014). https://doi.org/10.1016/j.sbspro.2014.07.164
34. Nasiopoulos, D.K.: Implementation stages and risks of Logistics Companies reengineering. Business Intelligence and Modelling. **6**, 197–206 (2020a)
35. Nasiopoulos, D.K.: Web search applications in Logistics companies in the context of evolution of Management Information Systems. Business Intelligence and Modelling. **6**, 241–254 (2020b)
36. Dimitrios, N.K., Sakas, D.P., Vlachos, D.S.: The role of open source leadership in developing high technology companies. Key Eng. Mater. **543**, 402–405 (2013)

Creation of Knowledge-Based Supply Chain Firms: Which Factors Matter?

Nikos Kanellos

1 Introduction

The concept of entrepreneurship is strongly heterogeneous because of the many different processes involved in starting up a supply chain firm. Creation factors vary by firms' characteristics like the founders' motivations and the different environments where they are growing. As a result, the heterogeneity of entrepreneurship is difficult to explain by embracing a general model. It requires a classification of supply chain firms according to their specific characteristics, indicating certain models of entrepreneurial behavior. In the research, we classified supply chain firms in various high-tech industries based on their knowledge-based entrepreneurship characteristics.

Although these firms are technology-related, we expected most of them to have a high stock of knowledge and significant innovation performance. This is not the case for the majority of supply chain firms we would expect because a large group of supply chain firms is characterized by relatively low-intensity innovation and knowledge. Therefore, the fact that a supply chain firm can operate in a high-tech sector in the Greek area does not necessarily ensure that this supply chain firm will be distinguished for its basic characteristics related to knowledge-intensive entrepreneurship. That is the existence of an important stock of knowledge and innovation.

N. Kanellos (✉)
Department of Agribusiness and Supply Chain Management, School of Applied Economics and Social Sciences, Agricultural University of Athens, Athens, Greece
e-mail: nikos.kanellos2@aua.gr

D. P. Sakas et al. (eds.), *Computational and Strategic Business Modelling*, Springer Proceedings in Business and Economics,
https://doi.org/10.1007/978-3-031-41371-1_27

2 The Characteristics of Knowledge in Supply Chain Firms

Supply chain management is considered a core corporation or process integrator to be the firm's backbone, with every based service supplier to be a complete supply chain. This addresses core functions like service-oriented acquisition, service manufacturing and deployment, and service merchandise revenues, as well as the layout, capacity requirement, and workmanship of service demand from responsive service providers to delivery integrators, as well as client efficiency reviews [1, 2]. There remains a shortage of skilled management and information processing employees throughout the business, leading to the logistics sector's low industry competition and potential growth. The sector's operational effectiveness is poor [3]. To achieve efficient supply chain joint creativity throughout the big data context, big data analysis and handling technologies must be used to gather, evaluate, extract, and utilize diverse vast data assets [4].

The knowledge-based entrepreneurship is a kind of high-potential entrepreneurship [5] and can be a driving force for economic growth; its targeted promotion can be achieved by focusing on those factors that can influence it. Such factors may be the knowledge and skills of founders, the size of supply chain firms, their ability to recognize and exploit opportunities in the market, their export orientation [6], and whether they were established before or after the economic crisis. The simple assumption that creating new supply chain firms in high-tech sectors will provide multiplier effects and benefits for the other economic sectors is not entirely correct.

To make sense in the search for such supply chain firms in high-tech industries, we should consider whether a supply chain firm is distinguished by the technical knowledge of its founders, whether it is sufficiently large or has the potential to grow in a short time, whether it can adapt to technological changes and produce diverse products and services, whether it is extroverted, and whether it was founded by taking advantage of a market opportunity rather than the need that may have been caused by the economic crisis or the pandemic crisis (COVID-19).

By the term new supply chain firms, we mean the firms established during the decade (from 2000 to 2010) in the context of this research. The choice of this time range allows us to see the evolution of many firms after the critical period for their survival of the first 3.5 years. This period includes the period when the economic crisis occurred both at the international and national levels (2008 and beyond).

3 The Role of the Founding Team

Although many empirical studies have highlighted the advantages of founding a supply chain firm from a group of founders over a single individual, little has been written about how the relationships between aspiring entrepreneurs and the structure of the founding team influence launching a new venture. Forster [7] describes the difference between the founders of a supply chain firm and the members of a new venture team and their founding partnerships with other project teams. In the first case, the founders have a deeper relationship with their firms. The number of other members in a new venture team can change, as opposed to the number of founders that is predetermined from the start, despite the reclassifications that can occur after a business starts. In the latter case, while the partnerships developed at the founding resemble other forms of working groups, they are different because of the greater risk taken by the founders compared to the members of a task group, who may even be part of a large organization that already exists. Organizations are described as complex systems composed of relationships between people and subgroups. The human interaction that can exist within an organization can be unlimited but the number of bottlenecks.

The relationships a person can effectively manage have some limits [8]. Thus, a complex system such as an organism is most likely to consist of compact subsystems with strong bonds within them. The composition of the founding team and its influence on the performance of the supply chain firm are very important in these subsystems. Opinions differ in the literature on whether there is a positive or negative dependence between the number of founding members and the performance of a firm. The advantages of having a large number of founding teams [9] because of the amount of knowledge, experience, and wisdom that exists is also argued that the stronger and more stable the relationship between the founders, the more likely it is to achieve the formation of a new venture. In terms of the negative aspects of a large founding team, these relate primarily to the conflict that can arise from the different goals of each founder. Many people consider the binary relation to be unique [10]. In principle, a relationship between two people is the "basic unit of social behavior" and there is some degree of trust compared with larger groups. This trust stems from the fact that the goal of each entrepreneur is the success of the venture, so the relationship between the two entrepreneurs becomes stronger, thus contributing to the success of the whole venture. Conversely, this trust can be lost when a third party is involved. In general, the tendency to form a coalition between the two reduces the stability of the founding team.

As a result, the third person in the group left out of the alliance will sooner or later give up the venture. However, when there are more than three founders, similar results emerge, and binary or ternary alliances exist in larger groups. Relationships among founding members influence the start-up and the course of a business; for example, William Hewlett and Dave Packard, founders of the well-known company HP, simply wanted new engineers to build a business together without knowing its scope in advance. The bonds between the pair were strong enough to set the business up against all the difficulties that arose in the beginning and to achieve significant performance during its development. The two are very strong and are made up of relatives, for example, a married couple. The marital relationship is the longest effort to achieve the project, whereas when the founding team is not connected to marital relations, it is easier to interrupt the process of creating the project if the results appear unsuccessful. However, although partnerships between spouses are distinguished for the strong bonds they develop, the other partnerships are usually more heterogeneous, as the choice of the duet is not simply made in the context of a relational relationship but only in the effort of creating a supply chain firm.

Beckman [11] examines the role that the composition of the founding team plays in the behavior of the supply chain firm; when the founders of the firm were previously working in the same business, then the new business tends to have an exploitation behavior. When the founders come from different companies, then exploration qualifies as the most likely behavior. When both occur, the business is noted for its ambidexterity. The concept of exploration includes radical innovation, new product and market creation, broad search, and discovery, while the concept of exploitation involves incremental innovation, application, detail, routine creation, local search, and efficiency. In particular, the experience of the founders plays a very important role in the behavior of the supply chain firm (explorative or exploitative). There is a view that we should not only study what experience it is but also where it was acquired; for example, by comparing two three-member founding groups where on the one hand we have people coming from Apple while on the other one is from Apple, the other from Intel, and the third from HP, then in the first case, the founding members share a common language and perceptions while, in the second, experience and sources of information are heterogeneous.

Teams with common linkages from a previous business tend to use operating behaviors, improving existing processes, and quickly bringing new products or processes to the market. On the other hand, groups with different bonds are primarily associated with exploration behaviors such as discovering new ideas and technological leadership. We then say that the business is ambidextrous, supporting both the implementation of various ideas and practices as well as the discovery of new ones. Whether the founders share a common set of perceptions or unique knowledge, they will define the strategy of the business; the members of a founding group with common links from a previous business share three key things: (1) common language (shared perspective and reliability); (2) common organizational culture (shared visions and goals); and (3) common narratives and examples of behaviors to imitate or avoid.

According to Schoonhoven et al. [12], when founding groups have previously shared work experience, growth levels are higher than in the groups in which it occurs they don't share the experience. In general, these shared perceptions and knowledge shared by members of a founding team reinforce the behavior of exploitation. Previous shared work experience also improves time-to-market or strategic decision-making as roles and expectations are about to be known. Of course, the common starting point such as the founder's supply chain firm restricts the discovery of innovations because of the limited range of experience and knowledge. For supply chain firms, whether or not there are previous links between entrepreneurs, to have as much information as possible, they should develop links with the external environment (alliances, professional associations, etc.). Also, the fewer the common links between the founders, the greater the external social capital for the firm (due to the heterogeneity that exists).

We then say that the supply chain firm is ambidextrous, supporting both the implementation of various ideas and practices as well as the discovery of new ones.

Another important part that different linkages of the business bring is innovation. These firms develop exploration behaviors because of their greater range of knowledge and innovativeness. Of course, the risk that supply chain firms face with such behavior is what is called the "exploration trap," which essentially refers to the fact that new ideas are continually searched without the previous ideas being fully exploited first. The organizational structure of a firm is important for ambidexterity. Both exploration behaviors and exploitation behaviors can be run in parallel, but the pattern of affiliations in the founding group needs to be looked at to see which supply chain firms are capable of applying both behaviors. When the founding teams have both kinds of links from a previous business, then they can share perceptions and effectively transfer knowledge and support innovation. These paradigms can perpetuate over time, as founding teams with two types of ties tend to hire managers who also have two types of ties. So, firms explore and exploit in the long run, and this has benefits captured in business performance. Shared perceptions facilitate execution and improve speed, while unique knowledge is more about innovation and change [13].

The behavior of a supply chain firm ultimately depends on the "history" of its founders. It is the composition of the team that will determine whether the business will pursue exploration or exploitation or whether it will be ambidextrous. In the latter case, it follows that the founding groups should have significant shared and diverse experience in founding.

4 Creation Factors

The founders started their degree of agreement regarding business creation factors, answering a question that contained 5-point answers. According to the literature relating to the reasons why a new venture can be launched and discussed in the theoretical part, creation factors can be distinguished according to whether they are related to market knowledge and professional experience (cognitive factors), the need to work or the expectation of an increase in income (special incentives), the recognition and exploitation of technological, and other opportunities and, finally, general incentives associated with job independence, social recognition, etc.

As we observe from the elements in Table 1, four distinct factors are formed. It is characteristic that almost all variables in each factor exhibit a very high degree of correlation.

The first factor refers to general motives and consists of the need for personal creation, social recognition, and job independence. We are talking about incentives that are not directly linked to the individual's knowledge, his need to work, or the use of some opportunity. These are general factors that led to the establishment of the business and are related to the expression of personal creation, the perception of society for personal success, and the independence that a supply chain firm owner enjoys. Cognitive factors are the second distinct factor. This is the knowledge the

Table 1 Factor analysis of creation factors

Creation factors	Factor 1	Factor 2	Factor 3	Factor 4
Existence of need for personal creation	**0.802**	−0.044	−0.039	0.184
Social recognition	**0.764**	0.115	0.037	−0.043
Work independence	**0.645**	0.142	0.363	−0.097
Market knowledge	−0.013	**0.873**	0.095	0.011
Work experience in the same field	0.221	**0.821**	0.020	0.053
Job search for livelihood	−0.055	0.004	**0.862**	−0.170
Expectation for income increase	0.254	0.097	**0.776**	0.167
Recognition of a technological opportunity	0.045	0.071	−0.149	**0.818**
Recognition of a market opportunity	−0.121	0.375	−0.039	**0.660**
Utilization of scientific research results	0.094	−0.201	0.134	**0.554**
Eigenvalues	*2.28*	*1.71*	*1.36*	*1.13*
(variance explained %)	*22.86*	*17.17*	*13.65*	*11.3*

founder has about the market and the work experience they have in the industry they have founded and the current business.

Therefore, the key characteristic of this factor is knowledge. The next factor summarizes some special motivations. A key feature of this factor is the need for work and the expectation of income growth. Therefore, we can claim that there is more of a need for entrepreneurship here. On the contrary, the final factor is the recognition and exploitation of an opportunity. This opportunity might be either a technological opportunity or another one that was present in the market and was picked up by the founder who built the supply chain firm. Here are the results from scientific research as the basis for the establishment of the firm.

5 Sources of Knowledge

The sources of business knowledge are measured by a question that is formulated on a Likert scale with possible answers arranged on a 5-point scale. That is, participants state the degree of agreement regarding the exploitation of knowledge sources by their business for the exploration of new business opportunities, evaluating the responses on a 5-level ordered scale (1: none, 5: The dimensions highlighted by EFA are as follows: (1) Value chain: Competitors – Customers – Suppliers and (2) Scientific: Academic institutions and research institutes – Research programs – Scientific journals. That is, the prime factor can be defined as a value chain, as it includes all of the variables that make it up such as competitors, customers, and suppliers. The other factor concerns the scientific sources of knowledge that a supply chain firm can exploit since they contain academic institutions and research institutes, research programs, and scientific journals. We see, however, a clear distinction between these two factors. On the one hand, there is a strong element of the market (value chain), while on the other hand, it is the scientific part with what this entails

Table 2 Knowledge source factors analysis to explore new supply chain firm opportunities

Knowledge Sources	Factor 1	Factor 2
Universities and research institutes	**0.857**	−0.076
Research programs	**0.855**	−0.023
Scientific journals	**0.741**	0.120
Competitors	−0.007	**0.753**
Clients	−0.028	**0.709**
Suppliers	0.061	**0.708**
Eigenvalues	*2.01*	*1.59*
Variance explained %	*33.65*	*26.52*

Table 3 Analyzing factors in supply chain firm networks

Networks	Factor 1	Factor 2
Opportunities for the exploitation of new technologies and knowledge	**0.777**	0.118
Development of new products/services	**0.757**	0.211
Exploring opportunities for export activity	**0.636**	−0.115
Hiring highly qualified/skilled staff	**0.561**	0.107
Settlement of tax and other legal issues	−0.047	**0.884**
Production and operations management	0.246	**0.837**
Eigenvalues	*2.21*	*1.31*
Variance explained %	*36.84*	*21.89*

(scientific sources of knowledge). Table 2 details the results from driver analysis in the question examining business sources of knowledge the degree of correlation is quite high here.

6 Networks

The surveyed founders assessed the extent to which their networks contribute to day-to-day business operations. Two factors emerged from the analysis. The first is about exploring opportunities and creating something new. In other words, they are opportunities to exploit new technologies, knowledge, and opportunities for export activity. This also includes developing new products, services, and hiring the right staff to create them. The second factor concerns day-to-day supply chain firm operations such as operations, operations management, tax, and legal arrangement. Explore opportunities and create something new: Opportunities to exploit new technologies and knowledge—Developing new products/services – Exploring opportunities for export activity – Hiring highly qualified/skilled staff management of daily operations and obligations: Tax and legal settlement—Operations and operations management (Table 3).

Therefore, we note that the two factors differ in the quality of business. One is mainly about an opportunity while the other is about simple handling of everyday issues.

7 Barriers to Starting Up

Another aspect that has been studied is the examination of the factors that made the establishment of supply chain firms difficult and the obstacles that arise during their operation. Starting from the barriers the aspiring entrepreneur encountered at start-up, we measure the extent to which they influenced them with a 5-point Likert (1: none, 5: to a large extent). From EFA, we could distinguish the following factors: (1) Inadequate networking: Inadequate networking from the founder's professional/educational career to find partners; inadequate networking from the founder's relationships to find partners. (2) Initial investment and bureaucracy: High initial capital – Difficulty in accessing sources of finance – Bureaucracy. The first factor relates to the inadequate networking that exists on the part of the founder to set up the business. These networks concern their professional and educational environment and the personal relationships they develop. The second factor relates to the financial requirements that exist for starting up, which are the high-potential initial capital needed by the industry, and difficult access to various sources of finance. However, red tape is included as a major barrier to creating a supply chain firm (Table 4).

8 Conclusions

High-tech sectors are seen as pioneers in terms of innovation and have the potential to make an economy more competitive by creating new quality jobs and attracting new investment funds. However, the Greek supply chain companies operating in these sectors are highly heterogeneous and, in several cases, do not meet the basic conditions for their classification as knowledge-intensive. The main conclusions that emerge from the link between the concept of knowledge-based entrepreneurship and supply chain firm innovation reflect the factors that separate different types of businesses, depending on the dynamics they present. Firms that are closer to the concept of knowledge-based entrepreneurship have more important drivers of

Table 4 Factor analysis of start-up barriers

Obstacles to starting	Factor 1	Factor 2
Insufficient networking from the professional/educational career of the founder to find partners	**0.921**	0.073
Insufficient networking from the personal relationships of the founder to find partners	**0.921**	0.101
High initial capital	−0.038	**0.809**
Difficulty accessing funding sources	0.256	**0.739**
Bureaucracy	0.050	**0.682**
Eigenvalues	*2.08*	*1.36*
Variance explained %	*41.60*	*27.38*

opportunity recognition and exploitation than other businesses. They leverage their networks primarily to explore opportunities and develop new products and services. However, some supply chain firms have cognitive factors as the most important creators. They have founders with more job experience in the industry and have a better knowledge base by exploiting the value chain partners more than any other knowledge-intensive supply chain firm to explore new opportunities, but also by making greater use of the important networks they develop to manage day-to-day operations and explore opportunities [14]. But, in addition to the knowledge factors, these two categories of supply chain firms also stand out for the special incentives they have to form. The search for a job and the expectation that there may be income growth are more suited to supply chain firms that do not show strong innovative results. Therefore, state policies concerning the promotion of this type of entrepreneurship, which usually refer to sectors with technological content, should be done under other conditions and, above all, be targeted at the specific features of Greek supply chain firms. Other scientists with similar cases are studying similar problems using modeling and simulation [26–29].

9 Future Work

More future research could focus on digital innovation in other main high-tech sectors of the economy [15, 16], as well as financial technology [17], cryptocurrency trade organizations [17–20], and of course the wider scope of logistics and supply chain enterprises [21–25].

References

1. Hau, L.L., Shen, Z.J.: Supply chain and logistics innovations with the Belt and Road Initiative. J. Manag. Sci. Eng. **5**(2), 77–86 (2020)
2. He, L.F., Xue, M., Gu, B.: Internet-of-things enabled supply chain planning and coordination with big data services: Certain theoretic implications. J. Manag. Sci. Eng. **5**(1), 1–22 (n.d.)
3. Reza, K.M., Reza, F.S., Mark, G.: Joint analysis of eco-efficiency and eco-innovation with common weights in two-stage network DEA: A big data approach. Technol. Forecast. Soc. Chang. **144**(7), 553–562 (2019)
4. Gao, B.: Construction of Knowledge Service Model of Guizhou Supply Chain Enterprises Based on Big Data. Int. J. Inf. Syst. Supply Chain Manag. **15**(3) (2022). https://doi.org/10.4018/IJISSCM.290016
5. Garavaglia, C., Grieco, D.: Hand in hand with entrepreneurship. A criticaloverview from entrepreneurship to knowledge-based entrepreneurship. In: KEINS, Knowledge Based Entrepreneurship: Institutions, Networks and Systems. EU project n° CT2–CT–2004–506022, deliverable 0 (2005)
6. Groen, A.: Knowledge intensive entrepreneurship in networks: towards a multi-level/multi dimensional approach. J. Enterp. Cult. **13**(1), 69–88 (2005)
7. Forster, W. (2008). Co-founding New Ventures: founding Team Structure and Relationships in New Venture Creation, Working paper.

8. Folster, S.: Do entrepreneurs create jobs? Small Bus. Econ. **14**, 137–148 (2000)
9. Gartner, W.: A conceptual framework for describing the phenomenon of new venture creation. Acad. Manag. Rev. **10**, 696–706 (1985)
10. McGrath, R.G., MacMillan, I.C.: The Entrepreneurial Mindset. Harvard Business School Press (2000)
11. Beckman, C.: The influence of founding team company affiliations on firm behavior. Acad. Manag. J. **49**(4), 741–758 (2006)
12. Schoonhoven, C.B., Eisenhardt, K.M., Lyman, K.: Speeding products to market: waiting time to first product introduction in new firms. Adm. Sci. Q. **35**(1), 177–207 (1990)
13. Rowley, J.: From learning organization to knowledge entrepreneur. J. Knowl. Manag. (2000)
14. Easterby-Smith, M., Lyles, M., Peteraf, M.: Dynamic capabilities: current debates and future directions. Br. J. Manag. **20**, 1–8 (2009)
15. Kanellos, N.S.: A conceptual framework for analyzing knowledge-based entrepreneurship. In: 4th Symposium on Business and Management and Dynamic Simulation Models Supporting Management Strategies (2011) http://hdl.handle.net/10797/13771
16. Sakas, D.P., Nasiopoulos, D.K., Reklitis, P.: Modeling and simulation of the strategic use of the internet forum aiming at business success of high-technology companies. In: Sakas, D., Nasiopoulos, D. (eds.) Strategic Innovative Marketing. IC-SIM 2017. Springer Proceedings in Business and Economics. Springer, Cham (2019). https://doi.org/10.1007/978-3-030-16099-9_22
17. Sakas, D.P., Kamperos, I.D.G., Reklitis, D.P., Giannakopoulos, N.T., Nasiopoulos, D.K., Terzi, M.C., Kanellos, N.: The effectiveness of centralized payment network advertisements on digital branding during the COVID-19 crisis. Sustainability. **14**(6), 3616 (2022a). https://doi.org/10.3390/su14063616
18. Sakas, D.P., Giannakopoulos, N.T., Kanellos, N., Tryfonopoulos, C.: Digital marketing enhancement of cryptocurrency websites through customer innovative data process. Processes. **10**(5), 960 (2022b). https://doi.org/10.3390/pr10050960
19. Sakas, D.P., Giannakopoulos, N.T., Kanellos, N., Migkos, S.P.: Innovative cryptocurrency trade websites' marketing strategy refinement, via digital behavior. IEEE Access. **10**, 63163–63176 (2022c). https://doi.org/10.1109/ACCESS.2022.3182396
20. Sakas, D.P., Giannakopoulos, N.T., Reklitis, D.P., Dasaklis, T.K.: The effects of cryptocurrency trading websites on airlines' advertisement campaigns. J. Theor. Appl. Electron. Commer. Res. **16**(7), 3099–3119 (2021). https://doi.org/10.3390/jtaer16070169
21. Sakas, D.P., Giannakopoulos, N.T.: Harvesting crowdsourcing platforms' traffic in favour of air forwarders' brand name and sustainability. Sustainability. **13**(15), 8222 (2021a). https://doi.org/10.3390/su13158222
22. Sakas, D.P., Giannakopoulos, N.T.: Big data contribution in desktop and mobile devices comparison, regarding airlines' digital brand name effect. Big Data Cognit. Comput. **5**(4), 48 (2021b). https://doi.org/10.3390/bdcc5040048
23. Sakas, D.P., Reklitis, D.P.: Predictive model for estimating the impact of technical issues on consumers interaction in agri-logistics websites. In: Bochtis, D.D., Pearson, S., Lampridi, M., Marinoudi, V., Pardalos, P.M. (eds.) Information and Communication Technologies for Agriculture—Theme IV: Actions. Springer Optimization and Its Applications 185. Springer, Cham (2021). https://doi.org/10.1007/978-3-030-84156-0_14
24. Terzi, M.C., Sakas, D.P., Seimenis, I.: The contribution of the scientific committee in the development of conferences. Procedia Soc. Behav. Sci. **73**, 373–382 (2013). https://doi.org/10.1016/j.sbspro.2013.02.064
25. Nasiopoulos, D.K., Sakas, D.P., Vlachos, D.S.: Modeling the scientific dimension of academic conferences. Procedia Soc. Behav. Sci. **147**, 576–585 (2014). https://doi.org/10.1016/j.sbspro.2014.07.164
26. Nasiopoulos, D.K., Ioanna, D., Lida, G., Paraskevi, P., Giakomidou, D.S.: Modeling of B2C communication strategies in electronic commerce. Strategic Innovative Marketing, 531–537 (2017)

27. Nasiopoulos, D.K.: Implementation stages and risks of Logistics Companies reengineering. Bus. Intell. Model. **6**, 197–206 (2020a)
28. Nasiopoulos, D.K.: Web search applications in Logistics companies in the context of evolution of Management Information Systems. Bus. Intell. Model. **6**, 241–254 (2020b)
29. Dimitrios, N.K., Sakas, D.P., Vlachos, D.S.: The role of open source leadership in developing high technology companies. Key Eng. Mater. **543**, 402–405 (2013)

Knowledge-Based Business Skills and Dynamic Capabilities for Supply Chain Firms

Nikos Kanellos

1 Introduction

High-tech sectors are generally seen as pioneers in terms of innovation and have the potential to make an economy more competitive, creating new quality jobs and attracting new investment funds. However, the Greek supply chain firms, aiming to improve innovation levels, appear to be highly heterogeneous and, in several cases, do not meet the basic requirements for their classification as knowledge intensive. The main conclusions that emerge from the link between the concept of knowledge-based entrepreneurship (KBE) and supply chain business innovation reflect the factors that separate different types of supply chain firms, depending on the dynamics they present.

2 The Importance of Knowledge-Based Entrepreneurship in Supply Chain Sector

Every form of entrepreneurship is based on the exploitation of some kind of knowledge, even when it comes to arbitrage opportunities (investment opportunity that brings profit without risk). KBE is a special form of entrepreneurship and is linked to the so-called knowledge economy, which is characterized by the crucial role of ICT, the high proportion of knowledge-intensive activities, the largest capital

N. Kanellos (✉)
Department of Agribusiness and Supply Chain Management, School of Applied Economics and Social Sciences, Agricultural University of Athens, Athens, Greece
e-mail: nikos.kanellos2@aua.gr

D. P. Sakas et al. (eds.), *Computational and Strategic Business Modelling*,
Springer Proceedings in Business and Economics,
https://doi.org/10.1007/978-3-031-41371-1_28

of intangible resources about the corresponding capital of material resources in the whole stock capital, and increased R&D costs [1, 2]. KBE is related to activities in which the role of creating new knowledge is central to creating value [3].

Starting from a broader definition of entrepreneurship and incorporating the concept of knowledge into it, we consider that: "*entrepreneurship is the process of creating something new with value, devoting the necessary time and effort to knowledge-intensive economic activities, taking on economic, mental and social risks, and taking as a reward monetary and personal satisfaction and independence*" [4, 5]. This definition is based on three key characteristics of entrepreneurship. First is the creation process, and indeed the creation process of something new that is of value both to the entrepreneur and to the target audience. Second is the time and effort required to create this new one, which must also be operational. All-time and effort are estimated only by those involved in the business process. Finally, the third important aspect of entrepreneurship that stands out from this definition is the assessment of the risks needed to obtain the expected benefits. These risks, despite their diversity, focus on economic, psychological, and social areas. When studying the KBE phenomenon, it is important to take into account, in particular, the first of three key aspects of entrepreneurship. In the context of the article, we will adopt some key elements that Malerba and McKelvey [6] present for the concept of KBE. Therefore, KBE is linked to the following:

- New firms
- Innovative firms (in terms of knowledge-based processes)
- Firms with significant knowledge intensity in their activities
- Firms that exploit innovative opportunities not only in various sectors
- Firms in collaborative networks [7, 8]

Therefore, KBE is not just about start-ups. It is more than just new firms. These are new and innovative firms with a high knowledge intensity in their activities, involved in a process of transforming knowledge into innovation. That is, these firms are transforming knowledge into new or significantly improved products and services that can be brought to market. They are also firms that exploit innovative opportunities in various sectors and achieve through a strategy that pursues sustainable competitive advantage.

Yang and Lin [9] corroborated the significance of knowledge in the supply chain for innovation and discovered that regulatory restrictions, corporate management dedication, and societal acknowledgment are the primary motivating factors for sustainable innovation adoption. According to Khan et al. [10], the deployment of sophisticated knowledge and the formation of communication with the global setting can overcome many supply chain management constraints. Firms leverage their human value to produce extra strategic benefits via interaction in order to accomplish greater efficiency in sustainable innovation [11]. Song et al. [12] investigated the impact of knowledge transfer in regulating the relationship between sustainable knowledge-based skills and innovativeness and proposed that enterprises increase their permeable ability to accomplish sustainability initiatives. Nevertheless, it has been shown that studies on knowledge management in more sustainable knowledge-based business skills and manufacturing have gotten limited consideration [13].

3 Variables of Analysis

Some of the key features of knowledge-based entrepreneurship are the knowledge that is significant in the supply chain firms' activities, the innovations they create, and the networks they form and leverage in their operations. In addition, another feature of interest to us in the context of KBE is the "age" of supply chain firms and their sector. The focus of the investigation is new supply chain firms operating in supply chain sectors, and therefore, the selection of the sampled firms was based on their date of establishment and the economic sector in which they operate. Therefore, for the first three traits of KBE, it is necessary to have indicators that will allow us to measure the specific sizes, while for the other two traits, the initial selection of the supply chain firm population was sufficient to satisfy. Below are all the variables that were used to determine the characteristics of KBE.

To classify supply chain firms based on specific traits that they have and which are related to the concept of knowledge entrepreneurship, we have used three categories of variables. The first category is the knowledge intensity of supply chain firms. We distinguish this category into two subcategories of variables: (a) sources of knowledge and (b) the business knowledge stock. The sources of knowledge contain the internal dimension of the supply chain firms as well as two ratios derived from factor analysis: the value chain (competitors – customers – suppliers) and the scientific sources of knowledge (academic and research institutes – research programs – scientific journals). The firm knowledge stock includes the educational level of the founding group (1 = Primary-Secondary-Vocational education, 2 = Higher education, 3 = Master, 4 = PhD), the previous professional experience of the founders (Experience in founding – Experience in the field), and the educational level of employees (same categorization as founders, but considering whether it exceeds 75%). The second category is business networking, where here too there are two indicators indicated by factor analysis. These are exploring opportunities and creating something new (opportunities for exploiting new technologies and knowledge – developing new products/services – exploring opportunities for export activity – hiring highly qualified staff/skills) and managing day-to-day operations and obligations (addressing tax and other legal issues – operations and operations management). The creation or introduction of innovation in the last 3 years, the introduction of a product from abroad, the introduction of ready technology (e.g., equipment) from abroad, the adaptation of technology from abroad, the exploitation of results of research activity, either by own means or through cooperation, the exploitation of business practices, the modernity of innovations (new for the enterprise – new for the market – new for the international market), and the protection of intellectual property in the last 3 years (patents, copyrights, trademarks, and confidentiality agreements).

After classifying the supply chain firms into clusters, we selected variables with which comparative analysis can be performed between them. Here, we have four categories of variables, where the indexes contained in each have been indicated by factor analysis. The first category is the factors of firm creation. This category includes the general incentives (need for personal creation – social recognition – job independence), the knowledge factors (market knowledge – work experience in the same industry), the specific incentives (job search for a livelihood – the expectation of income growth), the recognition and exploitation of opportunity (recognition of a technological opportunity – recognition of a market opportunity – exploitation of scientific research results). The second category, which is the obstacles to the operation of enterprises, always includes, according to the analysis of the effects, the risks and costs of developing technology (high costs for developing technology – technological risks – high costs of skilled labor) and the economic crisis (market uncertainty – limited domestic demand). The third category, which is concerned with business strategy, focuses on new products and new markets (increasing sales by offering new products and services – entering new niche markets). The last category is innovation barriers, which include innovation uncertainty (uncertainty associated with the commercial success of the innovative project – uncertainty in demand for innovative products/services – uncertainty associated with the technical part of the innovative project) and research requirements (high research and development costs – lack of funding for innovation – lack of qualified staff). Comparative analysis allows us to confirm the differences between the four supply chain firm clusters by using variables that were not used in the classification process, so we try to confirm some logical differences that exist between the groups and are associated with the selected variables.

We propose a classification based on their specific characteristics linked to knowledge-intensive entrepreneurship (by cluster analysis technique). The types of variables selected for the analysis are divided into three broad categories (Knowledge Intensity – Networks – Innovation). The classification proposed by the results of the analysis is particularly useful for studying the differences and relationships that the supply chain firms have concerning the concept of knowledge entrepreneurship. It is not just a categorization of the supply chain firms into groups, as it allows us to trace the different levels of knowledge, networking, and innovation in firms, to draw useful conclusions about the theoretical framework of research.

The purpose of the cluster analysis approach is to group observations based on their similarity to several variables. In the context of the investigation, we have applied hierarchical approach methods, which like the other approach offered by the SPSS statistical package (K-Means) do not require the existence of a model and rely only on algorithmic solutions. The final choice of the number of groups is made after the smallest observations in the distance are united at each step (each observation is itself a group).

Based on the analysis carried out, four supply chain firm clusters arise: (a) innovative supply chain firms, (b) networking supply chain firms, (c) Research & Development (R&D), and (d) low innovation intensity (LII) supply chain firms. The name of each group is due to the prices of the three categories of variables received by the supply chain firms when compared. More specifically, innovative supply chain firms have the highest levels of innovation, networking supply chain firms exploit to a greater extent of their networks, either for exploring new opportunities or for settling everyday issues, R&D supply chain firms innovate, using exclusively the results of research carried out within them, and LIIs show almost zero rates of innovation.

The majority of firms (34.4%) belong to the fourth category, where there are low innovation-intensive firms. That is, a key characteristic of KBE, that is innovation, does not show or shows it with minimal percentages of a large part of the firms in the sample. This result is not at all in line with the fact that the sectors to which these firms belong are high-tech sectors, expecting high innovative performance. Next is the number of projects in the category of innovative firms with 29.7% and networking firms with 23.5%. The smallest group is R&D firms with only 12.4% in the sample.

Assessing the behavior of each cluster by applying the statistical technique of polynomial logistic regression, we consider how to differentiate the selected independent variables among the four different supply chain firms' groups. The selection of independent variables was made with five supply chain firms' components in mind. The first involves the founder of the founding group of firms, focusing on the most important areas of knowledge and skills. The founders had to select from five areas of knowledge and skills the two best areas for themselves in order of preference (1 and 2). We isolated the "Industry-Related Technical Knowledge" and "Product and Service Design" (where they were the first choice) options as the skills that best reflect their professional identity. In our model, we consider both supply chain firm resources and in particular their dynamic capabilities; as we have seen, factorial analysis has identified two factors that affect business capabilities:

- Ability to adapt to market changes: The supply chain firm follows market trends. The supply chain firm responds quickly to changes in demand. The supply chain firm immediately responds to the movements of competitors.
- Ability to adapt to technological changes and produce differentiated products: The supply chain firm has the flexibility to produce differentiated products and services. The supply chain firm can understand and adapt to technological change.
- About the human resources of supply chain firms, we have chosen the variable relating to the number of employees of enterprises as an average over the last 3 years:

  - 0 = Enterprises without employees
  - 1 = 1–9 employees
  - 2 = 10–49 employees
  - 3 = 50–249 employees
  - 4 = >250 employees

The proposed accounting model also includes the variables that reveal the extroversion of supply chain firms and the year of their foundation, using as a starting point 2008, the year when the first signs of the financial crisis in the market began to appear. The general equation discussed can be attributed as follows:

Firm Group = f{TECHNICAL KNOWLEDGE – PRODUCTS, NUMBER OF EMPLOYEES, ADAPTABILITY
– MARKET CHANGES, ADAPTABILITY
– TECHNOLOGICAL CHANGES, EXPORTS, YEAR OF ESTABLISHMENT}.

The results of the analysis are presented in detail in the tables below. Note that the reference category is the group of LIIs.

The first comparison is made between innovative and LIIs supply chain firms. We observe that four independent variables show statistically significant differences between the two groups. The first is exports, with a positive factor B, which essentially expresses the positive relationship that exists between this variable and innovative supply chain firms. That is, when an enterprise is export-oriented, it is more likely to be in the category of innovative enterprises than in that of LIIs. In contrast, the greater the ability of a supply chain firm to adapt to market changes is (keeping other variables constant), the greater the likelihood that it will be in the LII group.

On the other hand, as the ability of a business to adapt to technological change and to produce differentiated products increases, the more likely this business is to be innovative. The last variable that identifies a statistically significant difference between the two groups is the year when the supply chain firms were founded and, more specifically, whether supply chain firms were founded before or after 2008. We observe that this variable adversely affects innovative businesses so that the businesses that were founded more recently and after 2008 are more likely to be LIIs than innovative ones (Table 1).

The next comparison is made between networking supply chain firms and LIIs. Here, we observe that an independent variable shows a statistically significant difference. It is the ability to adapt to technological change and produce differentiated products and services. The relation is positive, that is, the larger the size of this variable, keeping the values of the others stable, the greater the likelihood that a business with this feature belongs to the networking supply chain firms rather than to the LIIs (Table 2).

Table 1 Logistic regression results in the supply chain firm group – Case "Innovative supply chain firms" (Reference category: LIIs)

Variables	Coefficient (B)	Exp β	Confidence limits	
CONSTANT	−2.14 (2.96)			
Technical knowledge/product design	0.53 (1.77)	1.701	(0.77)	−3.71
Exports	**0.90* (5.00)**	**2.479**	**(1.11)**	**(5.49)**
Ability to adapt to market changes	**−0.61* (4.52)**	**0.540**	**(0.30)**	**(0.95)**
Ability to adapt to technological change	**0.76* (8.44)**	**2.147**	**(1.28)**	**(3.59)**
Number of employees	0.26 (1.35)	1.299	(0.83)	(2.01)
Year of establishment	**−1.00* (4.90)**	**2.720**	**(1.12)**	**(6.59)**

* indicates statistical significance (p) below 5% in parentheses in the column of coefficients B are the values of Wald

Table 2 Accounting regression results in the enterprise group — "Networking supply chain firms" case (reference category: LIIs)

Variables	Coefficient (B)	Exp β	Confidence limits	
CONSTANT	−4.16 (7.92)			
Technical knowledge/product design	0.56 (1.83)	1.767	(0.77)	(4.02)
Exports	0.29 (0.49)	1.349	(0.58)	(3.10)
Ability to adapt to market changes	0.28 (0.75)	1.324	(0.70)	(2.49)
Ability to adapt to technological change	**0.71* (6.59)**	**2.046**	**(1,18)**	**(3.53)**
Number of employees	−0.28 (1.42)	0.753	(0.47)	(1.20)
Year of establishment	−0.91 (2.13)	0.403	(0.11)	(1.36)

* indicates statistical significance (p) below 5% in parentheses in the column of coefficients B are the values of Wald

Table 3 Accounting regression results in the enterprise group – "R&D supply chain firms" Case (Reference category: LIIs)

Variables	Coefficient (B)	Exp β	Confidence limits	
CONSTANT	−5.55 (8.53)			
Technical knowledge/product design	**1.50* (4.72)**	**4.515**	**(1.16)**	**(17.58)**
Exports	0.81 (2.49)	2.259	(0.82)	(6.21)
Ability to adapt to market changes	−0.36 (0.82)	0.696	(0.31)	(1.52)
Ability to adapt to technological change	**0.83* (4.70)**	**2.314**	**(1.08)**	**(4.93)**
Number of employees	**0.68* (5.21)**	**1.992**	**(1.10)**	**(3.59)**
Year of establishment	**−20.01** (0.00)**	**2.027E-09**	**(2.027E-09)**	**(2.027E-09)**

**, * denote statistical significance (p) below 1% and 5%, respectively in parentheses in the column of coefficients B are the values of Wald

The last comparison is made between R&D supply chain firms and LIIs. A total of four independent variables show statistically significant differences between the two clusters. Three of them have a positive relationship with the R&D group, while one has a negative relationship (negative factor B). In particular, technical knowledge and product design as the most important knowledge and skills areas of the founders, the ability of the supply chain firm to perceive technological change and have the flexibility to produce differentiated products, and the size of the supply chain firm, positively influence a supply chain firm's R&D integration. Finally, the variable showing a negative factor from the data of the analysis is the one related to the year of establishment of the supply chain firms and, more specifically, whether a supply chain firm was founded before or after the onset of the economic crisis (before or after 2008). This parameter suggests that when a supply chain firm has been founded more recently and after 2008, it is more likely to belong to the LII group. Moreover, as we have seen from the descriptive results of the four clusters, no R&D Company has been established since 2008, whereas, on the contrary, 20% of LIIs were set up in 2008–2010 (Table 3).

4 Conclusions

Regarding the factors that shape knowledge-based entrepreneurship in supply chain sectors in the Greek area, it can be seen that the analysis split the supply chain firms into four clusters. Three of these four groups present characteristics that are inextricably linked to the concept of knowledge entrepreneurship such as the strong body of knowledge supply chain firms can possess and its significant innovative performance. These groups of supply chain firms are R&D supply chain firms, Innovative supply chain firms, and networking supply chain firms, whereas LIIs do not have all the prerequisites of a knowledge-based enterprise.

The sector-related technical knowledge of the founders as well as their ability to design products are two important factors that can differentiate a supply chain firm and integrate it into a different business group. More specifically, according to the results of the statistical analysis, in the comparison between the clusters of R&D and LIIs, we saw that this factor positively affects the supply chain firms belonging to the first group. That is, when a supply chain firm is made up of founders or founders whose technical knowledge and product design are the knowledge and competence areas that best reflect their professional identity, then this supply chain firm is more likely to be in the R&D business group than in the LIIs. Therefore, the factor relating to knowledge and skills of the founders plays an important role in the formation of knowledge-intensive entrepreneurship as more specifically this factor of differentiation makes the R&D business group from the LII business group.

In addition, the factor relating to the number of employees of supply chain firms has a positive impact on a company's R&D integration. In other words, the higher the number of employees, the more likely a company is to be distinguished for its better innovation performance and the higher stock of knowledge, with the result that it is more likely to join the group of R&D companies than the group of LIIs. Therefore, the size of the business influences and shapes knowledge entrepreneurship, as there is a positive relationship between the number of employees and the characteristics of knowledge-based entrepreneurship.

A supply chain firm's ability to recognize and seize opportunities, as statistical analysis suggested, can be divided into two factors: (a) the ability to adapt to changes in the market and (b) the ability to adapt to technological changes and to produce differentiated products. We observed from the results of the analysis that these two factors differ in how they affect knowledge entrepreneurship. In particular, by comparing the groups of innovative enterprises and LIIs, we saw that when a supply chain firm has the flexibility to produce differentiated products, perceive technological changes, and adapt to them (producing "ability to adapt to technological changes and produce differentiated products"), then it is more likely to belong to innovative enterprises than to LIIs. Conversely, when a supply chain firm simply follows market trends, it responds quickly to changes in demand and responds immediately to the movements of competitors (factor "ability to adapt to changes in the market"), then the chances of its inclusion in the LII group of supply chain firms increase. The corresponding comparisons made between LIIs and networking supply chain firms and R&D supply chain firms also showed that the factor "ability to adapt to technological changes and produce differentiated products" differentiates these groups of supply chain firms, since when a firm is distinguished for this capacity, it is more likely to be a networking or R&D supply chain firm than an LII supply chain firm. So, the ability to recognize and exploit opportunities influences knowledge-based entrepreneurship and differentiates it according to the forms it takes.

In most cases, the export activity of a supply chain firm is linked to the added value, quality, and innovation of the products it offers since they are also "tested" in international markets. By comparing the clusters of supply chain firms with each other to determine whether the export orientation of a company varies according to the group to which it belongs, we saw that this factor has a positive relationship with the innovative supply chain firms. First of all, we should mention here again that the groups that export the most are the innovative supply chain firms and also the R&D supply chain firms, as 1 in 2 or so are extroverts. By contrast, 61% of networking supply chain firms and about 73% of LIIs do not carry out any export activity. Therefore, in the comparison between innovative and LIIs supply chain firms where there is a clear strategic difference in export orientation, we see that the factor "exports" positively affects the former. That is, as the export activity of an enterprise increases, it is more likely to be placed in the group of innovative supply chain firms than in the group of LIIs.

About the year of establishment of enterprises, based on 2008, when the first signs of the economic crisis were visible, we can observe the following: The supply chain firms in the sample are new firms that were established between 2000 and 2010. However, it was important to consider whether the typology of businesses changed before or after the onset of the economic crisis (before or after 2008). From the descriptive results related to business clusters, we saw that no R&D firms were established after 2008, only 7% of networking firms were established during this time, while many more are the innovative and LII firms that were founded when the signs of the economic crisis we are now emerging (31% and 20%, respectively). In the comparison we made between R&D and LIIs, we saw that the factor relating to the year of establishment (before or after 2008) has a negative relationship with the former, that is, as the results of the analysis show, the supply chain firms established since 2008 are more likely to be LII than R&D. The same factor also positively affects LIIs as compared to innovative supply chain firms in their comparison, concluding that enterprises established in 2008–2010 are more likely to be LIIs than innovative ones [25, 27]. Therefore, whether a business has been set up before or after the economic crisis has an impact on knowledge-intensive entrepreneurship, and, indeed, from the comparison made between the groups, we have seen that supply chain firms that were set up from 2008 onward are more likely to be LIIs [26, 28].

5 Future Work

Future studies concerning the clarity of the factors affecting supply chain knowledge-based business skills [14, 15] should be enhanced and expanded into other sectors like cryptocurrency [16–18], centralized payment networks [19], and logistics related [20–24].

References

1. Foray, D.: The Economics of Knowledge. MIT Press (2004)
2. Stam, E., Garnsey, E.W.: Entrepreneurship in the Knowledge Economy Centre for Technology Management (CTM) Working Paper, No. 2007/04 (2008). https://doi.org/10.2139/ssrn.1923098
3. Garavaglia, C., Grieco, D.: Hand in hand with entrepreneurship. A critical overview from entrepreneurship to knowledge-based entrepreneurship. In: KEINS, Knowledge Based Entrepreneurship: Institutions, Networks and Systems. EU project n° CT2–CT–2004–506022, deliverable 0 (2005)
4. Hisrich, R., Peters, M., Shepherd, D.: Entrepreneurship, 6th edn. McGraw Hill (2005)
5. Bosma, N.: Overview of theory, statistical evidence, and case studies on knowledge-based entrepreneurship, clusters and economic growth. AEGIS project, del. 2.4.1. (2010)
6. Malerba, F., McKelvey, M.: Conceptualizing knowledge-intensive entrepreneurship: concepts and models. AEGIS project, del. 1.1.1. (2010)
7. Groen, A.: Knowledge intensive entrepreneurship in networks: towards a multi-level/multi dimensional approach. J. Enterp. Cult. **13**(1), 69–88 (2005)
8. Radosevic S., Savic M. & Woodward R. (2008). Knowledge-based Entrepreneurship in Central and Eastern Europe: Myth and Reality. Economics Working Paper No. 93.
9. Yang, Z., Lin, Y.: The effects of supply chain collaboration on green innovation performance: an interpretive structural modeling analysis. Sustain. Prod. Consumption. **23**, 1–10 (2020). https://doi.org/10.1016/j.spc.2020.03.010
10. Khan, S.A.R., Godil, D.I., Jabbour, C.J.C., Shujaat, S., Razzaq, A., Yu, Z.: Green data analytics, Blockchain technology for sustainable development, and sustainable supply chain practices: evidence from small and medium enterprises. Ann. Oper. Res. (2021). https://doi.org/10.1007/s10479-021-04275-x
11. Chen, P.-C., Hung, S.-W.: Collaborative green innovation in emerging countries: a social capital perspective. Int. J. Oper. Prod. Manag. **34**(3), 347–363 (2014). https://doi.org/10.1108/ijopm-06-2012-0222
12. Song, M., Yang, M.X., Zeng, K.J., Feng, W.: Green knowledge sharing, stakeholder pressure, absorptive capacity, and green innovation: evidence from Chinese manufacturing firms. Bus. Strateg. Environ. **29**(3), 1517–1531 (2020). https://doi.org/10.1002/bse.2450
13. Wang, F., Li, H., Cao, Y., Zhang, C., Ran, Y.: Knowledge sharing strategy and emission reduction benefits of low carbon technology collaborative innovation in the green supply chain. Front. Environ. Sci. **9**, 783835 (2022). https://doi.org/10.3389/fenvs.2021.783835
14. Kanellos, N.S.: A conceptual framework for analyzing knowledge-based entrepreneurship. In: 4th Symposium on Business and Management and Dynamic Simulation Models Supporting Management Strategies (2011) http://hdl.handle.net/10797/13771
15. Sakas, D.P., Nasiopoulos, D.K., Reklitis, P.: Modeling and simulation of the strategic use of the internet forum aiming at business success of high-technology companies. In: Sakas, D., Nasiopoulos, D. (eds.) Strategic Innovative Marketing. IC-SIM 2017. Springer Proceedings in Business and Economics. Springer, Cham (2019). https://doi.org/10.1007/978-3-030-16099-9_22

16. Sakas, D.P., Giannakopoulos, N.T., Kanellos, N., Tryfonopoulos, C.: Digital marketing enhancement of cryptocurrency websites through customer innovative data process. Processes. **10**(5), 960 (2022a). https://doi.org/10.3390/pr10050960
17. Sakas, D.P., Giannakopoulos, N.T., Kanellos, N., Migkos, S.P.: Innovative cryptocurrency trade websites' marketing strategy refinement, via digital behavior. IEEE Access. **10**, 63163–63176 (2022b). https://doi.org/10.1109/ACCESS.2022.3182396
18. Sakas, D.P., Giannakopoulos, N.T., Reklitis, D.P., Dasaklis, T.K.: The effects of cryptocurrency trading websites on airlines' advertisement campaigns. J. Theor. Appl. Electron. Commer. Res. ***16****(7)*, 3099–3119 (2021). https://doi.org/10.3390/jtaer16070169
19. Sakas, D.P., Kamperos, I.D.G., Reklitis, D.P., Giannakopoulos, N.T., Nasiopoulos, D.K., Terzi, M.C., Kanellos, N.: The effectiveness of centralized payment network advertisements on digital branding during the COVID-19 crisis. Sustainability. ***14****(6)*, 3616 (2022c). https://doi.org/10.3390/su14063616
20. Sakas, D.P., Giannakopoulos, N.T.: Harvesting crowdsourcing platforms' traffic in favour of air forwarders' brand name and sustainability. Sustainability. **13**(15), 8222 (2021a). https://doi.org/10.3390/su13158222
21. Sakas, D.P., Giannakopoulos, N.T.: Big data contribution in desktop and mobile devices comparison, regarding airlines' digital brand name effect. Big Data Cognitive Comput. **5**(4), 48 (2021b). https://doi.org/10.3390/bdcc5040048
22. Sakas, D.P., Reklitis, D.P.: Predictive model for estimating the impact of technical issues on consumers interaction in agri-logistics websites. In: Bochtis, D.D., Pearson, S., Lampridi, M., Marinoudi, V., Pardalos, P.M. (eds.) Information and Communication Technologies for Agriculture—Theme IV: Actions. Springer Optimization and Its Applications 185. Springer, Cham (2021). https://doi.org/10.1007/978-3-030-84156-0_14
23. Terzi, M.C., Sakas, D.P., Seimenis, I.: The contribution of the scientific committee in the development of conferences. Procedia Soc. Behav. Sci. **73**, 373–382 (2013). https://doi.org/10.1016/j.sbspro.2013.02.064
24. Nasiopoulos, D.K., Sakas, D.P., Vlachos, D.S.: Modeling the Scientific Dimension of Academic Conferences. Procedia Soc. Behav. Sci. **147**, 576–585 (2014). https://doi.org/10.1016/j.sbspro.2014.07.164
25. Nasiopoulos, D.K., Ioanna, D., Lida, G., Paraskevi, P., Giakomidou, D.S.: Modeling of B2C Communication Strategies in Electronic Commerce. Strategic Innovative Marketing, 531–537 (2017)
26. Nasiopoulos, D.K.: Implementation stages and risks of Logistics Companies reengineering. Bus. Intell. Model. **6**, 197–206 (2020a)
27. Nasiopoulos, D.K.: Web search applications in Logistics companies in the context of evolution of Management Information Systems. Bus. Intell. Model. **6**, 241–254 (2020b)
28. Dimitrios, N.K., Sakas, D.P., Vlachos, D.S.: The role of open source leadership in developing high technology companies. Key Eng. Mater. **543**, 402–405 (2013)

The Operating Obstacles and the Strategy Followed by Greek Innovative Firms in the Supply Chain Sector

Nikos Kanellos

1 Introduction

Large investments aimed at creating new knowledge, such as research and development activities, do not automatically lead to innovation and economic growth. According to Carlsson et al. [1], (a) the factors that facilitate or hinder the transformation of knowledge into economically usable knowledge and (b) the presence or lack of a mechanism that will enable this transformation, significantly determine the effectiveness of the transformation of knowledge into innovation and new economic activity. Such a mechanism may be knowledge-based entrepreneurship (KBE). This chapter examines the factors that influence the operation of a knowledge-based supply chain firm, the obstacles it faces in its effort to innovate, the strategy that it follows, as well as how it recognizes and exploits the opportunities that arise.

2 Literature Review

2.1 Supply Chain Firms and Challenges

By employing various advancements, the supply chain industry evolution alters corporate processes with increased effectiveness, adaptability, and production for a greater degree of resilience [2]. The notion of innovative supply chain strategies and its connections with corporate permeable capability and entrepreneur performance

N. Kanellos (✉)
Department of Agribusiness and Supply Chain Management, School of Applied Economics and Social Sciences, Agricultural University of Athens, Athens, Greece
e-mail: nikos.kanellos2@aua.gr

D. P. Sakas et al. (eds.), *Computational and Strategic Business Modelling*,
Springer Proceedings in Business and Economics,
https://doi.org/10.1007/978-3-031-41371-1_29

are experimentally investigated [3]. Other analyses are looking into how particular expenditures affect sustainable supply chain performance and the function of knowledge sharing and supply chain associates' accountability [4].

Although several industrial companies use supply chain management to increase operating effectiveness, there are regulations and industry constraints for reducing pollution and waste reduction in the production moving processes [5]. Policy requirements raise company regulatory and policy knowledge, which promotes sustainable supply chain collaboration in many industries [6]. Corporations must innovate to enhance their procedures, particularly those for supply chain management, in order to reduce costs and enhance operational performance. Researchers have addressed supply chain entrepreneurship in domains such as logistics, advertising, and technological advancement [7]. There exist many experimental investigations that explore the relationship between innovation and corporate growth. With additional operating areas throughout the world, the difficulty of organizing innovative supply chain operations grows [8].

2.2 *Resources and Capabilities*

Strategic choices of a business play an important role in its growth and development process and determine to a large extent whether it will lead to sustainable competitive advantage. Every new business venture is characterized by its uncertain evolution and the strategy of the entrepreneur will judge whether this uncertainty can be overcome [8]. A supply chain firm can be seen as a bundle of unique resources [9]. The approach of the firm with this view, although it came to the fore in the 1980s and, above all, in the 1990s is not new. The strategy formulated based on the resources the firm has at its disposal appears in the literature during the 1960s and 1970s with the work of Andrews, Christiansen, and Chandler. However, the most important resource-based view of the firm is Penrose's [10] contribution to the book: "The theory of the growth of the firm" [10], which essentially underpinned the theory of resources and capabilities. Penrose studies the principles underlying a firm's growth, how fast it can be and how far it can be reached. For Penrose [10], the firm is both an administrative organization and a collection of operations resources. The resources of the firm consist of the "natural things" of the business that it purchases, borrows, or produces for its use and its people, who are hired on such terms as to be an effective part of the firm. The heterogeneity of these resources and the outcome they produce (products and services) is what makes all businesses different. Several examples in the literature relate the evolution of a new venture to the impact of internal resources, for example, Westehead et al. [11] report that the human capital of a firm, such as the skills and knowledge or previous experience and knowledge of the founding team, can influence its competitive strategies and play a key role in the future success of start-ups.

2.3 Technology Resources

Technological (intangible) resources, such as patents, trade secrets, good reputation, and internal culture, can be important sources of competitive advantage, growth, and development. Technological knowledge is a resource that can create superior value in a supply chain firm, and the use of that knowledge is at the core of KBE. According to Strand [12], knowledge is one of the most important resources of the supply chain firm and its application is at the heart of value production. For this reason, a firm's patent portfolio is a fairly important aspect of the firm. On the same wavelength, Grant [13] sees knowledge as the most strategically important resource in the business, underlining the role of the individual, and the business in its possession and application. It also raises the issue of transfer of knowledge not only between supply chain firms but mainly within the firm. The ease with which knowledge is exchanged within a supply chain firm, and in the case of tacit knowledge, is a vital task, so communities of practice, groups of people who come together regularly to share information and learn about their shared interests, are useful for communicating experiences.

2.4 Organizational Resources

The organizational resources (or organizational capabilities) of the supply chain firm, refer to the complex combinations of tangible (most of which are listed on firms' balance sheets, e.g., facilities, land, and raw materials) and intangible resources, together with the people and processes that firms use to turn inputs into outputs. They include the coordination and control processes, the learning processes within the supply chain firm, and the links it develops with the external environment (e.g., networks with other actors such as enterprises, universities, and research institutes) for the exchange and dissemination of knowledge, which have a significant impact on its innovative performance and growth. Teece et al. [14] focused their study on the firm's organizational resources, distinguishing between static processes/routines that can be easily repeated in the same way and more dynamic ones that create new resources and capabilities in the supply chain firm.

2.5 External Environment

The influence of the external environment on new small supply chain firms is also important for the process of their growth and development. Begley et al. [15] state that factors such as the availability of finance, market, and immediate sectoral opportunities, the supply of skilled labor, and government regulation all influence the level of entrepreneurial activity. The role of government policies and the general institutional environment can be very important in fostering entrepreneurial activity and the emergence of new technologies and industries. In addition, networks, which can be seen either as a subset of factors within the external environment or as an autonomous factor, play an important role in knowledge transmission and are beginning to become of particular interest in strategic management. The proliferation of networking observed in high-tech industries is mainly due to the high costs of technology development and the uncertainty of emerging technologies. It is extremely important for a new supply chain firm operating in a high-tech industry to have access to resources to help it overcome the first difficult obstacles. The knowledge resources exchanged can be either tangible resources such as equipment and products, or intangible resources such as know-how or trademarks.

2.6 Dynamic Capabilities

In the last 20 years, the dimension of supply chain firm capabilities has appeared strongly in the literature and seems to have created a distinct current. Prahalad and Hamel [16] introduced the concept of core competencies, describing them as the capabilities or skills that permeate all parts of the firm like a strong thread, and "bind" them into a single and comprehensive whole. The concept of dynamic capabilities, whose founder is considered to be David Teece, is a related concept to core competencies, and combines elements from both evolutionary economics and strategic management, giving a more dynamic dimension to a strategy that focuses on survival and growth of a supply chain firm. It essentially refers to those competencies that enable companies to integrate, redefine and develop their internal and external structures to respond to their competitive environment. The theory of dynamic capabilities is quite close to the resource theory of the firm, as it focuses on the internal environment of the firm, and although it has begun to form its branch of thought, it is essentially an extension of resource theory. Teece et al. [17] typically state [18]: "*our view of the firm is to some extent richer than that of resource theory. It is not only the bundle of resources that is important, but also the mechanisms by which a firm learns and accumulates new skills and capabilities, and the forces that constrain the pace and direction of this process*".

It is therefore a dynamic theory of firm resources that goes beyond the critiques of the static and equilibrium nature of the "simple" model of resource theory [17], and helps us to understand how a supply chain firm's resources evolve over time and therefore, how competitive advantage is maintained. It is linked to several elements of KBE, such as innovation, organizational learning, and knowledge management, providing a knowledge-based perspective. Moreover, many argue that it is an extension of the knowledge-based approach. Also, the processes for developing dynamic competencies include methods for structuring R&D, information technology-assisted coding, problem-solving, knowledge sharing, knowledge development, and absorptive capacity. However, unlike other fields dealing with change-related issues (organizational learning, innovation, etc.), which we have largely discussed in the previous sections, dynamic capabilities theory focuses on how supply chain firms can change their valuable resources over time. In general, the exploitation of resources, including human capital (founders/managers and employees), technological capital, and knowledge-based capital, is an important part of dynamic capabilities [19]. Looking at the relevant literature, it can be concluded that the types of dynamic capabilities are diverse and mainly related to functional (zero-level) and dynamic capabilities [20], and first-tier and metacapabilities. Ambrosini et al. [21], based on these typologies, developed three levels of dynamic competencies: incremental, renewing, and regenerative. The first two exploit the existing resource base, while the third assesses and adapts the overall portfolio, "allowing the supply chain firm to move away from previous change practices toward new dynamic capabilities". Indeed, they infer the need to study the recreational dynamic capabilities in newer firms, and whether the use of dynamic capabilities differs across industries and in the perceptions of founders.

Teece [21] distinguishes three basic types of capabilities: (1) the ability to sense and shape opportunities and threats, (2) the ability to seize opportunities, and (3) the ability to maintain competitiveness by enhancing, combining, protecting, and, when necessary, redesigning the supply chain firm's tangible and intangible resources. Let us look at some examples of dynamic capabilities. Dynamic capabilities can be R&D activities, absorptive capacity [22], product innovation leading to organizational renewal [23], reconfiguration of organizational structure [24], etc. According to Aragon-Correa and Sharma [25], some factors in the environment affect the development of dynamic competencies such as uncertainty and complexity. Various forms of uncertainty can occur. When the founder or manager of the supply chain firm perceives the business environment as something unpredictable then we talk about the uncertainty of the environment. Organizational outcome uncertainty occurs when managers have difficulty understanding changes in their general business environment, while decision uncertainty refers to managers' perception that it is impossible to predict the consequences of a decision [26]. In terms of complexity, it is true that the more factors in the general business environment that the manager perceives and the more differences between these factors, the more complex the business environment is. When managers perceive the environment to be complex, they may find it difficult to identify which dynamic competencies to develop and may be unwilling to develop some. Therefore, a supply chain firm is affected differently by the factors of the external environment mentioned above, and the dynamic capabilities that are developed vary according to the perceptions and role of the founder, or generally, the person who manages the business.

3 Factor Analysis

We examine the obstacles that an entrepreneur encounters in the process of running the supply chain firm. The focus of the research is on new start-ups operating in high-tech sectors, so the sample companies were selected based on their date of establishment (between 2000 and 2010) and the economic sector in which they operate.

The factors are assessed with a 5-point Likert scale question (1: not at all, 5: to a great extent), and from the exploratory factor analysis (EFA) we can distinguish the following three factors:

- Risk and cost of technology development: high cost for technology development; technology risk; high cost of skilled labor
- Economic crisis: market uncertainty; limited domestic demand; limited domestic demand
- High taxation

The first factor relates to the cost and risk involved in technology development. The second factor relates to the consequences of the economic crisis and includes market uncertainty and limited domestic demand. Finally, the third factor is the high taxation faced by companies. We note that the first factor is directly related to the sectors in which the supply chain firms under investigation operate, while the other two are more related to the external environment of the supply chain firms and the economic crisis conditions prevailing at the time of the survey (Table 1).

3.1 *Barriers to the Capacity for Innovation*

In addition, we trace the barriers to supply chain firms' ability to innovate. And this is a 5-point Likert scale question (1: not an important barrier at all, 5: extremely important barrier), wherefrom the EFA we could distinguish the following factors:

Table 1 Factor analysis of operating barriers

Operating barriers	Factor 1	Factor 2	Factor 3
High costs for technology development	**0.737**	0.122	−0.079
Technology risk (uncertainty of emerging technologies)	**0.701**	−0.122	0.427
High cost of skilled labor	**0.651**	0.170	0.030
Market uncertainty/economic crisis	−0.030	**0.830**	0.283
Limited domestic demand High taxation	0.278	**0.800**	−0.159
High taxation	0.053	0.101	**0.926**
Eigenvalues	*1.90*	*1.16*	*1.02*
Variance explained %	*31.78*	*19.32*	*17.04*

Table 2 Factor analysis of barriers to innovation

Barriers to innovation	Factor 1	Factor 2
Uncertainty related to the commercial success of the innovative project	**0.845**	−0.009
Uncertainty in demand for innovative products/services	**0.813**	0.154
Uncertainty related to the technical part of the innovative project	**0.708**	0.326
High spending on research and development	0.023	**0.866**
Lack of funding for innovation	0.125	**0.811**
Lack of qualified staff	0.222	**0.541**
Eigenvalues	*2.47*	*1.29*
Variance explained %	*41.30*	*21.55*

- Innovation uncertainty: uncertainty related to the commercial success of the innovative project; uncertainty in the demand for innovative products/services; uncertainty related to the technical part of the innovative project.
- Research requirements: high expenditure on research and development; lack of funding for innovation; lack of qualified personnel.

The first factor is related to the uncertainty inherent in innovation, such as uncertainty about the commercial success of the innovative outcome, uncertainty about the demand for innovative products and services, and uncertainty about the technical part of the innovative project. The second factor relates to the requirements inherent in research for the development of innovations, such as the sometimes-high costs, the lack of financial resources that may be available, and the lack of qualified personnel to work on innovation projects (Table 2).

3.2 *Strategy*

As regards the strategy followed by supply chain firms, the results obtained from the EFA lead to the creation of three main factors. More specifically, from the question related to the extent to which firms adopt specific strategic choices (1: not at all, 5: to a large extent), we obtain the grouping of the following variables:

- New products and new markets: Increasing sales by offering new products/services; entering a new market.
- Increase in sales: Increase in sales by penetrating new markets by offering the same products/services.
- Increasing market share: Increasing market share in existing markets by offering the same products/services (Table 3).

The first factor is related to a strategy that aims to increase sales through the creation of new products and services and a strategy that focuses on specific markets. It can be said that this factor is more suited to supply chain firms that innovate and seek opportunities in other market segments. The second factor is again aimed at increasing sales, but this time by offering the same products and services in new

Table 3 Factor analysis in strategy

Strategy	Factor 1	Factor 2	Factor 3
Increase sales by offering new products/services	**0.913**	−0.026	0.037
Entry to new markets	**0.676**	0.441	−0.023
Increase sales by penetrating new markets by offering the same products/services	0.100	**0.944**	0.113
Increase market share in existing markets by offering the same products/services	0.015	0.096	**0.993**
Eigenvalues	*1.62*	*1.03*	*0.74*
Variance explained %	*40.53*	*25.79*	*18.63*

markets. Therefore, here too there is the characteristic of opportunity-seeking, but this does not, however, equate to the development of innovations. Finally, the third factor is linked to the strategy of increasing the share of one or more existing markets by offering the same products. This factor is therefore also not focused on creating something new, but rather on gaining a larger market share than competitors.

3.3 *Identifying and Exploiting Opportunities*

The last analysis relates to the ability of supply chain firms to identify and exploit opportunities. In response to the question on the extent to which firms identify and exploit opportunities (1: not at all, 5: to a large extent), we could define the following factors:

- Ability to adapt to market changes: The supply chain firm follows market trends. The supply chain firm responds quickly to changes in demand. The firm responds promptly to the movements of competitors.
- Ability to adapt to technological change and to produce differentiated products: The supply chain firm has the flexibility to produce differentiated products/services. The supply chain firm can perceive technological changes and adapt to them (Table 4).

The first factor shows the ability of a supply chain firm to adapt to changes that occur in the market, whether they relate to demand or the strategy pursued by its competitors. In addition, the firm may have the flexibility to follow trends in the market in which it operates. The second factor is linked to a firm's ability to adapt to technological changes and to its ability to produce differentiated products and offer customized services. The second factor also comes closer to the innovative performance of a supply chain firm.

Table 4 Analysis of factors in identifying and exploiting opportunities

Identifying and exploiting opportunities	Factor 1	Factor 2
The company follows market trends	**0.804**	−0.041
The company responds quickly to changes in demand	**0.766**	0.315
The supply chain firm responds immediately to the movements of competitors	**0.733**	0.206
The company has the flexibility to produce differentiated products/services	0.025	**0.868**
The company can understand and adapt to technological changes	0.269	**0.732**
Eigenvalues	*2.25*	*1.02*
Variance explained %	*36.89*	*65.53*

4 Conclusions

The obstacles that the supply chain companies in question may face are on the one hand those related to technology, due to the sectors in which they operate, and on the other hand, those related to the economic crisis, as we should not forget that the environment in which they operate is characterized by an economic recession. The first barrier contains, according to the results of the factor analysis, the risk, and high cost of developing technology, but also the high cost of skilled labor that is often required. This barrier is more important for supply chain firms that utilize technology and that employ people with higher qualifications and possibly higher wages. In trying to innovate, these firms take the technological risk required and the high cost of developing technology. The second obstacle is the consequences of the economic crisis, which are none other than the uncertainty in the market and the limited domestic demand. The second barrier affects more the companies that are looking for alternative ways to cope with it daily, for example, through innovation or exports.

The basic strategy followed by a company clearly shows its intentions regarding the way it moves within the market and the objectives it wants to achieve. Such a strategy can be expected to be followed by supply chain firms that seek to increase their sales through the production of innovative products and services, and that seek to identify opportunities by penetrating new markets or even very niche market segments. The results of the statistical analysis fully confirm this way of thinking. In other words, supply chain firms that rely on the innovation process are differentiated from other firms that do not focus on developing innovations. These firms also differ in terms of the strategy they choose and indeed in terms of offering new products and entering new market islands.

The obstacles that supply chain firms face in their efforts to innovate, the statistical analysis has revealed two important obstacles that firms face in their efforts to innovate. The first is innovation uncertainty, that is, uncertainty related to the commercial success of the innovation project, uncertainty in the demand for innovative products and services, and uncertainty related to the technical part of the innovation project. The second barrier is the requirements that accompany research, namely, high research and development costs, the frequent lack of research funding funds, and the lack of qualified personnel. Therefore, there are two major obstacles for supply chain firms trying to innovate, that they have to overcome to achieve their goal: the uncertainty inherent in innovation and the demands of research.

Finally, a supply chain firm's ability to identify and exploit opportunities, as the statistical analysis has shown, is divided into two factors: (1) the ability to adapt to market changes and (2) the ability to adapt to technological changes and produce differentiated products. From the results of the analyses, we observed that these two factors differ in the way they affect knowledge entrepreneurship. More specifically, when a firm has the flexibility to produce differentiated products and to perceive and adapt to technological changes (factor "ability to adapt to technological changes and produce differentiated products"), it is more likely to innovate. Conversely, when a firm simply follows market trends, responds quickly to changes in demand, and responds immediately to competitors' movements (the "ability to adapt to market changes" factor), it is more likely not to innovate.

5 Future Work

Similar studies could be performed in other sectors of the economy in order to improve the level of knowledge over the strategy and the operating obstacles of innovative firms [27, 28], apart from the supply chain sector. These sectors could refer to decentralized [29–31] and centralized payment networks [32], as well as other sectors [33–37]. Other scientists with similar cases are studying similar problems, using modeling and simulation [38–41].

References

1. Carlsson, B.O., Zoltan, J., Audretsch, D.B., Braumerhjelm, P.: The knowledge filter, entrepreneurship, and economic growth. Jena Economic Research Paper No 2007–057. (2007)
2. Bai, C., Dallasega, P., Orzes, G., Sarkis, J.: Industry 4.0 technologies assessment: a sustainability perspective. Int. J. Prod. Econ. **229**, 107776 (2020)
3. Hong, J.T., Zheng, R.Y., Deng, H.P., Zhou, Y.L.: Green supply chain collaborative innovation, absorptive capacity and innovation performance: evidence from China. J. Clean. Prod. **241** (2019)
4. Wu, A.H., Li, T.F.: Gaining sustainable development by green supply chain innovation: perspectives of specific investments and stakeholder engagement. Bus. Strateg. Environ. **29**(3), 962–975 (2020)
5. Sarkis, J., Zhu, Q.H., Lai, K.H.: An organizational theoretic review of green supply chain management literature. Int. J. Prod. Econ. **130**(1), 1–15 (2011)
6. Mondal, C., Giri, B.C.: Investigating strategies of a green closed-loop supply chain for substitutable products under government subsidy. J. Ind. Prod. Eng., 1–24 (2021)
7. Wong, D.T.W., Ngai, E.W.T.: Critical review of supply chain innovation research (1999–2016). Ind. Mark. Manag. **82**, 158–187 (2019)
8. Zhang, D.Y., Rong, Z., Ji, Q.: Green innovation and firm performance: evidence from listed companies in China. Resour. Conserv. Recycl. **144**, 48–55 (2019)

9. Feng, Y., Lai, K.-H., Zhu, Q.: Green supply chain innovation: Emergence, adoption, and challenges. Int. J. Prod. Econ. **248**, 108497 (2019)
10. Bishop, K.: Knowledge based entrepreneurship in The Czech Republic and Hungary: results from 4 case studies. Economics Working Paper No. 71. (2006)
11. Hitt, M., Ireland, R., Hoskisson, R.: Strategic Management: Competitiveness and Globalization. ITP (1999)
12. Penrose, E.: The Theory of the Growth of the Firm. Wiley (1959)
13. Westhead, P., Wright, M., Ucbasan, D.: The internationalization of new and small firms: a resource based view. J. Bus. Ventur. **16**(4), 333–358 (2001)
14. Strand, S.: Patents as a Source of Competitive Advantage Working Paper. Helsinki University of Technology (2006)
15. Grant, R.: Toward a knowledge based theory of the firm. Strateg. Manag. J. **4**, 109–122 (1998)
16. Teece, D.: The role of managers, entrepreneurs and the literati in enterprise performance and economic growth. Int. J. Technol. Learn. Innov. Dev. **1**(1), 1007 (2007a)
17. Begley, T., Tan, W., Schoch, H.: Politico-economic factors associated with interest in a starting a business: a multi country study. Entrepreneurship Theory and Practice, 35–55 (2005)
18. Prahalad, C., Hamel, G.: The core competence of the corporation. Harv. Bus. Rev. (1990)
19. Teece, D., Pisano, G., Shuen, A.: Dynamic capabilities and strategic management. Strateg. Manag. J. **18**(7), 509–533 (1997)
20. Ambrosini, V., Bowman, C.: What are dynamic capabilities and are they a useful construct in strategic management? Int. J. Manag. Rev. **11**(1), 29–49 (2009)
21. Easterby-Smith, M., Lyles, M., Peteraf, M.: Dynamic capabilities: current debates and future directions. Br. J. Manag. **20**, 1–8 (2009)
22. Helfat, C.E., Finkelstein, S., Mitchell, W., Peteraf, M., Singh, H., Teece, D., Winter, S.: Dynamic Capabilities: Understanding Strategic Change in Organizations. Blackwell (2007)
23. Ambrosini, V., Bowman, C., Collier, N.: Dynamic capabilities: an exploration of how firms renew their resource base. Br. J. Manag. **20**, 9–24 (2009)
24. Teece, D.: Explicating dynamic capabilities: the nature and microfoundations of (sustainable) enterprise performance. Strateg. Manag. J. (2007b)
25. Zahra, A., George, G.: Absorptive capacity: a review, reconceptualization, and extension. Acad. Manag. Rev. **27**(2), 185–203 (2002)
26. Danneels, E.: The dynamics of product innovation and firm competences. Strateg. Manag. J. **23**, 1095–1121 (2002)
27. Kanellos, N.S.: A conceptual framework for analyzing knowledge-based entrepreneurship. In: 4th Symposium on Business and Management and Dynamic Simulation Models Supporting Management Strategies (2011) http://hdl.handle.net/10797/13771
28. Sakas, D.P., Nasiopoulos, D.K., Reklitis, P.: Modeling and simulation of the strategic use of the internet forum aiming at business success of high-technology companies. In: Sakas, D., Nasiopoulos, D. (eds.) Strategic Innovative Marketing. IC-SIM 2017. Springer Proceedings in Business and Economics. Springer, Cham (2019). https://doi.org/10.1007/978-3-030-16099-9_22
29. Sakas, D.P., Giannakopoulos, N.T., Kanellos, N., Tryfonopoulos, C.: Digital marketing enhancement of cryptocurrency websites through customer innovative data process. Processes. **10**(5), 960 (2022a). https://doi.org/10.3390/pr10050960
30. Sakas, D.P., Giannakopoulos, N.T., Kanellos, N., Migkos, S.P.: Innovative cryptocurrency trade websites' marketing strategy refinement, via digital behavior. IEEE Access. **10**, 63163–63176 (2022b). https://doi.org/10.1109/ACCESS.2022.3182396
31. Sakas, D.P., Giannakopoulos, N.T., Reklitis, D.P., Dasaklis, T.K.: The effects of cryptocurrency trading websites on airlines' advertisement campaigns. J. Theor. Appl. Electron. Commer. Res. **16**(7), 3099–3119 (2021). https://doi.org/10.3390/jtaer16070169
32. Sakas, D.P., Kamperos, I.D.G., Reklitis, D.P., Giannakopoulos, N.T., Nasiopoulos, D.K., Terzi, M.C., Kanellos, N.: The effectiveness of centralized payment network advertisements on digital

branding during the COVID-19 crisis. Sustainability. **14**(6), 3616 (2022c). https://doi.org/10.3390/su14063616
33. Sakas, D.P., Giannakopoulos, N.T.: Harvesting crowdsourcing platforms' traffic in favour of air forwarders' brand name and sustainability. Sustainability. **13**(15), 8222 (2021a). https://doi.org/10.3390/su13158222
34. Sakas, D.P., Giannakopoulos, N.T.: Big data contribution in desktop and mobile devices comparison, regarding airlines' digital brand name effect. Big Data Cognit. Comput. **5**(4), 48 (2021b). https://doi.org/10.3390/bdcc5040048
35. Sakas, D.P., Reklitis, D.P.: Predictive model for estimating the impact of technical issues on consumers interaction in agri-logistics websites. In: Bochtis, D.D., Pearson, S., Lampridi, M., Marinoudi, V., Pardalos, P.M. (eds.) Information and Communication Technologies for Agriculture—Theme IV: Actions. Springer Optimization and Its Applications 185. Springer, Cham (2021). https://doi.org/10.1007/978-3-030-84156-0_14
36. Terzi, M.C., Sakas, D.P., Seimenis, I.: The contribution of the scientific committee in the development of conferences. Procedia Soc. Behav. Sci. **73**, 373–382 (2013). https://doi.org/10.1016/j.sbspro.2013.02.064
37. Nasiopoulos, D.K., Sakas, D.P., Vlachos, D.S.: Modeling the scientific dimension of academic conferences. Procedia Soc. Behav. Sci. **147**, 576–585 (2014). https://doi.org/10.1016/j.sbspro.2014.07.164
38. Nasiopoulos, D.K., Ioanna, D., Lida, G., Paraskevi, P., Giakomidou, D.S.: Modeling of B2C communication strategies in electronic commerce. Strategic Innovative Marketing, 531–537 (2017)
39. Nasiopoulos, D.K.: Implementation stages and risks of Logistics Companies reengineering. Bus. Intell. Model. **6**, 197–206 (2020a)
40. Nasiopoulos, D.K.: Web search applications in Logistics companies in the context of evolution of Management Information Systems. Bus. Intell. Model. **6**, 241–254 (2020b)
41. Dimitrios, N.K., Sakas, D.P., Vlachos, D.S.: The role of open source leadership in developing high technology companies. Key Eng. Mater. **543**, 402–405 (2013)

Reengineering Energy Sector SMEs to Marketing via Backlinks and Domains Strategy

Kriemadis Thanos and Despina S. Giakomidou

1 Introduction

Reengineering for marketing primarily allows for the description of firm assets, which are later employed to identify the best marketing concept [1]. Each marketing technique has its particular collection of methodology, linkages, interactions, and activities, which serve as the fundamental thread in market intelligence. As a consequence, the reengineering of the marketing process is synonymous with this integration step, and it emerges as a necessary consolidation of advantageous firm approaches, marketing automation, and trades.

Throughout the phase of transformation, each business selects a particular strategy or method to accomplish success and obtain a strategic edge. In volatile environments, reengineering might be beneficial. However, using only one operational instrument to succeed is no longer sufficient [2]. A mixture of successful methods may result in a considerable level of efficiency and triumph for the company. In reality, the corporate method of reengineering necessitates the concurrent development of operations, as well as the structure and maintenance of digital technologies, in order to obtain significant gains in speed, expense, reliability, and consumer attentiveness to goods and services [3].

The present surge of rigorous inquiry about the organizational process of marketing has corresponded with the advent of a revolutionary novel management strategy, the process of marketing reengineering [4]. Digital marketing, in its broadest sense, relates to promotion given via digital platforms like search engines, webpages, social networks, email, and mobile applications. Digital marketing is the

K. Thanos (✉) · D. S. Giakomidou
Department of Management Science and Technology, University of Peloponnese, Tripolis, Greece
e-mail: thanosk@uop.gr

D. P. Sakas et al. (eds.), *Computational and Strategic Business Modelling*, Springer Proceedings in Business and Economics,
https://doi.org/10.1007/978-3-031-41371-1_30

means by which businesses promote products, services, and trademarks via the use of various digital media platforms. Customers largely depend on digital methods to conduct marketing research.

The remainder of this chapter contains a reference overview and ramifications of related studies, statistical methods and rollout of the modeling concept, and relevant data about the findings and practical application of reengineering SMEs' marketing procedures.

2 Related Background

Applying the viewpoint that marketing should be primarily a corporate strategy that places consumer pleasure at the center of decision-making, the importance of reengineering in marketing may be seen from a very favorable perspective [2]. According to the assessment of data acquired with the knowledge that originated the impact of reengineering, corporate factors on the effectiveness of the enterprise were evaluated, and corporate climate was the factor that had greater importance to company performance [5]. As a consequence of international industrial developments, small and medium-sized firms' market sensitivity is steadily rising [5].

Many firms' goals are to expand with superior efficiency, produce exceptional jobs, reduce the expense of goods and services, and give worth to customers via a thorough grasp of their needs. As a result, in a new digital age world, with innovations, and powerful rivals, businesses must be successfully and continuously developed in order to achieve conceptual and functional performance. The reasons for organizational strategic failings include inefficient corporate procedures and a shortage of creativity, both of which have major consequences for firms and their profitability [6]. It has been shown that the majority of the employed designs struggle with several challenges following the deployment of company reengineering when it is customarily implemented without taking into account the modifications encircling the company's operations context [7].

Web page analytics data applications have already been extensively researched and utilized to evaluate the effectiveness of digital marketing approaches. Big Data, particularly user satisfaction information and technical aspects of organizations' websites, have been confirmed to be crucial parameters for improving digital marketing success in a variety of areas of the economy, ranging from centrally controlled and blockchain-based payment networks [8–10] to air carriers and logistics enterprises [11–13].

The researchers attempt to replace the gap in the literature considering the effectiveness of SMEs reengineering marketing operations to strengthen their backlinks and referring domains strategy, thereby increasing the enterprises' digital marketing performance by attracting more new visitors and increasing the time they spend on site. As a result, the research concentrated on energy SMEs, and the research hypotheses are as follows:

H1: *"With increased backlinks and connected domains, energy SMEs can attract more new visitors to their website."*

H2: *"The more energy SMEs increase/decrease the connectivity of their website, the more the time their visitors spend on it gets affected."*

3 Methodology

By utilizing regression analysis and the simulation tool of fuzzy cognitive mapping (FCM), we developed the framework of the methodology. The FCM analysis was performed based on Mpelogianni and Groumpos [14] context. In order to perform the FCM simulation, we capitalized on the variables' intercorrelations and regression coefficients. Throughout this chapter, the need for reengineering the marketing procedures of energy SMEs is promoted, via their websites' backlinks and domains strategy. For analysis purposes, the authors gathered analytic data from 5 innovative energy SMEs [15], regarding website visits, visits' duration, backlinks, and connected domains. Above big data gathering was carried out by using platform-based decision support systems [16].

4 Results

In the results section, as shown in Table 1, the linear regression models of the research were deducted, with dependent variables the new visitors and time on site metrics of energy SMEs' websites. As independent variables, the number of backlinks, referring domains, direct website traffic, and internal links were used.

Table 1 Energy SMEs' digital marketing regressions

Variables	Standardized coefficient	R^2	F	*p*-value
Dependent: New visitors				
Constant	–	0.996	52.490	0.104**
Website response	0.381			0.138
Backlinks	−1.899			0.146
Referring domains	3.268			0.141
Internal links	−1.145			0.123
Time on site	2.348			0.088
Dependent: Time on site				
Constant	–	0.965	13.613	0.070
Website response	−0.118			0.539
Backlinks	1.193			0.121
Referring domains	−2.158			0.040*
Internal links	0.650			0.051

* and ** indicate statistical significance at the 95% and 99% levels, respectively

These variables can affect significantly the performance of SMEs' websites and digital marketing, hence the need for an appropriate reengineering procedure of marketing activities.

It becomes clear that the first regression has p-value $= 0.104$ above $a = 0.05$ significance level, which means it is not overall verified, despite high $R^2 = 0.996$. Moreover, none of the independent variables of new visitors' regression have p-values below $a = 0.05$, so they do not affect the dependent variable significantly. This chapter's first research hypothesis (H1) is rejected, and thus increased backlinks and referring domains do not attract more visitors to energy SMEs' websites.

Time on site's regression is not overall verified since its p-value is above the significance level of 0.05 (0.070), despite the R^2 of 0.965. The independent variable of referring domains appears to significantly affect the time visitors spend on energy SMEs' websites, with p-value $= 0.040 < a = 0.05$. So, this chapter's second hypothesis is verified and the more energy SMEs use to vary the connectivity of their website, the more time their visitors spend on it gets affected.

5 Simulation Model

The development of the fuzzy cognitive mapping context was performed in the simulation section, where two FCM scenarios were executed. These scenarios aimed to simulate the outcomes of SMEs website metrics' modeling. Through the FCM, graphical presentations of diagnostic models are performed, deducting cause-and-effect relationships. For this reason, the MentalModeler [17] platform was capitalized, with blue and red arrows indicating the negative and positive correlations of variables, respectively. By using the FCM simulations, we aim to point out specific metrics and website strategies for SMEs in the energy sector that improve digital marketing performance. The author's intent is to show that, through observing and tuning specific website metrics, energy SMEs could attract more new visitors and extend the time they stay on their website. FCM outcomes are illustrated in Fig. 1.

Moving to the scenarios phase, depicted in Fig. 2, we can discern that in Scenario 1, in order energy SMEs to increase website new visitors and the time they spend on it, they should increase the direct traffic, the external and internal links, the response time of the website, and decrease the number of its of referral domains and backlinks. On the other hand, by increasing the number of their website' referral domains and backlinks, while reducing its direct traffic, the external and internal links, the response time, the new visitors, and the time they tend to spend on it are also decreased. From FCM, analysis becomes clear that in order for energy SMEs to increase their website's visibility and the engagement of their visitors, they could adjust the number of backlinks and referring domains of the website. So, a shift to this type of digital marketing activities could benefit SMEs, highlighting the importance of reengineering to marketing of corporate procedures.

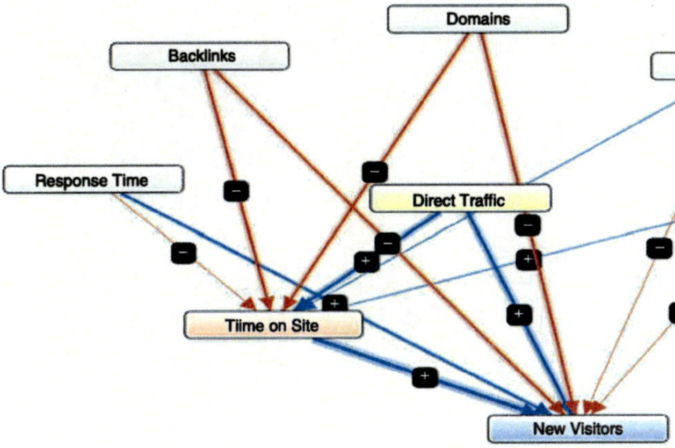

Fig. 1 Fuzzy cognitive mapping of energy SMEs' website

6 Conclusions

The results that emerged from this chapter relate to the need for SMEs to reengineer their core processes to marketing strategies for increased profit. This can be achieved by focusing more on marketing activities like increasing or decreasing the number of the website's backlinks and referring domains. Hence, through this research focus has been shifted to the marketing strategy of decreasing the number of energy SMEs website backlinks and referring domains, in order to increase the number of new visitors to their website and the amount of time they tend to spend on it. Other scientists with similar cases are studying similar problems, using modeling and simulation [18–24].

7 Practical Implications

SMEs in energy markets could potentially benefit from the outcomes of this chapter since the implication of backlinks and referring domains strategy increases the time their visitors spend on the website. This highlights the efficiency of strategies related to reengineering marketing processes.

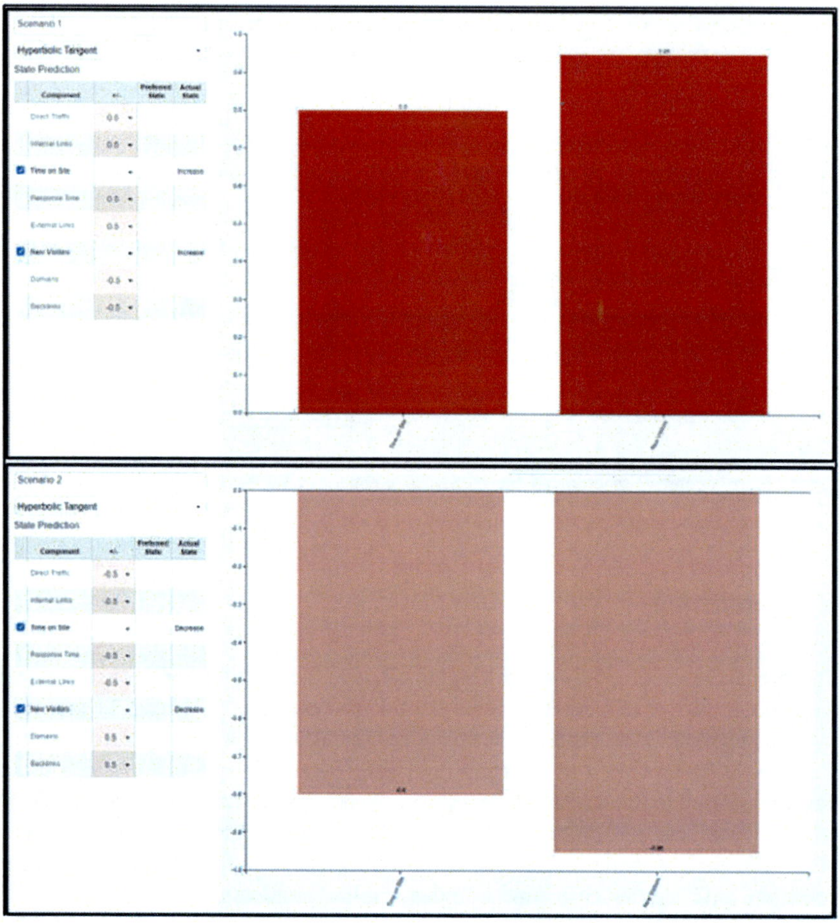

Fig. 2 Fuzzy cognitive mapping scenarios

References

1. Hadzhiev, B.I.: Re-Engineering Marketing (RM). Pecvnia Revista de la Facultad de Ciencias Económicas y Empresariales Universidad de León, 45–51 (2010). https://doi.org/10.18002/pec.v0i2010.763
2. Kocel, T.: Business Management *Beta Yayınları*, 15. Baskı, İstanbul (2014)
3. Omidia, A., Khoshtinata, B.: Factors affecting the implementation of business process reengineering: taking into account the moderating role of organizational culture (Case Study: Iran Air). Procedia Econ. Finance. **36**, 425–432 (2016)
4. Lynch, J.E.: Marketing and business process re-engineering. J. Mark. Pract. Appl. Mark. Sci. **1**(1), 45–53 (1995). https://doi.org/10.1108/EUM0000000003879

5. Akbaba, O., Altındağ, E.: The effects of reengineering, organizational climate and psychological capital on the firm performance. Procedia Soc. Behav. Sci. **235**, 320–331 (2016)
6. Rao, L., Mansingh, G., Osei-Bryson, K.-M.: Building ontology based knowledge maps to assist business process re-engineering. Decis. Support. Syst. **52**(3), 577–589 (2012). https://doi.org/10.1016/j.dss.2011.10.014
7. AbdEllatif, M., Farhan, S.M., Shehata, S.N.: Overcoming business process reengineering obstacles using ontology-based knowledge map methodology. Future Comput. Inf. J. **3**(1), 7–28 (2018). https://doi.org/10.1016/j.fcij.2017.10.006
8. Sakas, D.P., Kamperos, I.D.G., Reklitis, D.P., Giannakopoulos, N.T., Nasiopoulos, D.K., Terzi, M.C., Kanellos, N.: The effectiveness of centralized payment network advertisements on digital branding during the COVID-19 crisis. Sustainability. ***14***(*6*), 3616 (2022a). https://doi.org/10.3390/su14063616
9. Sakas, D.P., Giannakopoulos, N.T., Kanellos, N., Tryfonopoulos, C.: Digital marketing enhancement of cryptocurrency websites through customer innovative data process. Processes. **10**(5), 960 (2022b). https://doi.org/10.3390/pr10050960
10. Sakas, D.P., Giannakopoulos, N.T., Kanellos, N., Migkos, S.P.: Innovative cryptocurrency trade websites' marketing strategy refinement, via digital behavior. IEEE Access. **10**, 63163–63176 (2022c). https://doi.org/10.1109/ACCESS.2022.3182396
11. Sakas, D.P., Giannakopoulos, N.T.: Harvesting crowdsourcing platforms' traffic in favour of air forwarders' brand name and sustainability. Sustainability. **13**(15), 8222 (2021a). https://doi.org/10.3390/su13158222
12. Sakas, D.P., Giannakopoulos, N.T.: Big data contribution in desktop and mobile devices comparison, regarding airlines' digital brand name effect. Big Data Cognit. Comput. **5**(4), 48 (2021b). https://doi.org/10.3390/bdcc5040048
13. Sakas, D.P., Giannakopoulos, N.T., Reklitis, D.P., Dasaklis, T.K.: The effects of cryptocurrency trading websites on airlines' advertisement campaigns. J. Theor. Appl. Electron. Commer. Res. ***16***(*7*), 3099–3119 (2021). https://doi.org/10.3390/jtaer16070169
14. Mpelogianni, V., Groumpos, P.P.: Re-approaching fuzzy cognitive maps to increase the knowledge of a system. AI & Soc. **33**, 175–188 (2018). https://doi.org/10.1007/s00146-018-0813-0
15. Peters, A.: The 10 most innovative energy companies in 2022. (2022). https://www.fastcompany.com/90724394/most-innovative-companies-energy-2022
16. Semrush. (2022, March 16). https://www.semrush.com/
17. MentalModeler. (2022, March 16). https://dev.mentalmodeler.com/
18. NK Dimitrios, P Sakas (2019). Modeling and simulation of the strategic use of marketing in search engines for the business success of high technology companies, proceedings in Springer
19. Sakas, D.P., Nasiopoulos, D.K., Reklitis, P.: Modeling and simulation of the strategic use of marketing in search engines for the business success of high technology companies, pp. 217–226. Business and Economics (2019)
20. Nasiopoulos, D.K.: The role of digital marketing in the development of a distribution and logistics network of information technology companies. Bus. Intell. Model. **6**, 267–276 (2020)
21. Foteini, P.E., Dimitrios, N.K., Sakas, D.P., Vlachos, D.S.: The development of new ideas for IT products, through social media. Strategic innovative marketing, 399–404 (2017)
22. Plikas, J.H., Nasiopoulos, D.K., Sakas, D.P., Vlachos, D.S.: Modeling the promotion process of academic conferences through social media. Strategic Innovative Marketing, 499–505 (2017)
23. Papadopoulou, T., Nasiopoulos, D.K., Vlachos, D.S.: Merchandizing IT products via social networks. Modeling and simulation of the procedures. Strategic Innovative Marketing, 507–513 (2017)
24. Alexandros, N.K., Damianos, S.P., Dimitrios, N.K., Dimitrios, V.S.: Modelling the process of a web-based collaboration tool development. Strategic Innovative Marketing, 383–389 (2017)

SMEs' Reengineering to Marketing in Supply Chain Sector by Focusing on Website Improvement

Kriemadis Thanos and Despina S. Giakomidou

1 Introduction

Overall, digital advancement introduces all opportunities and risks to SMEs [1]. Since digital innovations and corporate concepts get quite prevalent, they impede aging outdated technologies, marketing strategies, process management, and relationships with suppliers, presenting the company with novel possibilities. Technological advances are "configurations of data, informatics, interaction, and interconnection innovations" that potentially evolve strategies and structure.

Reengineering for marketing initially provides the possibility to describe corporate resources, and then use them to determine the most suitable marketing orientation [2]. Every marketing tactic has its own set of methodologies, connectivity, conversations, and transactions, and they form the common post in market research. As a result, the reengineering to marketing process is synonymous with this unifying post, and it emerges as an essential accumulation of beneficial company strategies, marketing solutions, and transactions.

The rest of this chapter includes the related background review and implication of research topics, the statistical analysis and deployment of the simulation model and at last, information regarding the results and practical implications of reengineering to marketing process.

K. Thanos (✉) · D. S. Giakomidou
Department of Management Science and Technology, University of Peloponnese, Tripolis, Greece
e-mail: thanosk@uop.gr

D. P. Sakas et al. (eds.), *Computational and Strategic Business Modelling*, Springer Proceedings in Business and Economics,
https://doi.org/10.1007/978-3-031-41371-1_31

2 Related Background

The digital technology of SMEs is still understudied, especially the focusing switches to which emerging technologies promote innovation and disturbances in SMEs [3]. A company that has successfully created strategic flexibility could further establish yet another conceptual or greater dynamic activities through leveraging these skills. Emerging technologies have an exceptional function to perform in corporate entrepreneurship [4, 5], particularly in restructuring and changing the entrepreneurial process [6].

Technological advances have a beneficial impact on entrepreneurial companies. According to von Briel et al. [7], information technology behaves as exterior influence factors, playing a "possibly precedent character." SMEs accomplish things in this action plan by constructing strategic activities based primarily on digital technology acceptance via detecting, utilizing, and modifying [8].

Implication of website analytic data has been widely studied and used in order to assess the efficacy of digital marketing strategies. More specifically, Big Data, in the form of user engagement data and technical factors, has been proven as important metrics for enhancing digital marketing performance in various sectors, from centralized and decentralized payment networks [9–11] to airlines and supply chain firms [12–14].

From the earlier, the authors try to fill a gap regarding the utility of SMEs' reengineering to marketing processes to improve their website's technical characteristics, and thus increase the companies' digital marketing efficiency by attracting more organic visits. Hence, the study focused on supply chain SMEs and the research hypotheses are the following:

H1: "*The number of visitors a website of supply chain SMEs attract is mostly based on its technical characteristics.*"

H2: "*Specific characteristics of SMEs' website can affect the number of keywords that lead to their site.*"

3 Methodology

At this stage, the methodological context of this chapter is presented. The authors used regression and correlation analyses for data elaboration, followed by the simulation procedure of Fuzzy Cognitive Mapping (FCM) [15]. Regression and correlation coefficients were capitalized as inputs for the FCM. In this way, we aim to provide important data for highlighting, through simulation, the benefits of reengineering to marketing process for SMEs. The authors collected data from five innovative SMEs in the supply chain sector [16] during a 3-month period, regarding website technical characteristics and digital marketing analytics. These data were collected from onsite decision support applications [17] through daily collection and observation.

4 Results

Throughout the results section, in Table 1, we can see the results of the performed linear regressions. Firstly, organic traffic and then organic keywords were used as dependent variables and the most common technical characteristics of a website are used as independent variables. Some of these characteristics are the raw internal link rate, the crawl depth, the page load time, etc., factors that can affect the experience of a website user and indicate the need for a turning point in terms of marketing performance.

The first regression is not overall verified with p-Value $= 0.113 > a = 0.05$, despite high $R^2 = 0.974$. None of the individual variables of website technical characteristics appears to have a significant impact on organic traffic (significance levels above 0.05). So, the first research hypothesis (H1) is rejected, meaning that the number of visitors a website of supply chain SMEs attract is not entirely based on its technical characteristics.

Next, organic keywords regression is neither overall verified with p-Value $= 0.114 > a = 0.05$, despite also having a significant $R^2 = 0.973$. This time, from the individual levels of significance, the raw internal link rate variable affects significant the organic keywords in $a = 0.05$ level, with p-Value $= 0.039 < 0.05$. This means that chapter's second hypothesis (H2) is verified; hence, specific characteristics of SMEs' websites can affect the number of keywords that lead to their site.

Table 1 Supply chain SMEs' digital marketing regressions

Variables	Standardized Coefficient	R^2	F	p-Value
Dependent: Organic Traffic				
Constant	–	0.974	8.219	0.113
Raw internal link rate	0.107			0.823
Crawl depth	−0.088			0.719
HTTP Status Code	2.641			0.367
Page load time	−1.522			0.448
Incoming internal links	−0.373			0.149
Outcoming internal links	0.382			0.253
Size of JS_CSS	0.398			0.871
Webpage Issues	1.873			0.494
Internal Link Rate	0.948			0.160
Dependent: Organic Keywords				
Constant	–	0.973	8.127	0.114
Raw internal link rate	−2.079			0.039*
Crawl depth	0.172			0.506
HTTP Status Code	−1.170			0.661
Page load time	0.872			0.647
Incoming internal links	0.059			0.755
Outcoming internal links	−0.129			0.645
Size of JS_CSS	0.026			0.992
Webpage Issues	−1.608			0.553
Internal Link Rate	1.383			0.086

* and ** indicate statistical significance at the 95% and 99% levels respectively

5 Simulation Model

For simulation reasons, we developed a Fuzzy Cognitive Mapping framework and performed two scenarios regarding technical characteristics' values of SMEs websites. FCM is an exploratory model that uses graphical illustrations to depict inferential reasoning. This can be performed through the MentalModeler onsite platform [18]. FCM has many layers of cause-and-effect relationships among its variables, which are depicted by blue and red arrows that show the extend and the type (positive or negative accordingly) of the variables' relationship. In Fig. 1, we can observe the developed Fuzzy Cognitive Map for the simulation analysis of SMEs' websites. The purpose of the FCM Scenarios is to discern the factors that enhance the digital marketing performance of SMEs through evaluating their websites. This enhancement can be achieved via increasing organic traffic and decreasing organic keywords variation.

In the following scenarios, as shown in Fig. 2, we can see at Scenario 1 that an improvement in all of the technical characteristics and factors of supply chain SMEs' websites leads to higher organic traffic and lesser organic keywords variation. Scenario 2 shows that a decrease in all technical characteristics' values leads to a deterioration of organic traffic and to an increase in the variation of organic keywords. Since the improvement of SMEs website technical characteristics offers enhanced digital marketing efficiency, by attracting more organic visitors and limiting keywords' variation, a shift to analogous marketing strategies could benefit supply chain SMEs.

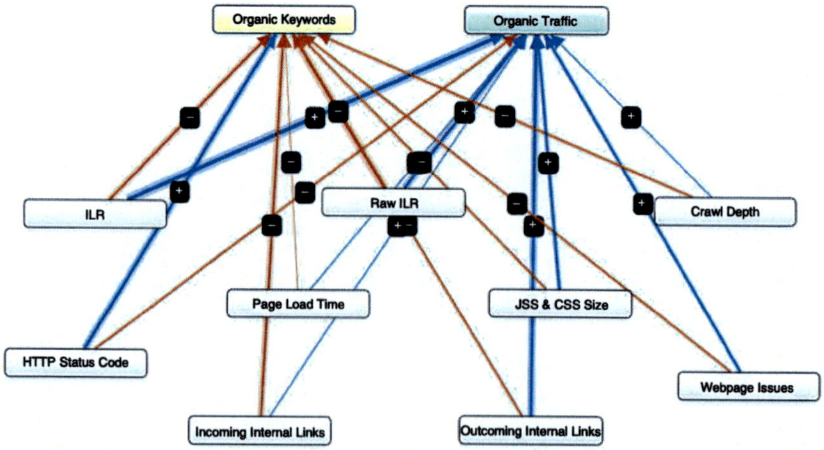

Fig. 1 Fuzzy cognitive mapping of supply chain SMEs' website

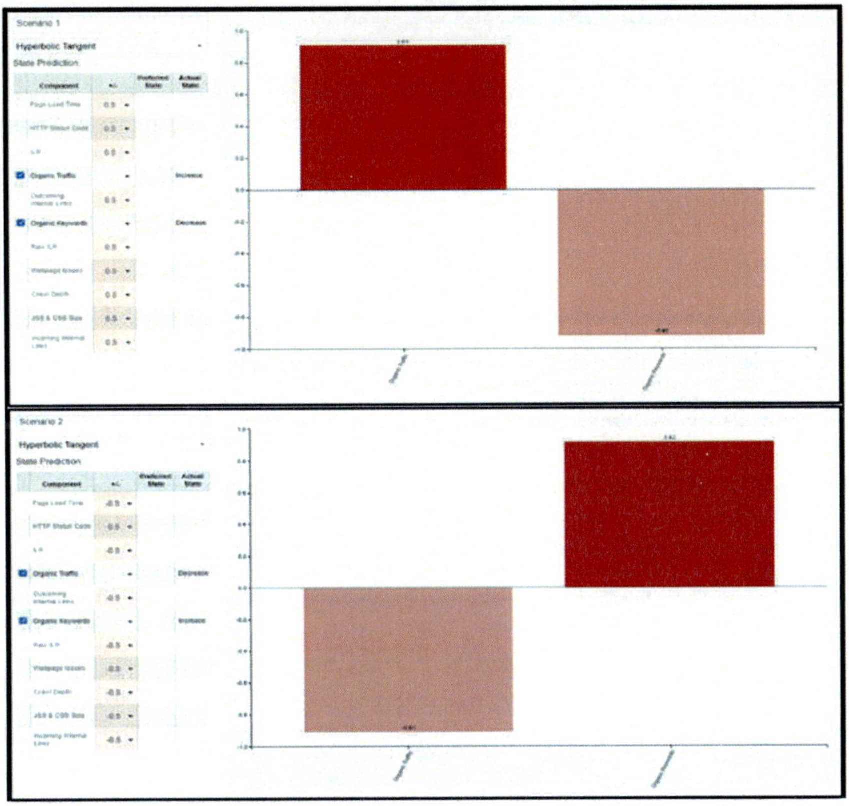

Fig. 2 Fuzzy cognitive mapping scenarios

6 Conclusions

Having analyzed the technical characteristics of SMEs' websites, the authors aim to highlight the importance of marketing reengineering strategy for SMEs in the supply chain sector. This chapter focused on whether the enhancement of specific technical characteristics of SMEs' websites can improve the results of their digital marketing performance. Hence, through regression and simulation analysis, there can be seen that organic visits of SMEs were increased and organic keywords variation was decreased, highlighting the requirement of reengineering to marketing for better corporate efficiency. Other scientists with similar cases are studying similar problems, using modeling and simulation [19–25].

7 Practical Implications

The practical implications for SMEs in supply chain sector could be the suggestion of their webpage technical characteristics that can attract more organic visits to the website, thus emphasizing on the benefits of reengineering to marketing strategies.

References

1. Roy, R., Lampert, C.M., Stoyneva, I.: When dinosaurs fly: the role of firm capabilities in the 'avianization' of incumbents during disruptive technological change. Strateg. Entrep. J. **12**(2), 261–284 (2018). https://doi.org/10.1002/sej.1278
2. Hadzhiev, B.I.: Re-engineering Marketing (RM). Pecvnia Revista de la Facultad de Ciencias Económicas y Empresariales Universidad de León, pp. 45–51. https://doi.org/10.18002/pec.v0i2010.763 (2010)
3. Berger, E.S.C., von Briel, F., Davidsson, P., Kuckertz, A.: Digital or not – the future of entrepreneurship and innovation: introduction to the special issue. J. Bus. Res. **125**, 436–442 (2021). https://doi.org/10.1016/j.jbusres.2019.12.020
4. Goerzig, D., Bauernhansl, T.: Enterprise architectures for the digital transformation in small and medium-sized enterprises. Procedia CIRP. **67**, 540–545 (2018). https://doi.org/10.1016/j.procir.2017.12.257
5. Proksch, D., Rosin, A.F., Stubner, S., Pinkwart, A.: The influence of a digital strategy on the digitalization of new ventures: the mediating effect of digital capabilities and a digital culture. J. Small Bus. Manag., 1–29 (2021). https://doi.org/10.1080/00472778.2021.1883036
6. Elia, G., Margherita, A., Passiante, G.: Digital entrepreneurship ecosystem: how digital technologies and collective intelligence are reshaping the entrepreneurial process. Technol. Forecast. Soc. Chang. **150**, 119791 (2020). https://doi.org/10.1016/j.techfore.2019.119791
7. von Briel, F., Davidsson, P., Recker, J.: Digital technologies as external enablers of new venture creation in the IT hardware sector. Entrep. Theory Pract. **42**(1), 47–69 (2018). https://doi.org/10.1177/1042258717732779
8. Khurana, I., Dutta, D.K., Ghura, A.S.: SMEs and digital transformation during a crisis: the emergence of resilience as a second-order dynamic capability in an entrepreneurial ecosystem. J. Bus. Res. **150**, 623–641 (2022). https://doi.org/10.1016/j.jbusres.2022.06.048
9. Sakas, D.P., Kamperos, I.D.G., Reklitis, D.P., Giannakopoulos, N.T., Nasiopoulos, D.K., Terzi, M.C., Kanellos, N.: The effectiveness of centralized payment network advertisements on digital branding during the COVID-19 crisis. Sustainability. **14**(6), 3616 (2022). https://doi.org/10.3390/su14063616
10. Sakas, D.P., Giannakopoulos, N.T., Kanellos, N., Tryfonopoulos, C.: Digital marketing enhancement of cryptocurrency websites through customer innovative data process. Processes. **10**(5), 960 (2022). https://doi.org/10.3390/pr10050960
11. Sakas, D.P., Giannakopoulos, N.T., Kanellos, N., Migkos, S.P.: Innovative cryptocurrency trade websites' marketing strategy refinement, via digital behavior. IEEE Access. **10**, 63163–63176 (2022). https://doi.org/10.1109/ACCESS.2022.3182396
12. Sakas, D.P., Giannakopoulos, N.T.: Harvesting crowdsourcing platforms' traffic in favour of air forwarders' brand name and sustainability. Sustainability. **13**(15), 8222 (2021). https://doi.org/10.3390/su13158222
13. Sakas, D.P., Giannakopoulos, N.T.: Big data contribution in desktop and mobile devices comparison, regarding airlines' digital brand name effect. Big Data Cogn. Comput. **5**(4), 48 (2021). https://doi.org/10.3390/bdcc5040048

14. Sakas, D.P., Giannakopoulos, N.T., Reklitis, D.P., Dasaklis, T.K.: The effects of cryptocurrency trading websites on airlines' advertisement campaigns. J. Theor. Appl. Electron. Commer. Res. **16**(7), 3099–3119 (2021). https://doi.org/10.3390/jtaer16070169
15. Mpelogianni, V., Groumpos, P.P.: Re-approaching fuzzy cognitive maps to increase the knowledge of a system. AI & Soc. **33**, 175–188 (2018). https://doi.org/10.1007/s00146-018-0813-0
16. Lindsky, D.: The 10 Most Innovative Logistics Companies of 2022. https://www.fastcompany.com/90724437/most-innovative-companies-logistics-2022 (2022)
17. Semrush.: 16 March 2022. https://www.semrush.com/ (2022)
18. MentalModeler.: 16 March 2022. https://dev.mentalmodeler.com/ (2022)
19. Dimitrios, N.K., Sakas, P.: Modeling and simulation of the strategic use of marketing in search engines for the business success of high technology companies. In: Proceedings in Springer. Springer, Cham (2019)
20. Sakas, D.P., Nasiopoulos, D.K., Reklitis, P.: Modeling and simulation of the strategic use of marketing in search engines for the business success of high technology companies. In: Business and Economics, pp. 217–226. Springer, Cham (2019)
21. Nasiopoulos, D.K.: The role of digital marketing in the development of a distribution and logistics network of information technology companies. Bus. Intell. Modell. **6**, 267–276 (2020)
22. Foteini, P.E., Dimitrios, N.K., Sakas, D.P., Vlachos, D.S.: The development of new ideas for IT products, through social media. In: Strategic Innovative Marketing, pp. 399–404. Springer, Cham (2017)
23. Plikas, J.H., Nasiopoulos, D.K., Sakas, D.P., Vlachos, D.S.: Modeling the promotion process of academic conferences through social media. In: Strategic Innovative Marketing, pp. 499–505. Springer, Cham (2017)
24. Papadopoulou, T., Nasiopoulos, D.K., Vlachos, D.S.: Merchandizing IT products via social networks. Modeling and simulation of the procedures. In: Strategic Innovative Marketing, pp. 507–513. Springer, Cham (2017)
25. Alexandros, N.K., Damianos, S.P., Dimitrios, N.K., Dimitrios, V.S.: Modelling the process of a web-based collaboration tool development. In: Strategic Innovative Marketing, pp. 383–389. Springer, Cham (2017)

Part V
Markets, Supply Chains and Value Chains

Giannis T. Tsoulfas

Description: This symposium attempts to explore contemporary developments in supply chains and value chains management and their implications in local, national, regional and international markets. The scope is to analyse the relationships among various stakeholders as well as the critical parameters and constraints that shape the efficiency, the performance and the competitiveness of businesses, organisations and sectors. Moreover, the drivers that lead to radical organisational transformation and evolution are investigated with the use of theoretical frameworks and empirical evidence. As advances in supply chains and value chains are increasingly used in order to achieve sustainable competitive advantage in global markets, they are nowadays regarded as of strategic importance for companies and organisations. Moreover, the contribution of various stakeholders is necessary not only for the narrow objectives of supply and value chains, but in order to join forces for addressing various sustainability challenges associated with the operations of those chains.

G. T. Tsoulfas
Agricultural University of Athens, Athens, Greece

The Implementation of Risk and Business Continuity Management Systems in Healthcare Supply Chain in Times of Crisis

Adriana Galoutzi, Panos T. Chountalas, and Giannis T. Tsoulfas

1 Introduction

The volatile business environment causes increased uncertainty with unexpected risks [1, 5]. Global supply chains, in particular, are more vulnerable due to unique internal operational and external physical or human risks [37]. However, most companies are not adequately prepared for the challenges they face in their supply chain, and managers do very little to prevent related risks [34]. Especially in the last 20 years, the major crises and disasters around the world have demonstrated how vulnerable businesses can become and how necessary it is to adopt a structured approach to risk management and business continuity [17, 35].

Supply-chain risk management can be defined as "the identification, assessment, treatment, and monitoring of supply-chain risks, with the aid of the internal implementation of tools, techniques, and strategies and of external coordination and collaboration with supply-chain members so as to reduce vulnerability and ensure continuity coupled with profitability, leading to competitive advantage" [10]. The standard dealing with risk management is ISO 31000. It is not certifiable but can be implemented and integrated into existing management systems and, more commonly, in ISO 9001 for quality management [4, 15]. From a business continuity perspective, supply-chain risk management manages exposure to severe business

A. Galoutzi
Department of Business and Economics, Metropolitan College, Athens, Greece

P. T. Chountalas (✉)
Department of Business Administration, University of Piraeus, Piraeus, Greece
e-mail: pchountalas@unipi.gr

G. T. Tsoulfas
Department of Agribusiness and Supply Chain Management, Agricultural University of Athens, Thiva, Greece

D. P. Sakas et al. (eds.), *Computational and Strategic Business Modelling*,
Springer Proceedings in Business and Economics,
https://doi.org/10.1007/978-3-031-41371-1_32

disruptions arising from risk within and outside the supply chain [10]. The business continuity management standard is ISO 22301; it is certifiable and provides a comprehensive framework for building business resilience.

The importance of the combined approach between business continuity and risk management processes has already been highlighted in the literature [2, 8, 12]. This approach is fully applicable in the vulnerable area of the supply chain, as shown by relevant research [9, 18, 24, 28]. In particular, when a supply chain implements risk management along with business continuity practices, it creates a competitive advantage as it continually develops skills to become more resilient [19, 20]. In this context, it is proposed for organizations to develop a business impact analysis first, then risk management and assessment, and, finally, the business continuity plan or the disaster recovery plan [3].

From the previous considerations, risk management and business continuity in the supply chain is an attractive and popular research area [35] and is of great interest to academics and researchers as an emerging and promising field [25]. Suresh et al. [26] recently argued a critical need to reconcile and synthesize both approaches to enable supply chains to adopt a coherent methodology. In line with this suggestion, the purpose of this chapter is to provide a deeper understanding of the combined approach between business continuity and risk management processes, specifically in supply chains, with a particular focus on procurement. A company dealing with the sale and distribution of medical devices was chosen as a research setting since its operation requires special attention to the objectives of business continuity and risk management.

The chapter is structured as follows: First, the methodological approach is presented. Next, the case study results are presented, focusing on how the combined implementation of the risk and business continuity management systems helped the company successfully handle the crises of capital controls and the COVID-19 pandemic. Finally, some concluding remarks are outlined.

2 Research Method

This section presents the research design selected for this study, including the procedures for case selection and data collection, coding, and analysis.

2.1 Research Design

This chapter aims to demonstrate the importance of the combined implementation of risk management and business continuity standards in the supply chain. The need to better understand such a complex phenomenon through direct observation in a natural setting led to the choice of a case study analysis as a basis for developing the research design [23, 36]. For the specific purpose of this study, a single case was selected since this is considered appropriate for exploratory investigations [7]. Furthermore, a single case provides a greater opportunity for in-depth observations [33] and thus enables a more careful study [6]. Several researchers used a similar research approach in their studies on business continuity and risk management [13, 21, 22, 29, 30].

2.2 Case Selection

For the case study analysis, a Greek company was chosen that holds a leading position in the Greek market of sales and distribution of medical devices. Operating as a hospital supply hub, it has undertaken all the relevant supplies of one of the largest groups in the Greek health sector. Hospitals are very complex supply networks, as they process thousands of materials per day, the orders response time must be very short, and the satisfaction of demand must be completely reliable.

The company is the only Greek company in this industry certified for business continuity management (ISO 22301), aiming to protect its activity from unforeseen risks and emergencies. The company is also certified to ISO 9001 (Quality Management System) and ISO 13485 (Quality Management System for Medical Devices). For the specific application of the risk management-related requirements of the ISO 9001 and ISO 13485 standards, the company uses the guidelines of the non-certifiable standard ISO 31000 (Risk management—Guidelines).

The company experienced the two major crises that hit the Greek companies during the last years: the capital controls imposed in Greece in 2015 and the COVID-19 pandemic that started in 2019. Capital controls were implemented on bank transfers from Greek to foreign banks in June 2015, after the European Central Bank's decision not to increase further the level of its Emergency Liquidity Assistance for Greek banks. To prevent the collapse of the Greek banking system, the Greek government decided to close the Greek banks for 20 days and impose limitations on cash withdrawals. This major crisis caused serious trouble to Greek businesses, especially those trading cross-border [32]. The COVID-19 pandemic started in Wuhan, China, in December 2019 and has since spread around the world. The first case in Greece was confirmed in late February 2020. Since then, more than 600,000 cases and 14,000 deaths have been officially reported (as of September 15, 2021). As a result, several lockdowns and restrictions were imposed at a national or local level, testing the resilience of all businesses worldwide, and especially SMEs [11, 27].

The abovementioned characteristics render the company an interesting field of study in which one can explore the combined implementation of business continuity and risk management standards in the supply chain under the pressure of two severe crises.

2.3 Data Collection Procedures

Multiple data collection methods were employed to ensure data triangulation. These included interviews with executives such as the CEO and Quality Manager, a review of the company's management system documentation like manuals and procedures, and field observations, particularly focusing on critical processes like ordering, storing, and distributing medical devices. In this study, triangulation occurs within a given methodology (i.e., case study); therefore, it pertains to within-method forms of triangulation [31]. The purpose of triangulation, as employed here, is the confirmation of data [14].

2.4 Data Coding and Analysis

Each interview was transcribed and analyzed with specialized software (i.e., QDA Miner). The researchers conducted iterative screening to identify the highlights of the interviews and subsequently categorize and code the results. At the same time, several other issues referenced multiple times by the interviewees were obtained. Furthermore, a qualitative comparison was made between the interviews and the documented procedures to verify that what was said was valid. Validation of these data (as derived from interviews and documents) was completed by further cross-checking them with the data gathered from observations at site visits.

3 Results

As a prerequisite for the effectiveness of business continuity management, a company should know how to mitigate the potential risks that threaten its business operations. That is why the company under study applies the Business Impact Analysis Framework based on the ISO 22301 standard in combination with the Risk Management and Assessment Framework based on the ISO 31000 standard. The outcomes of these two frameworks form the basis for the business continuity plan (as shown in Fig. 1), which the company implements in cases of threats. The above methodology results in the combined implementation of business continuity with risk management and thus enhances the company's resilience.

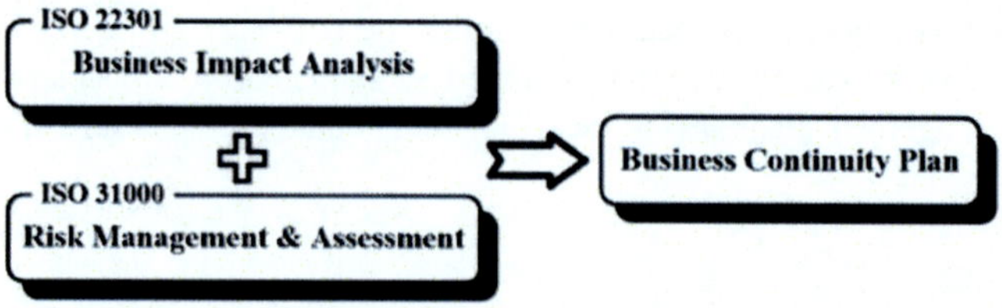

Fig. 1 The combined implementation of business continuity with risk management

In recent years, the company has faced two severe crises, which had never appeared in Greece: the imposition of capital controls in 2015 and the COVID-19 that started in 2019. The financial risks posed by these crises were not identified in time by the company. Therefore, they were not analyzed in the Risk Matrix nor subsequently documented in the business continuity plans. Thus, the company relied on some paradigms in the existing business continuity plans, which were deemed appropriate to deal with the two abovementioned unforeseen crises.

Specifically, when capital controls were imposed, an urgent meeting was held by the Crisis Management Team. Having experience from the existing business continuity plans, the team sought paradigms to deal with suppliers' inability to meet demand. These paradigms were similar to the situation the company was facing during the capital controls crisis. So, based on these paradigms, the company immediately took specific actions to continue its operation and ensure the uninterrupted service of its customers. Initially, their customers were categorized as critical, important, and minor through ABC Analysis. The company's information systems (i.e., Enterprise Resource Planning and Warehouse Management System) enabled the location of the materials and the specific quantities allocated in the customers' categories. This action was followed by an inspection of the materials inventory in the company's warehouse and an estimation of how long this inventory could cover the customers' demand under each category

From that moment on, when a customer placed an order, the company filtered the materials, and depending on the quantity requested by the customer, the available inventory, and the customers' category, one of the following alternatives was followed: send the entire ordered quantity; inform the customer that they would receive half the ordered quantity; inform the customer that they would receive one-third of the ordered quantity. The earlier process ran dynamically for the customers' categories.

However, the biggest problems would occur when the quantities of materials available in the company's warehouse ran out. To avoid this situation, the company initially increased the quantity per order to its suppliers so that the inventory would last more than previously planned. The warehouse, in its regular operation, used a safety stock of 20 days. To manage this new situation, the company proceeded with a larger quantity of orders that made the safety stock last for 30 days. It is important to note that the company orders from a wide range of suppliers in Greece and abroad. In the first days of the capital controls, the local suppliers smoothly supplied the company with the requested quantities. If one of the main suppliers did not have the ordered quantities available, the company used its pool of alternative suppliers, which was part of the existing business continuity plan. Furthermore, all the items distributed by the company have at least one alternative material, which the Quality Department has evaluated in advance. The employees of the Procurement Department were trained to use information systems to immediately find the alternative materials for each item if needed.

When the stocks of the local suppliers started to run out, and they could not supply the company anymore, the company made direct purchases from abroad. An issue that the company faced with foreign suppliers was that some demanded immediate payment, even if credit had been agreed beforehand, because they no longer trusted the financial situation in Greece. Fortunately, the company never had liquidity problems, so it could make direct payments to its suppliers.

The previous experience of crisis management in the case of capital controls was used 5 years later in the COVID-19 pandemic. In addition to the abovementioned actions, during the pandemic, there was a schedule of orders and deliveries depending on the needs and safety levels for some months ahead. The company had significantly increased the safety stock in line with the cooperating hospitals' demand. Additional materials were also added to the list of critical materials. This was a crucial element in the ordering of the coming months.

As with capital controls, the company had many alternative suppliers, especially for the critical materials. All potential suppliers (main and alternative) were pre-evaluated. Thus, if a major supplier could not meet the demand, the company could turn to alternative suppliers. In addition, all the required safety measures were used (temperature measurement, teleworking, occupational physician, etc.), ensuring the company's continued operation and the satisfaction of its customers' orders without delays or interruptions in deliveries. A particularly crucial element was the company's communication with the hospitals and its customers to immediately identify their needs and be directly informed about any problem occurring.

4 Conclusion

In this chapter, the role of the combined implementation of the ISO 22301 and ISO 31000 standards for business continuity and risk management, respectively, was highlighted, especially for the treatment of severe crises that could not be predicted in time.

Specifically, a case study was presented of a Greek company dealing with the sale and distribution of medical devices that experienced two major crises within a period of 5 years (i.e., capital controls and the COVID-19 pandemic). The financial risks posed by these two crises to the company's supply chain were unpredictable. According to Ojha et al. [18], it is not possible to implement a strictly standardized process for crisis treatment because every crisis is unique, and the same treatment will not be effective in all crisis scenarios. However, in the case study presented here, even though the specific risks were not identified in time, the company used the experience from the implementation of the ISO 22301 and ISO 31000 standards and immediately recognized the possible consequences and which of the predefined paradigms would be more effective to be used in the treatment of the new crises.

Following specific actions, the company managed to satisfy all customers adequately and at the same time survive under challenging conditions. This was mainly due to safety stock control practices and the use of alternative suppliers, especially in cases where the main suppliers could not supply critical materials. This finding confirms the research of Lavastre et al. ([16]), who identified these practices as highly effective for mitigating supply-chain risks. Future research could generate consensus around this issue among different business sectors. However, other scientists with similar cases are studying similar problems, using modeling and simulation [38–41].

References

1. Bakshi, N., Kleindorfer, P.: Co-opetition and investment for supply-chain resilience. Prod. Oper. Manag. **18**(6), 583–603 (2009)
2. Cerullo, V., Cerullo, M.J.: Business continuity planning: a comprehensive approach. Inf. Syst. Manag. **21**(3), 70–78 (2004)
3. Cha, S.C., Juo, P.W., Liu, L.T., Chen, W.N.: Riskpatrol: a risk management system considering the integration risk management with business continuity processes. In: 2008 IEEE International Conference on Intelligence and Security Informatics, pp. 110–115. IEEE, Taipei, Taiwan (2008)
4. Chountalas, P., Magoutas, A., Zografaki, E.: The heterogeneous implementation of ISO 9001: A multiple case study in service-oriented organizations. The TQM Journal. **32**(1), 56–77 (2019)
5. Craighead, C.W., Blackhurst, J., Rungtusanatham, M.J., Handfield, R.B.: The severity of supply chain disruptions: design characteristics and mitigation capabilities. Decis. Sci. **38**(1), 131–156 (2007)
6. Dyer Jr., W.G., Wilkins, A.L.: Better stories, not better constructs, to generate better theory: a rejoinder to Eisenhardt. Acad. Manag. Rev. **16**(3), 613–619 (1991)
7. Eisenhardt, K.M.: Building theories from case study research. Acad. Manag. Rev. **14**(4), 532–550 (1989)
8. Engemann, K.J., Henderson, D.M.: Business continuity and risk management: essentials of organizational resilience. Rothstein Publishing, Connecticut (2014)
9. Faertes, D.: Reliability of supply chains and business continuity management. Procedia Comput. Sci. **55**, 1400–1409 (2015)
10. Fan, Y., Stevenson, M.: A review of supply chain risk management: definition, theory, and research agenda. Int. J. Phys. Distrib. Logist. Manag. **48**(3), 205–230 (2018)
11. Fitriasari, F.: How do Small and Medium Enterprise (SME) survive the COVID-19 outbreak? J. Inov. Ekonomi. **5**(2), 53–62 (2020)
12. Graham, J., Kaye, D.: A Risk Management Approach to Business Continuity: Aligning Business Continuity and Corporate Governance. Rothstein Publishing, Connecticut (2015)
13. Jain, P., Pasman, H.J., Mannan, M.S.: Process system resilience: from risk management to business continuity and sustainability. Int. J. Bus. Contin. Risk Manag. **10**(1), 47–66 (2020)
14. Jick, T.D.: Mixing qualitative and quantitative methods: triangulation in action. Adm. Sci. Q. **24**(4), 602–611 (1979)
15. Lalonde, C., Boiral, O.: Managing risks through ISO 31000: a critical analysis. Risk Management. **14**(4), 272–300 (2012)
16. Lavastre, O., Gunasekaran, A., Spalanzani, A.: Supply chain risk management in French companies. Decis. Support. Syst. **52**(4), 828–838 (2012)
17. Mulla, G., Ademi, A.: Introduction to business continuity planning. Int. J. Sci. Innov. New Technol. **1**(13), 31–34 (2015)
18. Ojha, D., Gianiodis, P.T., Manuj, I.: Impact of logistical business continuity planning on operational capabilities and financial performance. Int. J. Logist. Manag. **24**(2), 180–209 (2013)
19. Pettit, T.J., Croxton, K.L., Fiksel, J.: Ensuring supply chain resilience: development and implementation of an assessment tool. J. Bus. Logist. **34**(1), 46–76 (2013)
20. Ponomarov, S.Y., Holcomb, M.C.: Understanding the concept of supply chain resilience. Int. J. Logist. Manag. **20**(1), 124–143 (2009)

21. Rashid, A.H.M., Loke, S.P., Ooi, K.B.: Strengthening supply chain risk management for business continuity: a case study approach. Int. J. Manag. Enterp Dev. **13**(3–4), 278–301 (2014)
22. Sawalha, I.H.S.: Organisational performance and business continuity management: a theoretical perspective and a case study. J. Bus. Contin. Emerg. Plan. **6**(4), 360–373 (2013)
23. Siggelkow, N.: Persuasion with case studies. Acad. Manag. J. **50**(1), 20–24 (2007)
24. Skipper, J.B., Hanna, J.B.: Minimizing supply chain disruption risk through enhanced flexibility. Int. J. Phys. Distrib. Logist. Manag. **39**(5), 404–427 (2009)
25. Sodhi, M.S., Son, B.G., Tang, C.S.: Researchers' perspectives on supply chain risk management. Prod. Oper. Manag. **21**(1), 1–13 (2012)
26. Suresh, N.C., Sanders, G.L., Braunscheidel, M.J.: Business continuity management for supply chains facing catastrophic events. IEEE Eng. Manag. Rev. **48**(3), 129–138 (2020)
27. Thukral, E.: COVID-19: Small and medium enterprises challenges and responses with creativity, innovation, and entrepreneurship. Strateg. Chang. **30**(2), 153–158 (2021)
28. Thun, J.H., Hoenig, D.: An empirical analysis of supply chain risk management in the German automotive industry. Int. J. Prod. Econ. **131**(1), 242–249 (2011)
29. Torabi, S.A., Soufi, H.R., Sahebjamnia, N.: A new framework for business impact analysis in business continuity management (with a case study). Saf. Sci. **68**, 309–323 (2014)
30. Tsogkas, A., Tsoulfas, G.T., Chountalas, P.: Risk management for business process reengineering: the case of a public sector organization. In: Sakas, D.P., Nasiopoulos, T., Y. (eds.) Business Intelligence and Modelling. Springer's Proceedings in Business and Economics, pp. 279–288. Springer, Heidelberg (2021)
31. Turner, S.F., Cardinal, L.B., Burton, R.M.: Research design for mixed methods: a triangulation-based framework and roadmap. Organ. Res. Methods. **20**(2), 243–267 (2017)
32. Ungerer, J., Ziaka, L.: Reflections on the Greek capital controls: How the rescue of the national economy justifies restricting private business. Legal Issues of Economic Integration. **44**(2), 135–149 (2017)
33. Voss, C., Tsikriktsis, N., Frohlich, M.: Case research in operations management. Int. J. Oper. Prod. Manag. **22**(2), 195–219 (2002)
34. Wagner, S.M., Bode, C.: An empirical examination of supply chain performance along several dimensions of risk. J. Bus. Logist. **29**(1), 307–325 (2008)
35. Wieland, A., Wallenburg, C.M.: Dealing with supply chain risks: linking risk management practices and strategies to performance. Int. J. Phys. Distrib. Logist. Manag. **42**(10), 887–905 (2012)
36. Yin, R.K.: Case study Research: Design and Methods, 5th edn. Sage, London (2014)
37. Zeng, B., Yen, B.P.-C.: Rethinking the role of partnerships in global supply chains: a risk-based perspective. Int. J. Prod. Econ. **185**, 52–62 (2017)
38. Plikas, H., Nasiopoulos, D.K., Giakomidou, D.S.: The contribution of social media in the management of social relations with customers. Modelling and simulation of the problem. Strategic Innovative Marketing, 515–521
39. Kanellos, N., Sakas, D.P., Kamperos, I.D.G., Reklitis, D.P., Giannakopoulos, N.T., Nasiopoulos, D.K., Terzi, M.C.: The effectiveness of centralized payment network advertisements on digital branding during the COVID-19 crisis. Sustainability. **14**(6)
40. Sakas, D.P., Nasiopoulos, D.K., Reklitis, P.: Modeling and simulation of the strategic use of marketing in search engines for the business success of high technology companies. Bus. Econ., 217–226
41. Nasiopoulos, D.K., Sakas, D.P., Vlachos, D.S.: A Mavrogianni: simulation of generation of new ideas for new product development and IT services. AIP Conf. Proc. **1644**, 60–68 (1644)

Green Supply Chain Management in the Greek Construction Industry

Eleni Marinaki, Anastasia Constantelou, and Giannis T. Tsoulfas

1 Introduction

The growing emphasis on sustainability and the shift towards greening have been noticeable trends in recent decades and have brought changes in many sectors of human activity. Supply chains (SCs) are an integral part of organizations as they combine, coordinate, and integrate their entire spectrum of functions and processes. This applies not only within but also across different organizations, which form parts of wider, even global value chains. Thus, SCs could not have been left intact from the global shift towards a greener, more sustainable future.

Since the 2010s, the greening of the SCs has only been meaningful through a holistic and integrated approach. Green practices, processes, projects, and goals cannot be dispersed and divided anymore; rather, organizations have recognized that the entire SC must become greener [13]. GSCM has emerged because of pressures from institutions, governments, stakeholders, and citizens. Simply monitoring organizations' environmental management programs on a reactive basis is no longer adequate. Organizations must now take proactive measures to avoid pollution

E. Marinaki
MBA Programme, School of Social Sciences, Hellenic Open University, Patras, Greece

A. Constantelou (✉)
MBA Programme, School of Social Sciences, Hellenic Open University, Patras, Greece

Department of Financial and Management Engineering, University of the Aegean, Chios, Greece
e-mail: a.konstantelou@fme.aegean.gr

G. T. Tsoulfas
MBA Programme, School of Social Sciences, Hellenic Open University, Patras, Greece

Department of Agribusiness and Supply Chain Management, Agricultural University of Athens, Thiva, Greece

D. P. Sakas et al. (eds.), *Computational and Strategic Business Modelling*, Springer Proceedings in Business and Economics,
https://doi.org/10.1007/978-3-031-41371-1_33

rather than just ignore it. According to Srivastava [18], these proactive methods are accomplished through multiple Rs: reduce, reuse, refurbish, rework, reclaim, remanufacture, recycle, and reverse logistics. As a result, the green component in SCM incorporates environmental concerns across the whole SC, from product design and material procurement through operations and manufacturing, transportation and delivery to end users, and end-of-life management.

Ahi and Searcy [1] reported 22 unique definitions of GSCM in their comparative literature review of definitions for green and sustainable supply chain management, focused on environmental, long-term, economic, and stakeholder aspects of corporate sustainability. According to Srivastava's seminal paper [18], GSCM is defined as 'integrating environmental thinking into SCM, including product design, material sourcing and selection, manufacturing processes, delivery of the final product to the consumers as well as end-of-life management of the product after its useful life'. Green operations refer to an organization's operational life cycle, which incorporates green thinking in material acquisition and sourcing, production/manufacturing, and product distribution, as well as reverse logistics methods. Finally, waste reduction in SCs is also included in GSCM. This is accomplished by waste management and organizational methods such as the different Rs [6].

This study aims to investigate the major GSCM practices followed by large construction companies in Greece. The emphasis lies on the fundamental factors that prompt for the adoption of such practices, the major barriers that construction companies face in the adoption of GSCM, and the metrics that quantify their influence on performance. The findings are derived from a questionnaire-based survey addressed to key GSCM-related staff of eligible Greek construction companies. Overall, the empirical study demonstrates that prominent Greek contractors understand the significance of GSCM and the shift towards greener, environmentally conscious and sustainable constructions' building. However, more effort needs to be put to make green practices part of contractors' day-to-day operations. This goes beyond mere interventions in the legislative framework to include a cultural shift and individual inclination towards greening.

2 GSCM in the Construction Industry

The SC in the construction business has several traits that are not found in other industries. In order to properly accomplish a project, developers, architects, contractors and subcontractors, suppliers, consumers, and end-users must seamlessly communicate upstream and downward a complicated SC. The number of different organizations that make up a construction project's SC might range from hundreds to thousands, depending on the nature and complexity of the project [4]. As a result, there is a need to structure and manage this complex network from the perspective of stakeholders, with the goal of incorporating the benefits of SCM, which have been recognized and acknowledged in other industries. Furthermore, SCM adoption in the construction sector may increase performance and contribute to time and cost savings for those companies who use it [6].

According to the 2020 Global Status Report for Buildings and Construction [20], the building and construction industry used 35% of global energy and emitted 38% of global CO_2 emissions in 2019. Since 2016, the industry has been steadily losing ground in relation to the Paris Agreement's goal of attaining net-zero emissions by 2050. Furthermore, according to Durdyev et al. [9], the building sector consumes 40% of raw materials and 25% of timber globally, creates 40% of solid waste globally, and uses 16% of the world's water.

At the same time, due to rising environmental consciousness and legislative limits, efforts to "go green" in construction have incorporated the green concept into SCM. As a result, the idea of GSCM arose and began to find application in both research and practice, with the potential for continued evolution in the area.

The ongoing COVID-19 epidemic has led to a significant reduction in worldwide building activity in 2020 compared to 2019 [20]. For the coming years, this might have a short-term positive influence on the industry's energy use and total environmental impact. However, as economies attempt to recover and flourish, this brief window of opportunity can be viewed as a chance for governments and construction industry practitioners to reevaluate their approach towards a greener and more sustainable business to stay up with global mandates.

3 The Construction Sector in Greece

In Greece, construction is one of the most important economic activities in the country, although it faced a significant decrease during the period of the economic crisis. The Greek construction sector is responsible for the planning and implementation of numerous projects and activities, boosting the productivity of the Greek economy and promoting growth and development in the country. From residential buildings to roads, railways, ports, and industrial installations, and from the development of major building projects to specialized construction activities, such as demolitions, refinishing works, electrical, mechanical, and other installations, the Greek construction sector is responsible for the planning and implementation of numerous projects and activities.

The broader construction sector is a highly significant keystone of the Greek economy. Although the total investment cost for construction projects shrank from €34.1 billion to €9.6 billion over the period 2007 to 2017, the sector still presents significant facts and figures, according to the 2019 Report of the Foundation For Economic & Industrial Research titled 'The Development Perspectives of Constructions in Greece'. The value added of the construction sector was €10.8 billion in 2017, accounting for 5.2% of Greece's GDP.

According to Greece Country Profile of the European Construction Industry Observatory, published in October 2020 [8], the sector showed some signs of recovery between 2017 and 2019, but statistics are still substantially below those of 2010. In 2019, the wide construction sector employed over 280,000 people, with the narrow construction sub-sector employing around 55% of the total workforce. In

2019, the whole construction industry employed 5.6% of the country's total workforce, including architectural and engineering activities, manufacturing, and the real estate sub-sectors. It had a total turnover of €20.3 billion and a total value added of €6 billion, with the narrow construction sector accounting for 54.8% (€3.3 billion) of the total.

Despite the sector's total size and contribution to the Greek economy, its volume appears to be unevenly distributed across the participating businesses. The great majority of construction organizations are micro-sized, consisting of either self-employed individuals or businesses with less than ten employees.

Although the sector appears to be lagging behind that of other European countries and has yet to overcome the slowdown and complications brought in by the previous decade's economic crisis, the resumption of procedures for long-stalled large investments, combined with the real estate activity's growth, could signal positive prospects. However, the COVID-19 pandemic has brought severe delays in these plans. Although building in Greece did not halt during the lockdowns, the current scenario has harmed construction owing to contract delays and higher implementation costs due to force majeure or safety concerns.

With regard to environmental protection, the industry looks to be ill-prepared for the implementation of the country's climate and energy strategy plan. According to the European Construction Sector Observatory [11], despite significant advancements, the sector continues to rely on traditional landfilling and waste disposal processes. The large fines levied on companies in the sector are explained by the fact that they dump 80% of solid waste, compared to the European average of 24%, even utilizing illegal disposal sites. Furthermore, the restricted construction sector sub-emissions of 269 thousand tonnes of GHG in 2018 remained high and above the EU average levels, diverting the country away from the Paris Agreement aims and goals and its transition to sustainability.

4 A Comprehensive Assessment Framework for GSCM in the Construction Sector

Balasubramanian and Shukla [5] proposed a comprehensive assessment framework for GSCM in the construction sector. Nine constructs make up the framework: core and facilitating GSCM practices, external and internal drivers, external and internal barriers, environmental, economic, and organizational performance. This section gives a quick overview of all of these constructs.

According to Balasubramanian and Shukla [5], Green practices are those "taken out by enterprises to reduce the negative environmental consequences connected with their activities". Green practices in the construction industry are divided into two categories: core and facilitating, which include all stages of building except the operational stage, which is primarily covered by green design. Table 1 illustrates these practices.

Table 1 Core and facilitating practices for GSCM in the construction industry

Core practices	Facilitating practices
Green building design	Environmental management systems
Green purchasing	Environmental training
Green transportation	Environmental auditing
Green construction/ manufacturing	Cross-functional cooperation for environmental improvements
End-of-life management	Green-related research and development

Table 2 Drivers for GSCM implementation in the construction industry

External drivers	Internal drivers
Government green-related regulation	Environmental commitment
Stakeholder pressure	Enhancement of reputation
Competitor pressure	Cost reduction
End-consumer pressure	Entering foreign markets

Table 3 Barriers for GSCM implementation in the construction industry

External barriers	Internal barriers
Lack of qualified staff in the market Shortage of green suppliers No extra time is being allocated to new environmental initiatives Lack of stakeholder engagement Lack of government incentives	High implementation costs Lack of green awareness and knowledge Lack of management commitment

Firms are implementing green practices because of the pressure applied by these drivers, which might come from inside the business, such as an environmentally conscious top management, or from outside the organization, such as from relevant legislation or stakeholder demand. Table 2 shows the different types of drivers identified in the literature [5, 9, 12, 15, 16, 21].

Similarly, there are issues that obstruct the adoption of green building techniques, which are referred to as barriers to GSCM implementation. These external and internal barriers to GSCM adoption by businesses are listed in Table 3 according to their source [3, 5, 7, 9, 14, 17, 21].

According to Ahi and Searcy [2], performance measurement in SC is determined by factors such as analysing, controlling, and verifying an organization's development, understanding in depth its key practices and courses of action, and identifying prospective modifications in future plans and practices. We can distinguish three types of GSCM performance metrics: environmental, economic, and organizational [5].

5 Research Method

5.1 Research Design

The questionnaire was developed using the GSCM assessment framework described in the preceding part as a basis for assessing GSCM in the Greek construction sector. The questionnaire was divided into five sections:

- *Personal and organizational information*: Respondents were asked to fill out information about the company they work for, such as the name, size, class, and scope of the organization, as well as their position, field of expertise, and experience in the company.
- *Implementation of green practices*: Respondents were asked to rate on a 5-point Likert scale (1 corresponding to 'not adopted at all' and 5 corresponding to 'fully adopted') which of the questionnaire-specific green practices their firm employs in its building projects. In order to make the questionnaire as complete and inclusive as possible, 20 GSCM practices were offered as alternatives.
- *Drivers of green practices*: Respondents were asked to assess the significance they ascribe to different GSCM drivers in their firm on a 5-point Likert scale (1 corresponding to 'not important at all' and 5 corresponding to 'extremely important'). A total of 20 alternatives were available in the questionnaire based on Table 2.
- *Barriers to green practices*: Respondents were given a list of barriers to GSCM practices and asked to rank them on a 5-point Likert scale (1 being 'not important at all' and 5 being 'extremely important'). There were 15 alternatives for respondents to evaluate.
- *Awareness and perceptions regarding GSCM impact on performance*: In the final part, we attempted to identify the experts' concerns about the influence of GSCM methods on their organizations' performance. Respondents were asked to rate their level of agreement with performance-related statements, which included all types of performance impact for companies – environmental, economic, and organizational – on a 5-point Likert scale (1 corresponding to 'strongly disagree' and 5 corresponding to 'strongly agree').

5.2 Data Collection

The Greek construction sector, including all of its stakeholders and SC stages, appears to be extremely diverse and fragmented. Therefore, it was necessary to confine the research to a much more specific and predefined sample in order to be feasible in terms of implementation, specific in terms of objectives, and practical in terms of results. To this end, only five firms had a legitimate seventh class registration in the competent authority's official contractors' registry during the course of this study. The current study focuses on the GSCM practices of this class of Greek contractors because of their importance in defining the trajectory of the industry.

Questionnaires were distributed to officials whose field of competence related to GSCM. Managers of health, safety and environment, project managers, quality managers, procurement managers, and site managers were chosen as the target group for obtaining a competent sample that could provide input based on field experience. The sampling was purposive, which does not require a predetermined number of participants, as the selection of participants was not random, but deliberate – owing to particular attributes they possess, by virtue of their expertise and competence in the subject [10].

Following a series of emails, in February 2021, the first batch of emails including the web-based questionnaire link was sent to key recipients inside the organizations in question, with the request that they would be sent to GSCM-related staff and other representatives as they saw fit. In the event of a non-response or a poor answer, more correspondence and reminders via email or phone calls were required. At the same time, an effort was made to broaden the list of receivers without jeopardizing the necessary main traits, so that the questionnaire would be distributed to as many relevant persons as possible.

6 Results

As previously stated, only officials from large Greek construction companies were approached and asked to participate in this study. All 48 responders worked for one of the firms registered in the seventh upper-class. The majority of respondents worked on all three types of projects, which span a wide variety of activities and operations, with infrastructure projects accounting for 45 of the 48 responses (94%), buildings for 43 (90%), and industrial projects for 41 (85%). Table 4 shows the respondents' experience and specialization, while Table 5 shows their job positions.

Table 4 Respondents' experience and specialization

Experience in the firm	%	Field of specialization		%
Less than 5 years	10%	Civil	Civil engineering	52%
		Mechanical/electrical	Mechanical engineering	21%
5–10 years	25%		Electrical-mechanical engineering	2%
			Electrical engineering	6%
11–15 years	23%	Environmental	Environmental engineering	6%
		Management	Management	6%
16–20 years	19%	Finance and administration	Finance	2%
			Administration	2%
More than 20 years	23%	Other	Mining-metallurgical engineering	2%
Totals	**100%**		**Totals**	**100%**

Table 5 Respondents' job position

Position in the firm		%
Quality, health, safety, and environment	HSE director/manager	13%
	Environmental manager	2%
	Quality manager	8%
	Quality engineer	4%
Upper management in construction	Project/site/contract manager/director	25%
	Construction manager	4%
	Technical director	2%
Construction	Site/operations/construction/supervisor engineer	29%
Procurement and planning	Procurement manager	6%
	Planner	2%
Finance, administration, and HR	Financial controller	2%
	Head of HR and administration	2%
	Totals	**100%**

The descriptive statistics for all constructs of the assessment framework for GSCM are presented in Tables 6, 7, 8 and 9. Overall, the measurement instrument's Cronbach's alpha estimations are good, indicating that the questionnaire is reliable, which might lead to more accurate assessments and evaluations [19].

The statement 'Employees are encouraged to use shared and public transportation' had the lowest mean among the 20 items describing GSCM practices, indicating that it was the least accepted practice. The statement 'There is a waste management plan for project sites', on the other hand, had the highest mean, making it the most often used GSCM practice.

External drivers tend to be the most essential for the acceptance and implementation of green practices in construction, according to the drivers for GSCM implementation. In fact, of all the constructs, 'Provisions of standard quality certification (e.g. ISO 14001)' had the highest mean value. 'Employees' engagement in GSCM practices', on the other hand, has the lowest mean value among all GSCM drivers. Individual-level drivers tend to be the least important in terms of GSCM adoption by Greek contractors, which is not surprising.

When it comes to GSCM implementation barriers, it is clear that the value participants place on each item of GSCM barriers is lower than the importance placed on the related drives. All aspects of both external and internal constructs play a very important role to Greek contractors. The item with the lowest mean value is 'disbelief in environmental advantages', implying that a sufficient proportion of practitioners believe that environmental attention in construction has a positive impact. The item with the greater mean value, on the other hand, is another internal barrier item, this time the 'high implementation cost'. This might be considered as a sign of short-sightedness and/or lack of awareness and acceptance by practitioners of the long-term economic benefits associated with implementing green practices.

Table 6 GSCM practices: descriptive statistics

GSCM practices: descriptive statistics			
Construct	Item	Mean	Standard deviation
Green design	Materials with low embodied energy are considered during design	3.58	0.90
	Materials with high recycled content are considered during design	3.63	0.73
	Reduction of hazardous materials is considered during design	4.10	0.75
Green purchasing	Material purchase decisions incorporate environmental criteria	3.90	0.75
	Tendering processes incorporate environmental criteria	3.81	0.82
Green transportation	Employees are provided with accommodation near project sites	4.08	0.92
	Employees are encouraged to use shared and public transportation	2.58	1.18
	Transportation of materials is implemented under full truckload precondition	3.73	0.82
	Transportation of materials is implemented in fuel-efficient vehicles	2.88	1.04
Green construction	Prefabricated components are used in projects	3.56	0.90
	Materials with high recycled content and low embodied energy are used in projects	3.58	0.85
	Hazardous materials are less used in construction projects	3.94	0.84
	There is a provision for waste water recycling at project sites	3.90	0.99
	There is a waste management plan for project sites	4.15	0.87
	Fuel-efficient equipment is used on project sites	3.42	0.92
End of life management	The end-of-life demolition environmental impact of projects is assessed	3.54	0.92
	The end-of-life demolition materials of projects are recycled	3.75	0.93
Facilitating green practices	Your company adopts environmental training programs for its employees	3.83	0.91
	Your company performs audits for environmental compliance	4.02	0.91
	Your company focuses in green-related research and development	3.50	0.77

In terms of performance, it is clear that the mean values of the environmental performance items are the highest of the three constructs, with the most significant item being 'The decrease in the number of environmental accidents'. Respondents, on the other hand, do not appear to firmly believe that the adoption of green practices has resulted in a 'increase in sales price'. Overall, the items of organizational

Table 7 Drivers for GSCM implementation: descriptive statistics

GSCM drivers: descriptive statistics			
Construct	Item	Mean	Standard deviation
External drivers	Government green-related regulations	4.40	0.71
	EU's environmental legislation	4.38	0.79
	Provisions of standard quality certification (e.g. ISO 14001)	4.65	0.56
	Pressure from supply chain stakeholders (e.g. developers, consultants, etc.)	3.73	0.79
	Pressure from competitors	3.69	0.88
Project-level drivers	Cost reduction	4.15	0.87
	Decreased construction time	4.08	0.94
	Waste reduction	3.81	0.98
	Reduced on-site worker health and safety risks	4.38	0.70
	New kinds of partnerships and project stakeholders	3.98	0.86
Corporate-level drivers	Environmentally friendly corporate culture and vision	4.00	0.80
	Corporate social responsibility	4.08	0.74
	Enhancement of brand image and reputation	4.23	0.83
	Achievement of economic benefits	3.98	0.89
	Entering into foreign markets	4.02	0.81
Individual-level drivers	Top management environmental commitment	4.00	0.80
	Employees' involvement in GSCM practices	3.63	0.87
	Knowledge, awareness, and information	3.75	0.89
	Education and training	3.65	1.02
	Moral imperative or social conscience	3.71	0.99

performance have the lowest mean values, while the items of economic performance are regarded somewhat in the middle of the other two constructs.

Because of the poor response in certain cases, the data set was restricted, making it impossible to examine the importance of factors. Future research might enhance the sample size to study correlations between factors and draw appropriate findings. It might also include statistics and information about the sector's various stakeholders, including developers, consultants, suppliers, and subcontractors, as well as small- and medium-sized contractors: who are key and indispensable participants in the construction SC.

7 Conclusion

Construction operations degrade the natural environment globally, resulting in massive energy consumption, solid waste generation, nonrenewable resource depletion, and land, air, and water contamination. As a result, introducing GSCM into construction is critical for reducing its environmental impact. This study examined

Table 8 Barriers in GSCM implementation: descriptive statistics

GSCM barriers: descriptive statistics			
Construct		Mean	Standard deviation
External barriers	Shortage of green professionals	3.35	0.96
	Shortage of green suppliers	3.46	0.85
	Shortage of green technology, materials, and processes	3.38	0.91
	Tight and inflexible stakeholder deadlines	3.56	0.80
	Lack of stakeholder engagement/collaboration	3.56	0.90
	Lack of government incentives	3.33	0.86
	Inadequate enforcement of environmental regulations	3.46	1.03
Internal barriers	High implementation cost	3.83	0.91
	Lack of knowledge and awareness	3.46	1.13
	No extra time allocated to new environmental initiatives	3.48	0.85
	No extra resources allocated to new environmental initiatives	3.40	0.94
	Perception of 'out-of-responsibility' zone	3.42	1.03
	Disbelief about environmental benefits	3.13	1.21
	Lack of top management involvement	3.23	1.19
	Lack of environmental training among employees	3.31	1.11

the application of GSCM by the major Greek construction companies. The analysis concentrated on significant Greek contractors because of their size and importance to the whole sector, as well as their central location in the SC and their capacity to influence corporate activity upstream and downstream.

Overall, prominent Greek contractors tend to understand the significance of GSCM and the unquestionable necessity for a shift towards greener, more sustainable building. As a response to expanding global concerns and competitiveness, their conduct demonstrates a shift towards environmental consciousness and sustainability. However, more work has to be done to make greener practices a part of contractors' day-to-day operations. Laws and regulations already exist and are being implemented. However, in order to overcome persistent impediments and for GSCM to become the instrument for construction practitioners to further boost their companies' performance in the context of environmental respect, individual drivers must be strengthened and a cultural shift must occur.

The current study is the first to look at the sector's use of the GSCM's complete assessment methodology, since there has been little evidence of GSCM application by Greek construction professionals till now. Its uniqueness makes it valuable in terms of increasing awareness among practitioners and providing a foundation for a future wider and more fine-tuned empirical research. In addition, studies by other scientists who could contribute to the above study, examine the supply chain beyond its ecological context, using modelling and simulation [22–25].

Table 9 Performance repercussion of GSCM implementation: descriptive statistics

GSCM performance: descriptive statistics			
Construct	Item	Mean	Standard deviation
Environmental performance	Number of environmental accidents has declined	4.04	0.90
	Greenhouse gas emissions have decreased	3.92	0.74
	Water consumption has decreased	3.79	0.80
	Energy consumption has decreased	3.85	0.77
	Landfill waste has decreased	3.90	0.81
	Material use has decreased	3.60	0.82
	Hazardous material use has decreased	3.88	0.94
Economic performance	Material expenses per unit constructed have decreased	3.40	0.94
	Water expenses per unit constructed have decreased	3.48	0.80
	Energy expenses per unit constructed have decreased	3.50	0.80
	Waste management cost per unit constructed has decreased	3.42	0.94
	Environmental fines and penalties per unit constructed have decreased	3.79	0.90
Organizational performance	Sales have increased	3.33	0.88
	Sales price has increased	3.23	0.95
	Market share has increased	3.44	0.82
	Return on investment has increased	3.35	0.86
	Profits have increased	3.35	0.86

References

1. Ahi, P., Searcy, C.: A comparative literature analysis of definitions for green and sustainable supply chain management. J. Clean. Prod. **52**, 329–341 (2013)
2. Ahi, P., Searcy, C.: An analysis of metrics used to measure performance in green and sustainable supply chains. J. Clean. Prod. **86**, 360–377 (2015)
3. AlSanad, S.: Awareness, drivers, actions, and barriers of sustainable construction in Kuwait. Procedia Eng. **118**, 969–983 (2015)
4. Badi, S., Murtagh, N.: Green supply chain management in construction: a systematic literature review and future research agenda. J. Clean. Prod. **223**, 312–322 (2019)
5. Balasubramanian, S., Shukla, V.: Green supply chain management: an empirical investigation on the construction sector. Supply Chain Manag. **22**(1), 58–81 (2017)
6. Benachio, G.L.F., Freitas, M.C.D., Tavares, S.F.: Green supply chain management in the construction industry: a literature review. IOP Conf. Ser.: Earth Environ. Sci. **225**(1), 012011 (2019)
7. Bey, N., Hauschild, M.Z., McAloone, T.C.: Drivers and barriers for implementation of environmental strategies in manufacturing companies. CIRP Ann. **62**(1), 43–46 (2013)
8. European Commission: European Construction Sector Observatory - Country profile Greece (2020). Available at https://ec.europa.eu/docsroom/documents/44164

9. Durdyev, S., Zavadskas, E.K., Thurnell, D., Banaitis, A., Ihtiyar, A.: Sustainable construction industry in Cambodia: awareness, drivers and barriers. Sustainability. **10**(2), 392 (2018)
10. Etikan, I., Musa, S.A., Alkassim, R.S.: Comparison of convenience sampling and purposive sampling. Am. J. Theor. Appl. Stat. **5**(1), 1–4 (2016)
11. European Construction Sector Observatory: European Construction Sector Observatory: 2020 Report for Greek construction sector – SEGM (2020). Available at https://segm.gr/language/en/european-construction-sector-observatory-2020-report-for-greek-construction-sector/
12. Holt, D., Ghobadian, A.: An empirical study of green supply chain management practices amongst UK manufacturers. J. Manuf. Technol. Manag. **20**(7), 933–956 (2009)
13. Jaggernath, R., Khan, Z.: Green supply chain management. World J. Entrepreneurship Manag. Sustain. Dev. **11**(1), 37–47 (2015)
14. Kaur, J., Sidhu, R., Awasthi, A., Chauhan, S., Goyal, S.: A DEMATEL based approach for investigating barriers in green supply chain management in Canadian manufacturing firms. Int. J. Prod. Res. **56**(1–2), 312–332 (2018)
15. Oluwole Akadiri, P., Olaniran Fadiya, O.: Empirical analysis of the determinants of environmentally sustainable practices in the UK construction industry. Constr. Innov. **13**(4), 352–373 (2013)
16. Sarkis, J., Zhu, Q., Lai, K.: An organizational theoretic review of green supply chain management literature. Int. J. Prod. Econ. **130**(1), 1–15 (2011)
17. Shi, Q., Zuo, J., Huang, R., Huang, J., Pullen, S.: Identifying the critical factors for green construction – an empirical study in China. Habitat Int. **40**, 1–8 (2013)
18. Srivastava, S.K.: Green supply-chain management: a state-of-the-art literature review. Int. J. Manag. Rev. **9**(1), 53–80 (2007)
19. Tavakol, M., Dennick, R.: Making sense of Cronbach's alpha. Int. J. Med. Educ. **2**, 53–55 (2011)
20. UNEP GlobalABC: 2020 Global Status Report for Buildings and Construction. The Global Alliance for Buildings and Construction (GlobalABC) (2020). Available at https://globalabc.org/resources/publications/2020-global-status-report-buildings-and-construction
21. Walker, H., Jones, N.: Sustainable supply chain management across the UK private sector. Supply Chain Manag. **17**(1), 15–28 (2012)
22. Nasiopoulos, D.K., Sakas, D.P., Vlachos, D.S., Mavrogianni, A.: Modeling of market segmentation for new IT product development. AIP Conf. Proc. **1644**, 51–59 (1644)
23. Dimitrios, N.K., Sakas, D.P., Vlachos, D.S.: The contribution of dynamic simulation model of depiction of knowledge, in the leading process of high technology companies. Key Eng. Mater. **543**, 406–409
24. Chaldezos, C., Sakas, D., Dimitrios, N.K., Giakomidou, D.S.: Calculating handling of company resources for establishing an effective project team. In: Kavoura, A., Sakas, D., Tomaras, P. (eds) Strategic Innovative Marketing. Springer Proceedings in Business and Economics. Springer, Cham (2017). https://doi.org/10.1007/978-3-319-56288-9_49
25. Nasiopoulos, D.K., Reklitis, P., Giakomidou, D.S.: Build and Approach of Logistics Companies Reengineering. Modeling and Simulation of the Problem. In: Sakas, D.P., Nasiopoulos, D. K., Taratuhina, Y. (eds) Business Intelligence and Modelling. IC-BIM 2019. Springer Proceedings in Business and Economics. Springer, Cham (2021). https://doi.org/10.1007/978-3-030-57065-1_20

Prioritizing Supply Chain Services Features in Online Supermarkets: Evidence from Greece

Anastasia Mitropoulou and Giannis T. Tsoulfas

1 Introduction

Several studies have identified supply chain efficiency as a key factor in delivering low-cost [15, 16, 42] or high-quality [5, 6, 9, 12] items with short lead times [1, 11, 40]. Bowersox et al. [8] pointed out that there are three viewpoints that might provide value for consumers, namely, economic, market, and relevancy value. As a result, the overarching goal of supply chain system must be to maximize profitability [14, 41].

Various authors have explored, emphasized, and debated essential issues of supply chain efficiency from a variety of perspectives [3, 13, 28, 29, 35]. Supply chain efficiency must guarantee that the customers' expectations are met and that any non-value-adding processes are eliminated. The measure of supply chain quality is having the appropriate product, in terms of quality, at the right time and in the right location at the lowest cost. Similarly, the most effective supply chain meets customer service objectives, such as distribution precision and delivery time, at the lowest feasible cost. There are some primary deciding variables for supply chain efficiency: resource sustainability, enhanced customer satisfaction, greater value, lowest cost, shorter delivery time, better quality, profitability, and waste reduction.

The fast emergence of e-commerce has fueled the growth of the retail and logistics industries [32]. Retailers can attract more customers, extend the distribution

A. Mitropoulou
MBA Programme, School of Social Sciences, Hellenic Open University, Patras, Greece

G. T. Tsoulfas (✉)
MBA Programme, School of Social Sciences, Hellenic Open University, Patras, Greece

Department of Agribusiness and Supply Chain Management, Agricultural University of Athens, Thiva, Greece
e-mail: giannis@aua.gr

D. P. Sakas et al. (eds.), *Computational and Strategic Business Modelling*,
Springer Proceedings in Business and Economics,
https://doi.org/10.1007/978-3-031-41371-1_34

chain even further, and harness existing resources thanks to the widespread usage of the Internet. At the same time, customers may participate in the service encounter at any moment and quickly compare offerings. The last mile of a supply chain is regarded as the most demanding element, and customers' expectations are now being pushed to their limits. In particular, customers' expectations for service quality are rising as they seek improved service with a variety of delivery alternatives in terms of time and location, flexible payment terms, and quicker ways to retrieve and return their items. To keep their customers content, businesses must pay attention to a variety of elements and consider ongoing examination and improvement of their operations [46].

The goal of this study is to assess the logistical performance of online supermarkets from the perspective of their customers. It aims to improve our knowledge of customers' expectations for the relevant service and how they view the final delivery. The Kano model is used to assess customers' perceptions regarding the presence and absence of attributes related to supply chain service quality in the case of Greek supermarkets.

2 The Impact of COVID-19 in Supermarkets' Online Shopping

During the COVID-19 epidemic, and especially during the quarantine period, the necessity for internet purchasing for basic groceries and other things became critical [2, 4].

Because traffic in electronic supermarkets remained low before the pandemic, the great majority of them lacked the appropriate infrastructure to handle the increasing amounts of orders in the days after the lockdowns. In a relatively short amount of time, a substantial number of conventional customers have joined the ever-growing population of online customers, and this rapid change in demand has generated various concerns, on top of the difficulties posed by the present pandemic crisis. They did, however, respond quickly and planned for the future. Some of them updated their online stores, but the majority resorted to emergency hiring and cooperation with third-party service providers. At least 15% of the recruiting during the epidemic was done to satisfy the increased demands provided by the Internet channel, which is noteworthy. The increased use of new sales channels, such as electronic supermarkets and telephone orders, has resulted in the necessity for extra people to oversee, execute, and deliver orders, as detailed in IELKA's research [21]. Furthermore, the Greek supermarkets have committed to investments in the formation or growth of their online businesses. These investments cover both the configuration, operation, and support of online stores, as well as the processes that support them: logistics, including equipment purchases and the establishment of facilities and pickup points in stores. The growth of the industry stabilized at considerably high rates after the first wave of the pandemic and still evolves, which means that this channel might claim a higher market share in the near future. At the end of August 2020 in Greece, IELKA and ELTRUN conducted a rolling consumer survey with a sample of 750 people, the results of which show that the significant increase in the use of e-commerce during the lockdown period due to COVID-19 is consolidating and becoming the new normal [22]. Another IELKA survey shows that consumers who shop at online supermarkets now account for 25% of all purchases, up from 1% to 2% in 2019, and that one out of every four consumers is unaware of the benefits of online shopping. This indicates that the market has significant potential for growth [23].

3 The Role of Supply Chains in Customer Satisfaction and Service Quality: The Kano Model

As opposed to the financial crisis brought on by the COVID-19 pandemic, grocery earnings have increased significantly. The majority of them have also gone through an urgent update of their services, with the development of their online shops being one of their main goals, resulting in a gain in market share for their online shopping. Because of the abovementioned increase in the contribution of online stores' revenues to total supermarket profitability, assessing customer satisfaction in this area is critical. The research findings may be used by supermarket managers and administrators who want to develop their online services even further.

Companies' interest in the economics of customer retention has grown dramatically, since the American Consumer Association demonstrated in the late 1980s that maintaining an existing client is five times less expensive than acquiring a new one [20]. When consumers are happy with a product or service, they build faith in the firm and become loyal to it, as well as the likelihood of recommending the product or service to others.

Customer satisfaction is the most studied concept in marketing literature [31] and is defined in various ways [48]. It is regarded as a customer's pledge to return for the same product or service in the event of a repurchase, implying that the same brand will eventually be repurchased regardless of the environment or other marketing strategies [36].

However, in order for a business to improve customer satisfaction, it is critical that consumers provide constant feedback so that all relevant inputs may be gathered. This procedure is significantly easier if the company has previously developed a trusting connection with its clients. A "customer-oriented" strategy may be the source of this trust [24, 51].

Several studies have looked at the link between customer satisfaction and service quality. Several publications [18, 25, 47, 49] have concluded that service quality influences customer satisfaction, a connection that also works the opposite way.

A thorough understanding of consumers' wants and expectations is a prerequisite for customer satisfaction [17, 43]. For many years, researchers saw the link between customer satisfaction and customer requirement compliance as linear and one-dimensional. However, just because customer expectations are met to a larger extent does not always imply that customer satisfaction is met or enhanced [30].

On the basis of Herzberg's notion that the conditions that promote job satisfaction differ from those that cause job dissatisfaction, Kano and Takahashi established the motivator-hygiene attribute of quality in 1979 [19, 26]. Except for the one-dimensional element, Kano offered a nonlinear function that contradicted the initial proportional interpretation [27]. Kano's methodology proposes categorizing different consumer criteria according to how much they may influence customer satisfaction. The "must be," "one-dimensional," and "attractive" are the three main sorts of customer criteria.

The "must be" criteria may be characterized as the essential qualities of service quality in terms of customer satisfaction. As a result, if these parameters are not reached, the customer will be extremely dissatisfied. As a result, completing the "must be" constraints would only result in a "not dissatisfied" condition. To put it another way, it is important to be fulfilled, but it is not sufficient to satisfy the consumer [10].

When "one-dimensional" criteria are present, they ensure consumer satisfaction; when they are not, they guarantee discontent (Redfern and Davey 2003). As the functionality of these needs improves, so does client pleasure. As a result, the higher the perceived service quality, the higher the client satisfaction, and vice versa. As a result, they are regarded as both required and adequate to satisfy the customer [10]. These characteristics are often used to gain a competitive advantage through distinction and consumer popularity [50].

"Attractive" criteria are those that can provide consumer happiness when they are present but do not cause customer dissatisfaction when they are absent. As a result, these service qualities might influence the level of customer satisfaction with a given service [30]. Customers do not expect these features, but when they are supplied correctly, they bring satisfaction. In contrast to the "must be," the "attractive" ones are sufficient, but not required, for fulfillment [10]. As part of an aggressive marketing strategy, appealing characteristics should be employed to attract customers.

According to Kano et al. [27] and Berger et al. [7], attribute classification may be performed by consumers' responses to the following two situations: superior value attributes (good question) and low-performance attributes (negative question). Depending on how consumers respond to both questions, they are assessed as "dissatisfied," "I can live with," "neutral," "must be that way," and "satisfied."

Using Kano's approach, it should be highlighted that the service provider should not only understand current client wants that are critical to success but also prospective customer requests that might thrill them [45].

4 Research Method

4.1 Research Design

According to Saunders et al. [44], when choosing a research technique, there are two options: utilize a single data collection method and the associated analysis processes (mono method) or use numerous data collection methods and analysis processes (multiple method).

A quantitative research approach was used in this study. As a result, a single quantitative data collection approach (questionnaire) was utilized, together with the accompanying quantitative data analysis tools that generate or utilize numerical data.

The first section of the survey asks about the respondents' demographic profile, such as their gender, age, level of education, and employment. The second and third sections of the questionnaire help to determine the features that the service must have in order to satisfy customers. The questions pertain to the SERVQUAL model's tangibility, dependability, responsiveness, assurance, and empathy dimensions [37]. They were designed in this way to extract information on the degree of satisfaction in the presence of and the level of dissatisfaction in the absence of the features indicated in the SERVQUAL model's 22 questions, which were modified to suit the subject under investigation. Table 1 lists these characteristics.

The answers were based on the following Likert scale: 1, I like it; 2, I expect it; 3, I am neutral; 4, I can tolerate it; 5, I dislike it.

4.2 Data Collection

The questionnaire was disseminated online in order to collect the necessary data. Following some investigation, it was decided to use Google Forms, a free and web-based survey administration software, for the objectives of this study. As a result, relevant questions and multiple-choice responses were constructed, and the survey was saved in the online application.

A trial questionnaire was emailed to nine recipients before the questionnaire was sent out to assess the clarity and grasp of the questions. Following the respondents' input, it was decided to make some small adjustments to the material in order to make it more understandable to the audience.

The questionnaire was sent to citizens of Athens and Thessaloniki, as these cities are the key hubs for online supermarkets. Furthermore, it is self-evident that the questions could only be answered by online supermarket customers.

The sampling approach used in this study is a mix of convenience and snowball sampling. Because the questionnaire was sent via a variety of means, including emails, texts, and forwarded to others, it is hard to estimate the total number of people who eventually got it.

Table 1 SERVQUAL features assessed with the Kano model

Code	Feature
F1	Modern and easy to use website
F2	Tidy distributing vehicles
F3	Neat appearance of drivers and delivery personnel
F4	Proper packaging to product specification
F5	On time delivery of the order
F6	Employees with sincere interest to solve customers' problems
F7	Delivery of all products at once
F8	Reasonable compensation for damaged package
F9	Accurate records of the delivery
F10	Information provision to the customers about the exact status of the order
F11	Possibility to change the delivery time after order
F12	Employees' willingness to help customers
F13	Employees' availability to respond to customers' requests'
F14	Employees' behavior that promotes customers' confidence
F15	Website with a sense of safety to perform transactions
F16	Employees consistently courteous with customers
F17	Knowledgeable employees to answer customers' questions
F18	Individual attention/personalized promotions to customers
F19	Convenient delivery time frames to all customers
F20	Convenient payment methods to all customers
F21	Customers' best interests at heart
F22	Understand specific needs of the customers

5 Results

The demographic characteristics of the respondents are presented in Table 2.

Respondents were asked to rate their feelings about whether each attribute was present or not. Functional questions are used to assess presence, whereas dysfunctional questions are used to assess absence. Every attribute was categorized into one of the Kano model's six categories by integrating the replies to both functional and dysfunctional questions:

- Must be (M) – requirements that if they will not be met, the product/service will fail
- One-dimensional (O) – features that increase the satisfaction the more they are fulfilled
- Attractive (A) – attributes that are not expected by customers, but their presence makes the customer happy
- Indifferent (I) – customers who do not really care about these features
- Questionable (Q) – corresponds to not logical answers
- Reversal (R) – adverse effect of the existence of an attribute, as it increases dissatisfaction when it is implemented

Table 2 Respondents' demographic characteristics

Participants	147				
Gender			Education		
Men	61	41.5%	Secondary education	25	17.0%
Women	86	58.5%	Technical studies	15	10.2%
Age			Bachelor degree	55	37.4%
−25	8	5.4%	Master degree	52	35.4%
26–35	48	32.7%	Student	8	5.4%
36–45	42	28.6%	Employment		
46–55	19	12.9%	Employee (public sector)	19	12.9%
56–65	15	10.2%	Employee (private sector)	57	38.8%
65-	15	10.2%	Freelancer	32	21.8%
			Unemployed	12	8.2%
			Other	19	12.9%

Table 3 Kano evaluation table

		Negative questions				
Customer requirement		1	2	3	4	5
Positive questions	**1**	Q	A	A	A	O
	2	R	I	I	I	M
	3	R	I	I	I	M
	4	R	I	I	I	M
	5	R	R	R	R	Q

This categorization may be seen in Table 3, which takes into consideration both functional and dysfunctional questions.

We conducted a discrete analysis for categorization using the Kano model based on the responses collected from all respondents. As a result, we determined the average percent of the answers given for each of the possible classes for each attribute. Then, for each feature, (a) the relative highest score was used, and (b) the two higher scores were used. Table 4 summarizes the findings.

From the above results, we can see that most features are categorized as "must be." This implies that clients are expecting to see these elements, and if they are not there in the perceived service, they will be disappointed. A contemporary and easy-to-use website, on the other hand, and an employee's eagerness to assist clients are two traits that fall into the "one-dimensional" category. This is a category that mostly pertains to personal preferences. It indicates that the more of this feature there is, the happier the consumer is. If we use two categories, then the "must be" and "one-dimensional" categories prevail.

Table 4 Features categorization as per Kano's model

Feature	M	O	A	I	R	Q	Single category	Two categories
F1	29%	43%	12%	9%	0%	8%	O	O-M
F2	46%	24%	7%	17%	1%	4%	M	M-O
F3	35%	15%	16%	28%	1%	5%	M	M-I
F4	44%	27%	12%	10%	0%	7%	M	M-O
F5	43%	35%	10%	4%	1%	6%	M	M-O
F6	36%	31%	9%	16%	2%	7%	M	M-O
F7	35%	27%	17%	13%	1%	7%	M	M-O
F8	52%	25%	8%	5%	1%	8%	M	M-O
F9	41%	21%	14%	15%	1%	7%	M	M-O
F10	35%	25%	15%	16%	3%	5%	M	M-O
F11	35%	19%	16%	1%	8%	0%	M	M-O
F12	11%	15%	1%	9%	0%	0%	O	O-M
F13	37%	25%	13%	16%	2%	6%	M	M-O
F14	37%	25%	14%	16%	1%	7%	M	M-O
F15	48%	28%	7%	9%	2%	6%	M	M-O
F16	35%	29%	12%	18%	0%	7%	M	M-O
F17	36%	22%	16%	18%	0%	7%	M	M-O
F18	27%	20%	20%	24%	1%	8%	M	M-I
F19	29%	29%	16%	19%	0%	7%	M	M-O
F20	41%	28%	8%	13%	1%	9%	M	M-O
F21	33%	23%	18%	19%	1%	6%	M	M-O
F22	31%	22%	18%	18%	1%	8%	M	M-O

6 Conclusion

This study looks at how satisfied customers are with the services offered by online supermarkets and what they anticipate from such businesses. Managers and administrators of supermarkets may use the survey's findings in order to improve their services and claim more market share and profits. However, there are corresponding studies that, with the method of modeling and simulation, can help in this purpose [34, 38].

The survey highlighted areas where online supermarkets have to address to meeting customer expectations. Almost all investigated features are regarded prerequisites for a satisfied consumer, although the attributes of a modern website and helpful employees might increase satisfaction when recognized. Also, other scientists approach corresponding subjects in their studies [33, 39].

References

1. Anand, N., Grover, N.: Measuring retail supply chain performance: theoretical model using key performance indicators (KPIs). BIJ. **22**(1), 135–166 (2015)
2. Anastasiadou, E., Chrissos Anestis, M., Karantza, I., Vlachakis, S.: The coronavirus' effects on consumer behavior and supermarket activities: insights from Greece and Sweden. Int. J. Sociol. Soc. Policy. **40**(9–10), 893–907 (2020)

3. Ang, S., Zhu, Y., Yang, F.: Efficiency evaluation and ranking of supply chains based on stochastic multicriteria acceptability analysis and data envelopment analysis. Int. Trans. Oper. Res. **28**(6), 3190–3219 (2021)
4. Badenhop, A., Frasquet, M.: Online grocery shopping at multichannel supermarkets: the impact of retailer brand equity. J. Food Prod. Mark. **27**(2), 89–104 (2021)
5. Beheshtinia, M.A., Feizollahy, P., Fathi, M.: Supply chain optimization considering sustainability aspects. Sustainability. **13**(21), 11873 (2021)
6. Ben-faress, M., Elouadi, A., Gretete, D.: Framework and development of a collaborative supply chain model. Int. J. Supply Chain Manag. **8**(3), 833–837 (2019)
7. Berger, C., Blauth, R., Bolster, C., Burchill, G., DuMouchel, W., Pouliot, F., Richter, R., Rubinoff, A., Shen, D., Timko, M., Walden, D.: Kano's methods for understanding customer-defined quality. Center Qual. Manag. J. **2**(4), 3–36 (1993)
8. Bowersox, D.J., Closs, D.J., Stank, T.P.: Ten mega-trends that will revolutionise supply chain logistics. J. Bus. Logist. **21**(2), 1–16 (2000)
9. Brandenburg, M.: Supply chain efficiency, value creation and the economic crisis – an empirical assessment of the European automotive industry 2002-2010. Int. J. Prod. Econ. **171**(3), 321–335 (2016)
10. Busacca, B., Padula, G.: Understanding the relationship between attribute performance and overall satisfaction: theory, measurement and implications. Mark. Intell. Plan. **23**(6), 543–561 (2005)
11. Chen, W., Li, B., Song, D., Li, Q.: Decentralized supply chain decisions on lead time quote and pricing with a risk-averse supplier. Manag. Decis. Econ. **38**(4), 565–580 (2017)
12. Chen, B., Xie, W., Huang, F., He, J.: Quality competition and coordination in a Vmi supply chain with two risk-averse manufacturers. J. Indust. Manag. Optimiz. **17**(5), 2903–2924 (2021)
13. Chopra, S., Sodhi, M., Lücker, F.: Achieving supply chain efficiency and resilience by using multi-level commons. Decis. Sci. **52**(4), 817–832 (2021)
14. Dmuchowski, R.: Methods of measuring the effectiveness of logistics activities. Contemp. Econ. **15**(3), 309–320 (2021)
15. Duong, B.A.T., Truong, H.Q., Sameiro, M., Sampaio, P., Fernandes, A.C., Vilhena, E., Bui, L. T.C., Yadohisa, H.: Supply chain management and organizational performance: the resonant influence. Int. J. Qual. Reliab. Manag. **36**(7), 1053–1077 (2019)
16. Ellinger, A.E., Natarajarathinam, M., Adams, F.G., Gray, J.B., Hofman, D.O., Marah, K.: Supply chain management competency and firm financial success. J. Bus. Logist. **32**(3), 214–226 (2011)
17. Gajewska, T., Zimon, D., Kaczor, G., Madzík, P.: The impact of the level of customer satisfaction on the quality of e-commerce services. Int. J. Product. Perform. Manag. **69**(4), 666–684 (2020)
18. Gopi, B., Samat, N.: The influence of food trucks' service quality on customer satisfaction and its impact toward customer loyalty. Br. Food J. **122**(10), 3213–3226 (2020)
19. Herzberg, F., Mausner, B., Snyderman, B.: The Motivation to Work, 2nd edn. Routledge, London (1959)
20. Hill, N., Brierley, J., MacDougall, R.: How to Measure Customer Satisfaction, 2nd edn. Routledge, London (2003)
21. IELKA: Online supermarket: High growth rates despite low sales levels (2018). http://www.ielka.gr/?p=2339. Last accessed 2021/8/1
22. IELKA: Consumers' online shopping habits are consolidated (2020). http://www.ielka.gr/?p=2742. Last accessed 2021/8/1
23. IELKA: 1 in 4 Internet users orders food remotely (2021). http://www.ielka.gr/?p=2819. Last accessed 2021/8/1
24. Jeng, S.: Enhancing the creativity of frontline employees: the effects of job complexity and customer orientation. Int. J. Logist. Manag. **29**(3), 387–408 (2018)

25. Joon Choi, B., Sik Kim, H.: The impact of outcome quality, interaction quality, and peer-to-peer quality on customer satisfaction with a hospital service. Manag. Serv. Qual. **23**(3), 188–204 (2013)
26. Kano, N., Takahashi, F.: On MH property of quality. In: Nippon QC Gakka, 9th Annual Presentation Meeting, Abstracts, Japanese Society of Quality Control, 21–26 (1979)
27. Kano, N., Seraku, N., Takahashi, F., Tsuji, S.: Attractive quality and must-be quality. Hinshitsu: J. Jpn. Soc. Qual. Control. **14**(2), 39–48 (1984)
28. Ketokivi, M., Mahoney, J.T.: Transaction cost economics as a theory of supply chain efficiency. Prod. Oper. Manag. **29**(4), 1011–1031 (2020)
29. Kim, I., Kim, C.: Supply chain efficiency measurement to maintain sustainable performance in the automobile industry. Sustainability. **10**(8), 2852 (2018)
30. Matzler, K., Hinterhuber, H.H.: How to make product development projects more successful by integrating Kano's model of customer satisfaction into qualify function deployment. Technovation. **18**(1), 25–38 (1998)
31. Miao, M., Jalees, T., Zaman, S.I., Khan, S., Hanif, N.-U.-A., Javed, M.K.: The influence of e-customer satisfaction, e-trust and perceived value on consumer's repurchase intention in B2C e-commerce segment. Asia Pac. J. Mark. Logist. **34**(10), 2184–2206 (2022)
32. Moagăr-Poladian, S., Dumitrescu, G., Tănase, I.: Retail e-commerce (E-tail)—evolution, characteristics and perspectives in China, the USA and Europe. Glob. Econ. Obs. **5**(1), 167–178 (2017)
33. Nasiopoulos, D.K, Sakas, D.P., Reklitis, P.: The interaction of users' behavior to the development of a distribution and logistics network of the information technology companies. Bus. Intell. Model. **6**, 219–227 (2021)
34. Nasiopoulos, D.K., Sakas, D.P., Vlachos, D.S., Mavrogianni, A.: Modeling of market segmentation for new IT product development. AIP Conf. Proc. **1644**(1), 51–59, (2015)
35. Negi, S.: Supply chain efficiency framework to improve business performance in a competitive era. Manag. Res. Rev. **44**(3), 477–508 (2021)
36. Oliver, R.L.: Whence consumer loyalty? J. Mark. **63**(4), 33–44 (1999)
37. Parasuraman, A., Zeithaml, V., Berry, L.: SERVQUAL: a multiple-item scale for measuring consumer perceptions of service quality. J. Retail. **64**(1), 12–40 (1988)
38. Plessias, A.A., Nasiopoulos, D.K., Giakomidou, D.S.: Modeling of market segmentation in social networks and media. In: Kavoura, A., Sakas, D., Tomaras, P. (eds) Strategic Innovative Marketing. Springer Proceedings in Business and Economics. Springer, Cham. (2017)
39. Plikas, J.H., Nasiopoulos, D.K., Giakomidou, D.S.: The contribution of social media in the management of social relations with customers. Modelling and simulation of the problem. In: Kavoura, A., Sakas, D., Tomaras, P. (eds) Strategic Innovative Marketing. Springer Proceedings in Business and Economics. Springer, Cham, (2017)
40. Ramos, A.L.F.A., Ferreira, J.A.V., da Costa, S.R.: The impact of the handling unit on logistics costs: the case of a Portuguese food retail supply chain. Int. J. Math. Eng. Manag. Sci. **5**(5), 835–850 (2020)
41. Sahu, R., Kumar, A., Kumar, M.: A framework for evaluating the status of supply chain integration. Int. J. Serv. Oper. Manag. **28**(2), 204–221 (2017)
42. Salmiah, S., Rusmardiana, A., Desfiandi, A., Aswin, F., Reni, A., Hasibuan, A.N., Idris, I., Rafiq, M.: Project supply chain management in product development: study of literature and its development. Int. J. Supply Chain Manag. **9**(2), 1023–1028 (2020)
43. Santouridis, I., Veraki, A.: Customer relationship management and customer satisfaction: the mediating role of relationship quality. Total Qual. Manag. Bus. Excell. **28**(9–10), 1122–1133 (2017)
44. Saunders, M., Lewis, P., Thornhill, A.: Research Methods for Business Students, 5th edn. Prentice Hall, Harlow (2009)
45. Shahin, A., Zairi, M.: Kano model: a dynamic approach for classifying and prioritising needs of airline travellers with three case studies on international airlines. Total Qual. Manag. Bus. Excell. **20**(9), 1003–1028 (2009)

46. Sivadas, E., Baker-Prewitt, J.L.: An examination of the relationship between service quality, customer satisfaction, and store loyalty. Int. J. Retail Distrib. Manag. **28**(2), 73–82 (2000)
47. Slack, N., Singh, G., Sharma, S.: The effect of supermarket service quality dimensions and customer satisfaction on customer loyalty and disloyalty dimensions. Int. J. Qual. Serv. Sci. **12**(3), 297–318 (2020)
48. Suchánek, P., Richter, J., Králová, M.: Customer satisfaction, product quality and performance of companies. Rev. Econ. Perspect. **14**(4), 329–344 (2014)
49. Ueltschy, L.C., Laroche, M., Eggert, A., Bindl, U.: Service quality and satisfaction: an international comparison of professional services perceptions. J. Serv. Mark. **21**(6), 410–423 (2007)
50. Witell, L., Löfgren, M.: Classification of quality attributes. Manag. Serv. Qual. **17**(1), 54–73 (2007)
51. Zhang, H.Y., Yang, F.: The impact of customer orientation on new product development performance: the role of top management support. Int. J. Product. Perform. Manag. **67**(3), 590–607 (2018)

Part VI
Financial Perspectives, Tax and Accounting Environment, Industries and Banks During Greek Financial Crisis

Kanellos Toudas

During financial crisis in Greece, various industries were positively or negatively affected. Tourism and Courier industry were among them. It also created conditions for fiscal interventions and tax evasion as well as mergers and acquisitions for banks. This workshop tries to analyze the tax and accounting impact of the financial crisis in Greek entrepreneurship highlighting the future perspectives, through discussion and research papers.

K. Toudas
Agricultural University of Athens, Athens, Greece

Application of VSM as a Process Management Model in a Seaport Container Terminal

Victoria Tsolidou, Dimitrios Folinas, and Thomas Fotiadis

1 Introduction

Ports play a vital role in supporting national and local economic development. The increase in world populations, the industrialization of nations, and a heavier reliance on world trade have accelerated the need for the efficient travel and transfer of goods between overseas ships and domestic trucks, trains, and barges [1]. Today, most ports are competing with one another on a global scale. This has generated the drive to improve port efficiency, lower cargo handling costs, and integrate port services with other components of the global distribution network. Moreover, they are now perceived to be the remaining controllable component in improving the efficiency of, first, logistics, and, second, administrative processes.

A port is a complex business system described by a complex business model of operation [2]. This business model was the subject of many research initiatives. Especially, the topic of logistics processes in ports has been studied from different angles and with varying degrees of detail [3–8]. However, as authors have concluded, there is a limited bibliography that focused on the administrative processes, especially those that involve customers (e.g. shipping companies, freight forwards, 3PL's, etc.). By the administrative business processes of a seaport, we mean the

V. Tsolidou (✉)
International Hellenic University, Department of Organisation Management, Marketing and Tourism, School of Economics and Management, Sindos, Greece

D. Folinas
International Hellenic University, Department of Supply Chain Management, School of Economics and Management, Katerini, Greece

T. Fotiadis
Democritus University of Thrace, Department of Production and Management Engineering, Xanthi, Greece

D. P. Sakas et al. (eds.), *Computational and Strategic Business Modelling*, Springer Proceedings in Business and Economics,
https://doi.org/10.1007/978-3-031-41371-1_35

whole range of operations and related work performed in the container terminal that is required to release cargo. In the administrative business processes:

- Time is critical for customers and time is money. Any delay in one step/task in the administrative procedures creates turmoil in the implementation of the tasks of the entire supply chain.
- Administrative processes are heavily dependent on the customs authorities.
- Most of the processes are paper-based. Even if the digitalization of ports is a necessity, many ports today still use paper for the management of business documents. One of the major challenges is that in order for ports to go fully digital, all the supply chain members must be digitalized too.
- Transportation and waste of motion are critical. This is a major problem in container terminals where the distances between the various offices are rather long.

For the administrative processes, the authors strongly agree with [2], who argue that "the target characteristics of process management in ports and especially in a container terminal include increasing the transparency of management, standardizing business processes, reducing the degree of influence of the "human factor" on the operations of business processes, optimizing key performance indicators of processes, increasing the key factor parameters of success and customer focus of ports, improving the manageability of the port systems as a whole".

These processes in a port are a perfect area for applying lean thinking techniques. Womack, Jones, and Rose (1990) coined the term "lean thinking" as a "way to do more and more with less and less-less human effort, less equipment, less time, and less space while coming closer and closer to providing customers with exactly what they want" [9]. Lean is a methodology that tries to reduce or eliminate defects, costs, waste, and coming closer to this goal leads to coming closer to the customers' requirements [10, 11].

There isn't a single approach to implementing lean, and every organization has its own goals and culture. There are numerous tools of lean but the key part is to implement the right strategy, for example, a tool should be used when it is suitable under the circumstances, and not all tools can be used in the same situation. These are very useful in helping to identify waste and are necessary for analysing and diagnosing processes. When used correctly, these tools show the way to eradicate waste and enable turn the philosophy of lean into action. These differ in characteristics as some identify value, some prepare for flow, others are used for mapping, and some are used to reduce defects [12–14].

This paper is focused on the administration and operation processes in the modern container terminal. It suggests a thorough analysis of lean and value stream mapping methodology under the port perspective in order to understand the hidden waste. The rest of the paper starts with a discussion of lean thinking and the issues relative to sustaining improvement efforts as well as suggestions for additional focuses of improvement for implementing lean enterprise in port operations. It then examines a specific administrative process: the delivery of import full containers, which is a common business process in seaports. It analyses the current business process through VSM related to import procedures at container port terminals and proposes an optimized VSM model.

2 Literature Review

A thorough analysis of lean and value stream mapping methodology is required under the port logistics perspective to understand the waste that is hidden in the administration and operation processes in the modern container terminal.

A process model is a graphical illustration of a process, which documents the existing flow and knowledge. Process models contribute to the process assessment, controlling, compliance, and optimization, since they visually illustrate optimization capabilities such as replicating process steps or bottlenecks [15]. VSM is a lean process model that was first introduced by James Womack et al. in *The Machine That Changed the World* book (1990). A value stream map is a visual customer service-oriented map, where all main process activities are mapped in sequence end to end, along with flows of material and information. VSM map does not add value to the customer in either of the seven distinct categories (Transport, Inventory, Motion, Waiting, Overproduction, Overprocessing and Defects) [16].

In the literature, there are many existing research initiatives regarding the application of lean thinking in organizations throughout various industries. In the port logistics environment, lean methodology has been applied in ports successfully with positive results [17]. A summary of the research papers reviewed during the literature review is detailed in Table 1.

Furthermore, many research initiatives have emphasized a lot of ports under digital transformation initiatives have launched paperless cargo release service processes and employed different types of technologies such as blockchain with digital platforms for data sharing, such as the TradeLens ecosystem [18] or GSBN (Global Shipping Business Network) [19, 20].

3 Creation of the Current State of VSM

In this study, the discharging (unloading) of the containers in the terminal has finally been selected. This process at container terminals can be described with the following process steps: Vessels are arriving at container terminals and are allocated to a berth based on availability and capacity. Upon mooring, vessels are discharged with quay cranes according to the discharge plan. Containers are dispatched to straddle carriers, which transport them to the storage yard where they are temporarily stored. Cargo release is being performed between shipping lines, customs brokers, port, and customs authorities, and containers are loaded to trucks to reach the destination [21].

After the mapping of the key current process steps regarding the delivery of import full containers and the measurement of cycle times, processing times,

Table 1 VSM cases in the port logistics industry

Year	Authors	Title
2021	Sadeghi, Moein; Bagheri, Morteza; Pishvaee, Mir Saman	Evaluation of rail terminals in container ports using simulation: A case study
2021	Saini, Mohan; Efimova, Anastasia; Chromjaková, Felicita	Value stream mapping of ocean import containers: A process cycle efficiency perspective
2020	Kusrini, Elisa; Parmasari, Anggita Noviyanti	Productivity improvement for unit terminal container using lean supply chain management and single minute exchange of dies (SMED): A case study at Semarang port in Indonesia
2020	Sliwczynski, Boguslaw	Value of ICT integration model of e-booking system and intelligent truck traffic management system in the sea port of TEN-T corridor
2020	Fathurrahman, K; Hakim, IM	Improving the loading and unloading process efficiency with lean manufacturing approach using value stream. Mapping in Jakarta container yard
2019	Amrina, Elita; Kamil, Insannul; Rahmad, Dodi	Waste assessment using a lean approach in receiving process of container terminal: A case of Teluk Bayur port
2017	Andreadis, Eleftherios; Garza-Reyes, Jose Arturo; Kumar, Vikas	Towards a conceptual framework for value stream mapping (VSM) implementation: An investigation of managerial factors
2017	Tobias Meudt, Joachim Metternich, Eberhard abele	Value stream mapping 4.0: Holistic examination of value stream and information logistics in production
2017	Wibowo, Agus tri; Handayani, Naniek Utami	Design implementation of lean supply chain management: A case study on loading process of fertilizer at PT Petrokimia Gresik port
2016	Koley, Subhra; Datta, Biplab; Mukherjee, Srabanti	Reducing dwell time related to clearing and forwarding of export and import goods at Kolkata Sea and air ports
2015	Sánchez, Maria Jose González; Buiza, G; Beltran, J; Stanley, MT; Veloqui, M; Cerbán, MM; Turias, I	Logistic management optimization in the ports

waiting, and idle times of each process step from the receipt of the delivery order until the release of the cargo, the below VSMs were created. In order for the container terminal to release the import full container, port administration office should receive a valid delivery order from the respective shipping agent or line. In addition, the container will have to be cleared by relevant authorities. Then, relevant truck or train appointments will be created in order container terminal to issue the relevant permit documents and release cargo at the port gate accordingly.

Table 2 AS-IS value stream map analysis of import documentation cargo release process

Process steps	Description	LT (sec)	PT (sec)	Resources (FTEs)
Permit	Issuing permits for the receipt and delivery of containers from/to trucks/trains and vessels	320	120	2
Invoicing	Issuing and collecting invoices for all services provided	300	160	2
Treasury	Accounting and final settlement of all operations and services provided by the container terminal	240	150	1
Archive	Archive all records and their appropriate registers while sending copies of container receipts electronically to the supervising customs office	14,440	220	2
Total		*15,260*	*650*	*7*
Activity ratio	AR = PT/LT × 100			4.26%
Rolled % C&A[a]	Rolled %C&A = (%C&A)1 × (%C&A)2 × (%C&A)n × 100			33.53%

[a]C&A is complete and accurate

In this study, different attributes were used as indicators to increase the efficiency and improve the process, lead time, process time, and activity ratio (Table 2).

The following figure (Fig. 1) illustrates the current discharging cargo release process using the VSM approach.

We analysed the current process to identify the different types of waste (muda) and other inefficiencies in the process. Sub-processes are disconnected with incomplete and unclear information that causes bottlenecks. The workload is uneven (mura) in waves, which leads to muda. In the current process (muri), it is also present, since operators are utilized for more than 100% capability to complete different tasks. The following table presents the inefficiencies of the examined process indicating the type of waste (Table 3).

4 Creation of the Future State of VSM

Once the AS-IS map was created and analysed, the creation of the proposed future stated was designed and formulated (Table 4).

The following figure (Fig. 2) illustrates the optimized discharging cargo release process using the VSM approach.

Based on the proposed ("to-be") process, a number of benefits can be emerged in the key metrics that are provided in Table 5.

In order to apply the proposed (to-be) actions and improve the efficiency and effectiveness of the subject process, the authors have made some assumptions by leveraging technology:

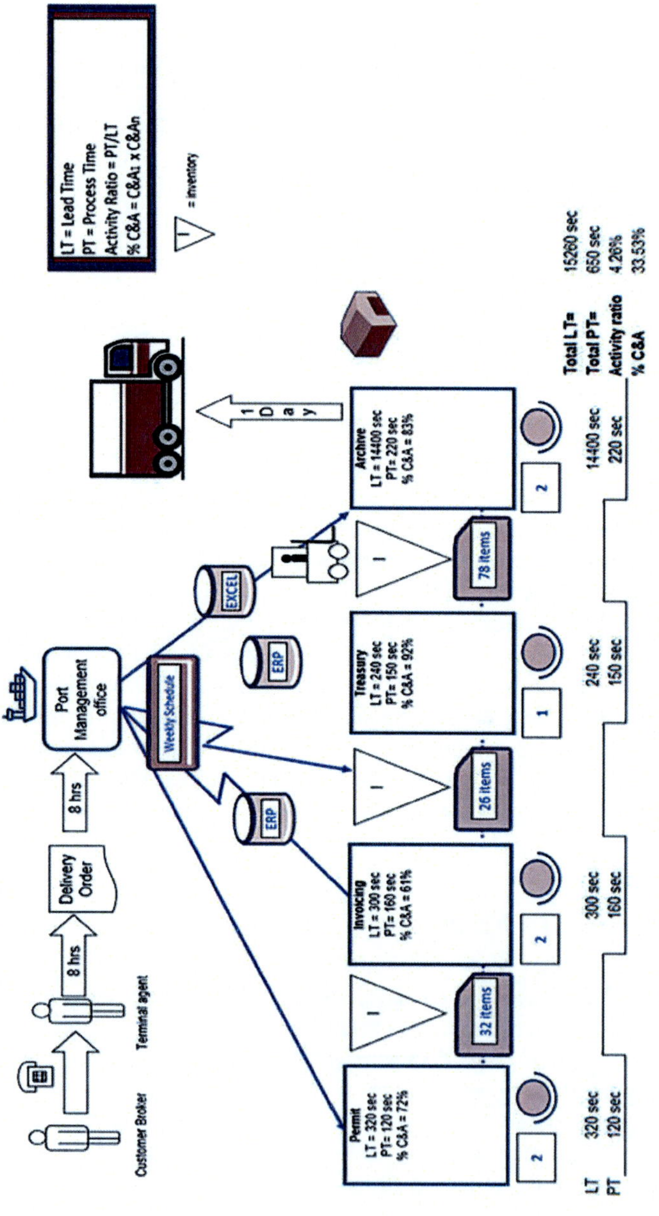

Fig. 1 AS-IS value stream map import documentation cargo release process

Table 3 AS-IS value stream map assessment

Process steps	Type of waste	Inefficiency
Permit	Overproduction	Permits are printed in multiple copies
Permit	Motion	Customer moves from one building to another in order to move to the next step.
Permit	Waiting	Long delays are experienced
Permit	Defects	Customer experience is affected by defects in the permit documents
Permit	Over-processing	Permits information is stored in multiple files and is printed in different type of reports and documents
Invoicing	Overproduction	Invoices are printed manually in multiple documents
Invoicing	Motion	Customer has to move to another building in order to complete the next process step
Invoicing	Defects	Errors in invoicing are producing credit invoices
Invoicing	Waiting	Customers are waiting invoicing to be produced in order to proceed to the next step
Invoicing	Over-processing	Invoicing records are stored in multiple files
Treasury	Waiting	Cashier is waiting for the previous process step in order to proceed with payment
Treasury	Defects	Defects to the payment party are experienced
Treasury	Over-processing	Treasury recordings are data entered into multiple systems
Archive	Overproduction	Archiving of all paper documents of previous process steps is tedious and cumbersome and takes a significant amount of time
Archive	Inventory	Office space and warehouse spaces are occupied for all the archiving records
Archive	Motion	Archiving is performed in another area, so employees have to move from one building to another
Archive	Waiting	Archiving documents are stored temporarily in each process step and then they are gathered all together per vessel

- First, a port community system (PCS) should be implemented to handle electronic communication between the container terminal and the transport operators, the importers, and their brokers. PCS will reduce inefficiencies in the processes and will facilitate a smooth flow of electronic data. Delivery order information will be sent through PCS to the port management system and permit issuance will be done automatically.
- Electronic exchange of custom declarations and custom responses through blockchain technology can be used to speed up the cargo release process. Data and resource exchange with industry chain partners will simplify and streamline operations.
- Throughout the PCS and by deploying e-invoicing, the customers will be able to view and generate their invoices days before vessel berthing, giving customers more options and convenience in terms of invoice generation and payments.

Table 4 AS-IS value stream map analysis of import documentation cargo release process

Process steps	Description	LT (sec)	PT (sec)	Resources (FTEs)
Permit	Issuing permits for the receipt and delivery of containers from/to trucks/trains and vessels	120	60	1
Invoicing	Issuing and collecting invoices for all services provided	120	30	1
Treasury	Accounting and final settlement of all operations and services provided by the container terminal	60	45	1
Archive	Archive all records and their appropriate registers whilst sending copies of container receipts electronically to the supervising customs office	60	30	0
Total		*360*	*165*	*3*
Activity ratio	AR = PT/LT × 100			45.83%
Rolled % C&A	Rolled %C&A = (%C&A)1 × (%C&A)2 × (%C&A)n × 100			78.25%

- Real-time information flow across all stakeholders and multiple systems (ERPs, TOS systems, and port community systems that are not communicating with each other) will contribute to the automated cargo release process and savings are significant.
- Cloud Digital archiving with document management systems (DMS) will save time, money, and even space and will prevent data loss while also ensuring that records are maintained and can be retrieved anytime.

5 Conclusion

Following the recommendation found in the literature, VSM was applied to evaluate the administration import cargo release process of a container terminal. Authors argue that the VSM method can be an effective and efficient tool for several improvements in the identification of the wastes in the operation of the discharging cargo release – a common and key daily task – in a seaport.

A number of to-dos (improvement) practices are proposed. Each of the improvement activities is automated with the use of a technology (such as the blockchain, digital archiving, port community systems technology, etc.) aimed at either eliminating non-value adding steps to reduce the length of the value stream, which subsequently would contribute to the reduction of the total process, lead, and customer query cycle times. A number of proposed actions have been proposed and introduced. After their deployment, the following achievements have been realized:

A critical success factor of the proposed action plan is the digitalization of the administrative processes. According to a [22] World Bank report (2021), "the impact of digitalization would extend far beyond the maritime sector. Given the strategic

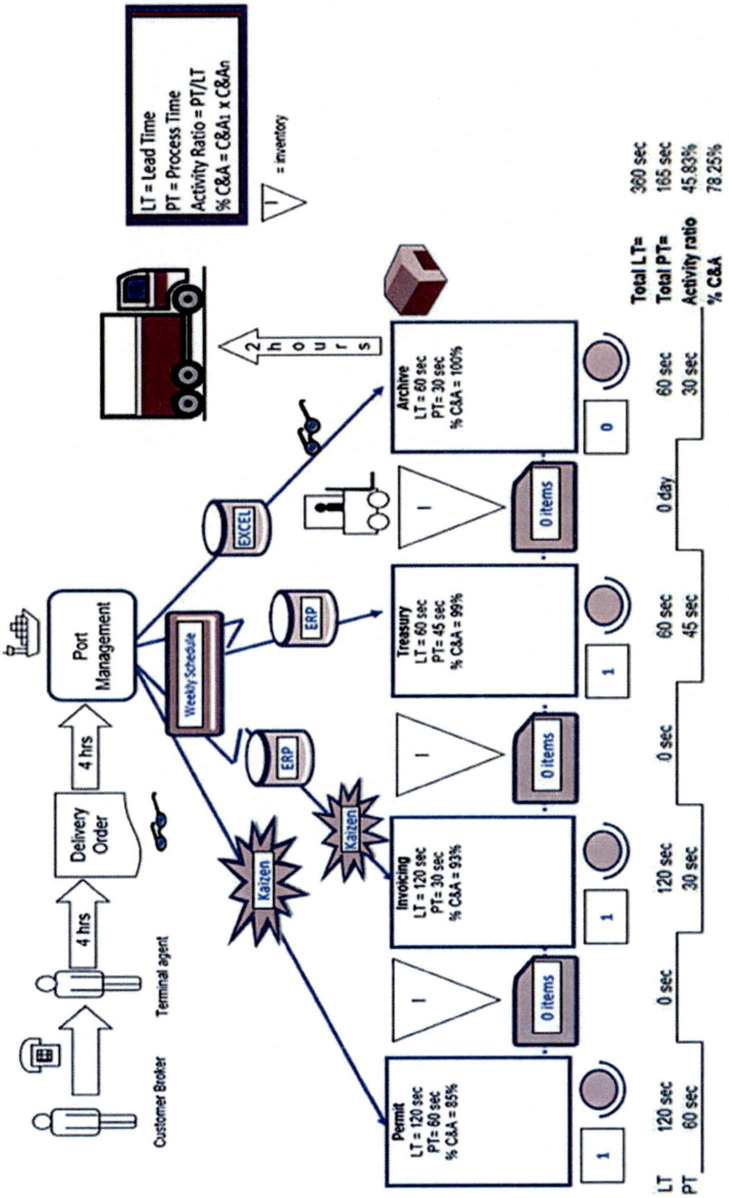

Fig. 2 TO-BE value stream map import documentation cargo release process

Table 5 Basic value stream: current vs. projected future state performance metrics

Metric	Current state	Projected future state	Projected % improvement
Lead time	15,260	360	97.6409% decrease
Process time	650	165	= 74.6154% decrease
Activity ratio	4.26%	45.83%	= 975.822% increase
Rolled % complete and accurate	33.53%	78.25%	= 133.373% increase

role of the shipping industry, digital transformation has the potential to bring wide-ranging economic benefits and contribute to a stronger, more sustainable recovery, particularly in low and middle-income countries". And this is the case in the examined seaport as it is a successful digital transition.

Considering the findings that were observed following the implementation of the pilot project at the examined seaport, the research project provided viable evidence that these lean techniques and principles have a positive impact and that VSM was a workable technique in the administrative processes in the container terminal. Many challenges need to be considered for future study regarding the examined sector. It requires sustained political commitment, adequate regulation, and effective collaboration between the public and private sectors, along with a concerted effort on education and skills training of the seaport staff. Other scientists, however, with similar cases study similar problems, using modelling and simulation [23–25].

References

1. Loyd, N., Jennings, L.C., Siniard, J., et al.: Application of lean enterprise to improve seaport operations. Transp. Res. Rec. **2100**(1), 29–37 (2009). https://doi.org/10.3141/2100-04
2. Bagirov, E., Pantina, T., Borodulina, S.: Seaport development management based on business process modeling. E3S Web Conf. **258**, 02002 (2021). https://doi.org/10.1051/e3sconf/202125802002
3. Elbert, R., Pontow, H., Benlian, A.: The role of inter-organizational information systems in maritime transport chains. Electron. Mark.., Springer; IIM University of St. Gallen. **27**(2), 157–173 (2017). https://doi.org/10.1007/s12525-016-0216-3
4. Besri, Z., Boulmakoul, A.: Framework for organizational structure re-design by assessing logistics' business processes in harbor container terminals. Transp. Res. Procedia. **22** (2016). https://doi.org/10.1016/j.trpro.2017.03.023
5. Heilig, L., Voss, S.: A scientometric analysis of cloud computing literature. Cloud Comput. IEEE Trans. **2**(3), 266–278 (2014). https://doi.org/10.1109/TCC.2014.2321168
6. Marlow, P., Casaca, A.: Measuring lean ports performance. Int. J. Transp. Manag. **1**, 189–202 (2003). https://doi.org/10.1016/j.ijtm.2003.12.002
7. González Sánchez, M.J., Buiza, G., Beltran, J., et al.: Logistic management optimization in the ports. J. Bus. Econ.., ISSN 2155-7950. **6**(11), 1910–1916 (2015). https://doi.org/10.15341/jbe(2155-7950)/11.06.2015/009
8. Amrina, E., Kamil, I., Rahmad, D.: Waste assessment using a lean approach in receiving process of container terminal: a case of Teluk Bayur Port. In: Conference on Innovation in

Technology and Engineering Science, IOP Conference Series: Materials Science and Engineering, vol. 602. Published under licence by IOP Publishing Ltd (2018)
9. Womack, J.P., Jones, D.T., Roose, D.: The Machine That Changed the World. Rawson Associates, New York (1990)
10. Womack, J.P.: Value stream mapping. Manuf. Eng. **136**, 145–156 (2006)
11. Womack, J.P., Jones, D.T.: Lean Thinking: Banish Waste and Create Wealth in Your Corporation. Simon & Schuster, New York (2003)
12. Goodridge, D., Westhorp, G., Rotter, T., et al.: Lean and leadership practices: development of an initial realist program theory. BMC Health Serv. Res. **15**(1), 362 (2015). https://doi.org/10.1186/s12913-015-1030-x
13. Burgess, N., Radnor, Z.: Evaluating lean in healthcare. Int. J. Health Care Qual. Assur. **26**(3), 220–235 (2013). https://doi.org/10.1108/09526861311311418
14. Nicholas, J.M.: Lean Production for Competitive Advantage: A Comprehensive Guide to Lean Methodologies and Management Practices. CRC Press, New York (2011)
15. Saragiotis, P.: Business process management in the port sector: a literature review. Maritime Bus. Rev. **4**(1), 49–70 (2019). https://doi.org/10.1108/MABR-10-2018-0042
16. Fathurrahman, K., Hakim, I.: Improving the loading and unloading process efficiency with lean manufacturing approach using value stream mapping in Jakarta container yard. IOP Conf. Series: Mater. Sci. Eng. **1003**(1), 012042 (2020). https://doi.org/10.1088/1757-899x/1003/1/012042
17. Saini, M., Efimova, A., Chromjaková, F.: Value stream mapping of ocean import containers: a process cycle efficiency perspective. Acta Logis. **8**(4), 393–405 (2021). https://doi.org/10.22306/al.v8i4.245
18. Louw-Reimer, J., Nielsen, J.L.M., Bjørn-Andersen, N., Kouwenhoven, N.: Boosting the effectiveness of containerised supply chains: a case study of TradeLens. In: Lind, M., Michaelides, M., Ward, R., Watson, R.T. (eds.) Maritime Informatics. Progress in IS. Springer, Cham (2021). https://doi.org/10.1007/978-3-030-72785-7_6
19. Balci, G., Surucu-Balci, E.: Blockchain adoption in the maritime supply chain: examining barriers and salient stakeholders in containerized international trade. Transport. Res. Part E: Logist. Transp. Rev. **156**, 102539 (2021). https://doi.org/10.1016/j.tre.2021.102539
20. Caliskan, A., Özkan Özen, Y.D., Ozturkoglu, Y.: Digital transformation of traditional marketing business model in new industry era. J. Enterp. Inf. Manag. **34**(4), 1252–1273 (2021). https://doi.org/10.1108/JEIM-02-2020-0084
21. Carlo, H., Vis, I., Roodbergen, K.: Transport operations in container terminals: literature overview, trends, research directions and classification scheme. Eur. J. Oper. Res. **236**(1), 1–13 (2014). https://doi.org/10.1016/j.ejor.2013.11.023
22. The World Bank: Accelerating digitalization across the maritime supply chain (2021). Available online at https://www.worldbank.org/en/topic/transport/publication/accelerating-digitalization-across-the-maritime-supply-chain. Accessed at 23 Apr 2022
23. Nasiopoulos, D.K., Sakas, D.P., Vlachos, D.S., Mavrogianni, A.: Modeling of market segmentation for new IT product development. AIP Conf. Proc. **1644**, 51–59 (1644)
24. Nasiopoulos, D.K.: Build and approach of logistics companies reengineering. Modelling and simulation of the problem. Bus. Intell. Model. **6**, 189–196 (2021)
25. Papadopoulou, T., Nasiopoulos, D.K., Vlachos, D.S.: Merchandizing IT products via social networks. Modeling and simulation of the procedures. In: Kavoura, A., Sakas, D., Tomaras, P. (eds) Strategic innovative marketing. Springer Proceedings in Business and Economics. Springer, Cham (2017). https://doi.org/10.1007/978-3-319-33865-1_61

Warehouse Stocks and Supply Chain Performance: The Case of Food Businesses

Georgios Tsimpoukas and Panagiotis Reklitis

1 Introduction

1.1 Warehouse

Warehouse is a complex concept and several definitions have been given from time to time to approach this concept. In order to fully understand its significance, some of them are formulated below. According to Sifnioti [20], the warehouse can be defined as "the space of the enterprise through which the products acquired or sold by the enterprise pass and are temporarily stored. The warehouse receives the work of receipt, storage and shipment, work necessary to reach the product from production to consumption, in the right condition, at the right cost" to Sifnioti [20].

We could also define the concept of a warehouse as an organized space that aims at adequate storage of goods. Alternatively in the term "warehouse", we could use the term "distribution center." Warehouses are an integral part of the supply chain [11]. The efficiency of any supply chain depends on the operation of parts of its network, such as warehouses. Warehouses are a key part of modern supply chains and play a major role in the success or failure of businesses today [10]. Warehouse is any space in which you carry out the process of receiving goods, as well as any space where raw materials and semifinished or finished products are stored or maintained in a certain order, which are called stocks. The warehouse is one of the most important functional elements of a company, its purpose is the proper management and preservation of the produced or purchased products as well as their disposal time [20].

G. Tsimpoukas · P. Reklitis (✉)
Department of Agribusiness and Supply Chain Management, Agricultural University of Athens, Thebes, Greece
e-mail: preklitis@aua.gr

D. P. Sakas et al. (eds.), *Computational and Strategic Business Modelling*, Springer Proceedings in Business and Economics,
https://doi.org/10.1007/978-3-031-41371-1_36

1.2 Inventories

The term stocks refers to any inert, financial instrument or resource (product or material) that the company acquires and stores it to use or to resell it at some future time [5].

The accumulation of assets and wealth and the preservation of stocks are a hallmark of human civilization. However, until the end of the first 50 years of the twentieth century, there was a passive attitude of business management towards stocks. Since then, this situation has been reversed, along with the changing business environment and rapid technological progress. Inventory management is now an important responsibility for the management of a productive system. Stocks are mostly created intentionally to smooth out time gaps between supply and demand. In other words, the necessity of the existence of the stock lies in the uncertainty regarding the supply and demand of the good. Inventory control is a scientifically based technique that aims to monitor the quantity of goods currently stored and make relevant decisions, such as the time point and quantity of the order placed with suppliers and which replenishes the stocks upon receipt of the relevant products by the company.

The problem of inventory management is very important for all companies as inventories usually freeze a large percentage of their capital and have significant maintenance costs. There are categories of businesses such as supermarkets, where their stocks cover about 50% of their assets. Therefore, they are directly related to the liquidity of the company's capital, as the reduction of the stock leads to the release of current capital and consequently reduces its lending needs. For example, Goodyear announced in 2009 that its inventory reduction program had led to savings of $ 1 billion dollars [12].

1.3 Supply Chain Performance

Nowadays, for the administrative sector of a company to function effectively, it must not only consider the measurement of the performance of each company but also carry out an evaluation of its other activities [14].

With the development of technology and the modernization of their systems, the largest percentage of companies now have the ability to quickly monitor and evaluate the situation and performance that prevails within them. Key performance indicators (KPIs) are also important business tools [19].

Performance metrics are a very useful tool for organizational change that is used by both employees and business executives. Employees use performance metrics as a means of setting priorities in order to accomplish the goals set, while executives use performance metrics to set personalized strategic goals for each employee [8].

However, every company must be very careful with the use of performance measurements as incorrect measurements can lead to undesirable consequences, such as reduced productivity and customer service levels, undermining employee morale and disorganizing processes. Performance metrics are a key component of performance management, which should be aligned with the business strategy. Through performance management, information technology is used in order to monitor the execution of the business strategy, thus achieving the defined goals of each company. Modern management uses some key performance indicators (KPIs) in the strategy formulation stage, in order to measure effectiveness not only from a qualitative but also from a quantitative point of view. In this case, the two categories of indicators facilitate the evaluation of performance [6].

2 Materials and Methods

The questionnaires were sent to dairy cooperatives (who filled in the respective questionnaire on their behalf) and then distributed the remaining questionnaires to dairy companies which are members of the cooperatives, in Thessaly (Greece).

Fifty (50) completed questionnaires collected by both the owners of the companies and their managers.

There was difficulty in getting face-to-face interviews from business owners and managers due to covid-19 (questionnaires were completed between September and December 2021). These questionnaires were sent via email and telephone instructions were given where necessary. The completed questionnaires were received via email emails.

The questionnaire used for the present survey was a structured questionnaire with closed-ended questions. The first category of questions refers to demographics and contains questions concerning gender, age, level of education, total length of service, position, and occupation. The second category of questions refers to the repository and contains 15 questions that include factors that shape its operation. The third category of questions refers to inventory management in the company and examines the factors that shape inventory management. The fourth category of questions refers to the performance of the supply chain in relation to the time required along the entire length of the supply chain, the cost along the supply chain, the improvement of the quality of the products/services provided, and finally the increase of the quantity shipments and receipts.

3 Analysis and Results

We conclude at the three hypotheses listed below:

There is a relationship between warehouse and supply chain performance.

Warehouse affect supply chain performance. Receiving, storage, and shipping occur at the warehouse to transport the product from production to consumption in the right condition and at the right cost. The warehouse receives, sorts, puts, maintains, stores, picks, and loads-ships. Warehouse equipment, warehouse systems, and informational systems are used for these tasks. All of the above must contribute to a competitive company operation and excellent supply chain performance in terms of service speed, product quality, and customer happiness.

There is a relationship between inventory and supply chain performance.

Inventory management's main aims are to ensure seamless item movement, accelerate product delivery from receipt, reduce storage, reduce material costs, lower per-item inventory, require less investment, and increase inventory turnover. When a firm manages its inventory well, it may provide appropriate storage space, produce and sell quality items, adapt to market demand, build customer confidence, reduce cancelled orders, and incur minimum losses. Bad inventory control can lead to significant financial commitments, negative cash flow, worse return on investment, and inability to expand due to lack of funds. Having a large inventory also increases storage costs, as more room is needed. The aforementioned factors influence (positively or negatively) the supply chain's ability to acquire a competitive advantage, given that this advantage depends on the supply chain's financial performance, speed of service, sustained product quality, and, ultimately, customer attitudes towards the firm's products.

There is a relationship between warehouse and inventory and supply chain performance.

Dependent variable:

Using the statistical program, the Cronbach coefficient was measured, which received the following values for the dependent variable:

Supply chain performance: Cronbach's alpha = 0.194

Independent variables:

The Cronbach coefficient, which received the following values for the independent variables, was also measured:

Warehouse: Cronbach's alpha = 0.707 (removes "available resources" and "warehouse accessibility" queries based on statistical program-scale if item deleted)

Inventory: Cronbach's alpha = 0.825

Correlation Factor

Then, the Pearson correlation coefficient was measured with the statistical program, with the following results (Table 1):

As shown in the table above:

Between warehouse-inventory, the correlation coefficient is $r = 0.599$ for $p < 0.001$

Between warehouse-performance, the correlation coefficient is $r = 0.402$ for $p < 0.005$

Between inventory-performance, the correlation coefficient is $r = 0.614$ for $p < 0.001$

Multiple Linear Regression

In this study, multiple linear regression was used using the statistical program to test the hypotheses:

Table 1 Correlations

		Warehouse	Inventory	Performance
Warehouse	Pearson correlation	1	.599[a]	.402[a]
	Sig. (2-tailed)		.000	.004
	N	50	50	50
Inventory	Pearson correlation	.599[a]	1	.614[a]
	Sig. (2-tailed)	.000		.000
	N	50	50	50
Performance	Pearson correlation	.402[a]	.614[a]	1
	Sig. (2-tailed)	.004	.000	
	N	50	50	50

[a]Correlation is significant at the 0.01 level (2-tailed)

Table 2 Multiple linear regression

					Change statistics				
Model	R	R square	Adjusted R square	Std. error of the estimate	R square change	F change	df1	df2	Sig. F change
1	.402[a]	.162	.144	.26.433	.162	9247	1	48	.004

[a]Predictors: (constant), warehouse

Table 3 Multiple linear regression

					Change statistics				
Model	R	R Square	Adjusted R square	Std. error of the estimate	R Square change	F change	df1	df2	Sig. F change
1	.614[a]	.377	.365	.22.777	.377	29.105	1	48	.000

[a]Predictors: (constant) inventory

A. *Warehouse and Supply Chain Performance Relationship* (Table 2)

Hypothesis: The variable "warehouse" affects the "performance of the supply chain."

The results of multiple linear regression were statistically significant F (1.48) = 9.247 for $p > 0.001$. The coefficient adjusted R square= 0.144, which means that 14.4% of the variance of the dependent variable is interpreted by the model with the participation of the independent variable; therefore, it is not confirmed in the hypothesis that has been made.

B. *Inventory and Supply Chain Relationship* (Table 3)

Hypothesis: The "inventory" variable affects "supply chain performance."

The results of multiple linear regression were statistically significant F (1.48) = 29.105 for $p < 0.001$. The coefficient adjusted R square = 0.365, which means that 36.5% of the variance of the dependent variable is interpreted by the model with the participation of the above independent variable, thus confirming to some extent the hypothesis that has been made (Table 4).

Table 4 ANOVA[a] [1] hypothesis 2

Model		Sum of squares	Df	Mean square	F	Sig.
1	Regression	1510	1	1510	29.105	.000[b]
	Residual	2490	48	.052		
	Total	4000	49			

[a]Dependent variable: performance
[b]Predictors: (constant) inventory
[1]The number of the model

As can be seen from the table above, the observed level of statistical significance is Sign = 0.000 < 0.05, so the conclusion to which one is led is that there is a linear relationship between the independent and the dependent variable. Thus, the assumption of linearity seems to be satisfied (Table 5).

We also observe from the coefficients table that the constant coefficient has a value of 2037, while the variable "stocks" has a fairly high positive value, StbBeta = 0.614 which is statistically significant since it has $p < 0.001$. Stocks are therefore positively and statistically significantly linked to supply chain performance.

C. *Warehouse, Inventory, and Supply Chain Performance Relationship* (Table 6)

The results of multiple linear regression were statistically significant F (2.47) = 14.35 for $p < 0.001$. The coefficient adjusted R square = 0.353, which means that 35.3% of the variance of the dependent variable is interpreted by the model with the participation of the above independent variables, thus confirming to some extent the hypothesis that has been made (Table 7).

As can be seen from the table above, the observed level of statistical significance is Sign = 0.000 < 0.05, so the conclusion to which one is led is that there is a linear relationship between the independent and the dependent variable. In consumptions, linearity seems to be satisfied (Table 8).

Hypothesis: The variables "warehouse" and "stocks" affect "supply chain performance."

We observe from the coefficients table that the constant coefficient has a value of 1932, while the variable "stocks" has a fairly high positive value, StbBeta = 0.583, which is statistically significant, since it has $p < 0.001$. Stocks are therefore positively and statistically significantly associated with supply chain performance. On the contrary, the warehouse that has a factor of StbBeta = 0.053 is not positively and statistically significantly related to the supply chain performance (StbBeta = 0.053) because $p > 0.001$

4 Discussion

The answers obtained from the research and processed by the relevant statistical tool show that from the case studies, (a) the warehouse does not affect the supply chain performance individually and (b) the stocks affect the supply chain performance, while in combination the warehouse and the stocks further affect supply chain performance.

Table 5 Correlation coefficient

Model		Unstandardized coefficients		Standardized coefficients			Correlations			Collinearity statistics	
		B	Std. error	Beta	t	Sig.	Zero-order	Partial	Part	Tolerance	VIF
1	(constant)	2037	.373		5463	.000					
	inventory	.531	.098	.614	5.395	.000	.614	.614	.614	1.000	1.000

[a]Dependent variable: performance

Table 6 Summary of the model 2, hypothesis 3

Model	R	R Square	Adjusted R square	Std. error of the estimate	Change statistics				
					R Square change	F change	df1	df2	Sig. F change
1	.616[a]	.379	.353	.22.984	.379	14.359	2	47	.000

[a]Predictors: (constant) inventory, warehouse

Table 7 ANOVA[a] [1]

Model		Sum of squares	df	Mean square	F	Sig.
1	Regression	1517	2	.759	14.359	.000[b]
	Residual	2483	47	.053		
	Total	4000	49			

[a]Dependent variable: performance
[b]Predictors: (constant), inventory, warehouse
[1]The number of the model

According to results from the Performance Improvement of Inventory Management System Processes by an Automated Warehouse Management System [2], which investigates the impact of a warehouse management system on the performance of the supply chain that provides less resource use, it is indicated that it is more efficient and reliable inventory management system. The supply chain processes performed in the warehouse were examined before adapting a software that can handle the necessary transactions. The software was tested to improve workflow and provide timely and efficient handling. The data was collected from the warehouse of a leading telecommunications service provider in Jordan. In addition, the layout of the facilities was studied, and we introduced a production station inside the warehouse, which resulted in better optimization/utilization of the warehouse space. The production station consists of three stages: grouping, labelling, and repackaging. The system handles three phases of the product life cycle: receiving, processing, and distributing SIM and prepaid scrapers. Each phase of the product life cycle was discussed in detail, and process gaps were identified. It also highlights the gap between theory and practice, in order to motivate researchers to develop and adapt to mitigate supply chain disruptions.

Evaluating Inventory Management Performance Using a Turnover Curve [3] also examines tools for better inventory management, which of course improve supply chain performance. A generalized formula (turnover curve) developed in previous research that mimics practical inventory control (shows the relationship between inventory and turnover) is used to control the performance of inventory regulation across multiple storage locations. The same turnover curve is used to estimate the impact of changing inventory control procedures or to set new targets for inventory levels. It is a simple but powerful tool for evaluating the performance of inventory management that can be developed from readily available corporate data. But also the work Logistics and Agri-Food: Digitization to Increase Competitive Advantage and Sustainability.

Table 8 Correlation coefficients

Model		Unstandardized coefficients		Standardized coefficients	t	Sig.	Correlations			Collinearity statistics	
		B	Std. error	Beta			Zero-order	Partial	Part	Tolerance	VIF
1	(constant)	1932	.471		4101	.000					
	Warehouse	.057	.153	.053	.370	.713	.402	.054	.043	.642	1559
	Inventory	.504	.124	.583	4061	.000	.614	.510	.467	.642	1559

[a]Dependent variable: performance

Literature Review and the Case of Italy [18] examines the current challenges facing the supply chain with a focus on the agri-food sector. The paper addresses two key research questions. First, how and with what tools can digitization help improve supply chain management and logistics sustainability? Secondly, what are the main management and strategic implications and implications for the agri-food sector in terms of efficiency, effectiveness, cost reduction, and supply chain optimization? Finally, the paper presents Italy as a case study, selected both for its particular internal differences in logistics infrastructure and business management between northern and southern regions (which could at least be partially overcome by the use of new technologies and frameworks), as well as the importance of the agri-food sector for the domestic economy (representing about 25% of the country's GDP), in which digitization is expected to have positive effects in terms of value creation and sustainability.

This approach aligns and synthesizes significant factors linked to supply chain performance measurement, according to " A Review of Performance Metrics for Supply Chain Management" [9]. The results of this study can assist managers and practitioners evaluate and measure their supply chains effectively and, more importantly, efficiently. Other scientists with similar cases are studying similar problems, using modelling and simulation [4, 15–17].

Furthermore, in "Supply Chain Performance Within the Agri Food Sector" [7], the article discusses the components of the agri-food supply chain, the main key performance indicators used to measure its performance, and the difference between a traditional supply chain and a sustainable supply chain, by analyzing the waste management component. A quantitative study has been designed in order to obtain organizational learning on key performance indicators.

Additionally, according to the findings of "Measuring Supply Chain Performance in the"Agri-Food Sector [1], some QAS criteria also seem to have a favorable influence on the performance of some supply chain members while having a negative impact on the performance of many others. Overall, the results show that all specified QAS standards are judged to have a positive influence on supply chain performance, while the total impact is very minor.

The purpose of this study, Role of Warehouse Characteristics in Supply Chain Warehouse Efficiency in Indonesia [13], is to investigate the role of warehouse attributes in supply chain warehouse efficiency. The study's results indicate that warehouse attributes have a positive impact on supply chain warehouse efficiency. Warehouse efficiency among Indonesian supply chain enterprises is increased through proper warehouse design and effective operations. Furthermore, MIS is discovered to be one of the mediating variables between warehouse variables and supply chain warehouse efficiency.

References

1. Aramyan, L.H.: Measuring supply chain performance in the agri-food sector. Thesis Wageningen University, p 164 (2007)
2. Atieh, A.M., Kaylani, H., YousefAl-abdallat, A.Q., Ghoul, L., Jaradat, L., Hdairis, I.: Performance improvement of inventory management system processes by an automated warehouse management system. Procedia CIRP. **41**(2016), 568–572 (2016)
3. Ballou, R.H.: Evaluating inventory management performance using a turnover curve. Int. J. Phys. Distrib. Logist. Manag. **30**(1), 72–85 (2000)
4. Chaldezos, C., Sakas, D., Dimitrios, N.K., Giakomidou, D.S.: Calculating handling of company resources for establishing an effective project team. In: Kavoura, A., Sakas, D., Tomaras, P. (eds) Strategic Innovative Marketing. Springer Proceedings in Business and Economics. Springer, Cham (2017). https://doi.org/10.1007/978-3-319-56288-9_49
5. Dervitsiotis, N.K.: Inventory Systems. Afon Kyriakidi Publications, Thessaloniki (in Greek) (1985)
6. Ding H., Benyoucef L., Xie X.: Key performance indicators "A modeling and simulation framework for supply chain optimization, Chapter 16, pp. 219–232 (2005)
7. Dinu, M.D.: Supply chain performance within agri food sector. Ekonomika Poljoprivrede. **63**(3), 919–928 (2016)
8. Eckerson, W.W.: Performance Dashboards: Measuring and Managing Your Buisness, 6th edn. Pearson Education, Inc (2006)
9. Elrod, C.: A review of performance metrics for supply chain management. Eng. Manag. J. **25**, 39–50 (2015)
10. Frazelle, E.: World-Class Warehousing and Material Handling. McGraw-Hill, NewYork (2002)
11. Gu, J., Goetschalckx, M., McGinnis, L.: Research on warehouse operation: a comprehensive review. Eur. J. Oper. Res. **177**(1), 1–21 (2007)
12. Iakovou, E.: Logistics for Greece of Tomorrow. Corfu Publications, Athens (in Greek) (2014)
13. Jermsittiparsert, K., Sutduean, J., Sriyakul, T.: Role of warehouse attributes in supply chain warehouse efficiency in Indonesia. Int. J. Innov. Creat. Change, 1–17 (2019)
14. Laios, L.: Supply Management, 1st edn. Humanantec Publications, Athens (in Greek) (2016)
15. Nasiopoulos, D.K.: The role of digital marketing in the development of a distribution and logistics network of information technology companies. Bus. Intell. Model.. **6**, 267–276
16. Nasiopoulos, D.K.: Build and approach of logistics companies reengineering. Modelling and simulation of the problem. Bus. Intell. Model. **6**, 189–196
17. Plessias, A.A., Nasiopoulos, D.K., Giakomidou, D.S.: Modeling of market segmentation in social networks and media. In: Strategic Innovative Marketing, pp. 523–529
18. Romandino, M., Zanin, A.: Logistics and agri-food: digitization to increase competitive advantage and sustainability. Literature review and the case of Italy. Sustainability. **14**(2), 787 (2022)
19. Schuh, C., Raudabaugh, L.J., Kromoser, R., Strohmer, F.M., Triplat, A., Pearce, J.: The Purchasing Chessboard 64 Methods to Reduce Costs and Increase Value with Supplies, 3rd edn. Springer, New York (2012/2017)
20. Sifniotis, K.: Logistics Management Theory and Practice. Papazisis Publications (in Greek) (1997)

Mergers and Acquisitions Between Systemic Banks in Greece and Their Impact on Concentration and Control

Apostolos Christopoulos, Ioannis Katsampoxakis, Ioannis Thanos, and Kanellos Toudas

1 Introduction

The financial crisis, which originated in the USA (2007) and spread to other countries, also affected Greece. In 2010, Greece sought bailout funding and signed an M0U with the so-called Troika, that is, the EU the ECB and the IMF. However, although the Greek crisis started as a sovereign debt crisis, it soon turned into a banking crisis, as banks were the main holders of Greek bonds.

On the other hand, liberalization allowed banks to undertake excessive risks, which played a significant role in the global financial crisis of 2007 and, to some extent, in the Greek financial crisis of 2010 [1, 2]. As a result, regulators started to reconsider their decisions emphasizing on the stability of the banking system. This

A. Christopoulos
Department of Business Administration, University of the Aegean, Chios, Greece

Hellenic Open University, Patra, Greece
e-mail: axristop@aegean.gr

I. Katsampoxakis
Department of Statistics and Actuarial—Financial Mathematics, University of the Aegean, Samos, Greece
e-mail: ikatsamp@aegean.gr

I. Thanos
School of Applied Mathematical and Physical Sciences, National Technical University of Athens, Athens, Greece

K. Toudas (✉)
Department of Agribusiness and Supply Chain Management, Agricultural University of Athens, Athens, Greece
e-mail: kstoudas@aua.gr

D. P. Sakas et al. (eds.), *Computational and Strategic Business Modelling*, Springer Proceedings in Business and Economics,
https://doi.org/10.1007/978-3-031-41371-1_37

new environment led researchers to turn their interests to topics such as (i) Concentration-Competition, (ii) Concentration-Stability and (iii) Competition-Stability.

2 Literature Review

The existing literature does not lead to clear results in terms of the relationships among market concentration, market power and the financial stability of the banking system. In fact, even empirical studies fail to make safe conclusions as to what is ultimately valid. This ambiguity is evident in the available literature as we demonstrate in the following section.

2.1 Concentration-Competition

The existing literature mainly researches the relationship between competition and concentration. There are two prevalent views on the relationship between these two components.

The first view is based on the Structure-Conduct-Performance relationship [3, 4], according to which increasing concentration will have a positive impact on the increasing market power of large banks (and a negative impact on competition in the banking industry); taking advantage of the increase in their market share and/or the eradication of their competitors, banks can more easily impose higher prices and record (abnormally) higher profits. On the other hand, the second relation is based on the effective structure Hypothesis [5, 6], which suggests that most efficient banks can increase their profitability and size, simultaneously increasing their concentration. Therefore, an increase in this concentration does not imply market power, which means that there is, not necessarily, a causal relationship between concentration and competition in the banking sector.

However, other studies have concluded that there is no significant relationship between concentration and competition [7–12].

2.2 Concentration-Stability

The first view emerging in the existing literature is that increasing concentration under certain conditions has a positive impact on the sector's stability. Such results can be found in cases of M&As occurring in the context of restructuring required of the sector (e.g., the acquisition of the 'Agricultural Bank of Greece' by 'Piraeus Bank' and the acquisition of the 'Emporiki Bank of Greece' by 'Alpha Bank'). The increasing degree of concentration led to an improved stability within the sector [2, 13].

This view holds that stability in the industry improves when the degree of concentration within the industry increases, whether this is due to new M&As or comes because of an increase in the market share of the bigger banks.

Other studies conclude that it is easier to monitor a system with a few big banks rather than one where many small banks operate, and therefore, more detailed and systematic monitoring is required.

The proponents of this view support that in the most concentrated markets, individual banks can charge higher interest rates for loans, which increases the likelihood of moral hazard as borrowers make risky decisions, and, as a result, banking portfolios become riskier, too [14].

In a more recent study, Shim (2019) [15] shows that high concentration leads to a more stable financial environment compared with less concentrated markets, whereas Azmi et al. (2019) [16] also argue that concentration is beneficial for banking stability, focusing on dual banking economies.

2.3 Competition-Stability

Perhaps the more intensely studied relationship is the relationship between Competition and Stability. It all began in the 1990s, when there was a tendency to reduce restrictions on the banking sector to obtain the benefits an increased level of competition might offer. These tactics following deregulation are also the main reason leading to the 2007 crisis [17, 18]. Thus, the study of the Competition-Stability relationship was considered by experts as a major issue.

As in the relationships discussed above, opinions in the literature differ. The two main relationships under discussion are Competition-Stability and Competition-Fragility.

Big banks were considered 'too big to fail' having a 'safety net' provided by the state can engage in activities with greater risk. In more competitive markets, interest rates are lower and the 'too big to fail' and 'safety net' parameters are of lesser importance. Thus, the moral hazard problem is mitigated, and we are led to stability [14].

In addition to the aforementioned studies that focused on the impact of concepts like concentration and competition on the stability of the financial system, researchers also studied the aftermath of the unconventional monetary policies that central banks applied as remediation measures regarding the recent economic crises [19–22]. The effect of the unconventional monetary policies by Central Banks in general is positive for the real economy [23, 24].

According to Acharya et al. (2019) [25], the ECB's Outright Monetary Transactions program indirectly recapitalized European banks through its positive impact on periphery sovereign bonds. Kenourgios, D., Christopoulos, A., and Dimitriou, D. (2013) [26] examined the returns on stocks, bonds, commodities, shipping, foreign exchange and real estate and found evidence of a correlated-information channel as a contagion mechanism between markets within different countries.

Yu (2017) [27] and Keddada and Schalckb (2020) [28] found that the correlation between sovereign and bank CDS spreads before the crisis remained small but had increased significantly until the end of the sample. Similarly, attempted to determine the extent to which European banks were vulnerable to sovereign credit risk from 2010 to 2013.

Thus, we conclude that for each case concerning a country, industry, system, or whether carried out for a different time-period, new research should be conducted to draw conclusions that are most likely to be valid for each case.

3 Data and Methodology

3.1 Data

The data in this research came from published consolidated financial statements, and specifically the balance sheets and profit and loss statements of the five largest banking groups in Greece for the period 2008–2018. Although only these five banks were included, our sample represents 97% of the industry's total assets. Considering country-level data, these were collected from the databases of the World Bank and the European Central Bank.

The Herfindahl-Hirschman Index as a measure of concentration

The Herfindahl-Hirschman index (HHI) is one of the most widely used indicators in the theoretical literature. It can often be used as a benchmark to assess other concentration indicators. This indicator is the sum of the squares of bank shares as shown in the following formula:

$$\mathrm{HHI} = \sum_{i=1}^{n} S_i^2$$

where S_i is the market share of the bank, and i and n are the number of enterprises in the sector. This indicator can take values from $1/n$ ($\mathrm{HHI} = n(1/n)^2 = 1/n$) where all banks are of equal size, and we have an indication of full competition up to10000 ($\mathrm{HHI} = 100^2$), when a bank has 100% of the shares and so we have an indication of a monopoly.

According to ECB 's instructions, market shares are calculated using the total assets.

The Lerner index as a measure of competition

This study uses the Lerner index, which has been commonly used in banking research, as a measure of competition (or market power). The Lerner index captures the capacity of price power by computing the disparity between price and marginal cost as a percentage of the price and ranges between 0 and 1. In case of perfect competition and monopoly, the index equals 0 and 1, respectively. The time-variant Lerner index at the bank-level is calculated as follows:

$$L = \frac{p_{it} - \text{mc}_{it}}{p_{it}}$$

where the p_{it} is the price of total assets proxied by the ratio of total revenues (interest and non-interest income) to total assets for bank i at time t. mc_{it} is the marginal cost for bank s at time t.

The marginal costs are obtained by differentiating a translog cost function with respect to one output. We use a stochastic frontier model to estimate a translog cost function with one output (total assets) and three inputs (labour, funding and physical capital). The cost function for a given bank s at time t can be specified as follows:

$$\begin{aligned} \ln\text{TC}_{it} &= \alpha_0 + \alpha_Q \ln Q_{it} + 0.5\alpha_{QQ}(\ln Q_{it})^2 + \sum\nolimits_{k=1}^{3} \alpha_k \ln W_{k,it} \\ &+ \sum\nolimits_{k=1}^{3} \alpha_{Qk} \ln Q_{it} \ln W_{k,it} + 0.5 \sum\nolimits_{j=1}^{3} \sum\nolimits_{k=1}^{3} \alpha_{jk} \ln W_{j,it} \ln W_{k,it} \ + \varepsilon_{it} \qquad (1) \end{aligned}$$

$$\varepsilon_{it} = v_{it} + u_{it}$$

where C is the total costs, Q is the output (total assets) and W is a vector of input prices (price of labour, price of funds and price of physical capital), v represents standard statistical noise and u captures inefficiency. Following Turk-Ariss (2010) [29], the total costs and prices of funds and labour are scaled by the price of physical capital to correct for heteroscedasticity and scale biases.

$$\begin{aligned} \ln(\text{TC}_{it}/W_{3,it}) &= \alpha_0 + \alpha_Q \ln Q_{it} + 0.5\alpha_{QQ}(\ln Q_{it})^2 + \sum\nolimits_{k=1}^{2} \alpha_k \ln(W_{k,it}/W_{3,it}) \\ &+ \sum\nolimits_{k=1}^{2} \alpha_{Qk} \ln Q_{it} \ln(W_{k,it}/W_{3,it}) \\ &+ 0.5 \sum\nolimits_{j=1}^{2} \sum\nolimits_{k=1}^{2} \alpha_{jk} \ln(W_{j,it}/W_{3,it}) \ \ln(W_{k,it}/W_{3,it}) + \varepsilon_{it} \end{aligned}$$

To obtain the marginal cost, Eq. (1) is differentiated with respect to Q:

$$\frac{\partial \ln\text{TC}_{it}}{\partial \ln Q_{it}} = \left(\frac{\text{TC}_{it}}{Q_{it}}\right)\left(\alpha_Q + \alpha_{QQ} \ln Q_{it} + \sum\nolimits_{k=1}^{2} \alpha_{Qk} \ln(W_{k,it}/W_{3,it})\right) = \text{MC}_{it}$$

The Bank Z Score as a measure of bank stability

The Z-score is a widely used measure of bank stability in the related literature and can be interpreted as the number of standard deviations by which returns would have to fall from the mean to deplete the equity capital. Hence, it determines how many standard deviations in return on assets a bank is from insolvency. The score is computed as follows:

$$\text{Bank } Z_{it} = \frac{\text{ROA}_{it} + \left(\frac{E}{\text{TA}}\right)_{it}}{\sigma(\text{ROA})_{it}}$$

where ROA is the return on assets, *E*/TA represents the equity to total assets ratio and σ(ROA) denotes the standard deviation of return on assets. A higher Z-score implies a lower probability of insolvency (failure), providing a more direct measure of soundness compared to other measures of risk.

3.2 Control Variables

In our econometric models apart from the basic variables, we use a list of bank-specific and macroeconomic control variables.

In this section, we present each one of them with its definition and divided according to the relationship studied.

Table 3 reports the descriptive statistics of the variables used in the regression. Taking a glance at the basic variables (Lerner, HHI, LnBank Z score), we can make a first assumption of the results we might find for the three relationships of this study. The mean value, as well as the median of the Lerner index, is relatively low and quite close to zero, while, on the other hand, the ones of the HHI index and LnBank Z score have higher values. So, this could be a first sign that there might be a negative relationship between Market Power and Concentration, as well as Stability, while there might be a positive between Concentration and Stability.

3.3 Methodology

Before the econometric study, using the HHI index and its yearly change ΔHHI, an analysis on the level of Concentration of the Greek Banking industry is carried out and subsequently we estimate, using the same indices, the change in the sector's Concentration the possible systemic banks merger scenarios might cause.

Then we proceed to the basic econometric study. Our first model has the form:

$$\begin{aligned}\text{COMP}_{it} = {} & a_0 + a_1\text{COMP}_{it-1} + a_2\text{CONC}_{it} + a_3\ \text{LLP}_{it} + a_4\ \text{Liquidity Risk}_{it} \\ & + a_5\ \text{Interest Rate risk}_{it} + a_6\ \text{Lending to Tot.Assets}_{it} \\ & + a_7\ \text{Deposits to Tot.Liabilities}_{it} + a_8\ \text{Other Earning Assets to Tot.Assets} \\ & + a_9\ \text{Fee Based Activities}_{it} + a_{10}\ \text{ATMs to Branches}_{it} + a_{11}\ \text{GDP}_{it} + \varepsilon_{it} \qquad (2)\end{aligned}$$

where COMP_{it} is the dependent variable for competitiveness approached through the Lerner index for bank *i* and year *t*; COMP_{it-1} is the first lag of the Lerner index and is used to examine not only the importance of whether we should consider the previous values of the dependent variable but also the factor a_1, which represents the rate at

Table 1 In the following table, we list the control variables that are used in the econometric model of concentration-competitiveness relationship

Variable	Definition
LLP	The ratio of loan loss provisions to total assets
Liquidity risk	As a proxy for liquidity risk, we employ the ratio 'liquid assets/short term funding'. Liquidity risk is expected to affect bank margins positively [27].
Interest rate risk	Computed as the difference between the interbank market (three months) rate and the interest rate for customer deposits [28].
Inefficiency	Computed as the ratio 'operating costs/gross income'. Higher operating costs imply increased operating inefficiency [29].
Capital to assets ratio	A proxy of banks' solvency, computed as the ratio 'capital and reserves/ total assets'. Capital requirements represent a premium on bank margins [30].
Lending/total assets	Customer and interbank loans as a ratio of total assets.
Deposits/total liabilities	Total deposits as a proportion of total liabilities.
Other earning assets/ total assets	Total earning assets other than loans as a ratio of total assets.
Fee-based activities	The Boyd and Gertler (1994) [30] estimator is a proxy for bank fee-based activities, which is computed as fee income/total revenue fee income.
ATMs/branches	A proxy for technical change in delivery channels. The ratio is computed using national data for bank ATMs and branches.
GDP	The relationship between bank margins and growth will depend on the correlation among prices, costs and the business cycle [31].

which competitiveness converges at a future level; and CONC_{it} is the concentration variable that is approached through the HHI index. The rest are as described in Table 1 and ε_{it} is the error term.

In the second analysis, we calculate two models, where in the first, we consider a linear relationship between competition and stability. In the third model, based on the research of Berger et al. (2009) [31], we add the square term of competition to test for non-linear relationships between Competition and Stability [32].

$$\begin{aligned}\text{RISK}_{it} = {} & b_0 + b_1\text{RISK}_{it-1} + b_2\text{COMP}_{it} + b_3\text{CONC}_t + b_4\text{SIZE}_{it} + b_5\text{RGDP}_t \\ & + b_6\text{LR}_{it} + b_7\text{NIM}_{it} + b_8\text{CRISIS}_t + b_9\text{LLP}_{it} + \varepsilon_{it}\end{aligned} \tag{3}$$

$$\begin{aligned}\text{RISK}_{it} = {} & c_0 + c_1\text{RISK}_{it-1} + c_2\text{COMP}_{it} + c_3\text{COMP}_{it}{}^2 + c_4\text{CONC}_t \\ & + c_5\text{SIZE}_{it} + c_6\text{RGDP}_t + c_7\text{LR}_{it} + c_8\text{NIM}_{it} + c_9\text{CRISIS}_t + c_{10}\text{LLP}_{it} + \varepsilon_{it}\end{aligned} \tag{4}$$

where RISK_{it} is the dependent variable for stability for Bank I and year t and is proxied by the Bank Z score index's natural logarithm; RISK_{it-1} is the lag of the stability variable and, as previously, is used to examine not only the importance of whether we should consider the previous values of the dependent variable but also the factor a_1, which represents the rate at which stability converges at a future level;

Table 2 In the following table, we list the control variables used in econometric models for Concentration-Stability and Competition-Stability relationships

Variable	Definition
SIZE	The size of the bank is measured as the natural logarithm of the total assets.
LLP	The ratio of provisions for loan losses to total assets (LLP) is used to measure the quality of the exported result and how managers invest in high-risk assets.
NIM	The net interest margin (NIM) is used to monitor the profitability of a bank's investment and lending activities.
Capital requirements	Indicates the minimum capital requirement that is interpreted as an entry barrier indicator.
RGDP	The conditions of the economic cycle are controlled by the introduction of the annual real GDP growth rate.
CRISIS	An indicator variable that takes a value of one for crisis years, and zero in the others, and is included as a control variable.
LR (Loans Ratio-Asset Composition)	Also, for the control of heterogeneity relating to banking-level factors, the composition of assets is included in the regressions. The composition of the assets is measured as the ratio of loans to assets.

and $COMP_{it}$, as before, for competition with the Lerner index, but, in addition, here, based on Berger et al. (2009) [31] and Jimenez et al. (2013) [33], we also use a square term for the competition measure to capture a possible non-linear relationship between competition and risk with $COMP_{it}^2$ in Eq. (4). $CONC_t$ again is for concentration with the HHI index; the rest are as described in Table 2, and finally e_{it} is the error term.

The two models used in this survey are calculated based on the generalized method of moments (GMM) to solve any endogenous problems that may arise. In the first case, the model used is based on the one used in the research by Carbo and Rodriguez-Fernandez (2007) [7], where we examine the relationship between Concentration and Competitiveness, and the second on models of Kasman and Kasman (2015) [34] and Fu et al. (2014) [35] to study the relationship between Concentration and Competitiveness with Stability. In both models, we used panel data from the five banks and the Greek Banking system for the period 2008–2018, which was also the most critical for the industry in Greece. Variables for stationarity reasons are at first difference. To address correlation and possible endogenous problems, Arellano and Bond (1991) [36] propose the use of the lags of the explanatory variables as instruments (Table 3).

Table 3 Descriptive statistics of the variables used in the study

Variable	Full Sample				
	Mean	St. Dev.	Median	Min	Max
Lerner	0,210	0,433	0,280	−1762	1107
HHI	0,181	0,051	0,214	0,117	0,233
LnBankZ	1163	0,552	1193	−1523	1920
LLP	0,026	0,030	0,017	0,005	0,171
Liquidity risk	0,204	0,079	0,201	0,035	0,363
Interest rate risk	−0,017	0,009	−0,018	−0,038	0,006
Inefficiency	0,627	0,247	0,576	0,279	1803
Capital to assets ratio	0,084	0,050	0,087	−0,039	0,185
Lending/total assets	0,699	0,096	0,707	0,433	0,888
Deposits/total liabilities	0,633	0,113	0,628	0,424	0,909
SIZE	24,456	1199	24,925	21,932	25,517
NIM	0,027	0,006	0,027	0,011	0,041
Capital requirements	13,543	2880	13,508	9569	17,100
RGDP	−0,024	0,036	−0,004	−0,091	0,019
CRISIS	0,455	0,503	0,000	0,000	1000
LR	0,664	0,089	0,663	0,408	0,831
Other earning assets/total assets	0,205	0,069	0,213	0,031	0,320
Fee-based activities	6212E+09	3893E+09	6860E+09	1578E+08	1596E+10
ATMs/branches	1967	0,642	1892	0,732	3589
GDP (bil.€)	255,192	52,921	239,862	195,222	354,461

4 Empirical Results

4.1 Merger Scenarios

In this section, we present the level of Bank Concentration in Greece as well as its yearly change measured by the HHI and ΔHHI indices, respectively, for the period 2004–2018.

Table 4 shows that up to 2012 the HHI index in column 4 is below the limit of HHI $>$ 2000 set by the ECB. However, in 2012, the change in the index as seen in column 5 comes quite close to the limit of ΔHHI $>$ 150, due to the forced acquisitions that followed the debt crisis in that period. In 2013, however, we see that the index combined with its change (ΔHHI) far exceeds the limits. The fact that regulators allowed these acquisitions is because they were used as a measure to restore stability for the Greek banks, instead of injecting further capital coming from Greek taxpayers [2, 13]. So, we see that this index acts more as a guideline and is combined with other factors characterizing its situation, by regulators, to allow or not a M&A deal.

Over the next few years, mergers and acquisitions did not cause any significant change in the index, but the previous events had already raised the index to very high

Table 4 Greek banking sector HHI index for the period 2004–2018

(1)	(2)	(3)	(4)	(5)	(6)
YEAR	HHI	ΔHHI	HHI*10000	ΔHHI*10000	CONCENTRATION SUSPICION
2004	0,107		1070		SLIGHTLY PROBABLE
2005	0,1096	0,0026	1096	26	SLIGHTLY PROBABLE
2006	0,1101	0,0005	1101	5	SLIGHTLY PROBABLE
2007	0,1096	–0,0005	1096	−5	SLIGHTLY PROBABLE
2008	0,1172	0,0076	1172	76	SLIGHTLY PROBABLE
2009	0,1183	0,0011	1183	11	SLIGHTLY PROBABLE
2010	0,1214	0,0031	1214	31	SLIGHTLY PROBABLE
2011	0,1278	0,0064	1278	64	SLIGHTLY PROBABLE
2012	0,1487	0,0209	1487	209	SLIGHTLY PROBABLE
2013	0,2136	0,0649	2136	649	YES
2014	0,2195	0,0059	2195	59	SLIGHTLY PROBABLE BUT ALREADY CONCENTRATED
2015	0,2254	0,0059	2254	59	SLIGHTLY PROBABLE BUT ALREADY CONCENTRATED
2016	0,2332	0,0078	2332	78	SLIGHTLY PROBABLE BUT ALREADY CONCENTRATED
2017	0,2307	–0,0025	2307	−25	SLIGHTLY PROBABLE BUT ALREADY CONCENTRATED
2018	0,2304	–0,0003	2304	−3	SLIGHTLY PROBABLE BUT ALREADY CONCENTRATED

Source: European Central Bank

Note: To present the results at the scale presented by the ECB guidelines, we multiply the index by 10,000. Also, based on these guidelines, we characterize the level of concentration for each year according to the value of the index and its yearly change

Table 5 Merger scenarios of the four systemic Greek Banks for the year 2019

Merger	Results		
	HHI AFTER MERGER	ΔHHI	SUSPICIOUS
Ethniki Bank – Alpha Bank	3502	1198	YES
Ethniki Bank – Piraeus Bank	3519	1215	YES
Ethniki Bank – Eurobank	3443	1139	YES
Alpha Bank – Piraeus Bank	3443	1139	YES
Alpha Bank – Eurobank	3371	1067	YES
Piraeus Bank – Eurobank	3386	1082	YES

levels, causing the sector to be considered very concentrated, even though we cannot be certain at this time if this result is mostly negative. Next, based on the HHI index, we test the possible merger scenarios for the year 2019 for the four systemic banks in Greece representing more than 90% (EBF, 2020) of the sector's market share.

As we can see from Table 5, the results for each merger go far beyond the limits set by the ECB, i.e., $H > 2000$ in conjunction with $\Delta\text{HHI} > 150$. Thus, based on this indicator and the limits set, we can conclude that for reasons of creating a monopoly, or price cartel conditions, none of the above scenarios can be realized. However, as mentioned, this index and its yearly change act as guidelines. So, if the sector needs to be further consolidated (as recently stated by the ECB), due to the benefits of integration within EU banking, those other factors might play a much more important role than the index and foster M&As of systemic banks. If this is the case and the merger scenarios illustrated in Table 4 could be realized, based only on the above results, we can propose that the most preferable ones are those of Alpha Bank – Eurobank and Piraeus Bank – Eurobank, as they raise the overall sector's Concentration index less, compared to the other scenarios. This is a sign that they are less probable to create a monopoly in the Greek sector, with whatever drawbacks this situation might cause.

5 Econometric Models

In our econometric study, we examine separately the relationships of Concentration-Competition (Market Power) and Concentration and Competition-Stability based on the models mentioned in Sect. 3.

5.1 *Concentration-Competition*

We estimating model (2) using the GMM, considering the HHI index as the basic variable. The results are summarized in Table 6:

As we see above the variable HHI is not significant to our model. This leads us to rule in favor of the theory of effective structure as it is more possible to explain the competitive conditions within the Greek banking industry. These results agree with the ones found by Claessens and Laeven (2004) [8], Casu and Girardone (2006) [9], Carbo and Rodriguez-Fernandez (2007) [7], Yeyati and Micco (2007) [10], Efthyvoulou and Yildirim (2014) [11], and Rakshit and Bardhan (2019) [12].

On the other hand, we have control variables that are significant to the model. Initially, we see the paradox that there is a positive relationship with the 'Provisions for Losses' from 'Loans to Total Assets', whereas we would expect there to be a negative relationship with market power.

Another variable is inefficient. As we see, it has a negative relationship, which seems perfectly reasonable. As banks do not enjoy market power, this negative element may affect the dependent variable this way.

The 'Fee Based Activities' variable is also negative. In this case, it makes sense as the more banks specialize in their primary activities, the more they will gain a competitive advantage, since they become more efficient and consequently gain more market power.

Table 6 GMM regression results for Eq. (2)

Variable	Coefficient	Std.Error	t-Stat.	Prob.
Lerner $_{-1}$	−0,434,157	0,190,977	−2,273,346	0,0283
HHI	−566,899	6,468,742	−0,876,367	0,3859
LLP	5,594,816	2,308,396	2,423,681	0,0199
Interest rate risk	−479,904	13,83,778	−0,346,807	0,7305
Inefficiency	−1,518,091	0,523,648	−2,899,071	0,006
Liquidity risk	−0,57,804	3,169,756	−0,182,361	0,8562
Capital to assets	−0,72,641	2,576,925	−0,28,189	0,7794
Lending/tot. assets	−3,537,447	2,994,225	−1,181,423	0,2442
Deposits/tot. liabilities	247,783	1,620,892	1,528,683	0,134
Other earning assets/tot. assets	2,564,012	3,155,856	0,812,462	0,4212
Fee-based activities	−1,04E-10	4,71E-11	−2,204,215	0,0332
ATMs/branches	−0,131,112	0,303,287	−0,432,304	0,6678
GDP	−6,74E-13	5,58E-12	−0,120,806	0,9044
Const.	0,065709	0,103,166	0,636,924	0,5277
R-squared	0.303239			
J-statistic	6.722881			
Prob (J-statistic)	0.875370			

Finally, the only other important variable is the 'lag of the dependent'. This states the importance of considering previous index values. But the fact that it is negative shows us that the previous results affect the index in reverse, which means that we will have constant ups and downs.

5.2 *Concentration-Competition-Stability*

In the second analysis, we consider the effect of two variables on stability. For the first, we consider the relationship between competition and stability, and for the second, the effect of concentration. Using again a GMM estimation on model (3) and having as basic explanatory variables, the Lerner and HHI indices, we gain the following results (Table 7):

For the first basic variable, the Lerner index, we see that it has a marginal significance to our model. Nonetheless, in our investigation, we accept it to extract conclusions. Thus, we see that it has a positive relationship with the stability index. This makes us lean toward the Competition-Fragility view (consistent with, among others: Yeyati & Micco, 2007 [10], Turk-Ariss, 2010 [29], Leroy & Lucotte, 2017 [37]). This means that the increase in the market power of banks (within reasonable limits) is expected to bring greater stability.

The second variable is HHI. For this variable, we see that it is not significant to the model. Therefore, we find no relation with Stability. As a result, we cannot rule in

Table 7 GMM regression results for Eq. (3)

Variable	Coefficient	Std.Error	t-Statist.	Prob.
LnBankZ-1	0,135,644	0,124,842	1,086,524	0,2902
LERNER	0,114,471	0,05717	2,002,304	0,059
HHI	9,113,786	9,124,209	0,998,858	0,3298
RGDP	20,61,677	2,931,706	7,032,344	<0,0001
SIZE	−0,578,229	0,477,173	−1,211,781	0,2397
LLP	−3,981,513	1,215,019	−3,276,914	0,0038
LR	1,194,722	0,886,229	1,348,097	0,1927
CRISIS	1,865,079	0,153,508	12,1497	<0,0001
NIM	−26,82,579	16,94,398	−1,583,205	0,1291
CAPITAL REQ.	0,832,779	0,043131	19,30,833	<0,0001
Const.	−0,376,809	0,053155	−7,088,878	<0,0001
R-squared	0.979992			
J-statistic	8.640407			
Prob(J-statistic)	0.279514			

favor of any of the views Concentration-Stability, or fragility, for which we have not found similar results in the literature.

Of the control variables that are important, initially, GDP's annual growth rate is positively related to the stability variable. The next expected result is the negative relationship of the 'Provisions for Loan Losses' to 'Total Assets' with Stability. However, the unexpected result for which we have not been able to provide a satisfactory explanation is the positive relationship between the indicative variable 'CRISIS' and Stability.

The last significant variable positively affecting stability is the amount of capital requirements.

For a robustness test following Jimenez et al. (2013) [33] and Fu et al. (2014) [35], we also use a quadratic term of the measure of competition, to capture a possible non-linear relationship between competition and risk, as illustrated in model (4). The results are as follows (Table 8):

Because both the Lerner index and its square are not important to the model, we can deduct that the Competition-Stability relationship is linear, which is in line with previous studies including the quadratic term but supporting the linearity.

6 Conclusions

This chapter examined the impact of Systemic Greek Bank M&A on the financial sector's major concepts: Concentration, Competition and Stability, as the subject is highly topical these days given the various reports circulating in financial news sources regarding upcoming systemic bank mergers in Greece.

Table 8 GMM regression results for Eq. (4)

Variable	Coefficient	Std. error	t-Statistic	Prob.
LnBankZ-1	0,129,482	0,133,476	0,970,076	0,3442
LERNER	0,144,587	0,101,608	1,422,989	0,171
$LERNER^2$	0,025094	0,058573	0,428,426	0,6732
HHI	9,281,579	9,398,323	0,987,578	0,3358
RGDP	21,56,667	2,755,268	7,827,433	<0,0001
SIZE	−0,388,827	0,440,332	−0,883,031	0,3883
LLP	−3,604,646	120,143	−3,000,297	0,0074
LR	1,577,162	0,905,025	1,742,674	0,0976
CRISIS	1,835,308	0,158,131	11,60,628	<0,0001
NIM	−30,92,965	18,19,624	−1,699,783	0,1055
CAPITAL REQ.	0,829,995	0,045954	18,06162	<0,0001
Const.	−0,356,842	0,051624	−6,912,338	<0,0001
R-squared	0.979456			
J-statistic	8.638275			
Prob(J-statistic)	0.373723			

For this purpose, we tested the relationships of Concentration-Competition, Concentration-Stability and Competition-Stability.

Thus, as a final view, based on our results, it could be said that while a merger between two Greek systemic banks seems, at first sight, dissuasive, on the contrary, it can have positive effects on the stability of the system.

For the Greek economy, these results mean that measures such as forced mergers and acquisitions to save the banks, and consequently the system, constitute a positive development. As we have observed from our results, the increase in concentration following mergers and acquisitions did not have any major negative impact on the stability and did not lead to the sector becoming a monopoly; on the contrary, the results of these moves by the banks helped reverse the situation and save the banking system. Opposite results would have resulted in bankruptcy for the banks, and consequently the collapse of the economy, as the system's overall money flow relies on the banks. Also, our results cannot explain extreme situations of concentration and competition, but the conclusions are made based on values given to the variables not far exceeding our sample.

Our last limitation concerned our models. The variables within our models were based on previous ones presented in similar studies dependent on a previously developed theoretical framework. Even though the variables used in our model emerged from a thorough study of the existing literature and might not be the optimal ones to test the pertinent relationships without the shadow of a doubt, testing every possible variable for suitability is rather impractical. As a result, we relied on those variables that are most probable and most appropriate for the research conducted.

Closing this chapter, we shall refer to possible expansions of this research. One relevant expansion might be the inclusion of the Covid-19 crisis in the sample, testing dynamically by modeling and simulating the sample [38–42]. Whether M&A are moves that banks or other kind of businesses might use to gain positive results and cover their losses. Finally, another future research development might be to examine other advantages that mergers and acquisitions might bring considering where most synergies are coming from.

References

1. Ozturk, S., Sozdemir, A.: Effects of global financial crisis on Greece economy. Proc. Econ. Financ. *23*, *568–575* (*2015*)
2. Provopoulos, G.: The Greek economy and banking system: recent developments and the way forward. J. Macroecon. **39**, 240–249 (2014)
3. Mason, E.S.: Price and production policies of large-scale enterprise. Am. Econ. Rev. **29**, 61–74 (1939)
4. Bain, J.S.: Relation of profit rate to industry concentration. Q. J. Econ. **65**, 293–324 (1951)
5. Demsetz, H.: Industry structure, market rivalry, and public policy. J. Law Econ. **16**(1), 1–9 (1973)
6. Peltzman, S.: The gains and losses from industrial concentration. J. Law Econ. **20**, 229–263 (1977)
7. Carbó, V.C., Rodríguez, Fernández, F.: The determinants of bank margins in European banking. J. Bank. Financ. **31**, 2043–2063 (2007)
8. Claessens, S., Laeven, L.: What drives bank competition? some international evidence. J. Money Credit Bank. **36**(3), June part 2, 563–June part 2, 583 (2004)
9. Casu, B., Girardone, C.: Bank competition, concentration and efficiency in the single European market. Manchester School. **74**(4), 441–468 (2006)
10. Yeyati, E.L., Micco, A.: Concentration and foreign penetration in Latin American banking sectors: impact on competition and risk. J. Bank. Financ. **31**, 1633–1647 (2007)
11. Efthyvoulou, G., Yildirim, C.: Market power in CEE banking sectors and the impact of the global financial crisis. J. Bank. Financ. **40**, 11–27 (2014)
12. Rakshit, B., Bardhan, S.: Bank competition and its determinants: evidence from Indian banking. Int. J. Econ. Bus. **26**(2), 283–313 (2019)
13. Sompolos, Z., Mavri, M.: Estimating the efficiency of Greek Banking System during the last decade of world economic crisis: an econometric approach. Benchmark. Int. J. **25**(6), 1762–1794 (2018)
14. Boyd, J.H., De Nicolo, G.: The theory of bank risk-taking and competition revisited. J. Financ. **60**, 1329–1343 (2005)
15. Shim, J.: Loan portfolio diversification, market structure and bank stability. J. Bank. Financ. **104**, 103–115 (2019)
16. Azmi, W., Ali, M., Arshad, S., Rizvi, S.A.: Intricacies of competition, stability, and diversification: evidence from dual banking economies. Econ. Model. **83**, 111–126 (2019)
17. Beck, T., De Jonghe, O., Schepens, G.: Bank competition and stability: cross-country heterogeneity. J. Financ. Intermediat. **22**(2), 218–244 (2013)
18. González, L.O., Razia, A., Búa, M.V., Sestayo, R.L.: Competition, concentration, and risk taking in Banking sector of MENA countries. Res. Int. Bus. Financ. **42**, 591–560 (2017)
19. Lenza, M., Pill, H., Reichlin, L.: Monetary policy in exceptional times. Econ. Policy. **25**, 295–339 (2010)
20. Eser, F., Schwaab, B.: Evaluating the impact of unconventional monetary policy measures: empirical evidence from the ECB's securities markets programme. J. Financ. Econ. **119**(1), 147–167 (2016)
21. Steeley, J., Matyushkin, A.: The effects of quantitative easing on the volatility of the gilt-edged market. Int. Rev. Financ. Anal. **37**, 113–128 (2015)

22. Beetsma, R., de Jong, F., Giuliodori, W.D.: Realized (co) variances of eurozone sovereign yields during the crisis: the impact of news and the securities markets programme. J. Int. Money Financ. **75**, 14–31 (2017)
23. Fratzscher, M., Duca, M.L., Straub, R.: ECB unconventional monetary policy: market impact and international spillovers. IMF Econ. Rev. **64**(1), 36–74 (2016)
24. Kenourgios, D., Papadamou, S., Dimitriou, D.: Intraday exchange rate volatility transmissions across QE announcements. Financ. Res. Lett. **14**, 128–134 (2015)
25. Acharya, V.V., Eisert, T., Eufinger, C., Hirsch, C.: Whatever it takes: the real effects of unconventional monetary policy. Rev. Financ. Stud. **32**(9), 3366–3411 (2019). https://doi.org/10.1093/rfs/hhz005
26. Kenourgios, D., Christopoulos, A., Dimitriou, D.: Asset markets contagion during the global financial crisis. Multinatl. Financ. J. **17**(1/2), 49–76 (2013)
27. Yu, S.: Sovereign and bank interdependencies – evidence from the CDS market. Res. Int. Bus. Financ. **39**, 68–84 (2017)
28. Keddada, B., Schalckb, C.: Evaluating sovereign risk spillovers on domestic banks during the European debt crisis. Econ. Model. **88**, 356–375 (2020)
29. Turk-Ariss, R.: On the implications of market power in banking: evidence from developing countries. J. Bank. Financ. **34**(4), 765–775 (2010)
30. Boyd, J.H., Gertler, M.: Are banks dead? or are the reports greatly exaggerated? Federal Reserve Bank of Minneapolis. Q. Rev. **18**, 2–23 (1994)
31. Berger, A.N., Klapper, L.F., Turk-Ariss, R.: Bank competition and financial stability. J. Financ. Serv. Res. **35**(2), 99–118 (2009)
32. Angbanzo, L.: Commercial bank net interest margins, default risk, interest-rate risk and off-balance sheet banking. J. Bank. Financ. **21**, 55–87 (1997)
33. Jimenez, G., Lopez, J., Saurina, J.: How does competition affect bank risk-taking? J. Financ. Stabil. **9**, 185–195 (2013)
34. Kasman, S., Kasman, A.: Bank competition, concentration, and financial stability in the Turkish banking industry. Econ. Syst. **39**, 502–517 (2015)
35. Fu, X., Lin, Y.: Molyneux P Bank competition and financial stability in Asia Pacific. J. Bank. Financ. **38**, 64–77 (2014)
36. Arellano, M., Bond, S.: Some tests of specification for panel data: Monte Carlo evidence and an application to employment equations. Rev. Econ. Stud. **58**, 277–297 (1991)
37. Leroy, A., Lucotte, Y.: Is there a competition-stability trade-off in European banking? J. Int. Finan. Markets. Inst. Money. **46**, 199–215 (2017)
38. Kanellos, N., Sakas, D.P., Kamperos, I.D.G., Reklitis, D.P., Giannakopoulos, N.T., Nasiopoulos, D.K., Terzi, M.C.: The effectiveness of centralized payment network advertisements on digital branding during the COVID-19 crisis. Sustainability. **14**(6), 3616 (2022)
39. Alexandros, N.K., Damianos, S.P., Dimitrios, N.K., Dimitrios, V.S.: Modelling the Process of a Web-Based Collaboration Tool Development. In: Kavoura, A., Sakas, D., Tomaras, P. (eds) Strategic Innovative Marketing. Springer Proceedings in Business and Economics. Springer, Cham (2017). https://doi.org/10.1007/978-3-319-56288-9_51
40. Sakas, D.P., Nasiopoulos, D.K., Reklitis, P.: Modeling and Simulation of the Strategic Use of Marketing in Search Engines for the Business Success of High Technology Companies. In: Sakas, D., Nasiopoulos, D. (eds) Strategic Innovative Marketing. IC-SIM 2017. Springer Proceedings in Business and Economics. Springer, Cham (2019). https://doi.org/10.1007/978-3-030-16099-9_27
41. Sakas, D.P., Nasiopoulos, D.K., Reklitis, P.: Modeling and Simulation of the Strategic Use of the Internet Forum Aiming at Business Success of High-Technology Companies. In: Sakas, D., Nasiopoulos, D. (eds) Strategic Innovative Marketing. IC-SIM 2017. Springer Proceedings in Business and Economics. Springer, Cham (2019). https://doi.org/10.1007/978-3-030-16099-9_22
42. Sakas, D., Vlachos, D., Nasiopoulos, D.: Modelling strategic management for the development of competitive advantage, based on technology. J. Syst. Inf. Technol. **16**, 187 (2014)

Tax Evasion in the Greek Construction Sector

Boufounou Paraskevi, Ilias Episkopos, Kanellos Toudas, and George Georgakopoulos

1 Introduction

Examining the factors that influence economic growth, several studies have focused on the role of taxation and in particular on tax evasion. Tax evasion can lead to a distortion of capital accumulation, as fewer people bear the tax burden, leading to unbalanced economic growth [1–5]. Furthermore, tax evasion reduces the public resources available to finance the provision of productive inputs provided by the government in the form of public goods, resulting in reduced endogenous growth [1, 4, 5]. In addition to the above, tax evasion reduces citizens' trust in the rule of law [4, 6–8], which leads more people to this illegal activity [6, 7].

Tax evasion constitutes one of the most important obstacles to economic growth, and it is very important to examine individual aspects of tax evasion practices in Greece [9, 10]. This study focuses on analyzing tax evasion in technical companies; the construction sector contributes significantly to the country's employment and GDP. The study is organized as follows:

Initially, the concept, causes and effects of tax evasion are presented, and the phenomenon of tax evasion with reference to Greece is further discussed. A thorough literature review follows. Then, the research method is applied, and the main findings are presented. Finally, the main conclusions are summarized, and suggestions for practical application are put forward accompanied by recommendations for further research.

B. Paraskevi · I. Episkopos
National and Kapodistrian University of Athens, Athens, Greece

K. Toudas (✉) · G. Georgakopoulos
Agricultural University of Athens, Athens, Greece
e-mail: kstoudas@aua.gr

D. P. Sakas et al. (eds.), *Computational and Strategic Business Modelling*,
Springer Proceedings in Business and Economics,
https://doi.org/10.1007/978-3-031-41371-1_38

2 Tax Evasion

2.1 The Importance of Tax Evasion

Tax evasion is an illegal activity, defined as the concealment of a taxable activity declaration. Tax evasion should be distinguished from tax avoidance, defined as the reorganization of economic activity, possibly at some cost, to reduce tax payments. Tax evasion refers to an illegal reduction of tax payments (for example, by not reporting revenues) [11]. According to the Greek legislation, any act aimed at the total or partial avoidance of a tax liability in violation of tax law is defined as tax evasion. Tax evasion takes various forms, the main ones being: "(a) concealment of income, (b) incurring excessive or fictitious expenses, (c) failure to submit or the inaccurate submission of tax returns, (d) non-reporting of VAT, and (e) falsification of books and records" [12].

The concept of "tax evasion" is often confused with that of the "informal economy". These two concepts do not necessarily coincide, although tax evasion is partly related to the informal economy (defined as economic activity that is not recorded in national accounts, although it should be).

Overall, tax evasion and avoidance exist in all economies. Although both entail a reduction in taxpayers' taxes, tax evasion is an illegal activity, while tax avoidance behavior is legal. Since the returns to tax evasion and tax avoidance are different in nature, they should be introduced into an economic model in a different way. The fact that one is likely and the other is not is in itself a big difference between them [13].

According to Kirchler et al. [14] that tax avoidance and tax evasion have similar effects, economists suggest not to distinguish between them but to analyze their effects together. Tax evasion is an important problem in several countries, such as Greece. The emphasis on this problem stems from the fact that it is an obstacle to growth and balanced growth, undermines the rule of law, prevents the maximization of social welfare and redistributive policies by governments, threatens social cohesion and reduces citizens' trust in the state [4]. For the welfare state to survive in the long run, tax compliance and reduction of tax evasion are crucial [7].

2.2 The Greek Case Study

The European Commission [15] found that tax evasion, combined with the Covid-19 pandemic, resulted in a decrease in VAT revenue to €8.28 billion in Greece, which (along with Denmark) recorded the largest losses (5.5 points increase in losses). More specifically, VAT revenue losses amount to 36.9% of expected revenue, which is the highest since 2000. On 19 July 2018, the European Commission referred Greece to the Court of Justice of the European Union for failing to transpose the Anti-Money Laundering Directive into their national law [16].

Table 1 Causes of tax evasion in Greece

Category	Description
Legislative/political factors	Complexity of the tax system, legal uncertainty, continuous increase of the tax burden and lack of political will to tackle the phenomenon
Technological factors	Lack of technological/organizational infrastructure of the tax administration
Organizational/bureaucratic factors	Inefficient administration, poor public sector organization, fixation on following procedures rather than achieving results
Structural factors	Excessively high number of self-employed and micro-enterprises which makes tax controls difficult. Greater tax evasion in certain professional categories of self-employed, and between the two extreme income categories (very rich and very poor), but also in certain regions of the country's periphery compared to urban centers
Cultural factors	Citizens' beliefs about the state and its responsiveness but also about their rights and obligations toward it

Source: Dianeosis [9], pp. 169–170

Violations of tax legislation in Greece are classified into three main groups: "(a) acts of tax evasion, i.e. misleading or inaccurate acts or omissions in relation to tax debts with the purpose of not paying the legal tax; (b) acts of tax avoidance, i.e. aggressive tax planning practices to artificially reduce tax debts; and (c) taxpayer's default, i.e. delaying the payment of overdue tax debts" [12]. The criminal fraud of tax evasion is provided for in Article 66 of Law 4174/2013 "Income Tax Code" (ITC) and is punishable by a minimum of two (2) years imprisonment.

Tax evasion has been one of the key challenges facing the Greek authorities in their efforts to achieve fiscal consolidation for decades [5]. According to Dianeosis [17], the revenue foregone from tax evasion for individuals ranges from 1.9% to 4.7% of GDP per year, while for legal entities tax evasion (combined with tax avoidance) amounts to 0.15% of GDP. The estimation of the GDP is done by collecting data indicators from the World Bank [18]. In addition, VAT evasion is estimated to cause a loss of 3.5% of GDP. Overall, based on the report, the size of tax evasion in Greece can be estimated at between 6% and 9% of GDP, i.e., between €11 billion and €16 billion per year. A later survey by Dianeosis [9] states that tax evasion by individuals and legal entities is estimated to be between 1.96% and 4.85% of GDP, which implies that the lost tax revenue ranges from €3.4 billion to €8.5 billion, stating that there are a variety of factors that lead to tax evasion as summarized in Table 1.

The tax policy itself should be examined as to whether it is an incentive for tax evasion. There are multiple tax rates and additional unanticipated taxes on already declared income, while there is favorable tax treatment of specific groups of the population. Examining the Greek tax system, Rapanos & Kaplanoglou [10] state that its structure is not growth friendly. The increase in withholding tax as part of the reforms did not lead to incentives for firms to invest. The complexity of the legislation, the inefficiency of the tax administration and the corruption that

Table 2 Key economic indicators of the construction sector in Greece, 2015–2018

Years	2015	2016	2017	2018
Number of firms	74,337	77,229	61,833	59,843
Number of employees	132,071	145,060	124,261	147,195
Number of paid employees	70,807	86,906	68,685	96,748
Salaries and wages	788,168,415	900,237,976	851,248,880	922,089,429
Production value	9,730,983,710	8,793,198,274	9,392,559,381	9,242,510,507
Total purchases of goods and services	7,930,753,717	7,445,137,973	8,500,397,008	8,801,422,797
Value added at factor cost prices	2,698,393,079	2,033,006,991	2,451,485,802	1,829,989,963
Gross investment in tangible goods	437,319,405	703,848,061	587,470,653	210,451,254
Turnover	10,368,013,395	9,249,303,087	9,886,605,666	10,033,327,191
Staff costs	1,016,069,003	1,141,333,892	1,104,244,197	1,177,622,211

Source: ELSTAT [19, 20]

characterizes the public administration, the unequal distribution of the tax burden, the feeling of non-reciprocity of taxes paid, the level of tax rates and the tax burden, social perceptions of tax evasion and, finally, the large number of self-employed persons are additional factors leading to tax evasion.

The construction sector is very important for the Greek economy, thus this study focuses on researching tax evasion on this sector. The construction sector experienced rapid growth from the early 1990s until 2007 but then suffered a dramatic contraction during the crisis years. Cumulatively, the decline in construction value added in the period 2008–2019 reached 87.7%, while employment in the sector reached 61.8%. However, the sector has been on the rise in the following years. The construction sector contributes a significant share to Greece's GDP. The key economic indicators of the construction sector in Greece are presented in Table 2.

More specifically, in 2017, the total contribution of the construction sector in terms of GDP amounted to 19.9 billion Euros in 2017, which corresponds to 11% of GDP. The construction sector's value in 2018 had approached 18.1 billion, or 8.3% of GDP, while combined with the other functions within the sector, the total direct contribution of construction was 20.2 billion Euros, or 9.3% of GDP. In 2019, the construction sector contributed 2.222 billion Euros of gross value added to GDP, which when combined with engineering services, amounted to 3.028 billion Euros.

In terms of employment, and taking into account multiplier effects, the total contribution of construction in 2017 is estimated at more than 500,000 jobs. Construction employed around 151,600 people in 2019, generating more than 2.22 billion Euros of value added to GDP [21]. For every one Euro spent in the construction sector, 1.8 Euros are added to the country's GDP, of which 0.4 Euros are government revenues, 39 jobs are created and 13 are related to the sector.

Greece's property market had a positive year in 2017, as sales rose by 2% and the market capitalization grew at a rate of 78%. The growth of the construction industry worldwide positively affected Greece, which recorded a 2% increase in annual sales. The period 2017–2019 saw an increase in demand for residential real estate due to short-term rentals (e.g., Airbnb) and the introduction of new players (Russia and China) into the market.

Three Greek companies are included among the top 100 construction companies worldwide: Hellaktor (rank 81), GEK Terna 9 (rank 91) and JP-AVAX (rank 98). It also estimates that the construction sector in Greece is expected to recover within 2018–2022, with an estimated compound annual growth rate of 4.7% compared to −2.4% in 2013–2017 (these estimates were made before the start of the pandemic). It is estimated that this sector could provide a significant boost to the country's economic growth after the pandemic. The combination of NSRF 2021–2027 program funds (estimated at more than €20 billion Euros), to funds from the Recovery Fund (estimated at €32 billion Euros) and public-private partnerships will boost the construction sector and ultimately contribute to Greece's economic growth.

3 Literature Review

Relevant literature reviewed suggests that tax evasion is a crucial factor, affecting economic growth. Indicatively:

- Bitzenis et al. [22] report "shadow economy" (i.e., tax evasion) in combination with undeclared work as determinants of economic growth. Important determinants of the "shadow economy" are factors related to macroeconomic conditions, such as unemployment and GDP growth, as well as institutional factors, such as tax morality and the rule of law.
- Cerqueti and Coppier [23] examine the relationship between tax evasion and economic growth, developing a Ramsey model of economic growth with a competing but non-exempt public goods financed by taxes that can be avoided through a corrupt tax inspector. The authors demonstrate that growth rates – both income and tax revenue – decline as the tax rate increases, but there are differences in how the growth rate declines as the tax rate increases: the rate of decline is higher in countries with low shame than in countries with high shame.
- Varvarigos [24] stresses, however, that concerns about the impact of tax evasion should not be limited to moral considerations. He conducted a study to examine the impact of tax evasion on economic development by focusing on the cultural aspects of tax compliance and their impact on tax evasion. A cultural norm that determines the defiance of tax avoidance practices links past tax evasion effects to taxpayers' current incentives to conceal sources of income. This dynamic can lead to multiple equilibria in the evolution of tax evasion. Because of the latter's effect on capital accumulation, this multiplicity can lead economies to divergent growth paths, in the case of differing initial levels of tax evasion. This is the case even though economies may, at the outset, be identical in terms of capital stock and structural characteristics, including those applying the tax.

- Bethencourt and Kunze [25] also incorporate a dynamic cultural perspective on tax evasion. Their aim is to examine how the effect of social norms on tax compliance can explain the positive relationship between marginal tax rates and tax evasion, the reduction in the tax evasion/GDP ratio and the positive effect of GDP per capita on tax compliance. Their model does not generate multiple equilibria in tax evasion and does not take into account the causal effect of tax evasion on capital accumulation due to the absence of a productive role for government spending.
- Blackburn et al. [26] study the relationship between the informal economy and financial development in a model of tax evasion and bank intermediation. The lower the stage of economic development, the higher the level of tax evasion and its effects, and the larger the size of the informal economy. Overall, the authors find bidirectional causal effects between corruption and economic growth. Individuals choose how much of their income to declare, trading off their incentives to provide security against their disincentives to comply with tax obligations.
- Roubini & Sala-i-Martin [27] use a monetary growth model in which tax evasion motivates the government to suppress the financial sector as a means of increasing the demand for money, expanding the tax base for inflation and, hence, covering the tax revenue shortfall. Consequently, tax evasion leads to a reduction in the rate of economic growth because the process of capital accumulation is hindered.
- Chen [28] in his study finds a negative impact of tax evasion on economic growth because tax evasion induces policy makers to increase the tax rate more than what is needed to maximize economic growth, thereby offsetting the loss of tax revenue.
- Caballé and Panadés [29] analyze how tax compliance policy affects the rate of economic growth. The authors show that tax compliance effects depend on the relative productivity of private and public capital. Even if higher enforcement leads to a reduction in savings, since individuals will have less disposable income, the overall effect may be incremental. It is the imposition of very high penalties with a very low probability of inspection that allows the government to best approximate the highest growth rate among all policies that induce honest behavior.
- Ameyaw et al. [30] investigate the impact of personal tax evasion on economic growth in Ghana using quantitative questionnaire survey. The results of the study revealed that tax evasion in Ghana has a negative effect on economic growth. Specifically, the study identified high tax rates as the determinant of tax evasion, which in turn has a negative significant impact on the socio-economic development of the country.
- Bostina [31] assesses the impact of tax evasion on economic growth in the European Union for the period 1997–2010. The main hypothesis that tax evasion negatively affects economic growth was not rejected, which leads to the conclusion that as tax evasion increases, economic growth is likely to decrease.
- Tax evasion was also found to have a negative impact on economic growth in the study by Jugurnath et al. [32]. The objective of the study was to examine the impact of tax evasion on economic growth in Mauritius. It was found that high tax

rates, trust in government and regulations were directly related to tax evasion, while more stringent and strict government measures directly influenced attitudes toward tax evasion.

The extent of tax evasion cannot be measured directly but must be inferred from economic variables that can be observed. In general, the definition and measurement of tax evasion is subject to theoretical and practical limitations stemming from the complexity of the economy, the tax framework and the tax collection mechanism [5].

The standard analysis of tax evasion assumes that the offender is the sole perpetrator of the tax evasion. However, this assumption is incomplete and misleading. The act of tax evasion sets in motion a range of general equilibrium adjustments as individuals and firms react to changes in incentives created by tax evasion.

The existence of factors (legislative, economic, political) that do not reduce the initial tax advantage of tax evasion is an important factor leading to tax evasion [1].

The study by Islam, Rashid et al. [33] attempted to investigate the impact of economic and non-economic public policies on tax evasion using panel data of seven SAARC countries covering the period from 1998 to 2015. The result of the study implies that the higher the degree of economic freedoms, the lower the tax evasion.

Using Swedish data, Hammar et al. [7] analyze which factors influence the perception of tax evasion. The analysis is carried out on ten different taxes (various forms of income tax, corporate tax, gift tax, property tax, estate tax, property tax, vehicle tax, carbon tax and alcohol tax), and the results vary widely on the basis of these taxes. People who distrust politicians are more likely to believe that tax evasion is common, and this holds for most of the taxes studied. If people stop trusting politicians, social norms about tax compliance deteriorate and the potential for tax collection to maintain the welfare state is reduced. The authors focus on the importance of trust – whether citizens are generally perceived as trustworthy has little impact on perceived tax evasion.

Frey and Torgler [34] consider tax payment as an example of "pro-social" behavior. Like Hammar et al. [7], they suggest that taxpayers are willing to pay their taxes conditionally, depending on the behavior of others. That is, people are more willing to pay taxes when they believe that others pay them. Developing an index of tax morality and using survey data from several Western and Eastern European countries, Frey and Torgler [34] find a high negative correlation between perceived tax evasion and tax morality.

"Trust" is important because even if a government is expected to provide exactly what the citizen wants, it is usually reasonable for the individual to evade taxes if he knows that he cannot be detected and sanctioned. Therefore, trust in government and other citizens, i.e., the belief that other citizens will pay their taxes, are important determinants of tax evasion. In this respect, this approach can be linked to fiscal sociology and state capacity, which presents the development of the tax system and tax collection as the result of an ongoing dialog between the state and the wider population [6, 8].

Overall, the tax compliance literature clearly reveals that social or generalized "trust" is an important factor affecting payment of taxes. If an individual believes that most citizens pay their taxes, then they are more inclined to pay equally and not evade taxes. This is called "social taxpaying": social taxpayers are influenced by social rules, feel guilty when they evade taxes and ashamed when they are caught. If they perceive that others are paying taxes, they tend to pay them as well. On the other hand, reducing others' contributions reduces their willingness to contribute themselves [7].

Legislation and legality are considered very important too. The law expresses social values and therefore could influence individual behavior independently of the imposition of sanctions. In this context, legislation and legality can influence individuals' decisions. The dividing line between legal and illegal tax acts (tax avoidance and tax evasion, respectively) is often unclear and varies between countries. Blaufus et al. [35] examined how legality relates to tax evasion. Without any risk of negative economic consequences, classifying tax evasion as illegal can lead to a reduction in tax evasion. However, once potential negative economic consequences are introduced, legality does not affect tax evasion unless the concept of morality is taken into account.

Individual psychological factors, such as perceived fairness and the role of opportunity for tax evasion, are also determinants of this activity [11]. The desire to maximize expected net income is an important factor leading to tax evasion. The higher the gross income and marginal tax rates, the greater the likelihood of tax evasion. Thus, tax evasion is more common in the higher income group than in the lower income groups, while high marginal tax rates clearly imply higher tax evasion [7].

Problems that arise in the collection of taxes by a government that can potentially lead to tax evasion are as follows: a) disproportionate tax burden relative to other regions, but also in combination with income level; b) unbalanced tax structure, especially with an emphasis on indirect taxes, combined with the insufficient redistributive benefit of the tax system [4].

Demalijaj Ukaj [2] emphasizes the level of taxation is very important for the economic growth of a country and fiscal policy should always be considered as an important factor in the behavior and ability of participants to evade taxes. In addition to the level of tax rates, corruption of public administration officials including tax and customs officials, inadequate staff and weak legal system have been cited as additional reasons leading to tax evasion.

Varvarigos [24] focuses on the relevance of social and cultural dimensions as drivers behind the persistence of tax evasion. This is because the cultural norm that influences individuals' motivations for tax evasion acts as a dissemination mechanism that perpetuates such actions, shaping a particular attitude of society toward tax compliance or non-compliance. For this reason, he cites the example of Hong Kong, where the Hong Kong Independent Commission Against Corruption (ICAC) organized an education program at all levels of education.

Finally, Freire-Seren et al. [13] report that, according to some studies, higher levels of education reduce individuals' preference for tax evasion, while based on others higher levels of education lead to higher levels of tax evasion.

4 Methodology and Results

Aiming in studying tax fraud by the Greek technical firms in depth, a special questionnaire was designed and the replies of 154 accountants from 53 technical firms obtained were analyzed using SPSS. The sampling method was random. The Likert scale 1 (low) to 5 (very high) was applied in the questionnaire. The main findings, classified in three categories, namely (a) Tax Evasion Practices (b) Opportunities for Tax evasion on technical companies' value chain and (c) Effectiveness of potential measures against tax evasion, are critically assessed against the literature surveyed above.

As shown in Table 3:

It is evident that tax evasion in technical companies is mostly committed by companies registered in the country, and that a common method for tax evasion for companies established in countries with high VAT rate is to carry out technical activities in countries with low VAT rate.

As presented in Table 4:

Greek technical companies use a variety of tax evasion strategies, as also argued by Freire-Serene & Panades [13]. Transport services and suppliers' overbilling are mostly considered as providers of tax evasion opportunities. The exploit of tax evasion opportunities provided to technical companies through transport services, advertising business and short-term leases of real estate evidenced, is also reported [5, 12].

As illustrated in Table 5:

It is evident that the existence of further controls and write-offs of fines by the Ministry of Finance enhances tax evasion in technical companies, while regarding the countermeasures, the stabilization of the tax legislative framework is the most

Table 3 Tax evasion practices

Tax evasion practices	*N*	Average	Standard deviation
Extent to which tax evasion in technical companies is related to the receipt of fictitious invoices, e.g., in order to reduce tax liability and profits	154	3.93	0.789
Extent to which tax evasion in technical companies is based on the concealment of transactions	154	4.01	0.842
Extent to which tax evasion in technical companies is committed by a registered domestic company	154	4.15	0.829
Extent to which tax evasion in technical companies is committed by a non-registered third-country national company, in the Member State where the construction works and delivery of services take place	154	3.95	0.848
Extent to which technical companies that tax evade follow the method of physical establishment in a country with a high VAT rate and carry out the activity of technical works in a country with a low VAT rate	154	4.10	0.822

Table 4 Opportunities for tax evasion on technical companies' value chain

Opportunities for tax evasion across technical companies' value chain	*N*	Average	Standard deviation
Extent to which the education of workers and employees of technical companies provide opportunities for tax evasion	154	3.89	0.813
Extent to which transport services provide opportunities for tax evasion for technical companies	154	4.05	0.82
Extent to which advertising business provides opportunities for tax fraud for technical companies	154	3.97	0.855
Extent to which short-term leases and leases provide opportunities for tax evasion for technical companies	154	3.95	0.82
Extent to which subcontracting technical works provides opportunities for tax evasion for technical companies	154	3.84	0.745
Extent to which overbilling of suppliers provides opportunities for tax evasion for technical companies	154	4.07	0.804

Table 5 Effectiveness of potential measures against tax evasion

Effectiveness of potential measures against tax evasion	*N*	Average	Standard deviation
Extent to which high taxation is responsible for the existence of tax evasion in technical companies	154	3.97	0.813
Extent to which the reduction of tax rates, contributions, fees etc. would contribute to the reduction of tax evasion in technical companies	154	3.98	0.835
Extent to which the constant change in tax legislation is responsible for tax evasion in technical companies	154	3.97	0.828
Extent to which the stabilization of the tax legislative framework would contribute to the reduction and disappearance of tax evasion in technical companies	154	4.07	0.804
Extent to which the existence of further checks and fine write-offs by the Ministry of Finance enhances tax evasion in technical companies	154	4.00	0.835
Extend to which improvement of the tax control mechanism would contribute to reducing tax evasion in technical companies	154	3.98	0.762
Extend to which the tightening of fines, penalties etc. would contribute to the reduction of tax evasion in technical companies	154	3.93	0.774

important factor that would contribute to the reduction of technical companies' tax evasion, although the improvement of the tax control mechanism and the reduction of tax rates, contributions, fees, etc. and finally the tightening of fines, penalties, etc. can play a very important role too. The results obtained suggest tightening of the tax evasion framework, combined to improving the tax control mechanism. As discussed by Caballé & Panadés [29] the revenues from penalties imposed on tax evaders used to finance public capital enhance further growth and development as they enable public capital formation by (a) leading taxpayers to behave more honestly, therefore ending up paying more taxes, and (b) creating additional resources from the fines paid by tax evaders.

The above results indicating that improvement of tax evasion controls and rationalization of tax policy as critical factors are in line with those of Dianeosis [9], which also suggests the increase of use of plastic money and electronic invoicing for both individuals and legal entities, the modernization the tax collection mechanism through digitization of records and electronic interconnection of services, coupled with the enhancement of knowledge, skills and competences of human resources.

5 Concluding Remarks

As Alm [1] points out, tax evasion results in tax revenues' reduction, thus affecting the taxes faced by taxpayers and the public services received by citizens. In addition to revenue losses, tax evasion leads to misallocation of resources when people change their behavior to defraud tax services, such as their choice of working hours, and their occupation. Governments must spend resources to identify, measure and punish non-compliance. Non-compliance alters the distribution of income in arbitrary, unpredictable and unfair ways to society as a whole.

According to Gërxhani et al. [6], individuals and companies pay taxes only because they have to (i.e., because they believe that if they do not then they will be penalized); thus, the probability of being caught for tax evasion is a very important factor shaping their behavior [7]. Tax ethics provide a framework within which people can think of paying taxes as payment for government services or "political exchange". In this light, there is a relationship between concepts such as trust, social capital and solidarity. Tax evasion is related to the extent to which citizens trust the government to be honest and to provide the services promised [6].

According to Vasardani [5], a broad policy package should be put in place to effectively tackle the complex phenomenon of tax evasion, focusing on three main objectives:

(a) Simplification of Tax legislation in order to ensure a flexible and rational regulatory framework and reduce the need for explanatory circulars is a key prerequisite for effective tax policy design and implementation. Reviewing taxes with low yields and high costs (assessment costs, collection costs etc.) compared to expected returns should be reviewed. The existence of a multitude of tax systems and constant revision of the relevant legislation can de facto cause confusion.

(b) Reform and mobilization of the tax administration in order to be possible to cross-check and monitor information over time through a central integrated information system, eliminating tax corruption and reducing personal contact between tax auditors/collectors and taxpayers. Measures that may contribute to this goal include dismissal from service of auditors who accept bribes; complete separation of audit and collection services; decrease of the number of local revenue offices while increasing the powers of central services; and full computerization of basic operations such as taxpayer registration and monitoring the filing of tax returns.

(c) Improve taxpayers' understanding of the tax framework and of the different options available to them. This, in turn, will help to enhance voluntary tax compliance, to free administrative resources to target specific groups that represent a high risk of tax evasion or have a history of non-compliance.

Greece could attain these goals exploiting the opportunities the European program Fiscalis 2020 [36] in which participates, which "allows tax administrations to cooperate with each other in the exchange of information, experience and knowledge". Among the objectives of this program is to support the fight against tax fraud and tax evasion. The scope of the program covers VAT, excise duties, taxes on energy products and electricity as well as other taxes as far as they relate to the internal market and administrative cooperation between Member States.

The main strength of the present study is that it illustrates the opinions of the accountants of a significant group of technical companies, therefore providing a very in-depth review of tax evasion, still the practical utilization of its findings is limited by the fact that they only reflect technical companies' approach. An analogous study in the future, extended to all other major sectors of the Greek economy, would provide a more comprehensive view of tax evasion. Such study could include the use of modeling and simulation systems [37–41]. Furthermore, as tax evasion also exists among individuals, if an extensive survey would be conducted among individuals, interesting results would be obtained and could be compared to those of companies (especially as countermeasures for limiting tax evasion are concerned). A future study could also focus on the phenomenon of tax evasion based on different tax categories' analysis, as they are characterized by different levels of evasion and different incentives. Similarly, the views of individuals regarding their tax compliance on the bases of different types of taxes, and the factors that drive them to such behavior, could be examined. Such surveys could potentially lead to more concrete proposals for practical implementation.

References

1. Alm, J.: Tax evasion, labor market effects, and income distribution (2014)
2. Demalijaj Ukaj, S.: Tax evasion and the impact on economic growth. Acta Universitatis Danubius. **10**(6), 18 (2014)
3. EBRD: How we assess transition qualities (n.d.)
4. Gomez Sabaini, J.C., Jimenez, J.P.: Tax structure and tax evasion in Latin America (2012)
5. Vasardani, M.: Tax evasion in Greece: an overview. Bank Greece Econ. Bullet. **35**, 15–24 (2011)
6. Gërxhani, K., Wintrobe, R.: Understanding tax evasion: combining the public choice and new institutionalist perspectives. In: Douarin, E., Havrylyshyn, O. (eds.) Handbook of Comparative Economics, pp. 785–810. Springer, New York (2021)
7. Hammar, H., Jagers, S.C., Nordblom, K.: Perceived tax evasion and the importance of trust. J. Socio-Econ. **38**(2), 238–245 (2009)

8. Moore, M.: Revenues, state formation and the quality of governance in developing countries. Int. Polit. Sci. Rev. **25**(3), 297–319 (2004)
9. Dianeosis: Income taxation in Greece. Comparative analysis and reform proposals (2018) (in Greek)
10. Rapanos T., Kaplanoglou G.: Taxation and economic development. The case of Greece. Proceedings of the Hellenic Bankers Association: competitiveness for growth: policy proposals. pp. 609–638 (2014) (in Greek)
11. Schneider, F., Kirchler, E., Maciejovsky, B.: Social Representations on Tax Avoidance, Tax Evasion, and Tax Flight: Do Legal Differences Matter? Johannes Kepler University of Linz (2001)
12. Broumas, A.: Η Φοροδιαφυγή στην Ελληνική Νομοθεσία (2017). (in Greek)
13. Freire-Seren, M.J., Panades, I., Marti, J.: Tax avoidance, human capital accumulation and economic growth. Econ. Model. **30**, 22–29 (2013)
14. Kirchler, E., Maciejovsky, B., Schneider, F.: Everyday representations of tax avoidance, tax evasion, and tax flight: do legal differences matter?. SFB 373 Discussion Paper, No. 2001,43, Humboldt University of Berlin (2001)
15. European Commission: European economic forecast. Autumn (2020)
16. European Parliament: Report on financial crime, tax evasion and tax avoidance (2018/2121 (INI) (2018). (in Greek)
17. Dianeosis: tax evasion in Greece. Causes, extent and proposals to combat it (2016) (in Greek)
18. Plikas, J.H., Dimitrios, N.K., Delis, P., Vlachos, D.S.: Software protection and piracy focusing on the 2008 crisis: a comparative study and simulation modeling regarding the case of Greece, Germany, and England. In: Strategic Innovative Marketing, pp. 319–326
19. ELSTAT: Construction (Structural research) (2021) (in Greek)
20. ELSTAT: Production index in construction (2021) (in Greek)
21. Tsiobanou, M.: The construction sector today (2021) (in Greek)
22. Bitzenis, A., Vlachos, V., Schneider, F.: An exploration of the Greek Shadow economy: can its transfer into the official economy provide economic relief amid the crisis? J. Econ. Issues. **50**(1), 165–196 (2016)
23. Cerqueti, R., Coppier, R.: Economic growth, corruption and tax evasion. Econ. Model. **28**, 489 (2009)
24. Varvarigos, D.: Cultural norms, the persistence of tax evasion, and economic growth. Working Paper no 15/10, University of Leicester (2015)
25. Bethencourt, C., Kunze, L.: Tax evasion, social norms and economic growth. Working paper No. 48427, Munich Personal RePEc Archive. (2013)
26. Blackburn, K., Bose, N., Capasso, S.: Tax evasion, the underground economy and financial development. J. Econ. Behav. Organiz. **83**(2), 243–253 (2012)
27. Roubini, N., Sala-i-Martin, X.: A growth model of inflation, tax evasion, and financial repression. J. Monetary Econ. **35**, 275–301 (1995)
28. Chen, B.L.: Tax evasion in a model of endogenous growth. Rev. Econ. Dynam. **6**, 381–403 (2003)
29. Caballé, J., Panadés, J.: Tax evasion and economic growth. Publ. Financ. **52**, 318 (2000)
30. Ameyaw, B., Addai, B., Ashalley, E., Quaye, I.: The effects of personal income tax evasion on Socio-economic development in Ghana: a case study of the informal sector. Br. J. Econ. Manag. Trade. **10**(4), 1–14 (2015)
31. Bostina, F.: Impact of tax evasion on the economic growth in the European Union. Ann. Econ. Ser. Constantin Brancusi Univ. **6**, 163–169 (2017)
32. Jugurnath, B., Soondrum, H., Seedoyal, A., Kisto, H.: Tax evasion and economic growth: evidence from Mauritius. Int. J. Manag. Appl. Sci. **3**(7), 15–19 (2017)
33. Islam, A., Rashid, H.U., Hossain, S.Z., Hashmi, R.: Public policies and tax evasion: evidence from SAARC countries. Heliyon. **6**, e05449 (2020)
34. Frey, B.S., Torgler, B.: Tax morale and conditional cooperation. J. Comp. Econ. **35**(1), 136–159 (2007)

35. Blaufus, K., Braune, M., Hundsdoerfer, J., Jacob, M.: Does legality matter? The case of tax avoidance and evasion, arqus Discussion Paper, No. 193, Arbeitskreis Quantitative Steuerlehre (arqus), Berlin (2015)
36. AADE: Fiscalis 2020 (2021) (in Greek)
37. Dimitrios, N.K., Sakas, D.P., Vlachos, D.S.: Analysis of strategic leadership simulation models in non-profit organizations. Procedia Soc. Behav. Sci. **73**, 276–284 (2013)
38. Nasiopoulos, D.K., Sakas, D.P., Vlachos, D.S., Mavrogianni, A.: Modeling of market segmentation for new IT product development. AIP Conf. Proc. **1644**, 51–59 (2015)
39. Chantzaras, A., Dimitrios, N.K., Vlachos, D.S.: Mobile commerce and success factors. simulation and modeling of the problem. In: Strategic Innovative Marketing, pp. 349–355. Springer
40. Sakas, D.P., Nasiopoulos, D.K., Reklitis, P.: Modeling and simulation of the strategic use of marketing in search engines for the business success of high technology companies. In: Business and Economics, pp. 217–226. Springer
41. Plikas, J.H., Nasiopoulos, D.K., Giakomidou, D.S.: The contribution of social media in the management of social relations with customers, modelling and simulation of the problem. In: Strategic Innovative Marketing, pp. 515–521. Springer

Taxation and the Greek Financial Crisis

Boufounou Paraskevi, Vasiliki Athanasiou, Kanellos Toudas, and Lida A. Galani

1 Introduction

By imposing taxes, the state saves the resources needed to serve the needs of its citizens, such as public safety, public health, education and the administration of justice. The imposition of taxes, particularly income tax, serves to more fairly distribute wealth and to reduce economic and social inequalities. Furthermore, the imposition of taxes, especially indirect taxes and tariffs imposed on imported goods, aims to protect domestically produced goods and develop competition between economic units against other countries [3, 4]. The tax burden is intricately tied to the tax capability of each type of firm, and changes in taxes have the effect of increasing or lowering citizen income and governmental revenues. Its significance is even more critical in times of economic crisis.

The tax burden on corporate activity in Greece and its repercussions before and during the economic crisis are investigated in this study. In particular, an empirical study is conducted and the periods 2005–08 and 2009–12 are compared. The study is structured as follows: At the beginning, the issues addressed in this study are presented. This is followed by a discussion of the research methods and a critical

B. Paraskevi
National and Kapodistrian University of Athens, Athens, Greece

V. Athanasiou
University of Economics and Business, Athens, Greece

K. Toudas (✉)
Agricultural University of Athens, Athens, Greece
e-mail: kstoudas@aua.gr

L. A. Galani
Department of Economics, University of Peloponnese, Tripolis, Greece

D. P. Sakas et al. (eds.), *Computational and Strategic Business Modelling*,
Springer Proceedings in Business and Economics,
https://doi.org/10.1007/978-3-031-41371-1_39

analysis of the primary findings. Finally, the main conclusions are summarized and thoughts for further research are developed.

2 Taxation

The importance of taxes is even more critical in the period of the financial crisis. Table 1 shows the evolution of tax rates for legal entities in Greece for the period 2005–2012 examined in this study.

With a net taxable profit of 100,000 euros and the assumption that no regular reserve and advance tax is calculated and that the remaining profit after income tax is distributed to the shareholders, Table 2 provides an example of how the tax burden of a public limited company can be calculated using the rates in Table 1.

As illustrated in Table 3, there is a continuous downward trend in the tax rates on corporate profits in the Member States of the European Union, namely, the average since 2005 was 25.5%, in 2008 it reached 24%, after the peak of the crisis in 2010 it fell to 23.3% and in 2012 it reached 23%. The only country that is an exception is Malta. We note, however, that from 2005 to 2012, the average of the 17 countries of the European Union is higher than the average of the 27 countries. Due to the inclusion of Member States with relatively low tax rates, such as Bulgaria and Cyprus, the 27-country zone's average is 25.4 percent. It is also clear that Greece's tax rates are much higher than those in all of its surrounding nations. This makes it particularly challenging to draw in and keep investment, especially when combined with the bureaucracy and tax system's complexity.

Due to the diversity of the factors and the globalization of economic activity, member nations now compete with one another to draw in international investment. Due to tax rivalry, especially with the more recent members, particularly those in the Balkan region, who have much lower tax rates, the majority of member states have decreased tax revenues. The corporate tax rate in Bulgaria is 10%. This is one of the lowest rates in Europe together with that of Cyprus, which is also 10%. So that is why there is a relocation of businesses to Bulgaria and an increase in the activity of Greek businesses there. Entrepreneurs profit from the business-friendly environment and the streamlined processes for starting, registering and running a firm in addition to the lower tax rate. Besides, according to the World Bank, in the Doing Business 2015 ranking, Bulgaria ranks 38th in terms of ease of doing business compared to Greece, which ranks 61. Also, Romania has a business tax rate of 16%, Latvia and Lithuania have a tax rate of 15%, while the Czech Republic and Slovakia have tax rates of 19% and 23%, respectively. It is worth noting that in Slovakia, the rate was also 19% but last year it became 23%. Finally, in Slovenia, the tax rate is 17%. It is important to note that Greece, after Portugal, the Czech Republic and Poland, has implemented 20 reforms to boost entrepreneurship in the period 2005–2014 and ranks fourth among OECD countries in terms of the number of reforms aimed at stimulating entrepreneurial ideas and restoring economic growth, in addition to the low tax rate, which is an initial incentive to start a business.

Table 1 Tax rates by type of legal entity in Greece

Type	-General partnerships —Limited partnerships	—Civil law societies	—Joint ventures (Art. 2§2 CBR) —Civil-law partnerships —Joint stock companies —Silent partnerships	—S.A. companies —Companies listed in ATHEX	Dividend payout ratio	Board of Directors remuneration ratio	—Ltd companies	—OTHERS (as defined by article 109 Income Tax Code)
2012	20%	20%	25%	20%	25%	35%	20%	20%
2011	20%	20%	25%	20%	25%	35%	20%	20%
2010	20%	20%	25%	24%	21%	35%	24%	24%
2009	20%	20%	25%	25%	10%	35%	25%	25%
2008	20%	20%	25%	25%	10%	25%	25%	25%
2007	20%	20%	25%	25%	-	25%	25%	25%
2006	22%	22%	29%	29%	-	29%	29%	29%
2005	24%	24%	32%	32%	-	32%	32%	32%

Source: http://www.taxheaven.gr/pages/klimakes

Table 2 Tax burden of a company (case-study)

Year	2005	2006	2007	2008	2009	2010	2011	2012
Profit for the year	100.000,00	100.000,00	100.000,00	100.000,00	100.000,00	100.000,00	100.000,00	100.000,00
Tax burden	32%	29%	25%	25%	25%	24%	20%	20%
Income tax on taxable profits	32.000,00	29.000,00	25.000,00	25.000,00	25.000,00	24.000,00	20.000,00	20.000,00
Balance of profits to be distributed	68.000,00	71.000,00	75.000,00	75.000,00	75.000,00	76.000,00	80.000,00	80.000,00
Profit-sharing ratio	–	–	–	10%	10%	21%	25%	25%
Tax on distributed profits	0	0,00	0,00	7.500,00	7.500,00	15.960,00	20.000,00	20.000,00
Total tax burden	32.000,00	29.000,00	25.000,00	32.500,00	32.500,00	39.960,00	40.000,00	40.000,00
Tax rate	32,00%	29,00%	25,00%	32,50%	32,50%	39,96%	40,00%	40,00%

Table 3 Corporate income tax rates in the EU

Year	2005	2006	2007	2008	2009	2010	2011	2012	2013
Belgium	34.0	34.0	34.0	34.0	34.0	34.0	34.0	34.0	34.0
Bulgaria	15.0	15.0	10.0	10.0	10.0	10.0	10.0	10.0	10.0
Czech Republic	26.0	24.0	24.0	21.0	20.0	19.0	19.0	19.0	19.0
Denmark	28.0	28.0	25.0	25.0	25.0	25.0	25.0	25.0	25.0
Germany	38,7	38,7	38,7	29,8	29,8	29,8	29,8	29,8	29,8
Estonia	24.0	23.0	22.0	21.0	21.0	21.0	21.0	21.0	21.0
Ireland	12,5	12,5	12,5	12,5	12,5	12,5	12,5	12,5	12,5
Greece	**32.0**	**29.0**	**25.0**	**35.0**	**35.0**	**24.0**	**20.0**	**20.0**	**26.0**
Spain	35.0	35.0	32,5	30.0	30.0	30.0	30.0	30.0	30.0
France	35.0	34,4	34,4	34,4	34,4	34,4	34,4	36,1	36,1
Italy	37,3	37,3	37,3	31,4	31,4	31,4	31,4	31,4	31,4
Cyprus	10.0	10.0	10.0	10.0	10.0	10.0	10.0	10.0	10.0
Latvia	15.0	15.0	15.0	15.0	15.0	15.0	15.0	15.0	15.0
Lithuania	15.0	19.0	18.0	15.0	20.0	15.0	15.0	15.0	15.0
Luxemburg	30,4	29,6	29,6	29,6	28,6	28,6	28,8	28,8	29,2
Hungary	17,5	17,5	21,3	21,3	21,3	20,6	20,6	20,6	20,6
Malta	35.0	35.0	35.0	35.0	35.0	35.0	35.0	35.0	35.0
The Netherlands	31,5	29,6	25,5	25,5	25,5	25,5	25.0	25.0	25.0
Austria	25.0	25.0	25.0	25.0	25.0	25.0	25.0	25.0	25.0
Poland	19.0	19.0	19.0	19.0	19.0	19.0	19.0	19.0	19.0
Portugal	27,5	27,5	26,5	26,5	26,5	29.0	29.0	31,5	31,5
Rumania	16.0	16.0	16.0	16.0	16.0	16.0	16.0	16.0	16.0
Slovenia	25.0	25.0	23.0	22.0	21.0	20.0	20.0	18.0	17.0
Slovakia	19.0	19.0	19.0	19.0	19.0	19.0	19.0	19.0	23.0
Finland	26.0	26.0	26.0	26.0	26.0	26.0	26.0	24,5	24,5
Sweden	28.0	28.0	28.0	28.0	26,3	26,3	26,3	26,3	22.0
United Kingdom	30.0	30.0	30.0	30.0	28.0	28.0	26.0	24.0	23.0
Iceland	18.0	18.0	18.0	15.0	15.0	18.0	20.0	20.0	20.0
Norway	28.0	28.0	28.0	28.0	28.0	28.0	28.0	28.0	28.0
EU-27	**25,5**	**25,3**	**24,5**	**24.0**	**23,9**	**23,3**	**23,1**	**23.0**	**23,2**
EA-17	**28,1**	**27,7**	**26,8**	**26,3**	**26,2**	**25,6**	**25,3**	**25,4**	**25,9**

Source: Taxation trends in the European Union [8]

Businesses that cannot relocate to another country to enjoy lower tax rates often resort to tax evasion. In general, any act or omission by taxpayers intended to lessen or avoid paying the required taxes is considered tax evasion. The following traits are mentioned in another definition: It is the portion of income tax that people do not voluntarily declare to the tax authorities as they are required to do by law. In addition, it is also defined as that part of indirect and other direct taxes which, under the law, should be collected and voluntarily remitted to the state, but which is never collected by taxpayers [12].

The concept of tax avoidance or tax fraud, which can be found in the international literature, is another concept connected to tax evasion because they both have similar effects. As opposed to tax evasion, which is the circumvention of simply the letter of the law, tax avoidance is described as the complete or partial avoidance of tax [12]. Therefore, in order for many economists to be able to define tax evasion, they considered that an act of this kind must have the following characteristics: an element of dishonesty, fictitiousness or forgery to avoid paying tax; it must aim to reduce the tax burden due to the existence of loopholes, ambiguities in the legislation against the will of the legislator; and there must be complete secrecy as the emergence of the specific act means immediate action to combat it by the competent authorities.

It should be noted that tax evasion is not the same as the underground economy, but tax evasion is the main cause of the creation of the underground economy and is part of the overall economic activity [2]. Informal economy is defined as all economic activities that contribute to the creation of added value, and while they should be included in the national income, they are not counted due to the inadequacy of the relevant services [10]. Informal economy is also referred to as: unofficial, hidden, black, unrecorded, shadow, under-ground, grey, parallel, dual, moonlight, unobserved, etc.

Tax evasion varies from country to country and from taxpayer to taxpayer. Many economists have concluded that the main causes of tax evasion due to its complexity are as follows [12]:

1. The educational and cultural level. The lower the people's educational level, the greater the tendency to tax evasion. They do not understand the reason for the existence of taxes to cover public needs, and their individual needs prevail over collective needs.
2. The social justice perception of the distribution of the tax burden. Many taxpayers are consistent with their tax obligations, but seeing that someone who evades taxes survives better and with more profit creates competition and for the sake of defense and survival, the tax consciousness of the legitimate taxpayer stops and equates to a tax evader.
3. The extent of the tax burden, since tax evasion tends to increase as tax burden increases. The foregoing situation can be explained by the fact that there is less money available to meet and satisfy the requirements of taxpayers the higher the tax load is. In higher income brackets, where the advantage is greater, it becomes more obvious.
4. The management of public money. The better the quality of government spending to meet social needs, the lower the tendency to tax evasion as taxpayers remain happy with the use of public money, i.e. money contributed by the taxpayers themselves.
5. The functioning and structure of the tax system. The more the complex and unstable a tax system is, the greater the propensity for tax evasion. How the tax is levied is also an important factor indicating the tendency to tax evasion, as it would be different to levy a tax at the industrialization stage and different at the

retail stage, where there are more businesses with lower incomes, worse accounting records that make it difficult to control. Also, the categorization of taxation by the source of the origin of income creates flexibility in terms of preventing tax evasion, i.e. employees and pensioners are taxed differently and self-employed persons differently.
6. The organizational structure and proper utilization of tax and audit services for the proper functioning of the tax system greatly reduce tax evasion.
7. The organization of the economy is a basic prerequisite for reducing the tendency to tax evasion, as it is possible to record transaction data and use them by the competent audit bodies.

The consequences of tax evasion for both economic policy and the functioning of the economy are negative [12]:

1. There is an unfair distribution of the tax burden, as it leads to different tax burdens for taxpayers without having the same tax capacity. Due to the existence of tax evasion, the state, to keep its revenues at the desired level, does not reduce tax rates and imposes tax burden on the consistent taxpayers.
2. The state's tax revenues are reduced, and it is unable to finance its expenditures. Consequently, it turns to measures of a net revenue raising nature such as indirect tax increases, special levies, borrowing with negative effects on both the social and economic levels. Moreover, we are not forgetting the negative impact of the increase in debt, which leads to huge budgetary problems.
3. The effectiveness of the tax system and economic policy is reduced. For example, if the state promotes some tax measures such as tax incentives, those who evade taxes have more flexibility to adapt to this policy but with less benefit, which means inefficiency of this measure. Also, the existence of tax evasion mainly affects the low and middle income classes since it has a positive relationship with the maintenance of the tax burden.
4. It creates problems in the competition of enterprises, as the enterprises that evade taxes have a comparative advantage of survival, earning a higher profit margin, liquidity over those that do not evade taxes.
5. Finally, it creates a negative impact on the taxpayers' tax awareness and spreads with great speed, with the result that consistent taxpayers also become tax evaders.

Given that tax evasion and tax fraud is a global phenomenon, and its fight is an immediate need for the existence of a tax law, in a joint effort of all Member States of the European Union, the European Commission has proposed some European measures to combat these phenomena to ensure the efforts of fiscal consolidation and fairness of tax systems, with a series of actions that were harmonized and compatible with the multiannual financial frameworks 2007–2013 and 2014–2020 [1]. The concept of tax audit includes all audit procedures on the books and records kept by economic units to verify and check the declarations submitted by the taxpayers. Due to the complexity of the tax system, tax audit is an arduous process, with great importance as the accurate determination of the taxpayers' net taxable income implies the fairer distribution of tax burdens and tax justice for the existence of a favoured state [3]. According to Koromilas G. and Giannopoulou C., tax audit in the Greek territory is divided into three main categories [5, 6].

1. The preventive audit, which controls a company's actions or omissions in the current fiscal year. It creates the impression to taxpayers that there is continuous and effective control. The characteristic of the preventive audit is its short duration and its substantial nature.
2. The temporary audit, whereby a part of the taxpayers' income is partially attributed to the State Treasury. The head of the public financial service issues an audit order to examine one or more subjects and objects of the full and final tax audit. With the ability to produce temporary audit sheets without having to undertake a normal audit, the temporary audit combines the preventative and regular audits, which we shall discuss below [11]. For example, the head of a tax office may issue an audit sheet for companies that have not submitted returns or submit negative returns, as well as for companies for which there is information on tax evasion.
3. Finally, the regular audit under Article 66 of the Income Tax Code [11], which is aimed at the full and final control of the obligations of an enterprise and the attribution of any violations.

In particular, from 2005 to 2012, three significant orders—POL.1037/1.3.2005, POL.1072 /8.4.2011 and POL 1178/23.8.2011—were in force for the control of traders' tax affairs. These orders indicated a number of criteria for the selection of cases for audit through the risk analysis function of the ELENXIS system (it should be noted that POL 1178/23.8.2011 specifies 87 criteria for the selection of instances for audit), including (indicatively):

- Substantial violations of the VAT Code or information sheets on tax evasion, failure to submit declarations such as VAT, income tax, large VAT credit balances which are not justified by the scope of activity of each company, confiscated books and data and loss of books and data but also tax cash registers and mechanisms
- The large expenses that give rise to suspicion, the incorrect application of the anticipated net profit rate, the difference between supplementary income tax and VAT returns, the existence of a loss, the low gross profit rate and the use of development laws.
- The case of exemption from VAT under the provisions of POL 1262/2.8.1993
- Large intra-community transactions
- The existence of repeated requests for mutual assistance from other Member States
- Cases where the limitation period is about to expire
- The non-acceptance of a tax clearance certificate under paragraph 1 of Article 9 of Law 3888/2010
- The winding-up procedure

Despite the fact that numerous laws have been passed on the procedures and conditions for the control of economic units and the performance of the Greek tax system, all studies show that public revenues do not increase from taxation. These conclusions are supported by Eurostat data, proving the effectiveness of the tax collecting and control system. Greece's informal economy is larger compared to that of other EU countries, leading Greece to a higher rate of tax evasion. In order for the state's tax control mechanism to cope, it will have to address some of its weaknesses [12], including high administrative costs (paying the salaries of a large number of employees in the Ministry of Finance). The complexity and variety of the tax legislative framework, which is currently a patchwork of laws, ministerial decisions and circulars, add to the inefficiency of the control and revenue collection process. The complex Greek tax code is a fact that necessitates quick tax reform and is of interest to many academics.

3 Methods and Materials

For the present research, a special questionnaire was prepared and responses were collected from a sample of 120 professionals from the financial-accounting sector covering all the sectors active in the region of Epirus. The sample consists of 65.8% men and 34.2% women. Also, as presented in Fig. 1:

(a) The smallest percentage of participants (1.7%) belongs to the age group f18–25 years old, followed by the age group 25–34 years old (16.7%), which indicates that there are no tax incentives or investment incentives to boost youth

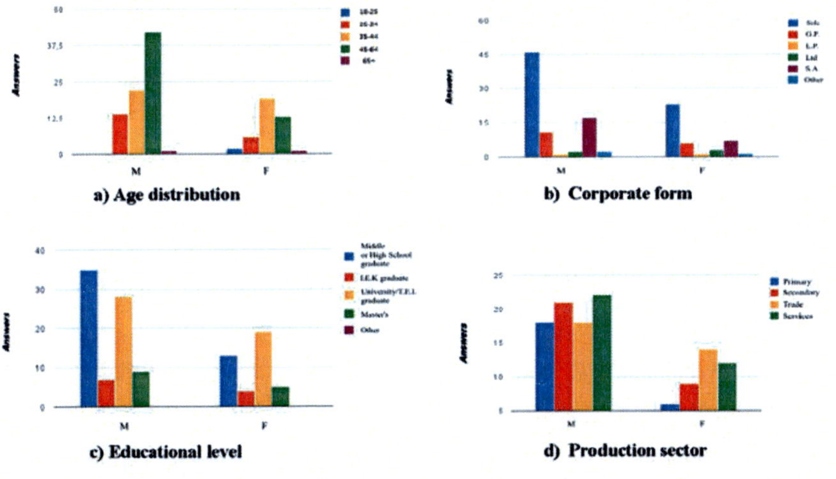

Fig. 1 Sample characteristics. **(a)** age distribution. **(b)** corporate form. **(c)** educational level. **(d)** production sector

entrepreneurship in the Greek economy. In addition, in the age group 18–25 the majority is female, in the age group 25–34 our sample is 70% male and 30% female, in the age group 35–44 it is 54% male and 46% female, in the age group 45–64 76% male and 24% female and finally in the age group 65+ it is equally divided between women and men. So we can see that female entrepreneurship has an advantage in the younger age bracket.

(b) The predominant legal form is that of the sole proprietorship (57.5%), followed by société anonyme (S.A) (20%). The legal form of the partnership – general partnership (G.P) comes third with 14.2%. Overall, 67% of the men and 33% of the women in the sample have a sole proprietorship, 65% of the men and 35% of the women have a general partnership, while the percentages of men and women for the legal form of the limited partnership (L.P) are 50% men and 50% women. The legal form of the Limited Liability Company (Ltd) appears to be used by 60% of women and S.A. by 29% of women.

(c) Overall, 40% are secondary school graduates, and almost four out of ten respondents are higher education graduates. Overall, 73% of men are secondary school graduates. Male IEK (Institutes of Vocational Training) graduates make up 64% of the sample, while female IEK graduates make up 36%. Male tertiary school graduates make up 60% of the sample, while female tertiary school graduates make up 40%. Finally, male postgraduate degree holders make up 64% of the sample, while female postgraduate degree holders make up 36%.

(d) The sample is almost evenly distributed across the four production sectors. 20% belongs to the primary sector, 25% belongs to the secondary sector, 26.7% belongs to trade and 28.3% belongs to services, as shown in Fig. 1. In the primary sector, the quota of men and women is as follows: 75% men and 25% women, in the secondary sector 70% and 30%, respectively, in trade 56% and 44%, respectively, and in services 65% and 35%, respectively.

4 Results

According to the study's findings, while turnover increased, businesses' financial performance during the crisis era declined while their tax burden increased, as illustrated in Fig. 2:

(a) The average turnover of the sample in the period 2005 to 2008 is lower than the average turnover in the period 2009 to 2012, i.e. turnover increased despite the economic crisis.

(b) However, the average accounting profit of firms in the period 2005 to 2008 is higher than the average accounting profit in the period of the financial crisis 2009 to 2012 (32.15%),

(c) The average off-balance sheet profits of enterprises in the period 2005–2008 (83.299,00 EUR) is much higher than the average off-balance sheet profits in the crisis period 2009–2012 (29.837,00 EUR) as in the precrisis period, many

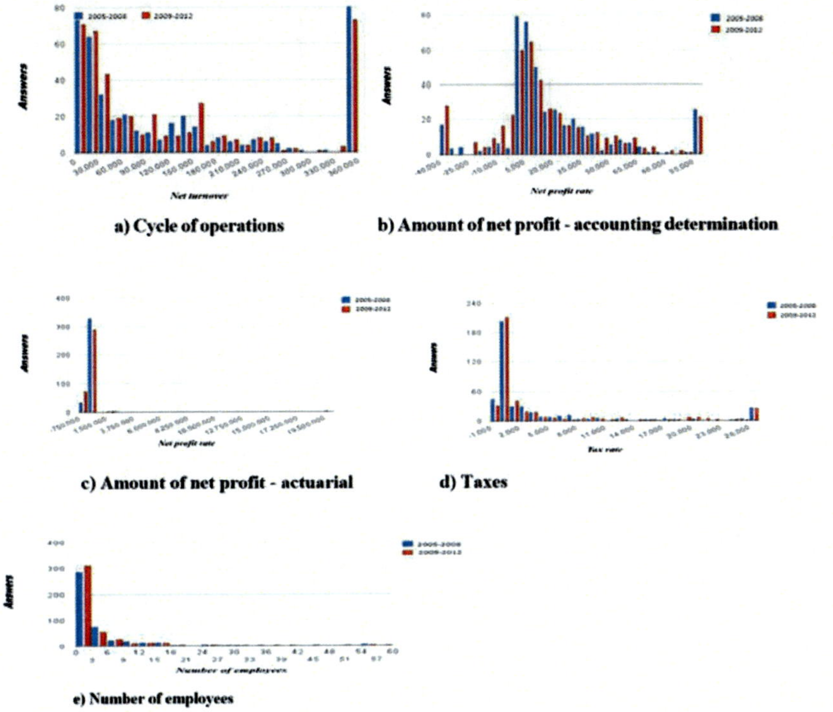

Fig. 2 Evolution of key financial indicators (turnover, earnings, employees). (**a**) Cycle of operations, (**b**) amount of net profit – accounting determination, (**c**) amount of net profit – actuarial, (**d**) taxes, (**e**) number of employees

enterprises took advantage of the so-called self-cleaning/self-control based on Law 3299/2004 in order to avoid any form of control. Between the two periods under consideration, there is a decrease in the average off-balance sheet profit of 64.18% in the crisis period. The average profit in both periods ranges between 5000.00 and 20,000.00 euros and in the crisis period 2009–2012 the losses of the companies increased.

(d) The average tax payable in the sample in the period 2005–2008 is higher than the average tax payable during the crisis period 2009–2012 by 7%. Thus combined with the aforementioned decrease in average accounting and off-balance sheet profits by 32.15% and 64.18% (i.e. the net income taxed by a firm), it becomes evident that the tax burden on firms in the crisis period was enormous.

(e) The average number of employees in the sample in the period 2005–2008 (4.06 persons) is higher than the average number of employees in the period 2009–2012 (3.81 people).

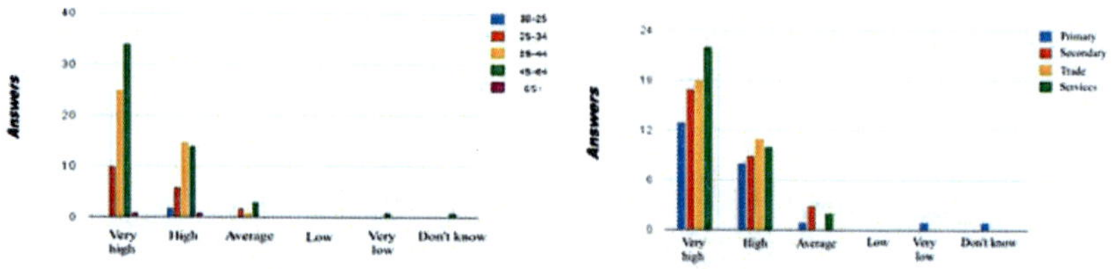

a) by age group b) by industry

Fig. 3 Assessment of the level of tax burden: (**a**) by age group, (**b**) by industry

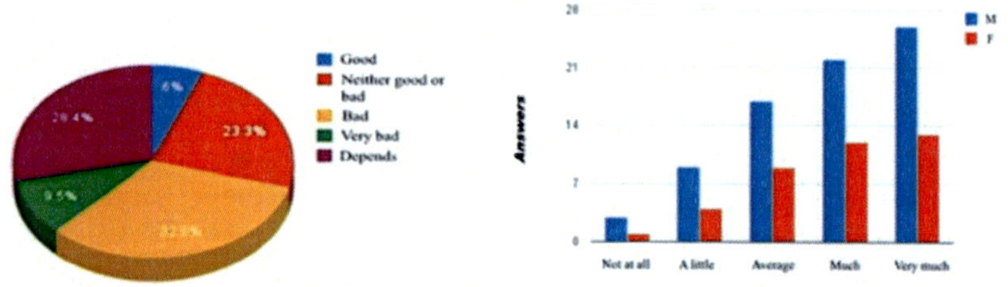

a) Assessment of revenue concealment b) Unjustified tax evasion

Fig. 4 Assessment of tax evasion. (**a**) Assessment of revenue concealment. (**b**) Unjustified tax evasion

As shown in Fig. 3, it is estimated that the general level of tax burden in Greece is very high across all age groups and all productive sectors, which is indeed confirmed by Table 3.

Survey respondents noted that the increase in tax rates during the financial crisis period increased the incentive to evade tax compared to the precrisis period, and this answer holds true regardless of gender, education level and age group. In fact, given that the tax burden in Greece during the financial crisis period is higher than in the pre-financial crisis period, and that a high tax burden is an incentive for tax evasion, it can be concluded that the incentive for tax evasion increased during the financial crisis. In particular, 78.5% of the sample believes that tax evasion would be reduced if tax rates were reduced and 25.2% do not agree that there would be such an impact. In similar surveys by Passadis [9] and Tatsos [12], the respective agree and disagree responses are 74.8% and 57% and the disagreeing responses are 25.2% and 43%, respectively. That is, with the financial crisis, there is an increasing perception that the measure of reducing tax rates will have a positive effect on reducing tax evasion. This may be an indicator of the psychological burden that Greek taxpayers have suffered from the general economic downturn.

As shown in Fig. 4,

(a) In our survey, three out of ten respondents said that a company hiding a small part of its income is bad and dependent, two out of ten neither good nor bad and one out of ten very bad. In the similar survey [12], four out of ten thought it was bad, two out of ten answered dependent and neither good nor bad, and one out of

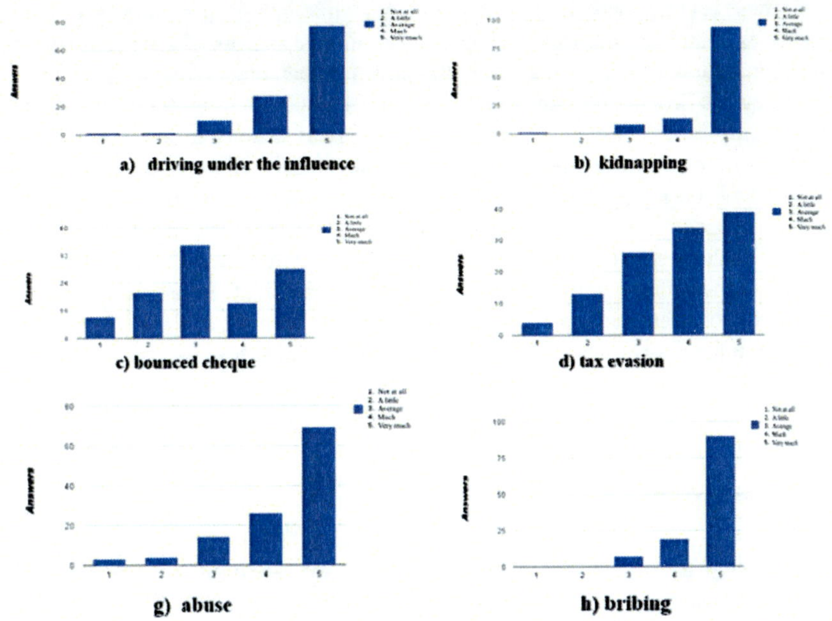

Fig. 5 Comparative assessment of criminal offence: (**a**) driving under the influence, (**b**) kidnapping, (**c**) bounced check, (**d**) tax evasion, (**e**) abandonment of victim in a traffic accident, (**f**) arson, (**g**) abuse, (**h**) bribing

ten answered good. This increase in the acceptance of tax evasion is estimated to be linked to the increase in tax rates during the crisis period in the country, as mentioned above. The higher the tax burden, the greater the tendency to tax evasion. In addition, in the relevant survey of the Economic Chamber of Greece, 23.4% intend to conceal part of their income and thus intend to evade taxes [7].

(b) It is noteworthy that only 33% consider tax evasion to be a criminal act and this answer is not influenced by gender.

Businesses also noted that the following measures would be useful to address the financial crisis:

- Simplification of the tax system (95.7%)
- Stabilizing the tax system for 5 years (91.4%)
- Introduction of special tax incentives for SMEs in the Region (77.6%)
- Simplification of bureaucratic procedures (69.8%)

and these rankings are not affected by gender, age and level of education

As seen on Fig. 5, tax evasion is not considered a significantly criminal offence compared to other offences such as driving while intoxicated and kidnapping. Similarly, it is considered less punishable than bribery and embezzlement. Only bounced check is considered a less serious offence of tax evasion. These findings are not affected by gender, age and educational level. In the study [12], the classification of offences is at the same level as in our study, while in a similar study by Passadis [9], tax evasion was considered a higher level offence as a tax offence compared to the current study but again at a lower classification than other offences. In a similar survey by the Economic Chamber of Greece (2011), 60.5% of respondents said that withholding income from the tax authorities was as serious an act as theft. It is therefore understandable that although during the financial crisis the penalties for tax evasion have been tightened (e.g. the procedure of self-incrimination for debts to the Treasury), tax evasion remains not only a financial problem but also a moral one.

Finally, the survey observed that the percentage of respondents who had not been tax audited was increasing with age, which means that a large number of entrepreneurs will complete their career without ever having been tax audited. Therefore an increase in the frequency of effective audits is estimated to lead to a reduction in tax evasion.

5 Conclusions

In conclusion, the study found that even though the average turnover of businesses in the Greek region increased during the financial crisis, their accounting and off-balance sheet profits decreased as their tax claims increased. This led to an increase (compared to the precrisis period) in the perception that tax evasion is "not reprehensible" and morally acceptable, regardless of gender, education level or age. It is predicted that tax evasion can be decreased by streamlining the tax code, stabilizing it for 5 years and firming up tax enforcement.

The findings of this study provide useful conclusions on policy implications for this Region. To extend them and provide a basis for country-wide implementation measures, it would be advisable to carry out a similar study in an urban area(s) and compare the results. Modeling and simulation systems could be used as tools in this study [13–16]. Also for the informed proposal of effective measures per geographical area, similar studies could be carried out for each of the 13 regions of Greece and the results could be compared. Finally, conducting a similar study covering the entire period of the financial crisis after its end could lead to useful policy conclusions.

References

1. Communication from the Commission to the European Parliament and the Council: an action plan to strengthen the fight against tax fraud and tax evasion (2016). https://eur-lex.europa.eu/legal-content/en/TXT/?uri=CELEX%3A52012DC0722
2. Vavouras, I., Kourtis, A.: Informal Economy: The Extent of the Phenomenon in Greece and Abroad. Papazisis eds, Athens (1991)
3. Kazantzis, C.: Auditing and Internal Control: A Systematic Approach to Concepts, Principles and Standards. Business Plus S.A, Piraeus (2006)
4. Karagiorgos, T.: Tax Practice. Germanos eds, Thessaloniki (2003)

5. Koromilas, G.L.: Income Tax Code, Law 4172/2013 (Journal of Laws 167/A) Analysis and Interpretation, Athens (2013)
6. Koromilas G., Giannopoulou C.: Seminar notes on business tax audit. http://www.epimetol.gr/aitniaimages /Forologikoselenchos_1_F9987.pdf
7. Economic Chamber of Greece: Study of taxpayer's behaviour and characteristics. Econ. Chron. **142**, 2–3 (2011) http://www.oecd.org/regreform/regulatorypolicy/Greece_2014_Measuring_administrative_burdens_Tax-Law-VAT.pdf
8. European Commission: Directorate-General for Taxation and Customs Union: Taxation Trends in the European Union: data for the EU Member States, Iceland, Norway and United Kingdom, 2021st edn. Publications Office of the European Union (2021) https://data.europa.eu/doi/10.2778/732541
9. Passadis S.: An empirical approach to the role of the informal economy and corruption in Greece: the prospect of Greece's remaining in the Economic and Monetary Union (2013) https://dspace.lib.uom.gr/bitstream/2159/15649/3/PassadisSpyridonMsc2013.pdf
10. Schneider, F., Enste, D.: Shadow economies: size, causes, and consequences. J. Econ. Literat. **XXXVIII**, 77–114 (2000) http://www.economics.uni-linz.ac.at/members/Schneider/files/publications/JEL.pdf
11. Stamatopoulos, D., Karavokyris, A.: Income Tax of Individuals and Legal Entities. ELFORIN, Athens (2010)
12. Tatsos, N.: Informal Economy and Tax Evasion in Greece. Papazisie eds, Athens (2001)
13. Dimitrios, N.K., Sakas, D.P., Vlachos, D.S.: Analysis of strategic leadership simulation models in non-profit organizations. Procedia Soc. Behav. Sci. **73**, 276–284 (2013)
14. Nasiopoulos, D.K., Sakas, D.P., Vlachos, D.S., Mavrogianni, A.: Modeling of market segmentation for new IT product development. AIP Conf. Proc. **1644**, 51–59 (2015)
15. Plikas, J.H., Dimitrios, N.K., Delis, P., Vlachos, D.S.: Software protection and piracy focusing on the 2008 crisis: a comparative study and simulation modeling regarding the case of Greece, Germany, and England. In: Strategic Innovative Marketing, pp. 319–326
16. Plikas, J.H., Nasiopoulos, D.K., Giakomidou, D.S.: The contribution of social media in the management of social relations with customers. Modelling and simulation of the problem. In: Kavoura, A., Sakas, D., Tomaras, P. (eds) Strategic innovative marketing. Springer Proceedings in Business and Economics. Springer, Cham (2017). https://doi.org/10.1007/978-3-319-33865-1_62

The Impact of the Greek Economic Crisis in Financing SMEs in the Tourism Sector: An Empirical Investigation

Kanellos Toudas, Aikaterini Gatze, Paraskevi Boufounou, and Lida A. Galani

1 Introduction

The economic crisis that has hit Greece in the past few years has been a particularly decisive factor in the economic and social life of the country over the last decade. The cumulative decline in GDP exceeded 25%, while thousands of jobs were lost. During the economic crisis, however, the country's most important sector, tourism, showed particular reflexes after the first years of the crisis and managed to grow further. However, despite the positive effects, the pressure on this sector was also considerable, as the increase in taxation and the general negative climate in the country were not negligible [1].

The aim of this chapter is to investigate the financing of tourist accommodation and hotels in the Halkidiki region by financial institutions and how these were affected by the financial crisis. For this reason, this chapter is structured as follows:

Initially, the basic elements of the theory of business financing are presented, while special emphasis is placed on the importance of their financing. Then, the importance of SMEs generally and for the Greek economy in particular is assessed, with emphasis on SMEs operating in the tourism sector critically discussing their difficulties in accessing finance. An appraisal of the impact of tourism on GDP and

K. Toudas (✉)
Agricultural University of Athens, Athens, Greece
e-mail: kstoudas@aua.gr

A. Gatze
Hellenic Open University, Athens, Greece

P. Boufounou
National and Kapodistrian University of Athens, Athens, Greece

L. A. Galani
Department of Economics, University of Peloponnese, Tripolis, Greece

D. P. Sakas et al. (eds.), *Computational and Strategic Business Modelling*,
Springer Proceedings in Business and Economics,
https://doi.org/10.1007/978-3-031-41371-1_40

an analysis of the data on tourism enterprises in Halkidiki follow. Then, the main findings on Greek SME financing opportunities, with an emphasis on the tourism sector, are summarized. Then, the methodological framework within which the research of the study takes place is presented in detail and a critical evaluation of the results obtained compared to those of previous relative studies follows. This chapter concludes with a critical assessment of the most important conclusions, and thoughts for further study are listed.

2 SMEs and Access to Finance

The management of business finance is undoubtedly among the main objectives of business financial management [2] and is broken down into sub-objectives according to Davies [3]:

- Securing regular and adequate financing
- Securing the optimal use of financing
- Ensuring investment security

The problem of financing is certainly an old problem dating back to the 1950s. Nevertheless, it is still valid today in both developed and emerging countries.

The SME market is important for national economies because it contributes significantly to employment and GDP and because its growth is linked to the formalization of an economy [4]. In many countries, the majority of jobs are provided by SMEs. From the 30 high-income countries of the Organisation for Economic Co-operation and Development [5], registered enterprises with fewer than 250 employees account for more than two-thirds of formal employment. In low-income countries, this number tends to be smaller, especially where the informal sector is large, but it is still important.

The contribution of SMEs to GDP also confirms their economic importance. In high-income and some middle-income countries, the sector accounts for more than half of the national output. Also, in low-income countries, the sector plays an important role, although the informal economy is more dominant.

The fact that the role of SMEs in an economy seems to increase with the income level of the country may suggest that they are the driving force of economic growth. Although this remains an open question, formalization has emerged as a potential channel through which a growing SME sector is linked to economic growth. Evidence shows an inverse relationship between the size of SMEs and informal sectors in a country [6].

For many decades, SME finance has been the main concern of researchers and policymakers worldwide. SMEs in general face serious barriers to obtaining bank finance and, in any case, more serious than large firms [7].

The challenges faced by SMEs in accessing bank credit are intended to arise from demand, supply, and country factors [8]. Basically, the difficulties faced by SMEs in accessing bank loans are caused by the serious problem of information asymmetry.

2.1 Tourism Business in Greece: The Case of Halkidiki

At a national level, Greece had 17.9 million international tourist arrivals (excluding cruise ship passengers), which amounted to €12.2 billion in 2013. This corresponds to 681 euro per arrival, 15.2% less than the global average. In any case, Greece has about 1.5% of the global tourism market and 2.9% of the European tourism market [9]. Moreover, when all the different impacts are taken into account, tourism contributes 16.4% to the Greek GDP and generates 18.3% of total employment in the country [9]. The importance of tourism for the Greek economy became even more recognized and understood by policymakers after the recent economic recession, which was the most severe one Greece has suffered in the last fifty years. In particular, Greece experienced six consecutive years of recession between 2008 and 2013 as a result of austerity measures introduced to address the sovereign debt crisis with devastating effects on the well-being of its citizens. In 2008, GDP per capita stood at €20,795, while in 2013, it fell to €16,302, a decrease of 21.6% in just five years. Moreover, the unemployment rate increased from 7.6% in 2008 to 26.9% in 2013, and the gross public sector debt as a percentage of GDP increased from 112.9% in 2008 to 175.7% in 2013 [10].

Moreover, and before introducing any grandiose tourism projects, it is very important to properly assess the impact of the recent economic crisis on tourism in Greece both at the national and regional level. In fact, in terms of supply, tourism in Greece is highly concentrated, as 40% of hotel beds (out of 773,445) were found in only two regions, namely, the island (and NUTS2 administrative region) of Crete with 166,370 beds and the Dodecanese (part of the NUTS2 administrative region of South Aegean) with 143,864 beds [11].

Halkidiki is a part of the geographical region of Macedonia, administratively part of the region of Central Macedonia, and Polygyros is its capital city. Halkidiki is a place of particular natural beauty of the Greek territory, which, traditionally, is a tourist destination for thousands of Greek and foreign tourists. It is mentioned that at least for 2014, there were 33 beaches with blue flags in Halkidiki, while in total there is a coastline of 550 km [12].

The graph below presents the updated data on the hotel capacity as of 30/6/2018, as provided through the official website of the Ministry of Tourism of Greece. According to the data of the following Diagram 1, Halkidiki has 34 out of 50 5-star hotels in the region of Central Macedonia, while in total it has almost 50% of the hotels in the region. A similar picture applies to the issue of beds with the largest participation of beds in Halkidiki in the whole region to be found in 5-star hotels, where 75% of the total number of beds in the region (Fig. 1).

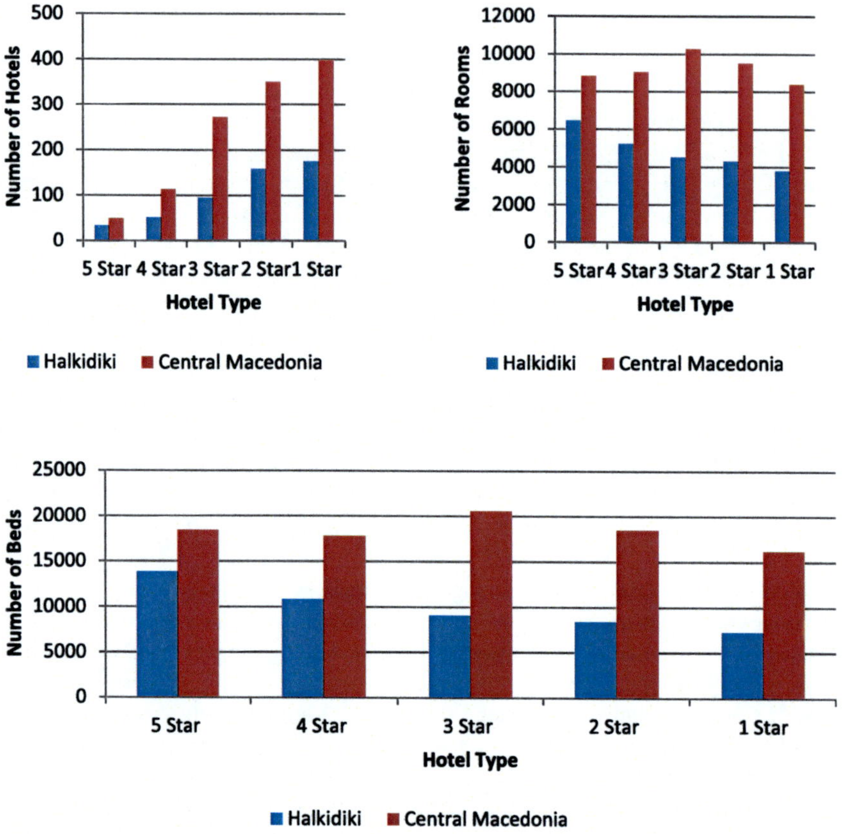

Fig. 1 Main tourism characteristics of Halkidiki and Central Macedonia. (Source: Hellenic Chamber of Hotels [13])

3 Literature Review

According to the European Commission survey [14], bank financing remains the first choice of financing for SMEs [15], while it is worth noting that according to the same survey, the approval rate of bank loans in the European Union for SMEs has increased to 70% of all applications, compared to the 60% recorded by the same survey in 2014. However, it is underlined that rejection rates are particularly high in Greece, reaching 20%, making Greece the first choice of financing for SMEs [16, 17].

According to Karatzias [18], the lack of liquidity is the biggest challenge that the Greek economy is currently facing in terms of access to finance for SMEs, while the relative lack of flexibility of existing measures to increase their liquidity is also noted, such as, for example, the inability to transfer European funds between regions in order to better meet the demand for these funds.

Karypidou [19] notes that although there are many ways of financing and the right choice depending on the needs of each business, bank loan is still the most commonly used method of financing. However, Karypidou [19] comes to another interesting conclusion, according to which, in order to cover the current needs of the business, apart from short-term bank lending, factoring is also appropriate, while on the other hand, in the case of expansion and upgrading of the business, the search for appropriate NSRF programs is a good option.

Moreover, in the earlier research by Moutsioulis [20], significant problems were identified, including the difficulty of access to financial resources, especially for enterprises in their early stages of operation, their low competitiveness, as well as information problems in relation to their support structures, and bureaucratic problems in their relations with the state and in the management of their financing programs. Moreover, it should be underlined that based on the results of his research, which were based on the PROMETHEE II method, it was found that SMEs with higher scores in their rating had lower levels of bank lending and higher levels of Equity Capital.

Kotsalu [21] aims to simplify the guarantee procedures and ensure sufficient liquidity for SME financing through the domestic banking system.

4 Methods and Materials

The aim of this chapter is to study the characteristics of the financing of tourist accommodation/hotels in the region of Halkidiki by financial institutions and their impact of the economic crisis. Questionnaires were sent to 190 enterprises and 165 responses were collected; therefore the degree of responsiveness of the survey was 86.84%. The sample of the survey consisted of 165 natural persons employed in tourism enterprises in Halkidiki (either in managerial positions in the case of limited liability companies, limited liability companies, limited liability companies, or as owners in the case of sole proprietorships and sole proprietorships), in various categories of accommodation. The questionnaire was completed face-to-face, and any queries raised by the survey participants were resolved on the spot. It is worth noting that a part of the questionnaires (36 in number) were answered by representatives/owners of businesses that had closed in the summer of 2020 due to COVID-19.

The main characteristics of the sample are as follows:

1. 67% of the sample is employed in enterprises that have been in business for a period of more than 5 years.
2. 40% of the sample was employed in sole proprietorships, followed by general partnerships with 28% and limited liability companies with 15%.
3. Almost 40% of the sample were employed in enterprises with up to 10 employees, while only 11% of the sample were employed in enterprises with more than 50 employees; almost 67% of the sample were employed in enterprises with a

turnover of up to 200,000 euros in 2019, while only 6% of the sample were employed in enterprises with a turnover of more than 1.5 million euros.

4. 34% of the sample was employed in rented rooms, and 21% of the sample was employed in four or five star hotels.
5. Most businesses focus on the Balkan countries (20.61%), but this does not mean that they do not attract a significant number of tourists from other countries (Central Europe 12.12%, Arab States 11.52%, Scandinavian 11.52%, Eastern Countries 11.52%, France 10.91%, United Kingdom 10.30%).
6. For 94% of the sample, the highest number of tourists is observed during the summer months.

The questionnaire created focused on the financing of the tourist SMEs and on the impact of the economic crisis on them. Data analysis was conducted in SPSS v.26 software. Cronbach's internal consistency coefficient α was calculated for all the questions, which obtained the satisfactory value of $\alpha = 0.598$ for the questionnaire as a whole, while for the financing section, the value was 0.658 and for the economic crisis impacts section was 0.586.

5 Results

Although Halkidiki is a promising area for tourism in Greece, there are no relative studies conducted so far; thus, the significance of the present study for providing implementation guidelines on the topics addressed becomes particularly important. The main findings of the research are presented below to further enhance the potential of their practical implementation, and they are critically evaluated against those of analogous previous studies conducted for other tourist regions of Greece presented above.

According to the results obtained, with *regard to access to finance*, the following emerged:

(a) For 51% of the sample, there was bank financing, while for 32%, the financing came from European funds. It should be noted that according to the research by Karypidou [19], the form of financing of companies was quite different. Specifically, only 14% of the sample received bank funding, while government funding was three times higher than in this survey. On the other hand, funding from European funds was at the same level in both surveys. It should be noted that Karypidou [19] carried out an analogous survey using a questionnaire of 60 employees of micro-, small-, and medium-sized enterprises in Thessaloniki and Athens. Therefore, it can be concluded that there has been an increase in bank financing of tourism SMEs from 2017 to 2019 [19].

(b) 50% of the financing was related to facilities (in the Karpydiou [19] survey, this percentage only reaches 10%), 25% was directed to the mechanical and technological equipment of the enterprises (with the Karpydiou [19] survey giving a percentage of 40%), and 12% was spent in advertising and financing of older needs (when in the Karpydiou [19] survey, these percentages were twice as high).

(c) Bank lending was chosen most often by 80% of respondents as the form of financing their business. Leasing follows with 13% while financing in the form of venture capital is also found with a percentage of more than 5%. It is worth noting that in a similar survey conducted earlier, bank lending constituted only 42% of the financing used by the sample when leasing and ventures were at the same level [22]. It should be noted that Kokodrouli [22] conducted an analogous study for micro-, small-, and medium-sized enterprises in Crete.

(d) 55% of the enterprises have used a public funding body in the past. In Kokolis' [23] survey, which was conducted using a questionnaire of 76 enterprises in Piraeus and Athens, only 24% of the sample had used a state financing institution. For most of the people who answered in the affirmative to the previous question, this institution was the ETEAN (now Hellenic Development Bank) and the Development Investment Law ($n = 44$, 48%), while 19% of the sample turned to the OAED and 33% of the sample used other European programs.

(e) 22% of enterprises have applied for any bank credit beyond the existing credit facilities. This percentage is similar to that noted in the Kokolis [23] survey, which was 27%. In 50% of the cases ($n = 18$), the result was 100% approval and in the remaining 50%, 75% approval. In the corresponding survey by Kokolis (2016), there was a 28% rejection rate, while 100% approval was only 28% of cases and 75% approval only 6%. In the remaining cases, approval involved smaller percentages.

(f) High interest rates and heavy bureaucracy are the biggest obstacles for businesses to obtain bank financing, with rates exceeding 44%, while the absence of sufficient collateral and guarantees is also noted at a rate of 6%. Regarding the first two deterrents, it should be noted that these coincide with the Kokolis (2016) survey with slight variations in frequencies.

(g) Financial programs (such as the NSRF) are the most common choice at almost 43%, followed by participation in local, national, or international exhibitions at almost 40%, and strategic programs and development actions at 16%. According to Veizi and Ioannou [24], strategic programs reach 40% when NSRF is at 33% and exhibitions at 22%. In other words, there is a variation in the way of approaching the companies participating in the surveys.

Regarding *the impact of the economic crisis on tourism*, the main findings are as follows:

(a) 79% of respondents think that financing has been affected by the financial crisis to a great and very great extent. The results are in line with the findings of Karypidou [19], as 67% of the sample states that they have been affected by the financial crisis "a lot" or "very much."

(b) Cost reduction (38%, $n = 63$) price reduction (24.2%, $n = 40$) and staff layoffs (37.6%, $n = 62$) are the most important actions that SMEs take in order to cope with the economic crisis and reduce their costs. A similar picture is observed in the results of the research of Lampropoulou, Mouzaki and Economou (2015) where their research was addressed to 40 small and medium hotel enterprises, where price reduction and cost reduction are the first choices for the enterprises of their sample.

(c) 53% of the sample noticed a decrease in bookings by Greeks when 36% noticed a decrease in bookings by both (Greeks and foreigners). However, in Lampropoulou, Mouzaki, and Economou's [25] survey, the decrease in bookings was equally between Greek and foreign tourists, which indicates an improvement in tourism conditions for foreign tourists.

(d) 70.3% of the participants in the survey made price reductions and other types of offers in order to maintain their clientele, while 29.7% of the sample answered negatively to this question, It is worth noting that in the survey by Lambropoulou, Mouzaki and Economou (2015), the entire sample had made a price reduction, so there was an improvement in the conditions of tourism over time.

(e) 44% of the sample considers that it was moderately affected by the economic crisis when almost 30% of the sample was little or not at all affected. On the other hand, in the survey by Lampropoulou, Mouzaki, and Economou [25], the impact was significantly more significant, as the majority of the sample stated that they were very (50%) or quite (25%) affected.

(f) Only 32.7% of the respondents consider that the general business strategies of the public sector contribute radically to addressing the economic crisis. The corresponding figure in the Vourneli [26] survey was significantly lower, at no more than 6%.

(g) 54.5% of the sample has thought about setting up an online shop "many times," while 38.2% of the respondents answered never or rarely. In Vourneli's [26] survey, which referred to a sample of 105 people working in small and medium enterprises in Kavala, the percentage of people who had not considered the Internet as a solution was significantly higher as it reached 80% of the sample. It is therefore clear that there is a shift in tourism SMEs toward digital.

(h) For 65% of the sample, was the situation of their business before the 2009 crisis characterized as satisfactory to excellent when for 20% the situation was characterized as moderate. On the contrary, according to the results of the survey by Veizi and Ioannou [24], 100% of the respondents stated that the situation of their business before the 2009 crisis was characterized as satisfactory to excellent.

(i) The main consequence of the crisis is a decrease in profits with a percentage of 51% when the corresponding percentage in the survey of Veizi and Ioannou [24] was 50%. This is followed by a decrease in customers (17%), which is, however, significantly lower than that identified in the survey, which was 40%.

(j) For 61% of the sample, turnover decreased 10–20%, while for 27% of the sample, turnover did not decrease at all. The results differ significantly from those of the Veizi and Ioannou [24] survey, as in this survey, the percentage of firms that experienced a zero or small decrease in turnover of up to 10% was only 16% of the sample, which highlights the regional differences of the regions (mainly due to the different countries attracting tourists to them).

(k) The crisis is an opportunity for businesses to develop tourism products or services that the respondents did not have until now, referring to green tourism (29%), sports and adventure tourism (28%) and conference tourism (13%). The results are quite similar to those of the Veizi and Ioannou [24] survey, according to which green tourism was boosted by 16%, conference tourism by 31% and religious tourism by 25%. This proves that the economic crisis has led to the search for exploitation of new opportunities in tourism segments that were previously untapped.

(l) The local or regional authority of Halkidiki has taken specific measures since 2008 to help SMEs to cope with the crisis at a rate of 33%, while in Poulas' (2013) survey of 40 SMEs operating in the region of Drama, the local or regional authority of Drama had taken significantly fewer measures, as the corresponding rate was only 10%, which highlights the particular importance of local authorities.

(m) Local authorities can show that they support SMEs by ensuring easier access to finance (24%), by improving access to new markets and enhancing international development (24%) and by improving access to the Internet (18%); in Pula's (2013) survey, ensuring easier access to finance was 30% with improving access to new markets and enhancing international development at 10% and improving access to the Internet (25%).

(n) Finally, the means of promoting the business were tourist guides at 19.4%, travel agencies (21.8%) and the Internet (15.2%) mainly. In Naki's [27], research, which had a sample of 156 people working in the field of tourism in businesses in Epirus, tourist guides, were used as a means of promoting the business at a rate of 42%, travel agencies at a rate of 76% and the Internet at a rate of 87%.

Finally, *the influence of the socioeconomic characteristics* of the sample on survey results was studied by using the univariate analysis of variance, and the t-tests for independent samples have found statistically significant the following:

A. Years of operation of the firm

- Firms with 5–10 years of operation are mainly financed by European funds, unlike the other firms which are mainly financed by banks as $F\ (3{,}161) = 15{,}172$, $P = 0{,}000 < 0{,}05$.
- New enterprises with less than 2 years of operation are believed to make moderate use of financing, in contrast to all other categories of enterprises for which the respondents' view is that they use this financing in a very good way as $F\ (3{,}161) = 7010$, $P = 0{,}000 < 0{,}05$.

B. Legal form of the firm

- People employed in sole proprietorships and in SMEs consider to a greater extent than other enterprises that the local or regional authority has taken specific measures since 2008 to help SMEs to cope with the crisis, as $F\ (4{,}160) = 4125$, $P = 0{,}000 < 0{,}05$.

- People employed in public limited companies are less likely than people employed in other enterprises to believe that the crisis has affected financing, as F (4,160) = 4785, P = 0,001 < 0,05.

C. Type of accommodation

- People employed in a four or five star hotel perceive to a lesser extent than employees in other firms that the crisis has affected financing as F (3,161) = 4417, P = 0,000 < 0,05.
- People employed in a four or five star hotel consider to a greater extent than employees in other businesses that the general business strategies of the public sector are making a radical contribution to the financial crisis as F (3,161) = 4588, P = 0,004 < 0,05.

6 Conclusions

The access to finance is certainly an old problem dating back to the 1950s. However, it remains significant for both developed and emerging countries. For many decades, SME financing has been (and remains) at the core of concern of researchers and policymakers worldwide.

SMEs in general face serious obstacles in accessing finance, more serious than bigger firms. Moreover, in a number of cases, it has been observed that banks, although having excess liquidity, are reluctant to grant loans to SMEs, especially in times of crisis.

In Halkidiki, for half of the tourism enterprises surveyed, bank lending is a reality, while the main area in which the money is directed is the facilities of the enterprise and the renewal and maintenance of its mechanical and technological equipment. It is worth noting that for the majority of respondents, the financing of the enterprise is used in the right way. Moreover, it should be highlighted that more than 50% of the respondents stated that they used a state institution for financing their business (e.g., the NSRF, OAED, or the Investment Law). At this point, it should be underlined that, the Greek state, in conjunction with the European Union, has offered several tools in the hands of SMEs to increase their funding levels in order to minimize the negative consequences of the economic crisis (such as the 4th Industrial Revolution Fund, the "Made in Greece" program, and the OP-Rise Fund). Moreover, the survey showed that high interest rates and high bureaucracy are the biggest obstacles to SME financing, which had also been noted by the study of relevant literature [28–30].

The results of the survey clearly indicated the negative influence of the economic crisis on the activities of the sampled tourism enterprises in Halkidiki. This was mainly reflected in the respondents' opinions as individuals (as more than 70% considered that the negative influence was high and very high). On the other hand, 70% of tourism enterprises indicated no to moderate influence. It is estimated that this reply combined with other answers (showing, e.g., a relatively small decrease in turnover of at most 20% for 30% of the sample surveyed) reflects the course of tourism in Halkidiki in recent years (which, as noted above, has shown resilience, turning mainly to tourists from the Balkan countries and developing sectors such as green tourism, which were not very popular in the last few years).

Furthermore, apart from the evidence that the years of a firm's operation are a crucial factor determining its access to finance, other sociocultural differences did not seem to significantly affect the questionnaire responses obtained.

Finally, the practical use of the present study would be enhanced if such surveys were to be conducted at the same time in the main tourist regions of Greece and the relative findings were to be comparatively analyzed, providing additionally for regional differences (islands/mainland destinations) and visitors' characteristics. Another way to improve the present study could be to conduct it in a dynamic way by modeling and simulating the data of research [31–34].

References

1. Bank of Greece: Developments in the Greek Travel Balance of Payments. Bank of Greece, Athens (2014)
2. Badulescu, D., Badulescu, A.: Financial constraints facing SMEs: some theory and evidence. Metalurgia Int. **15**(8), 49–62 (2010)
3. Davies, D.: The Art of Financial Management. PWN, Berlin (1993)
4. International Finance Corporation: The SME banking knowledge guide (2009)
5. OECD: Financing SMEs and entrepreneurs 2018: an OECD Scoreboard (2018)
6. Chrysomallis, A.: Small and medium-sized enterprises and economic crisis. Master Thesis. TEI of Kavala (2013) (in Greek)
7. Kenfack, K.: L'obtention du crédit bancaire par les PME au Cameroun. Review Afrique et Dévelopement. **41**, 21–158 (2016)
8. Lopez-Gracia, J., Aybar-Arias, C.: An empirical approach to the financial behaviour of small and medium sized companies. Small Bus. Econ. **14**, 55–63 (2000)
9. SETE: The importance of tourism for Greece (2014)
10. International Monetary Fund: World economic outlook database, Greece (2014)
11. Hellenic Chamber of Hotels: Developments in Tourism and the Greek Hotel Sector in 2013. Hellenic Chamber of Hotels, Athens (2014)
12. INSETE: Marketing plan of the tourist products of Central Macedonia. Road Map 2015–2020 (2015) (in Greek)
13. Hellenic Chamber of Hotels: Distribution of hotel potential 2018 by category and region (2018) (in Greek)
14. European Commission: Access to finance for small and medium-sized enterprises (2017) (in Greek)
15. Bank of Greece: Annual report for the year 2014 (2015)
16. European Commission: Annual report on European SMEs 2018/ 2019 (2019)
17. European Commission: Greece 2019 fact sheet (2020)
18. Karatzias, P.: The access of Greek SMEs to finance, programmes and financing tools developed in the last five years and the role of development laws. Master Thesis. School of Public Administration (2014) (in Greek)
19. Karypidou, I.: Sources of financing of small and medium enterprises. Master Thesis. School of Business Administration Sciences, University of Macedonia (2017) (in Greek)

20. Moutsioulis, K.: Sources of financing and evaluation of SMEs in Greece. Master Thesis. Department of Production and Management Engineering, Technical University of Crete (2010) (in Greek).
21. Kotsalu, E.: Small and medium enterprises, innovation and financing. Master Thesis. School of Applied Mathematics and Natural Sciences, National Technical University of Athens (2014)
22. Kokodrouli, E.: Financing of SMEs. Criteria & conditions. Master Thesis. TEI of Crete (2011) (in Greek)
23. Kokolis, G.: The economic crisis and the financing of SMEs in Greece. M.Sc. thesis. University of Piraeus (2016) (in Greek)
24. Veizi, I., Ioannou, I.: The management of SME tourism enterprises in the crisis period: strategic and business models for greater competitiveness. Master Thesis. TEI of Western Greece (2016) (in Greek)
25. Lampropoulou, M., Mouzaki, A., Economou, N.: Economic growth of small and medium enterprises in times of economic recession. Master Thesis. TEI of Patras (2015) (in Greek)
26. Vourneli, C.: Business strategies-tools for coping with the economic crisis. Master Thesis. ATEI of Kavala (2012) (in Greek)
27. Naki, K.: The contribution of tourism to the economic development of the Region of Epirus. Postgraduate Thesis. Harokopeio University (2016) (in Greek)
28. Cressy, R., Olofsson, C.: The financial conditions for Swedish SMEs: survey and research Agenda Robert. Small Bus. Econ. **9**, 179–194 (1997)
29. EBAN: European early stage market statistics 2015 (2016)
30. Freel, M., Carter, S., Tagg, S., Mason, C.: The latent demand for bank debt: Characterizing "discouraged borrowers". Small Bus. Econ. **38**, 399–418 (2012)
31. Nasiopoulos, D.K., Sakas, D.P., Vlachos, D.S., Mavrogianni, A.: Simulation of generation of new ideas for new product development and IT services. AIP Con. Proc. **1644**, 60–68 (1644)
32. Plessias, A.A., Nasiopoulos, D.K., Giakomidou, D.S.: Modeling of market segmentation in social networks and media. In: Strategic Innovative Marketing, pp. 523–529. Springer
33. Sakas, D.P., Nasiopoulos, D.K., Reklitis, P.: Modeling and simulation of the strategic use of marketing in search engines for the business success of high technology companies. In: Business and Economics, pp. 217–226. Springer
34. Plikas, J.H., Nasiopoulos, D.K., Sakas, D.P., Vlachos, D.S.: Modeling the promotion process of academic conferences through social media. In: Strategic Innovative Marketing, pp. 499–505. Springer

The Financial Perspective of the Courier Sector in Greece

Kanellos Toudas, Menexiadis Marios, Polyzois Mathaios, and Siouziou Ioanna

1 Introduction

Courier companies appeared and gradually began to grow in the early 1980s, while in the 1990s, they showed a boom that led to a remarkable growth rate. Development in the sector is understood by the fact that currently more than 700 companies operate given the obtained license by the National Telecommunications and Post Commission (EETT).

The top five companies in terms of annual turnover are ACS 127,5 mil €, Geniki Taxydromiki 72 mil €, Taximetafores ELTA 56 mil €, Speedex 42 mil € and Easy mail 6 mil € (2020 figures).

K. Toudas (✉)
Accounting and Financial Analysis, Agricultural University of Athens, Department of Agribusiness and Supply Chain Management, Athens, Greece
e-mail: kstoudas@aua.gr

M. Marios
National and Kapodistrian University of Athens, School of Economics, Athens, Greece

Aegean Airlines, Athens, Greece

P. Mathaios
National and Kapodistrian University of Athens, Financial and Accounting Department, Athens, Greece

Aramex Greece P.C, Athens, Greece

S. Ioanna
National Technical University of Athens, School of Electrical and Computer Engineering, Athens, Greece

D. P. Sakas et al. (eds.), *Computational and Strategic Business Modelling*, Springer Proceedings in Business and Economics,
https://doi.org/10.1007/978-3-031-41371-1_41

2 Financial Statements Analysis

The purpose of analyzing financial statements is to provide a complete picture of the business. It is considered to be a very important tool for stakeholders as they help them relying on specific data for decision-making, reducing as much as possible the uncertainty about the result. Thus, anyone who makes the analysis can aim at information about the ability of the company to meet its current obligations, prospects and profitability dynamics, the long-term stability and the structure or composition of the company's capital. Similar analyses have been made by various other scientists using modeling and simulation systems [9–11].

To complete the financial analysis, it is necessary to select some relationships, classify the data to calculate these relationships and, finally, evaluate, study and interpret them. As a general rule, analyses should have a standard size as a measure of comparison. Also, to evaluate the results of an analysis, they should be compared with the results of other companies in the same industry or with industry averages.

The most commonly used accounting analysis methods are the following:

1. Horizontal or comparative method
2. Vertical or layered method
3. Break even analysis
4. Analysis with ratios

Horizontal analysis is the analysis where the data of two or more years are compared and lead to conclusions, regarding the trend and the pace that a company has. Through this, one can understand the nature and trends of changes that occur in the company. The greater the number of financial years compared, the more the reliable are the conclusions. The comparison of data on an annual, semiannual, quarterly and monthly basis with a corresponding analysis of previous years can lead to useful conclusions about the course of business [1].

There are two main reasons for using the horizontal analysis. The first, as already mentioned, is the understanding of the course of the business in the period covered. The second is also important as it highlights the events that influenced the development of the business. If these facts are understood by the management, appropriate forecasting models can be created to design the correct treatment [6].

The *vertical analysis* makes it easy to compare the data of a company both in a number of years and with other companies of different sizes in the same industry. In fact, financial statements are prepared in common sizes so that comparisons can be made between them. Thus in the vertical method analysis, each item is divided by the total assets and liabilities and is presented as a percentage of the total. The same applies to the income statement where all data appear as a percentage of the total net sales [1].

Overall, the major advantage of this type of analysis is the convenience it provides in comparing companies in the same industry. Through this comparison, it can be distinguished whether the specific company has invested its funds correctly [3].

Break even analysis is the method used to find the amount of production or sales needed to achieve zero profit before taxes and interest. Break even is defined as the level of activity expressed in terms of sales value or product units, i.e. the point at which revenue equals costs and the firm has no profit or loss.

This analysis is more about the sales department and is used by management to make decisions in the proper way. It is understandable that every business has its own break even as both fixed costs and profit margins differ. To calculate the break even point, there are two equations, regarding the decision of the quantity to be sold or the value to sell [1].

Through the equations, it becomes clear that to change the break even point, one of the elements that determine it must also change. Therefore, break even is affected by sales volume, variable costs, selling price and fixed costs. It is therefore important for the management of the company to determine at all times the results of the above relationships to be aware of the possible impact on the profits of the company.

Ratio analysis is widely used today and is one of the most important methods in the world. It uses a series of relationships of balance sheet items and profit and loss results to calculate a company's financial position and performance to understand its position in the industry [8].

The ratios are grouped and during the analysis should be carefully selected, to provide substantial help when comparing companies. The most commonly used ratios are categorized as following:

1. *Liquidity ratios*: They study the ability of a company to meet its short-term liabilities. More specifically, through the liquidity that a company has, its ability to convert its assets into cash and with it to pay its short-term liabilities is distinguished. Often a liquidity problem signals the emergence of new issues [4].

The general liquidity ratio shows the degree of security that management will have in a potentially problematic situation for the company. Specifically, it reflects whether the receivables, stocks and assets of the company exceed the short-term liabilities. The surplus of assets enables the company to meet the payment of daily liabilities, without, however, being able to predict future liquidity levels. To be considered acceptable the general liquidity ratio must be close to 2 and certainly be much larger than 1. Otherwise, if it is smaller than 1, it means that the company cannot meet its obligations by showing borrowing as the only solution for its survival [7].

The special liquidity ratio is stricter than the previous one and shows whether the immediately liquid assets of the company cover its short-term liabilities. In essence, the difference between the two ratios is that special liquidity does not include stocks that are more difficult to liquidate. The higher the result, the more favorable the liquidation position of the company. Thus, a ratio higher than 1 shows the company's ability to meet its short-term liabilities and vice versa. If this ratio is around 1, it is satisfactory, provided that the doubtful receivables are not included in the requirements. Another important element to be taken into account is the absence of a large difference between the period of collection of receivables and the payment of liabilities [7].

2. *Capital structure ratios*: The capital structure is mainly related to the percentage of participation of each form of capital in the composition of the total capital of the company, while examining their proper use to increase the efficiency of the organization. Desirable by all organizations is, to a large extent, the balance between foreign capital and equity to balance the risk undertaken and the return desired by investors [1].

Lenders generally prefer a low credit score as they seek security in the event of a company bankruptcy. A steadily rising trend of the index is a negative sign and expresses the inability of the company to repay its loans and its shift to new forms of lending.

3. *Activity ratios:* Through the analysis of the activity ratios, the efficiency of the asset management carried out by the company is presented. This is generally a subjective analysis, and usually the rates are compared with the results of industry standards. It is important to understand operational performance as successful companies are usually effective in asset management, while less successful ones have low results that lead to problems in their operations [2].
4. *Profitability ratios*: A company's profitability corresponds to whether it achieves profits. Part of the efficiency of a business is the proper management of foreign capital [4]. The indexes of this group focus on the relationships of profits – capital as well as profits – sales [1].
5. *Investment ratios*: Investment ratios correlate the current stock price with items from the financial statements. This information is mainly used by shareholders when they want to buy, sell or hold shares in a company. Stock prices essentially show how attractive a stock is, while investment indicators reflect the current financial position of the company in relation to the stock price [7].

3 Financial Ratios Analysis

Given the financial statements between 2016 and 2020, the financial ratios have been calculated and analyzed.

Overall, general liquidity in the sector seems stable and fluctuates around one (1), but this can be also an indicator of future inability, since the higher than one, the better it is.

Year	ACS	Geniki Taxydromiki	Elta Courier	Speedex	Easy mail
2016	1.32	1.29	1.19	0.88	1.01
2017	1.62	1.28	1.18	0.89	1.05
2018	2.13	1.28	1.02	1.26	0.72
2019	2.17	1.16	0.70	1.24	0.94
2020	1.36	0.99	0.68	1.05	1.18

Special liquidity fluctuates around one (1), but again the same conclusions can be reached as the general liquidity, since in the sector there is no stock.

Year	ACS	Geniki Taxydromiki	Elta Courier	Speedex	Easy mail
2016	1.30	1.29	1.18	0.88	1.01
2017	1.60	1.28	1.17	0.89	1.05
2018	2.10	1.28	1.02	1.26	0.72
2019	2.14	1.16	0.69	1.24	0.94
2020	1.34	0.99	0.67	1.05	1.18

The net capital to assets ratio indicates the liquidity trend of each company. There have been years when some of the companies faced inability to meet short term liabilities; however, either by borrowing or increasing sales, this led to a reversed picture.

Year	ACS	Geniki Taxydromiki	Elta Courier	Speedex	Easy mail
2016	0.16	0.20	0.15	−0.13	0.01
2017	0.26	0.20	0.14	−0.11	0.04
2018	0.41	0.19	0.02	0.20	−0.24
2019	0.38	0.12	−0.33	0.16	−0.04
2020	0.14	−0.01	−0.37	0.05	0.13

The debt ratio indicates the risk of default and debts to third parties. The ratio presents a mixture since other companies reduce the ratio and others increase it.

Year	ACS	Geniki Taxydromiki	Elta Courier	Speedex	Easy mail
2016	0.57	0.72	0.83	1.15	0.78
2017	0.50	0.71	0.84	1.14	0.79
2018	0.44	0.69	0.97	1.08	0.84
2019	0.46	0.76	1.31	1.03	0.73
2020	0.51	0.90	1.36	1.04	0.74

The equity to loan ratio indicates creditworthiness, again with the minority exceeding the one (1).

Year	ACS	Geniki Taxydromiki	Elta Courier	Speedex	Easy mail
2016	0.74	0.39	0.21	−0.13	0.28
2017	0.98	0.40	0.19	−0.12	0.27
2018	1.28	0.46	0.03	−0.08	0.19
2019	1.18	0.32	−0.23	−0.03	0.36
2020	0.95	0.12	−0.26	−0.04	0.36

Current assets to total liabilities indicate that the company can meet its obligations in the long run. A mixture of this ability is presented at the table below.

Year	ACS	Geniki Taxydromiki	Elta Courier	Speedex	Easy mail
2016	1.14	1.26	1.11	0.84	1.01
2017	1.37	1.26	1.12	0.84	1.05
2018	1.76	1.26	0.96	0.87	0.72
2019	1.53	1.16	0.58	0.81	0.93
2020	1.06	0.97	0.58	0.86	1.18

The asset speed ratio determines the utilization of the company's assets to achieve sales. In the sector, it seems that the asset utilization is effective.

Year	ACS	Geniki Taxydromiki	Elta Courier	Speedex	Easy mail
2016	2.34	2.78	1.24	1.50	2.28
2017	2.24	2.65	1.18	1.60	2.89
2018	1.82	2.81	1.57	1.82	3.19
2019	1.60	2.15	2.13	1.79	2.49
2020	1.65	2.39	2.51	1.29	1.89

The Equity Turnover ratio reflects the efficiency of equity management, with very high rates indicating that the business is based on foreign capital – external borrowing.

Year	ACS	Geniki Taxydromiki	Elta Courier	Speedex	Easy mail
2016	5.49	9.88	7.22	−10.05	10.44
2017	4.52	9.25	7.44	−11.86	13.45
2018	3.24	8.94	57.42	−21.59	19.93
2019	2.95	8.87	−6.96	−53.27	9.38
2020	3.39	22.96	−7.04	−32.52	7.20

The fixed asset speed ratio indicates the connection between sales and fixed assets investment, where again the picture is a mix.

Year	ACS	Geniki Taxydromiki	Elta Courier	Speedex	Easy mail
2016	6.77	30.26	15.82	40.42	13.70
2017	7.23	26.47	20.20	38.40	21.09
2018	8.01	20.89	24.07	34.45	14.52
2019	5.36	17.24	8.84	11.13	13.18
2020	3.64	17.96	11.48	12.19	27.43

The Net Margin ratio indicates profitability. So above zero, the situation is better than below it; however, management has to find ways for higher sales and cost reduction.

Year	ACS	Geniki Taxydromiki	Elta Courier	Speedex	Easy mail
2016	0.12	0.02	0.02	0.04	−0.02
2017	0.12	0.02	0.03	0.05	0.00
2018	0.12	0.02	0.01	0.07	−0.02
2019	0.12	0.01	−0.14	0.05	−0.06
2020	0.11	0.03	−0.01	0.00	0.09

The equity yield ratio reflects the efficiency of utilizing the invested funds for the generation of profits. A mix in the picture as well as negative figures indicating losses that exceed the equity.

Year	ACS	Geniki Taxydromiki	Elta Courier	Speedex	Easy mail
2016	0.47	0.10	0.05	−0.08	−0.19
2017	0.35	0.10	0.06	−0.14	0.03
2018	0.28	0.09	−1.01	−0.74	−0.46
2019	0.26	0.04	1.11	−1.30	−0.59
2020	0.31	0.00	0.18	0.37	0.58

In the earnings per share ratio, profits are related to the share price increase.

Year	ACS	Geniki Taxydromiki	Elta Courier	Speedex	Easy mail
2016	23.57	1.10	0.49	0.36	−3.75
2017	23.46	1.19	0.64	0.56	0.66
2018	26.71	1.20	−1.54	1.69	−6.34
2019	29.42	0.55	−15.06	1.27	−12.66
2020	35.55	−0.01	−2.97	−0.62	23.47

The price earning (per share) ratio reflects the investor's confidence in the company and its ability to generate profits.

Year	ACS	Geniki Taxydromiki	Elta Courier	Speedex	Easy mail
2016	0.05	2.67	59.45	2.81	−8.00
2017	0.05	2.45	46.19	1.78	45.26
2018	0.05	2.45	−19.07	0.59	−4.73
2019	0.04	5.30	−1.95	0.79	−2.37
2020	0.03	−228.32	−9.87	−1.61	1.28

4 Viability Analysis Using Altman Z-Score Model

In the general situation of uncertainty over the last two decades, it is crucial to be able to determine the probability of failure of the business. Multiple discriminant analysis (MDA) aims to group data and simultaneously prevent participation in multiple classes, such as bankrupt and healthy businesses. Altman Z-Score MDA can detect a company going bankrupt at 95% the previous year and 85% 2 years before it happens. The variable used in the analysis is in the form of a numerator and variants of the model can use up to 33 numerators [5]. The advantages of the Altman Z-Score model include the simultaneous use of multiple indicators in an easy and simple calculation, thus taking into account a wide range of business activity. A disadvantage is the support of the model in some cases which violate the independence of the variables it uses.

To perform the analysis, five variables-numerators must be calculated, which, after multiplying by a factor, are added to give the result of each examined operation. This result can be positive or negative, even plus/minus infinity. The resulting price is used to rank the company in one of the three categories listed below:

(a) $Z < 1.81$ the company faces financial difficulties with a high risk of bankruptcy
(b) $1.81 < Z < 2.99$ grey zone
(c) $2.99 < Z$ the company is healthy

The Z-Score model was applied for the examined companies for the years 2018–2020, to study its reliability, with the below results:

Z-Score						
Year	ACS	Geniki Taxydromiki	Elta Courier	Speedex	Easy mail	Industry average
2018	4.58	3.82	0.95	2.21	2.54	2.82
2019	4.14	2.82	−0.79	2.11	1.75	2.01
2020	3.68	2.70	0.42	1.20	2.90	2.18
AVERAGE	4.14	3.11	0.19	1.84	2.40	
MINIMUM RATE	3.68	2.70	−0.79	1.20	1.75	
MAXIMUM RATE	4.58	3.82	0.95	2.21	2.90	

The below table presents the maximum prices per year and the average per year, as well as a statistical analysis and ranking of companies in the three categories as reported by the model.

	2018	2019	2020
Sample	5	5	5
Average/year	2.82	2.01	2.18
Maximum rate	4.58	4.14	3.68
Minimum rate	0.95	−0.79	0.42
$Z < 1.81$	1	2	2
% $Z < 1.81$	20%	40%	40%
$1.81 < Z < 2.99$	2	2	2
%$1.81 < Z < 2.99$	40%	40%	40%
$Z > 2.99$	2	1	1
%$Z > 2.99$	40%	20%	20%
% of model success	80%	60%	60%
Mistake type II	20%	40%	40%

5 Concluding Remarks

The analysis of the results obtained from the model in the courier sector indicates that the greatest result for the sector average is presented in the first year while fluctuations gradually appear. In general, the rates for the sector average remain in the grey zone; however, questions are raised about the fact that in 2018, the best result of 2.82 is very close to ranking the sector in healthy companies, while the following year the result is 2, 01 is close to the limit of 1.81. According to the Altman model, a company at risk may face operating issues in the future, find it difficult to raise funds from the banking system and may go bankrupt. It is clear that to draw clearer and safer conclusions, in-depth studies are needed, using a larger number of companies in the sector for a longer period of time. It is a fact that the Altman model relies heavily on the use of pointers and therefore gives an insight into the course of business. However, to obtain a complete picture and forecast, it is necessary to use extra statistical models as well. In addition, the current conditions in the country should be taken into account, such as the pandemic and the economic crisis that existed in previous years.

References

1. Vasileiou, D., Eriotis, N., Mpalios, D.: Principles of Financial Accounting – Financial Analysis and Decision making, 1st edn. Rosili, Gerakas (2019)
2. Fotopoulou, K.: Financial analysis of the banking sector with ratios, Nikosia (2020)
3. Georgopoulos, A.: Financial Statements Analysis, Athima, Hellas, 1st edn. Benou Publications (2014)
4. Goulas, N.: Ratio analysis of PLCs in the Athens Stock Stock Exchange, Thessaloniki (2021)
5. Ishmah, H., Solimun., Mitakda, M.B.T.: Multiple discriminant analysis altman Z-score, multiple discriminant analysis stepwise and K-means cluster for classification of financial distress status in manufacturing companies listed on the Indonesia Stock Exchange in 2019 (2022)
6. Kantzos, K.: Financial Statements Analysis, Athima, Hellas, 1st edn. Interbooks (2002)
7. Longaki, A.: Financial statement analysis of INTRUM AB, Patras (2021)
8. Palepu, K., Healy, P., Peek, E.: Business Analysis and Valuation (IFRS Edition), Palepu, K., P. Healy, E. Peek, (PHP), 4th edn, (2016). (Cengage)
9. Kanellos, N., Sakas, D.P., Kamperos, I.D.G., Reklitis, D.P., Giannakopoulos, N.T., Nasiopoulos, D.K., Terzi, M.C.: The effectiveness of centralized payment network advertisements on digital branding during the COVID-19 crisis. Sustainability. **14**(6), 23 (2022)
10. Nasiopoulos, D.K., Sakas, D.P., Trivellas, P.: Web search applications in logistics companies in the context of evolution of management information systems. In: Sakas, D.P., Nasiopoulos, D. K., Taratuhina, Y. (eds) Business intelligence and modelling. IC-BIM 2019. Springer Proceedings in Business and Economics. Springer, Cham (2021). https://doi.org/10.1007/978-3-030-57065-1_25
11. Plikas, J.H., Dimitrios, N.K., Delis, P., Vlachos, D.S.: Software protection and piracy focusing on the 2008 crisis: A comparative study and simulation modeling regarding the case of Greece, Germany, and England. In: Kavoura, A., Sakas, D., Tomaras, P. (eds) Strategic innovative marketing. Springer Proceedings in Business and Economics. Springer, Cham (2017). https://doi.org/10.1007/978-3-319-56288-9_42

The Role of Innovative Technologies in the Digitalization of the Modern Economy

Avdeeva Tatiana Viktorovna and Bart Tatyana Vyacheslavovna

In the modern economy, the problem of accelerating growth on the basis of the active introduction and use of various innovative technologies is becoming more acute. Many enterprises in conditions of tough competition are forced to use modern information technologies that allow them to quickly obtain information about the state of the external environment, analyze it and make appropriate management decisions. Modern strategic and operational management is no longer able to operate with "paperwork", data on the market, the behavior of competing companies, technological innovations, due to their large volume, the process of obsolescence and the inability of their timely qualitative analysis. In the light of the above, the use of modern electronic document management systems, electronic trading platforms corresponding to the level of development of information technologies of the latest generation, including software based on artificial intelligence, is becoming one of the most significant competitive advantages of the company. In general, electronic document exchange systems, including the stages of creating a document and endowing it with legal force, can significantly reduce costs, especially transaction costs, which are inevitably present in paper-based documents. At the same time, the system of executing tasks reporting by executors can drastically change, eliminating multiple duplications of both the reports themselves and the processes of entering the same data into different reports. An important advantage is also the acceleration of all processes, both managerial and production, which in turn significantly reduces

A. T. Viktorovna
The Russian Academy of Economy, The Institute of Management and Regional Development, Moscow, Prospect Vernadskogo, Moscow, Russia

B. T. Vyacheslavovna (✉)
Department of Marketing and Sport Business, Financial University under the Government of the Russian Federation, Moscow, Leningradsky Prospekt, Moscow, Russia
e-mail: tbart@mail.ru

D. P. Sakas et al. (eds.), *Computational and Strategic Business Modelling*, Springer Proceedings in Business and Economics,
https://doi.org/10.1007/978-3-031-41371-1_42

the cycles of product creation and its entry into the market, making high-risk investment decisions.

In 1995, Canadian expert in business and consulting, Donald Tapscott, proposed a new term to characterize the processes taking place in the global economy – "digital economy" – digital economy. In his work, Tapscott describes how people's lifestyles can change under the influence of information and communication technologies – ICT – and focuses on connections between users – primarily on the networking capabilities of people from different geographic locations and fields of activity. It is the avalanche-like growth of information links that is the basis of the new economy.

Digital is called the economy based on information and communication technologies – ICT. OECD identifies companies that create or provide processes for the electronic exchange of information, its processing, and so on. "Digital employment" includes any jobs in the digital sector and digital jobs in the non-digital sector.

According to Tapscott, it is the development of the digital economy that should be the main engine of globalization in the world economy. They are listed ten technological shifts that will help the development of the new economy and make the world economy much more efficient. Tapscott points out the need to introduce broadband Internet access, combining various data – audio, video, text in multimedia, the transition from proprietary to open source software, as well as the emergence of various aspects of virtual reality and artificial intelligence [1].

In 2019, we can conclude that all these shifts have already occurred. ICT had a tremendous impact on many sectors of the economy, but the depth of this influence and the speed of what is happening in each industry are very different.

The impact of digital technology is felt at both the global and local levels. You can consider the digital economy as a set of new industries representing a fast-growing part of the global economy in its traditional sense. In recent years, the rapid development of new technologies has led to the emergence of huge markets for cellular communications, Internet services, online gaming industry and other new products, without which humanity can no longer imagine existence.

New technologies have a transforming effect on some aspects of the activity of established economic entities, mainly it is the replacement of analog mechanisms of work – whether it be means of communication or industrial machines, with digital or digital elements, as well as further modernization – for example, existing production process.

The growth of the digital economy is due to the growth of a number of markets directly connected with digital and mobile technologies, with innovative developments in the military-industrial complex of states. The digital economy will still be at least 10–15 years old heavily dependent on the traditional one. At the current stage of technology development and the state of the markets, the digital economy should be considered a means of improving business efficiency. Modern digital economy involves new business models of human-resource-process-communication tools and emphasizes the need to transform management mechanisms to accommodate rapidly changing realities. There are industries in which the introduction of digital technology is particularly fast.

In the modern conditions of the spontaneous formation of "hotbeds" of the digital economy in different countries and regions, there is an acute problem of managing a phased transition to it, covering all economic entities. This is why in modern management theory and practice they are beginning to discuss the problems of identifying the essence of the digital economy, the allocation of its distinctive features, the systematization of the factors of its development, the definition of the resulting indicators.

The most common interpretations of the digital economy by Russian scientists are the following definitions: for example, O. Ivanov gives the broadest definition, treating the digital economy as a virtual environment that complements our reality [2].

R. Meshcheryakov believes that there are two approaches to the term "digital economy". The first approach is "classical": the digital economy is an economy based on digital technologies, and moreover, it is more correct to characterize exclusively the field of electronic goods and services. The second approach is the advanced one: "digital economy" is an economic production using digital technologies [3].

A. Engovatova treats the digital economy as an economy based on new methods of generating, processing, storing, transmitting data as well as digital computer technologies. She emphasizes that within the framework of this economic model, existing market business models undergo a radical transformation, the model of value-added significantly changes, the value of intermediaries at all levels in the economy is sharply reduced [4].

In the adopted strategy for the development of the digital economy in the Russian Federation, it is commonly understood as economic activity, the key factor in which production is digital data, and contributes to the formation of the information space, taking into account the needs of citizens and society in obtaining high-quality and reliable information, development of the information infrastructure of the Russian Federation, the creation and application of Russian information and telecommunication technologies, as well as the formation of a new technological basis for social and economic sphere [5]. Based on the above definitions, the essential distinguishing features of digital economy are defined. These include as follows:

- High level of automation
- Electronic document management
- Electronic accounting systems
- Electronic data storage
- Use GRM
- Creation of corporate social networks
- Use of ICT in production, management, communication, entertainment
- Electronic payment systems in the framework of e-commerce

Based on the opinions described, it can be argued that the digital economy can be viewed as a "platform" leading to the innovation and informational stage of economic development of systems of different levels, created on the basis of an effectively functioning information space, taking into account the needs of the economy and society, as well as the new technological basis business development and society as a whole. It is the foundation of the formation of a new postindustrial megacycle; it lays the potential for a transition to innovative and informational revitalization and neo-economic recovery.

Recently, the term "innovative economy" has been widely used in economic literature and in the practice of public administration (for example, the Russian government has adopted the state program "Economic Development and Innovative Economy"). In this regard, it is necessary to trace the relationship between the concepts of "globalization" and "neo-industrial development" with the term "innovative economy".

As the term "post-industrial development" does not adequately cover completely the changes taking place in the economy and, therefore, as we have proved, it is more expedient to apply the earlier term of John K. Galbraith "neo-industrial development" to the modern economy, and the now widely used terms "new economy" and the "knowledge economy" do not fully cover the events occurring in innovation [6].

Although the term "innovative economy" is relatively new, but innovative economies existed before, and the degree of "innovativeness" of the economy increased sharply in certain periods and in individual countries (for example, during the industrial revolution, first in England and then in a number of other countries). The stages of such bursts of innovation development are well characterized by S. Glazyev [7].

Experts of the international monetary fund consider the dilemma of globalization as "the growing economic interdependence of countries around the world occurring in the result of the increasing volume and diversity of cross-border flows of goods, services and capital, as well as a faster and wider diffusion of technologies" [8].

As for the "innovation economy", there are various definitions of it, all successful formulations directly or indirectly related to the concept of globalization.

A national innovative economy is, first, a type of economy based on borrowed and based on its own innovations, on continuous technological improvement, on the production and export of high-tech products with very high added value.

Second, it is an economy in which reliable government and corporate mechanisms are created and operate, ensuring innovation.

Third, an economy in which:

- Simple and effective innovations based on old knowledge are created and successfully implemented;
- New knowledge is not created, but innovations built on it are borrowed.

Fourth, a neo-industrial innovation economy can be built:

- On the basis of new knowledge created in the country
- On the provided opportunities of globalization to borrow new (modern) production technologies

Fifth, a rational combination is the rational option:

- Borrowed and own developments, built on both acquired abroad and on new knowledge created in the country; innovative development, the necessary traditional industries and the accelerated growth of several selected industries that are at the forefront of scientific and technological progress, which are at the present stage priority or critical.

In order to fit into the globalization processes in the country, in particular, there should be an intensification of innovation activity, and the directions of innovation activity should expand. In a small country, of course, borrowed innovations should take a particularly significant place, although the role of their own innovation and research sector is increasing, and in order to master borrowed technologies.

Now let us note the most important characteristics in which the national neo-industrial innovative economy should develop in the context of globalization of the world economy:

- The most important of the globalization factors contributing to the formation of a national neo-industrial innovative economy is the expansion of international technology exchange and cooperation in the implementation and achievement of final results. Foreign direct investment plays a significant role, but the priority should be investments that provide an influx of innovative technologies. Over the past twenty years, in many developed countries, the role of institutional investors (pension funds) and investment funds that actively invest in new technologies and innovation has increased by more than an order of magnitude. The state is obliged to create a coordination system to encourage businesses to increase spending on innovation and make it susceptible to innovation.
- An important factor of globalization is the fact that "in international trade at an accelerated rate, trade in services, technologies, and objects of intellectual property increases" – by what determines the content of a modern neo-industrial innovative economy.

The intensification of global competition necessitates the intensification of innovation activities and the provision of effective technological and sectoral structural shifts on this basis, achieving a structural effect due to the rapid growth of high-tech sectors of the economy and other priority sectors recognized by the country as well as the emergence of new types of activities.

Network structures play an increasingly important role in the modern economy, which we also drew attention to in our works.

The role of network structures in the modern economy is very successful; we can say that this detailed definition: "In the modern economy, network interconnections play a system-forming role, having a direct impact on the development of innovative activity, the very possibility of which is determined by the combination of direct and reverse links between different stages innovation cycle, knowledge producers and consumers, firms, the market, the state within national borders and, increasingly, on a global scale. The effectiveness of innovative development depends not only on how effective the activities of independent economic agents are, but also on how they interact with each other, how the knowledge system and their use interact" [9].

It is important to use in the process of neo-industrial innovative development of the factors of regionalization (cross-country). Regionalization is one of the directions of development of modern globalization and covers, as you know, a group of countries creating associations, unions, within the framework of which "there is greater or lesser liberalization of trade, pricing, capital movements, labor migration" [10].

The key factors in the digital transformation of the economy are shaping the near future. These factors will be focused on most innovative industrial clusters (both existing and planned).

Many studies have shown that an integral mechanism for increasing the competitiveness of economies of different countries of the world, accelerating innovation development processes in various sectors, encouraging local initiatives to enhance interaction between industrial enterprises, the state and end users, especially in the context of active use of opportunities and composite components economy, the so-called digital technology platforms, as well as a variety of infocommunications in a cluster approach, actively applied to industrial enterprises.

The cluster, as an object of economic agglomeration of interrelated enterprises in a certain territory, has been known since the days of handicraft production. In this case, the origin of the cluster approach (1830–1890), economics is obliged to A. Marshall, who himself called them "localized industries" (localized industry) [11]. Subsequently (1890–1950), the term industrial districts became widespread in the literature. The most popular term at the present time – "cluster" – appeared much later. Believe that the concept of "cluster" is introduced into the theory of economic relations by M. Porter (1998) [12].

M. Porter used the term "cluster" to analyze the structure and efficiency of economic relations in industries and organizations. Analyzing the competitiveness of the company, he focused on its economic environment. Currently, the cluster is considered a separate economic system, an integrated structure, which have a significant impact on the development of the state industrial policy and the economy as a whole.

Thus, the "Industrial Cluster" is a set of stakeholders in the field of industry and related organizations that collaborate and effectively solve production, marketing and innovation tasks [13].

Infocommunication technologies will coordinate and control the operation of production equipment based on digitalization. As a result, time spent on interoperable transitions will be reduced; there will be an opportunity for more flexible planning and use of available information and analytical resources.

In turn, innovation-active industrial cluster should be understood as a group of interrelated industrial, scientific and educational enterprises, as well as other organizations (financial, consulting, engineering and other companies), actively working together to achieve key performance indicators, as well as interacting with each other to develop, implement, commercialize innovations and innovative products.

That is, innovation-active clusters are clusters that actively use various innovations (process, structural, organizational, technical, etc.) in their activities and, accordingly, ensure the production of a significant proportion of high-tech competitive products with a high degree of added value. The assessment of the level of innovation activity should be carried out in conjunction with the assessment of the innovative potential of the enterprise, and the second should precede the first.

The problems of formation, distribution and use of labor resources, ensuring the balance of jobs and the professional and qualification structure of personnel, have always been the focus of economic science.

The current state of the labor market is focused on increasing the intensity of information processes of the productive forces of society, the active use of information resources and the need for their use in the modern economy. The problem of labor mismatch with the structure of workplaces is not in itself new for the world economy; it arose at different stages of economic development in different countries. At the same time, integrated approaches to research of the digital economy is not enough. Taking into account modern changes: technological, informational and new qualitative characteristics of the workforce, it is necessary to create a full-fledged mechanism for the formation of a quality workforce in the digital economy. The mechanism, in our opinion, should contain at least two stages:

1. Creation of integration links between the labor market and the market of educational services
2. Creation and strengthening of integration links between all institutions providing labor

In the conditions of the digital economic system, the training of labor resources must take into account the fact that training objectives may vary (training of labor resources-innovators and training of labor resources as a hired labor force capable of carrying out any activity with information technologies by industry). Impact factors are different (the system of training, the level of provision of new industrial equipment, the level of unemployment, investment in the labor force of the company, return on investment in the labor force, the process creating an innovation-idea or product, etc.). One of the ways to organize the preparation of highly skilled labor is the creation of clusters (including educational ones). Considering the process of clustering industry and education, it is necessary to understand that clusters can be intersecting and not intersecting depending on the industries involved in the clustering. In the process of clustering, vocational education will not blindly follow the demand of the current labor market; it will be able to develop according to the formula of anticipating need for labor resources.

To form human resources, it is necessary, first, taking into account the process approach of information industrial transformation, having analyzed the production chain of product creation, training should be carried out by introducing them to separate links of the industrial production chain. Second, given the sectoral transformation of the information economy, the specifics of the industry must be analyzed in the sectoral context of the need for personnel in all parts of the production chain. Third, given the emergence of various varieties of digital industrial markets, it is necessary to focus on the formation of managerial, engineering and other personnel, taking into account the technological communication of information and markets.

Thus, itself the system of formation of labor resources must also be significantly transformed.

Implementation of measures to establish an innovative economy in Russia, including through comprehensive modernization of production and increasing its competitiveness, should be provided by development institutions that support through financing business projects, infrastructure support and co-financing R&D.

The development of the digital economy in the future will depend on the success of a number of breakthrough technologies.

At the end of 2018, there are five such technologies:

- 5G communication
- 3D printing
- Blockchain
- Artificial intelligence (AI)
- The virtual reality

The listed technologies are among the most important development directions outlined in the strategic documents of most developed countries. Active attempts to develop them are also being undertaken in China and a number of other Asian countries, and in 2017, "digital" priorities also received wide attention from government officials of Russia.

The mentioned technologies are in the stage of intensive development; every year, they become more and more perfect and attract more and more young scientists. Each technology can have a disruptive impact on a number of traditional industries and on business as a whole.

Before turning to the consideration of leading industries that have already become digital or have fully felt the impact of new technologies, consider the prospects for selected breakthrough technologies.

5G communication – Popular mobile communication systems in large cities operate in LTE 4G – 2665 MHz networks at speeds up to 100 Mbit/s. Further progress is associated with the use of millimeter waves – 1-100 GHz, which have unique features and are widely used, for example, by the military. New networks will be able to provide speeds of up to 10 Gbps. It is an increase in the data transfer rate tenfold and is the opportunity that will lead to tremendous changes in the world. Already conducted tests show that the 5G era is near and, most likely, will come in the early 2020s after the introduction of a new unified communications standard – IEEE 5G. It is the new generation networks that will open the way for widespread industrial automation – m2m, unmanned vehicles and real breakthroughs in the field of the Internet of things. At first, the technology will be used in the largest megacities and can significantly change their economy.

3D printing – Additive production allows you to overlay thin layers of material at each other, creating the desired 3D model. Currently, this market is very small – about $6 billion in 2016. It is expected that the use of 3D-printing can reduce the cost of production, to give new highly skilled jobs and, together with other technologies, to transform the industry. In 2015–2016 in the 3D printing market, there was a slowdown and a complete mismatch with analysts' expectations; however, already in 2017, the sector is expecting a growth of about 16%, and the total number of 3D printers sold will grow by 39%. Estimates of consulting companies are very optimistic – in the next 5 years, the average annual growth of the market for 3D printers will reach 33% – according to Context. The greatest contribution is made by industrial devices.

Blockchain – The technology of distributed registries allows carrying out a transaction; it means transferring some valuable information, for example, currency or contractual obligations, without copying it and further comparing copies – by the method of distribution between multiple independent and anonymous users – nodes. It should be emphasized that first of all blockchain is relevant for those countries and sectors of the world economy where there is a low level of trust in established organizations – banks, governments, counterparties, etc., because the technology attracts with its transparency and security. It provides an opportunity to develop smart contracts – i.e., automatically executed when the specified conditions are reached, the fractional economy already held on the example of Uber and Airbnb, crowdfunding, electronic elections, new types of intellectual property protection and user identification, microgrids in the energy sector and can also find application in any other operations that require instant and safe data exchange. In addition, the blockchain technology allows you to create cryptocurrencies; the most famous of which is Bitcoin. Cryptocurrency is a digital currency circulating on a computer network. Cryptocurrencies have several advantages: high transaction speed, transparency of operations and security. Nevertheless, the future of cryptocurrency remains in question – the issuer of a cryptocurrency is simultaneously many participants, which can make such a system unstable. In any case, cryptocurrencies require rethinking the role and function of money within the economic system. Visa – Visa B2B Connect, Disney, IBM, the governments of Dubai, China and Korea are actively working on the blockchain transaction project. Until 2017, technology was almost entirely concentrated in the United States and Western Europe. In 2017, the blockchain has become particularly popular in Asia. For example, if in 2013, 78.7% of blockchain transactions accounted for the United States; then in 2016, the share of the United States amounted to 49.2%. Separately, it is worth noting the huge potential of the blockchain for Africa, where 70% of the population is deprived of banking services. A number of countries in the region have already expressed their interest.

Artificial Intelligence (AI) – Technologies of artificial intelligence include such developments as machine learning, image recognition and speech. AI is used in ICT, media industry, retail, health care and so on. According to McKinsey, the largest high-tech companies spend from 20 to 30 billion dollars on AI annually, and start-ups from 6 to 9 billion dollars. More than 90% of the costs go to research and development, 60% – in machine learning technology. Most of the applications of AI are at a pilot stage, so the estimates of the prospects for this market vary widely – from 0.6 to 126 billion dollars by 2025. Although investments are growing rapidly, it is still premature to talk about commercial use of AI. Best of all, AI is implemented in three industries: telecommunications, automotive and financial services.

The virtual reality – Virtual reality technologies are usually considered together with augmented and mixed reality technologies. Goldman Sachs estimates that by 2025, video games will be the leaders in introducing virtual reality – more than $ 11 billion and health care – more than $5 billion. The entire market may exceed $ 80 billion. By the end of 2016, the market is estimated at $2-7 billion. It should be noted that in sales of the corresponding software, Asia is ahead of North America. In the US, though, there is a rapid growth in sales of the corresponding software - from $ 97 million in 2016 to $ 403 million in 2017 - according to a report SuperData.

The competitiveness of states will be determined by the flexibility of management and the ability to anticipate the impact of new technologies and to respond quickly to changes. New technologies not only expand the capabilities of market participants but also involve a number of challenges. Special attention is required for the impact of digitalization on the competitive environment and on the labor market – the digital economy leads to automation of work processes, which leads not only to an increase in demand for highly and low-skilled jobs but also to a reduction in the number of workers with average qualifications.

Governments around the world are creating special Internet ministries, regulators are launching initiatives aimed at complying with uniform rules for digital business participants, special services are changing. Topical initiatives include the creation of e-government, the digitization of health-care systems/education/utilities, the preparation of laws governing new elements of the digital economy.

The greatest difficulty in the matter of regulating innovation processes is that technologies are developing faster than political regimes. However, many developed countries have already adopted some programs for regulating the digital economy. The objectives of regulation are: to stimulate new forms of communication and communications – the Internet, m2m, to stimulate competition both within the digital sector and in traditional industries, to protect the interests and safety of consumers – users. Solutions can have a reciprocal effect: for example, strengthening control over information can hinder the development of the Internet, and support for one or another format of digital communications can lead to the consolidation of a single business model, while creating obstacles to the emergence of new ones.

There are three challenges to the digital economy for the system of state regulation:

First, the digital sector, the Internet and new forms of communication are developing very quickly. There are new platforms and services that require the attention of regulators and new approaches. Accordingly, introduced legislation can quickly lose relevance and effectiveness, which increases the pressure on regulators.

Second, the media is converging. The Internet, television, mobile communications are increasingly interconnected. This entails the confusion of markets that have traditionally been regulated separately and also creates new challenges for antitrust authorities. The EU has already responded to this challenge by combining control of telecommunications and TV and broadcasting in one supervisory authority.

Third, the decentralized nature of the Internet and the digital economy, the lack of clear geographic boundaries in it, creates problems of jurisdiction and coordinated regulatory approaches in different countries. Some aspects of the Internet and digital technologies were standardized and coordinated at the international level – domains, mobile communication principles, some norms of intellectual property rights. However, there are more unsolved regulatory issues. These are proxy servers, pirated content, "darknet", transactions through third countries with different regulations. In addition, issues of taxation principles and the responsibility of the parties are acute.

Given these challenges, the choice of the regulatory approach becomes a central issue for regulators. The approaches range from self-regulation, in which users and companies have to solve problems between themselves before contacting the state regulator, and joint regulation, in which the public and private sectors work together in a single organization, to administrative and managerial regulation, with which the government establishes uniform rules.

Compared with the administrative and managerial regulation, self-regulation has advantages: decisions on technical standardization are made by industry representatives who are more familiar with how the industry functions. Such an approach seems relevant in a market as complex as the digital economy. Moreover, self-regulation can be more flexible than administrative and management regulation, which harmonizes well with a dynamic digital economy. However, there are also disadvantages: during self-regulation, representatives of the industry may not take into account the interests of other economic entities and consumers. Thus, co-regulation is the best approach.

The phenomenon of the digital economy emphasizes the patterns of modern technological development and the management of this development. Attempts by various countries to support innovation and economic growth have not yet revealed a single successful approach. The institutional environment in digital technology remains fragmented. Opposite principles of ICT regulation often operate in different countries. The most important task remains to develop a balanced approach that can maximize the positive effect of the digital economy and minimize the risks associated with the spread of new technologies – unemployment, social inequality, etc.

Based on the above, the following general conclusions can be made:

- Digital economy can be considered a "platform" for the innovation and informational stage of economic development of systems of different levels, created on the basis of an effectively functioning information space, taking into account the needs of the economy and society, as well as the new technological basis for business development and society as a whole;
- In accordance with the dual nature of the digital economy – dependence on the industrial and neo-industrial stage of development on the one hand and the ability to form the basis of the innovation-information economy – on the other, managing its development should also include two directions. The first of them focuses on the formation (ensuring the transition from the neoindustrial to the digital economy), the second – on the development (ensuring the transition from the digital to the innovative-information economy);
- At the present stage, within the boundaries of which an inter-cyclic transition is observed, management should focus on creating conditions to ensure:

1. The highest level of automation at all stages of economic transformations of the state
2. The introduction of electronic document management, the expansion of boundaries for inter-country cooperation
3. Large-scale use of electronic accounting systems, the use of innovative methods for the protection of electronic resources
4. Formation of electronic data warehouses, systematization and protection
5. Creation of corporate social networks
6. Use of ICT in production, management, communication, entertainment
7. Transition to electronic payment systems in the framework of electronic commerce. The accelerated formation of the digital economy requires the development and implementation of large-scale implementation tools with the participation of the state.

In conclusion, we note that the emergence and spread of ICT have had such a profound impact on the world economy that a new phenomenon has emerged – the digital economy. Nevertheless, its scale is still relatively small, and it's premature to talk about its decisive role in the development of the world economy. Other scientists with similar cases are studying similar problems, using modeling and simulation [14–18]. Continuously new technologies emerge that will continue to change the economic and institutional landscape.

References

1. Galbraith, J.K.: The New Industrial State. John Galbraith New industrial society. Translation into Russian: L. Ya. Rozovsky, Yu. B. Kochevrin, B. P. Likhachev, S. L. Batasov. - M. Electronic publication: Center for Humanitarian Technologies. - 03/29/2012 (2004) https://gtmarket.ru/laboratory/basis/5021
2. Dobrynin, A.P., Chernykh, K.Y., Kupriyanovskiy, V.P.: Digital economy - various ways to the effective application of technology. Int. J. Open Inf. Technol. **1**(4), 4–10 (2016)
3. Yu, G.S.: Possibilities and limitations of the technical and economic development of Russia in the context of structural changes in the global economy [Electronic resource]– Access mode: http://www.spkurdyumov.narod.ru/glaziev/htm
4. Marshall, A. Principles of economic science / A. Marshall. - M.: Izd. group "Progress", Moscow, (1993)
5. Porter, M.E.: The Competitive Advantage of Nations, p. 380. Free Press, New York (1990)
6. Michael Porter, M.: Competitive Strategy. Techniques for Analyzing Industries and Competitors, p. 456. Alpina Publisher, Moscow (2015)
7. The program of digital economy of the Russian Federation, 07.23.2017 [electr. resource] http://fond83.ru/novosti/federalnye-novosti/637-proektprogrammy-tsifrovoj-ekonomiki-rf-budet-predstavlen-na-pmef
8. Serbinovsky, B.Y.: About the content of the terms "innovative economy", "new economy" and "knowledge economy" [Electronic resource] / B. Yu. Serbinovsky, A. S. Zakharova // Sci. J. KubGAU. 61(07) (2010). https://cyberleninka.ru/article/n/o-soderzhanii-terminov-innovatsionnaya-ekonomika-novaya-ekonomika-iekonomika-znaniy

9. Solomatin, M.S., Yu, S.N.: The role of the digital economy in the development of the state // Institutional and infrastructure aspects of the development of various economic systems; collection of articles of the international scientific-practical conference: in 2 parts. –2017. pp. 137–139
10. Strategy is simple [electr. resource] https://www.kommersant.ru/
11. Tapscott, D., Williams, E.D., Wikinomics.: How Massive Collaboration Changes Everything, p. 392. M.: BestBusinessBooks, Moscow (2009)
12. Digital economy: how experts understand this term // RIA News – 2017 [E-mail https://ria.ru/science/20170616/1496663946.html]
13. Digital Russia: a new reality. July 19, 2017 McKinsey Global Institute [Email address: http://www.tadviser.ru/images/c/c2/Digital-Russia-report.pdf]
14. Nasiopoulos, D.K., Sakas, D.P., Vlachos, D.S., Mavrogianni, A.: Simulation of generation of new ideas for new product development and IT services. AIP Conf. Proc. **1644**, 60–68 (2015)
15. Sakas, D.P., Vlachos, D.S., Nasiopoulos, D.K.: Modeling the development of the online conference's services. Lib. Rev. **65**(3), 160–184 (2016)
16. Sakas, D.P., Nasiopoulos, D.K., Reklitis, P.: Modeling and Simulation of the Strategic Use of the Internet Forum Aiming at Business Success of High-Technology Companies. In: Sakas, D., Nasiopoulos, D. (eds) Strategic Innovative Marketing. IC-SIM 2017. Springer Proceedings in Business and Economics. Springer, Cham (2019). https://doi.org/10.1007/978-3-030-16099-9_22
17. Sakas, D.P., Nasiopoulos, D.K., Reklitis, P.: Modeling and Simulation of the Strategic Use of Marketing in Search Engines for the Business Success of High Technology Companies. In: Business and Economics, Springer Proceedings in Business and Economics, Springer pp. 217–226. https://link.springer.com/chapter/10.1007/978-3-030-16099-9_27
18. Foteini, P.E., Dimitrios, N.K., Sakas, D.P., Vlachos, D.S.: The development of new ideas for IT products, through social media. In: Strategic Innovative Marketing, Springer Proceedings in Business and Economics, Springer, pp. 399–404. https://link.springer.com/chapter/10.1007/978-3-319-56288-9_53

Analysis of the Current State of Socioeconomic Development of the Republic of Dagestan

Djamily M. Maksubova and Navrat M. Umargadzhieva

1 Introduction

The Republic of Dagestan is included in the group of regions of the country with a low level of socioeconomic development. To this end, large-scale work has been launched in the republic to develop a strategy for the socioeconomic development of the Republic of Dagestan until 2035. With the involvement of experts and the general public, strategic sessions are held to discuss problems and determine priorities for the long-term socioeconomic development of the republic [3].

One of the main instruments for the development of the economy and social sphere in Dagestan is the Republican Investment Program. For its implementation in 2021, 18.4 billion rubles were allocated, of which 11.2 billion are federal funds and 7.3 billion are republican funds. At this expense, 34 objects out of the planned 170 were commissioned, 344 km of roads were repaired and 27.3 km were built, 12 bridges with a length of 548 linear meters were repaired, and one bridge of 29.8 linear meters was built.

In addition, as part of the implementation of the 100 Schools project, 105 general education organizations were repaired in 2021. Under the Local Initiatives program, 84 projects were financed for 299 million rubles to modernize infrastructure in municipalities. Within the framework of the regional state program for the socioeconomic development of mountainous territories, 157 subsidies were issued with a total amount of 48 million rubles [5].

Within the framework of the pilot project, together with the Accounts Chamber, measures are being taken in the republic to reduce the volume of construction in progress. As of January 1, 2021, their number was 1604 objects for a total amount of

D. M. Maksubova · N. M. Umargadzhieva (✉)
Federal State Budgetary Educational Institution of Higher Professional Education Dagestan State University, Republic of Dagestan, Makhachkala, Russia

D. P. Sakas et al. (eds.), *Computational and Strategic Business Modelling*,
Springer Proceedings in Business and Economics,
https://doi.org/10.1007/978-3-031-41371-1_43

45.4 billion rubles, and as of January 1 of this year, 588 objects for the amount of 8.4 billion rubles were already excluded from the reporting. Of these, 500 – by issuing title documents necessary for commissioning, 88 – by writing off (due to irrelevance). Work in this direction, according to the minister, continues.

Revenues of the consolidated budget of the republic in 2021 amounted to 185.9 billion rubles, and expenses – 181.1 billion rubles. The largest volume of expenditures falls on education – 68.3 billion rubles and social policy – 51.7 billion rubles.

According to the Acting Minister of Economy and Territorial Development of the Republic of Dagestan Rustamov A.Z., "Budget funds provided for this year must be contracted before April 1, unless otherwise provided by the schedule for financing and implementing activities. Also, they must be mastered in full, and the planned activities must be implemented on time" [6].

In 2021, Dagestan received additional support for socioeconomic development at the federal level. The region has been approved a budget loan to finance the implementation of infrastructure projects in the amount of 10.5 billion rubles. These funds will be used to develop the water supply system, including the design of the Makhachkala collector.

The index of output of goods and services by basic types of economic activity, according to preliminary data, amounted to about 102.2%.

As part of the republican investment program, 389 objects were financed in 2021, of which construction work on 47 objects was completed.

A new impetus has been given to the development of a number of industries in the republic, including tourism showing good performance: in 2021, the tourist flow amounted to over one million people, which is 30% higher than in 2020. This indicates that the priorities identified jointly are correct. It is necessary to start working on fixing these priorities and creating normal conditions for the socioeconomic development of the republic [7].

At the same time, we note that it is impossible to talk about significant economic growth. According to the preliminary results of 2021, the index of output of goods and services by basic types of economic activity amounted to about 102.2% – and this, we believe, is not development.

In terms of agriculture and industry, the statistics are not bad – the industries are developing and there are a number of promising projects. There is potential, including in animal husbandry, but there is not a single full-cycle enterprise for the production and processing of lamb. We believe that it is worth looking for opportunities to develop both agriculture and industry by building processing enterprises, building a vertical of production in all directions [1, p. 328].

At the same time, the issue of employment of the population remains topical for the republic. The number of officially registered unemployed in 2021 amounted to about 34 thousand people. We believe that here it is necessary to change the system in such a way as to develop more dynamically to solve this problem. To do this, we must keep records. Based on statistics, conduct analysis, conduct planned and not chaotic work. To do this, it is necessary to revise approaches to the organization of labor, including in personnel policy – to stop appointing incompetent people to positions.

As for agriculture and food, according to the Minister of Agriculture of the Republic B.V. Battalov, "in terms of the performance of the industry, 450 thousand tons of grain were produced in 2021, which is 6.5% more than a year earlier. Rice was produced 118.9 thousand tons, which is a record for the entire history of rice cultivation in Dagestan. The main role in this was played by the unprecedented assistance to the land reclamation complex, for the development of which over the past three years 900 million rubles have been allocated from the budget of the republic. The rice cluster is developing in different directions – last year another plant for processing this crop was launched, which made it possible to produce 14,000 tons of rice groats" [4].

One fifth of the gross regional product of Dagestan is agriculture, which employs up to 30% of the economically active population and concentrates more than 12% of fixed production assets [2].

The volume of gross agricultural output at the beginning of 2022 in farms amounted to 140 billion rubles, including in crop production – 65.2 billion rubles, in animal husbandry – 74.8 billion rubles.

Gross agricultural output in 2021, despite the pandemic, grew by 1.3% and amounted to 141.5 billion rubles. At the same time, a positive dynamics of production volumes of the main types of crop and livestock products was ensured.

As of January 1, 2022, the number of cattle in farms of all categories decreased by 3% compared to the corresponding period last year and amounted to 949.1 thousand heads, sheep and goats increased by 0.6% (4720 thousand heads), the number of poultry decreased by 12.1% (4401.5 thousand heads), including in agricultural organizations – by 25.6% (611.9 thousand heads).

The main task of the social protection service is to support low-income, low-income families, and the disabled. According to the results of 10 months of 2021, the average monthly salary in the region increased by 6.6% and amounted to 31 thousand rubles, but we believe that its size remains one of the lowest in the country.

To ensure the growth of citizens' incomes in 2021, a special regional program to reduce poverty was adopted in the republic, within the framework of which the social contract mechanism is used. Over the past year, social security authorities signed 7.7 thousand contracts worth more than 871 million rubles, which allowed citizens to organize entrepreneurial activities, undergo vocational training, find a job, establish a subsidiary farm, and get out of a difficult life situation [5].

In order to provide additional family support, a new form of state support has been introduced since last year – a monthly payment for children from 3 to 7 years old in the amount of 5378 to 10,757 rubles for each child. In 2021, 195 thousand families received this assistance in the amount of more than 17 billion rubles.

In addition, last year the republic increased the amount of monthly social support for veterans of the Great Patriotic War (up to 20,000 rubles), for family members of fallen veterans of the war in Afghanistan (up to 10,000 rubles). In general, over 32 billion rubles were allocated for social security of the population in 2021 through the ministry, which were provided in the form of payments to war veterans, labor veterans, home front workers, families with children, and the disabled and other categories of citizens.

Another area of work in the republic is the implementation of state programs in the field of social services for the population, providing an accessible environment for the disabled, improving living conditions for certain categories of citizens. In 2021, the institutions of the Ministry of Labor of the region provided various social services to more than 120 thousand people. Last year, the number of non-profit organizations providing social services increased from the republican budget.

Over the past year, 100 facilities were equipped for the disabled and people with limited mobility under the accessible environment program in the amount of 82.6 million rubles. The housing conditions of 389 people on the waiting list among veterans, disabled people, families with disabled children and liquidators of the Chernobyl accident were improved for a total of 316.3 million rubles, which is 1.3 times more than in 2020.

In order to reduce the negative consequences on the labor market associated with the COVID-19 pandemic, a special set of measures was developed in the republic to restore the employed population in the region to the level of pre-pandemic values. As a result of the measures taken (employment, job creation, subsidies, and microfinance for entrepreneurship), over 156 thousand people managed to restore employment in 2021, in particular, over 51 thousand people were restored to permanent/temporary jobs. Approximately 750 people have been trained under employment assistance programs. Over 1300 citizens have mastered a new profession within the framework of the Demography national project [7].

Last year, a new mechanism for stimulating employers in hiring unemployed citizens was launched in the republic by subsidizing at the expense of federal funds in the amount of 32.3 million rubles (a total of 926 citizens were employed). According to this indicator, Dagestan ranks seventh among the constituent entities of the Russian Federation. In general, the work done made it possible to reduce the level of unemployed in the republic by 3 times from 106 to 34 thousand people.

In addition to the positive aspects in the socioeconomic development of the Republic of Dagestan, there are many problems.

Hence, there is no real information about the economic situation in the republic, which casts doubt on everything that is happening. The existing managed statistics are self-deception and complacency, convenient for managers of all levels, industries and are formed at their own discretion and by anyone, and, even worse, the lack of necessary information generally leads to an inaccurate analysis of the state and incorrect forecasting. This is a known situation. I don't want to blame anyone in particular, since the system and methodology have developed in such a way that no one bears any responsibility for distorting statistics. Therefore, based on a false analysis, ineffective management decisions are made, and everyone knows what this leads to. It hurts, but we need to put things in order in this matter so that the authorities know what we have and what we don't [1, p. 329].

There is a problem with the sale of manufactured products in all regions, but this is beneficial for someone, but not for society, but nothing is being done in this direction by the state and municipal authorities. It is necessary to manage and state support in creating a modern structure, with the participation of municipalities to create a network of enterprises and organizations, at least for the primary processing of raw materials: animal slaughterhouses with mechanized skinning, salting, extracts for standard drying, agricultural cooperatives for the purchase, processing, product storage, and distribution. But what is actually happening is what happened in the past – manufacturers were left with big losses, intermediaries and traders got rich quickly, got easy and big profits. It rejects workers in the manufacturing sector. Therefore, they flee to the cities to trade [2].

1.1 Ways to Improve the Strategy of Socioeconomic Development of the Republic of Dagestan

The socioeconomic development of the Republic of Dagestan should, above all, express the interests of the population and provide solutions to those problems that concern them. The basis of the SER strategy should be a focus on a person, an increase in his real income, the formation of a comfortable living environment, affordable and high-quality education and health care, housing and communal services and more [8]. The above analysis revealed a decline in the proportion of the young population in Dagestan, which is a challenge for regional economic development. Thus, the key challenge of the NER strategy for the coming years is the inevitability of population aging. Therefore, the strategy of the SIR of the Republic of Dagestan should be aimed at – increasing the measures of social support provided in cash to families with children; providing parents with the opportunity to combine the upbringing and care of children with economic activity. To reduce the mortality rate and increase the life expectancy of the population, the strategy should contain measures aimed at improving the republican healthcare system in terms of reducing early mortality and preventable mortality; formation of conditions for the growth of safe behavior of the population and an increase in the duration of an active and healthy life. In the Republic of Dagestan today, there is a significant outflow of the able-bodied population to work, and young people to study in other regions. In order to increase the migration attractiveness of the Republic, it is necessary to ensure – the organization of new jobs (of high quality); improvement of conditions for small business, development of housing and transport infrastructures. Theorists today distinguish education capital, health capital, labor capital and cultural capital as elements of the human capital of the region. An analysis of the system of preschool education in the Republic of Dagestan showed a low coverage of places in preschool educational institutions – 32%. Given the limited republican consolidated budget, it is necessary to stimulate the development of the non-state sector of preschool education, as well as public-private and social partnerships in this area. General education for each student provides not only educational achievements, but also socialization. An analysis of the general education system in the Republic showed an increase in the number of students with a decrease in the number of educational institutions and the number of teachers, which makes it necessary to organize the educational process in the Republic in two or even three shifts. The SER strategy should include the introduction of new places for students and an increase in the staff of teachers. The region needs a program for the construction of new and reconstruction of existing school buildings, the implementation of which will ensure the organization of education in one shift and the formation of an environment for children with disabilities in schools. An analysis of the system of higher professional education in the Republic revealed an imbalance in the areas of training specialists, a discrepancy between the HPE system and the strategy for building the republican innovative economy and the lack of connection between the strategies of universities and the development of the region. In the conditions of high demand for human capital, the region needs to take into account in the strategy of the NED of the Republic – development of partnerships with leading Russian universities; attraction of innovative enterprises in terms of partnership in the development of curricula of universities; expanding student internships at innovative enterprises; implementation of joint projects with innovative enterprises in terms of research and development. Preservation of health and longevity of the population should become priorities in the development strategy of the healthcare system of the Republic of Dagestan. To eliminate the shortcomings identified by the analysis, the Republic needs to – reduce mortality from cancer, diseases of the circulatory system, tuberculosis, for which purpose to form such models of behavior of the population that contribute to maintaining health; increase the motivation of medical personnel and strengthen human resources in this industry; modernize the healthcare system itself in order to increase its efficiency; to form employers' motivation to preserve the health of employees through the organization of working conditions by them that prevent the development of occupational diseases. Historically, the Republic of Dagestan among the subjects of the North Caucasus was distinguished by a large number of handicraft industries based on original technologies and centuries-old traditions. Folk arts and crafts are represented by the production of jewelry, carpet weaving, pottery, and other activities. Fisheries are an important component of the economy of entire regions of Dagestan. There is no doubt that the Republic has great potential for the development of ethnic tourism, which will provide new prospects for the effective use of cultural resources available in the Republic. The strategy of the SIR of Dagestan should contain such a direction as the development of ethno-tourism, since it promises great prospects for the growth of employment and economic benefits. The Republic of Dagestan ranks first in Russia in terms of subsidization of the republican budget due to low tax revenues. Thus, a vicious circle arises when, receiving funds from the federal center, the Republic is in no hurry to straighten out its economy. Therefore, the only way to improve the financial condition of Dagestan is to reduce its subsidization, to strengthen the republican revenue base. To increase the revenue base of the republican consolidated budget, it is necessary to increase the tax potential of Dagestan, which can ensure an increase in the collection of tax payments. The point of economic growth of the Republic of Dagestan may be the creation of a special economic zone in the Republic. The arguments are the following factors that speak of the difficult socioeconomic situation in the Republic – high subsidization of the republican budget; low standard of living; high unemployment; a significant decrease in the level of industrialization. The creation of the SEZ in Dagestan will be promoted by Dagestan is the southernmost border region of the Russian Federation; availability of labor potential; availability of natural resource potential. According to the type of SEZ in Dagestan, it can be a tourist and recreational one, capable of involving all available potentials in the economy. Dagestan is the most multinational republic of Russia; its territory was inhabited two million years ago. Due to its ancient history and ethnic diversity, the region has a unique cultural wealth. Each district, each village in Dagestan has its own rich history, culture, and folk crafts. There are more than five thousand objects of cultural heritage in the Republic. There are practically all conditions for the development of all types of tourism: agro-tourism, event tourism, ethno-tourism, culinary, and others. For the development of this industry in the Republic, it is necessary to stimulate the creation of modern infrastructure, hotels and guest houses with all amenities and cafes with traditional cuisine. It is necessary to develop routes, programs, produce souvenirs and more. The length of the Caspian coast with sandy beaches is 500 km (for comparison, the Crimean coast has a length of 750 km). In the Republic, it is realistic to receive hundreds of thousands of tourists in the summer season and not only from Russia, which can become an important source of income for the local population, stimulate the development of entrepreneurship and the creation of new jobs. An analysis of agricultural production in the Republic of Dagestan showed that the main producers are private households. For the normal development of the region's economy, large enterprises with a multi-structural form of agriculture should become producers. At the same time, the state sector (seed-growing and breeding farms) must also be preserved in the Republic. Seed production is on the verge of collapse, since personal farms are not financially able to buy seeds of higher reproductions and, accordingly, engage in their reproduction. For the same reason, no land improvement work is being carried out in the Republic; therefore, the soils are degrading, according to expert estimates, two thirds of the republican soils are saline. In this regard, the strategy of the SER of the Republic should contain measures to organize assistance to farms and peasants in terms of their specialization and optimization of the size of production, development of cooperation in the Republic [1, 8]. The development of the agricultural potential of Dagestan is especially important in the framework of import substitution. To implement the proposed measures in the Republic of Dagestan, it is necessary to carry out further land reform, allocating land primarily to farms and peasants; allocate financial resources to farming and peasant farms; increase investment in seed production and breeding farms; to form social infrastructure in villages; to organize a service for informing agricultural producers regarding the market conditions of the industry and marketing. The Republic of Dagestan has great potential for the development of foreign economic activity. However, the volume of foreign trade turnover in the region is extremely low, and the Republic of Dagestan is among the Russian entities with weak export-import ties. The tactics of conducting foreign economic activity is mainly determined by the subsidization of the republican consolidated budget [7]. So, in 2018, the volume of foreign trade turnover of the

Republic of Dagestan amounted to 172.6 million US dollars. A total of 284 economic entities participated in the foreign economic activity of the Republic. Goods produced in Dagestan are exported to China, the Czech Republic, Iran, Turkey, Belarus, and Azerbaijan. The basis of exports is food products. Thus, meat (mutton), rice, fish, and vegetables are supplied to Azerbaijan and Iran. Exported to a number of countries and sheet glass. The main products, according to experts, for increasing the export potential of Dagestan are rice and meat. As proposals, taking into account the actual state of affairs in the foreign economic activity of the Republic of Dagestan, we can recommend –in order to promote local goods on the world market, the executive power of the Republic to more actively carry out marketing research; to actively attract foreign investors to the Government of the Republic of Dagestan, under its auspices, create a bank with a guarantee of protection of investment funds during economic or political crises; create concessions for foreign investors for the extraction of minerals and for their processing; limit imports into the Republic of goods produced in the Republic itself, namely, fruits and vegetables, wine and cognac products, and confectionery. In order to improve the investment climate in the Republic, the following can be recommended. Develop cluster-type projects with a view to including them in the federal project financing program, since these are long-term projects with a large socioeconomic effect. The Government of the Russian Federation approved Decree No. 158 of February 15, 2018 "On the Project Finance Factory program". Participation in the program allows stimulating the investment activity of business entities by providing borrowed funds under syndicated loan agreements for up to 20 years in the amount of 3 billion rubles. Financing is being provided for projects implemented in non-commodity exports, innovations, infrastructure, and the transfer of defense enterprises to the production of civilian products. Back in 2014, JSC Development Corporation of the Republic of Dagestan was established in the Republic, the mission of which is as follows – ensuring a high rate of socioeconomic development of the region; formation of a favorable investment climate in the Republic; integrated development of the territory of the region through the creation of sites for industrial and residential construction, support for the implementation of investment projects in the field of infrastructure and transport; development of the principles of public-private partnership in order to attract technologies and investments. The corporation included 30 large enterprises whose economies "kept afloat" ("Aviagregat," "Dagdizel," "Rychalsu," and others), providing their sites for potential investors. The corporation has simplified the procedure for obtaining loans at low interest rates for its members. It was assumed that this would help the Republic of Dagestan in a few years to become not a subsidized, but an economically developed region. But the effective results of this cooperation are still missing. Taking into account the peculiarities of the entrepreneurial structure of the Republic of Dagestan, it can be recommended to reorient the Corporation's activities toward intensive development of small businesses in the region based on public-private partnerships. In order to accelerate the entrepreneurial layer in the Republic of Dagestan, it is recommended to organize the work of credit cooperatives, having studied domestic experience (e.g., the Volgograd region) [7]. The peculiarity of the credit-cooperative system is that cooperatives represent a savings-loan system, the main source of which is the savings of members of cooperatives; are autonomous financial structures that independently resolve issues of attracting sources of funds, interest rate policy, personnel issues and others; in rural areas, they are the closest financial structure to direct consumers; apply inter-cooperative loans – attraction of free funds of other cooperatives for issuing them to consumers. The introduction of such a system of financing small businesses will help to effectively develop this area, increase tax revenues to the consolidated budget of the Republic of Dagestan. The implementation of the above activities and the investment of financial resources in these activities, with constant monitoring of their targeted spending, will allow the Republic of Dagestan to move to a qualitatively new level of socioeconomic development, which will provide the population of the region with a decent standard of living.

2 Conclusion

The socioeconomic development of the region is a complex process aimed at solving the problems of the population, improving its standard of living through achieving a balance in the socioeconomic and environmental development of the region, implemented on the basis of the efficient and reasonable use of its resource potential, including geographical, economic, infrastructural features, and potential opportunities of the region. An effective management system for the socioeconomic development of the region must meet the following criteria:

- Focus on achieving the necessary results and democracy
- Possession of the necessary force of influence
- Efficiency of regulated processes
- Use of the system of social and psychological motivation
- Ease of perception
- Adaptability
- Stimulation of progressive development
- Other

Summing up the assessment of the current state of the socioeconomic development of the Republic of Dagestan, we note that the last year, 2021, due to previous omissions, only determined the main directions of socioeconomic development, to which the Republic of Dagestan is moving in 2022. This year, we believe, should be a breakthrough for Dagestan. Based on this, it is necessary to work closely with the federal center, with corporations. But this should be done together, helping each other. Other scientists with similar cases are studying similar problems, using modeling and simulation [9, 10].

References

1. Maksubova, D.M., Omarova, H.Sh.: Experience and problems of implementing socio-economic development programs in the Republic of Dagestan//Economics and Entrepreneurship. (2021).
2. Makhmudova, M.Z.: Modern socio-economic development of the Republic of Dagestan // UEPS: management, economics, politics, sociology (2019). https://cyberleninka.ru/article/n/sovremennoe-sotsialno-ekonomicheskoe-razvitie-respubliki-dagestan. Accessed 10 Apr 2022.
3. Long-Term Forecast of Socio-Economic Development of the Republic of Dagestan for the Period Up To (2035). http://minec-rd.ru/dolgosrochnyy-prognoz-sotsialno-ekonomicheskogo-razvitiya-respubliki-dagestan-na-period-do-2035-goda. Accessed 8 Apr 2022
4. The Government of the Republic of Dagestan Discussed the Results of Socio-Economic Development For (2021). http://minec-rd.ru/news/item/7370. Accessed 10 Apr 2022.
5. Results of the Socio-Economic Development of the Republic of Dagestan. http://minec-rd.ru/itogi-sotsialno-ekonomicheskogo-razvitiya-respubliki-dagestan. Accessed 11 Apr 2022.
6. Official site of the Ministry of Economy and Territorial Development of the Republic of Dagestan. http://minec-rd.ru/rukovoditeli. Accessed 10 Apr 2022.
7. Territorial Body of the Federal State Statistics Service for the Republic of Dagestan. https://dagstat.gks.ru/. Accessed 8 Apr 2022.
8. Umargadzhieva, N.M., Maksubova, D.M.: Evaluation of socio-economic indicators of innovative development of Russian regions // Economics and Entrepreneurship, pp.476–482 (2021).
9. Trivella, L., Nasiopoulos, D.K., Trivellas, P.: The Cultural Organizations' Revival Lies in the Use of Business Management Strategy. The case study of Greek local cultural organizations. Strategic Innovative Marketing, pp. 203–210.
10. Foteini, P.E., Dimitrios, N.K., Sakas, D.P., Vlachos, D.S.: The Development of New Ideas for IT Products, Through Social Media, pp. 399–404. Strategic Innovative Marketing

Part VII
Culture and Communication in the Modern Cross-Cultural World

Yulia Taratuhina

Description: In today's changing world, understanding of communication through culture and culture through communication becomes an important factor in the development of society. Taking into account the static and dynamic parameters of culture and cultural practices, both of institutional nature and everyday life, we can imagine the culture as a communication and semiotic system at different levels of interaction. At present, taking into account the influence of globalization on personality and communities, the attention should be focused on the study of the ways of communicative processes (through art, language, media, discourse) that are used to construct dominant and non-dominant social identities (race, class, sex, gender) and how these identities are reproduced, acquired, challenged, and/or appropriated by society.

Y. Taratuhina
Institute of Control of Sciences Russian Academy of Sciences, Moscow, Russian Federation

Tools for Stimulating Labor Efficiency in a Digital Environment

Avdeeva Tatiana Viktorovna and Panko Iuliia Vladimirovna

The current state of the Russian economy is characterized by the fact that enterprises operate in an environment of growing demands of various social groups. A special trend of the present time is the general course on digitalization and the processes of digital transformation. Since the middle of the twentieth century, information and computer technologies have been developed in society, which marked the beginning of a new era – Industry 4.0. For the first time, the need for wider use of information technologies in the production and development of smart factories technology was discussed in 2011 at an industrial exhibition in Hanover, and in 2016, the founder and permanent president of the World Economic Forum in Davos, Klaus Schwab, introduced the term "Industry 4.0" into mass use [1]. Starting from this time, it is possible to observe how the material world is increasingly connected with the virtual one, as a result of which new cyber-physical complexes are born, combined into a single digital ecosystem.

The problem of the current stage of digital transformation is the presence of many modern enterprises created long before the beginning of the active and comprehensive use of information technologies, which continue to function in the old way and do not have the resource potential for a quick transition to new conditions and full use of the digital environment. This slows down the processes of digital transformation and significantly slows down the overall development of the digital economy, the technological basis of which should be information and communication digital technologies. Another problem that hinders the development of the digital

A. T. Viktorovna
The Russian Academy of Economy, the Institute of Management and Regional Development, Moscow, Russia

P. I. Vladimirovna (✉)
Department of Economic Theory and Management, Russian University of Transport (MIIT), Moscow, Russia

D. P. Sakas et al. (eds.), *Computational and Strategic Business Modelling*,
Springer Proceedings in Business and Economics,
https://doi.org/10.1007/978-3-031-41371-1_44

economy at the macroeconomic level is the lack of understanding of the essence of the processes of digital transformation.

According to the officially accepted concept, the digital transformation of the economy involves a complex of four directions, implying a transition to a program-predictive model of economic management; a change in the economic structure, taking into account the restructuring of sectors and sectors of the economy due to the penetration of digital technologies into them; the formation of more efficient economic processes provided with digital infrastructures; the transition to digital models and processes [2].

As noted in their work, Gribanov Yu.I. and Shatrov A.A., "the content of the term 'digital transformation' has evolved along with the change and development of technologies. For a long time, digital transformation has been understood as the conversion to digital format or storage in digital format of traditional forms of data. This is also one of the directions of digital transformation. Its interpretation is in the 'narrow sense'. However, in the modern world, this concept is much broader than the translation of data into digital format. It is important to separate the definitions of digitization и digitalization" [3].

Due to a lack of understanding of the changes taking place, many enterprises are following the path of creating websites and electronic trading platforms, switching to new software products, and creating corporate groups in social networks, believing that these changes are enough for digital transformation. At the same time, neither the strategic management mechanisms (accompanied by beautiful reporting schedules) nor the operational management mechanisms (with all orders placed in a working chat) reflect the real changes behind the words about digital transformation.

It is necessary to understand that digital transformation implies deeper changes, which first of all must occur at the level of transformation of the approach to management – , the transition from traditional functional management to process management, in the context of which the meaning of dividing employees into departments and functional services is lost. It is replaced by the execution of tasks by the executor as separate projects implemented in the flow of time and information. The digital resource acquires the status of the basis of the economic system, being reflected not only in information tools that help the employee in the implementation of functions but also in rebuilding the mechanisms of the management system and the functioning of the enterprise [4].

As stated in the article by Rudenko M.N. and Gribanov Yu.I., "digital transformation is the introduction of modern digital technologies into the business processes of socio-economic systems at all levels. This approach implies not only the installation of modern hardware or software, but also fundamental changes in management approaches, corporate culture, and external communications. As a result, the productivity of each employee and the level of customer satisfaction increase, and the company acquires a reputation as a progressive and modern organization. In practice, this means creating a system of end-to-end business processes, which can be called a digital business ecosystem" [5].

All digital products with which it is customary to associate digitalization are technological solutions necessary for the organization of the company's work in new information conditions [6]. The digital transformation itself is an extensive activity aimed at transferring all aspects of the management and life of the enterprise into new conditions in which new patterns of interaction of departments with each other and with the external environment begin to manifest themselves (e.g., the use of blockchain technology in the system of production planning and contracting, IoT technologies in interaction with consumers and the external environment, comprehensive application of data analysis based on Big Data technology, and the introduction of artificial intelligence in particularly stressful areas of production), new forms of employee interaction (being in a 24/7 work chat, prompt exchange of information with all interested employees at once, the possibility of remote performance of work functions, etc.), the transformation of the product being created based on an operational analysis of market conditions and trends, the general shift of many processes exclusively into the digital field (electronic bidding, the use of EDS, market monitoring, controlling the internal processes of the enterprise).

Such global changes inevitably entail the need to revise the company's strategy, its mission and goals, and as a result, internal organizational restructuring of the team, adjustment of the corporate culture, the structure of the organization, and all mechanisms and management methods.

Thus, "digital transformation and new technologies are revolutionizing technological processes and value chains that have been established for a certain time. They are becoming the main source of economic growth due to increased productivity and competitiveness of products, the creation of new industries and the constant expansion of opportunities for the emergence of updated products and technologies" [7].

The survey conducted using the methods of analysis, observation, comparison, and deduction allowed us to formulate a number of limitations that hinder digital transformation in enterprises.

First, there is a different initial degree of readiness of the enterprise for transformation, which may require significant investments in the introduction of breakthrough technologies and new technological digital developments based on the characteristics of a particular enterprise and the lack of such developments in modern practice.

Second, the lack of ready-made solutions repeatedly increases the risks in the transformation of business processes – analytical support of the digital transition is necessary, and the class of business analysts has not yet fully formed, and the probability of making mistakes is very high.

Third, digital transformation cannot be carried out partially, "in a hurry," since they are built on the logic of end-to-end processes and require "putting the whole process in order" at once – partial adjustments can only lead to a significant increase in the costs of reengineering and restructuring of the functional units and executors of the enterprise in the future. Also, to implement a large-scale restructuring of the business in space and material embodiment, high coordination of these actions is necessary on the part of the management, which should fully own the project management tool. In practice, an enterprise rarely has a wide staff of project managers who are able to implement such transformational projects efficiently and in the shortest possible time in the mode of an operating enterprise. Thus, a modern manager (innovation or change management), necessary for an enterprise for digital transformation, must have a wide range of skills and abilities in the field of creating new business models, working with the latest technologies (artificial intelligence, robotics, AR and VR, blockchain, Internet of Things), data analysis, possession of advanced project management methods. Also, the manager needs skills and abilities for effective interaction with the team, partners, and clients (communication skills, emotional intelligence, design thinking, adaptability). The lack of a sufficient number of such specialists in the modern labor market and the prohibitively high cost of available specialists is another reason for slowing down the processes of digital transformation in modern enterprises.

Fourth, in the context of the already mentioned project approach, the availability of teamwork skills among the staff is of great importance. In addition, to create a team, it is necessary to have sufficient opportunity to choose highly qualified specialists from among their own employees or to have the opportunity to attract new specialists to form a team capable of implementing the processes of digital transformation of the enterprise. In the absence of an established team, the company, as a rule, expects another stage of team building when both the "lapping" of new performers to each other and their adaptation to the conditions, values, and organizational culture of the new enterprise will take place. This may take some time when the team will be ineffective and also lead to additional costs at the stage of selecting suitable team specialists.

Fifth, in an effort to implement digital transformation, the company will not be able to do without the involvement of "new knowledge" specialists and the formation of appropriate departments or working groups. It is necessary to create and maintain a new information infrastructure – IT maintenance and updating of software complexes, administration of websites, pages in social networks and management of other SMM tools, server management and cloud resources, information security, industrial communications and work with business analytics – to solve these tasks, which are integral in the digital environment, businesses need professionals in the relevant fields who need to be found, selected and attracted.

Sixth, an enterprise that has embarked on the path of digital transformation will need to maintain all digital processes and technologies at the appropriate level, which will require constant investment in technological improvements, taking into account high-speed progress in the IT field.

And finally, seventh, digital transformation requires a system of new knowledge and competencies from all employees of the enterprise – without this condition, the introduction and full use of modern information technologies cannot be ensured.

In the future, it will be necessary to constantly improve the level of technological IQ of other employees. In addition, it will be necessary to constantly implement adaptation measures for the entire team in connection with the flow of innovations and improvements generated by digital transformations to monitor the need for the introduction of new methods, models, and management style, and the execution of the work processes themselves.

It should be borne in mind that digital transformation concerns not only the real sector of the economy. In parallel, the processes of digitalization are taking place in the public administration system, social spheres (healthcare, education, culture, IT begins to act as an important tool of thinking), and this leads to a change in the perception of the environment of each member of society. New demands arise in the interaction of society and business, which requires businesses to accelerate digital transformation to meet the needs of a new environment and a new digital consumer in a timely manner. Enterprises that will not be able to adapt to the requirements of the digital consumer are likely to go bankrupt and cease to exist. They will be destroyed by new high-tech companies because consumers will have access to more advanced services.

Thus, first of all, digital transformation needs to be defined from the perspective of a radical restructuring of the system and management model of a modern enterprise. In these conditions, the creation of a system of incentives for employees, for their interest in digital transformations of the company's activities, becomes particularly relevant.

The effectiveness of such a system can be ensured by the implementation of a number of key principles: complexity, consistency, regulation, specialization, stability, and purposeful creativity [8].

The first principle finds its realization in the synergy of various factors, including organizational, legal, technical, material, social, moral, and social. The content of organizational factors is to establish a hierarchy of goals and objectives, the sequence of work, the definition of powers, and their boundaries. Organizational factors are directly related to legal factors, the implementation of which ensures the protection of the interests of the employee, on the one hand, and the compliance of his rights and obligations on the other hand. The combination of organizational and legal factors can be considered the basis for the formation of a system of incentives and employee interest. Technical factors are of particular importance in the context of digital transformation, as they are directly related to the level of providing the enterprise with digital technological solutions and equipping the workplace with computer equipment and a software package. The level of implementation of technical factors directly ensures the comfort and modernity of the employee's workplace. They are directly related to physiological factors that determine the composition of the environment and a set of measures aimed at improving and maintaining the health of the team, minimizing the influence of harmful factors.

The group of material factors traditionally includes a system of financial incentives, including the level and conditions for receiving bonuses and allowances, and the procedure for forming the basic salary.

Social and moral factors are closely intertwined and are associated with the sociopsychological microclimate of the organization, the formation of an organizational culture environment, the development of organizational values, and the awareness of the mission and goals of the organization. Over the past few years, the social component associated with corporate social responsibility has been developing in this context.

All of these factors should not be applied individually, but in combination, which guarantees good results. It is then that a significant increase in the efficiency and quality of work will become a reality [10].

Thus, the principle of complexity ensures the integration of interacting factors into a single incentive system for the employee. This principle should be supplemented with the following principle – the principle of consistency. The implementation of the second principle is aimed at identifying and eliminating contradictions between factors and linking them together. This ensures the formation of such an incentive system in which there is an internal balance of all components, their mutual coordination, which makes the incentive complex consistent and increases the effectiveness of its impact on the employee for the benefit of the organization.

Particular attention should be paid to the implementation of the third principle. The third principle – regulation – is to create the order of the company's activities by developing standards, instructions, algorithms, and orderly business processes and ensuring control over their implementation. In this regard, it is important to distinguish those areas of activity of employees that require strict compliance with instructions and control over their implementation from those areas in which the employee should be free in his actions and can take the initiative.

The regulation of the content of the work performed by the employees of the enterprise should solve the following tasks:

1. Determine the work and operations that should be assigned to employees.
2. Provide employees with the information resources they need to perform their assigned tasks.
3. Rationally distribute work and operations between the divisions of the enterprise.
4. Establish specific job responsibilities for each employee, taking into account his qualifications and level of education.

The regulation of labor content serves to increase the efficiency of the work performed and plays a very important role by streamlining the incentive system at the enterprise.

A significant role is also played by the fourth principle – specialization, which is reflected in the distribution and assignment of a limited number of tasks and operations to the executive link (a division of the organization or an individual employee), based on the principle of optimality and economic feasibility of the division of work. Specialization is an incentive to increase labor productivity, increase efficiency, and improve the quality of work.

Stability is the fifth principle and presupposes the presence of an established team, the absence of staff turnover, the presence of certain tasks and functions facing the team, and the order of their implementation. Any changes taking place in the work of the enterprise should take place without disrupting the normal performance of the functions of a particular division of the enterprise or employee. Only then, there will be no reduction in the efficiency and quality of the work performed.

The sixth principle, purposeful creativity, acquires special importance in an innovative environment since it is its implementation that allows an organization to achieve a competitive advantage through the creation of new, more advanced products, production technologies and designs of applied equipment or types of materials, and the search for new, more effective solutions in the field of production organization and management. Based on the results of the creative activity of the enterprise as a whole, the structural unit and each individual employee, measures of material, and moral incentives are provided.

Maximization of labor efficiency can be achieved by an optimal combination of various forms and methods of stimulation. To date, considerable experience has been gained in world practice in reflecting the qualifications of an employee in remuneration. The most promising is to replace the time-based form of remuneration with a more progressive approach using a grading system (or grading).

"The grading system is a salary scale (tariff grid) of the company, developed on the basis of expert assessments of positions on predefined factors. The essence of the method lies in the fact that the weight is taken into account not only for each position in the structure of the organization, but also for each employee holding the same position, but having additional characteristics. The basis of the grading system is most often based on such characteristics as the complexity of the work performed, the level of responsibility and independence, tension and working conditions, the level of qualifications required to perform the work. The grading system evaluates all types of jobs, which makes it an extremely valuable tool in shaping the structure of remuneration. The criterion for evaluating positions is the level of influence of the position of the position on the company as a whole and the type of impact on the final result." The grading technique was created by the American scientist Edward Hay (E. Hay). At the moment, due to the fact that it is based on a mathematical model, this technique can be easily adapted to the needs of any enterprise with a staff of more than 300 people, regardless of the industry and prevailing working conditions [10].

The transition to the grading system forms the employee's interest and responsibility in the performance of planned tasks and ensures the fairness and objectivity of remuneration, taking into account the real contribution of the employee to the company's activities. In addition to the role of grading as a tool for an objective assessment of employees' work, grading can become an effective mechanism for stimulating employees to improve their professionalism since, thanks to the transparency of criteria, each employee understands at any time how he can influence the level of his own remuneration (including by increasing and maintaining his professionalism).

A special advantage of the rating system can be considered the possibility of its adaptation to any enterprise strategy [11].

Hence, the conducted research made it possible to identify important problems that complicate the transition of enterprises to a new state, which is necessary for successful functioning in the digital economy. Seven factors were identified that have a negative impact on the ability of enterprises to digital transformation.

One of the identified factors that is of great importance is the stimulation of employees to work effectively and increase their interest in their own professional growth. To accomplish this task, we need a good incentive system that will be based on the principles of efficiency. The grading system is defined as the most appropriate for these conditions. Other scientists with similar cases are studying similar problems using modeling and simulation [12–15].

References

1. Schwab, K.: The Fourth Industrial Revolution: Translated from English./Klaus Schwab – text: direct – M.: Publishing House "E", 2018 – 208 p.– (Top Business Awards)
2. Kolesnikov, A.V., Zernova, L.E., Degtyareva, V.V., Panko Iu, V., Sigidov Yu, I.: Global trends of the digital economy development. Opción Rev. Cienc. Hum. Soc. **36**(26), 523–540 (2020)
3. Gribanov, Yu. I.: The essence, content and role of digital transformation in the development of economic systems/ Gribanov Yu. I., Shatrov A.A. – Text: electronic// Bulletin of the altai academy of economics and law- 2019. – № 3-1 – pp. 44–48. https://elibrary.ru/item.asp?id=37091654 (дата обращения: 11.10.2021) – (date of application: 11.10.2021) – Access mode: Scientific Electronic Library eLIBRARY.RU.
4. Panko, Iu. V: Trends and prospects of digital transformation in certain sectors of the economy — Text: electronic//Economics and entrepreneurship. **6**(131), 256–259 (2021)
5. Rudenko, M.N.: Trends of digitalization and servicization of the economy/Rudenko M.N., Gribanov Yu. I. – Text: electronic// Theory and practice of the service: economy, social sphere, technology. 2(40), 5–8, (2019). https://elibrary.ru/item.asp?id=39188912 (date of application: 12.10.2021) – Access mode: Scientific Electronic Library eLIBRARY.RU.
6. Avdeeva, T.V.: Strategic concept in the development of a modern enterprise – Text: electronic // In the collection: Modern problems of natural resources management and the development of socio-economic systems. materials of the XII International scientific conference: in 4 parts. pp. 174–182 (2016)
7. Panko, Iu. V: Tools and methods for improving the efficiency of personnel in a digital environment – Text: electronic//Science and business: ways of development. **6**(120), 123–125 (2021).
8. Dyusekeev, K.A.: Improving the mechanism of stimulating the work of employees of higher education organizations – Text: electronic//IV International Scientific and Practical Conference 'Mathematics, Statistics and information technologies in economics, management and education', pp. 131–136 (2015)
9. Makarov, V.M.: Theory of management: Textbook. Spb.: Publishing House of the Polytechnic University, 2012. 125 p. (Economics and Management in energy and environmental management). – the text is electronic. Access mode: https://elib.spbstu.ru/dl/2535.pdf/download/2535.pdf
10. Antonova, A.S., Patsuk, O.V.: The need to introduce and improve the system of stimulation and motivation of personnel at Russian enterprises – text: electronic// MODERN PROBLEMS
11. Avdeeva, T.V.: Digital economy and digital inequality: modern problems and prospects – Text: electronic //In the collection: Greater eurasia: development, security, cooperation. Yearbook. RAN. INION. Scientific department. cooperation; Ed. by V.I. Gerasimov. Moscow, pp. 416–420 (2021).
12. Dimitrios, N.K., Sakas, D.P., Vlachos, D.S.: Analysis of strategic leadership models in information technology. Procedia Soc. Behav. Sci. **73**, 268–275 (n.d.)
13. Dimitrios, N.K., Sakas, D.P., Vlachos, D.S.: The role of information systems in creating strategic leadership model. Procedia Soc. Behav. Sci. **73**, 285–293 (n.d.)
14. Chaldezos, C., Sakas, D., Dimitrios, N.K., Giakomidou, D.S.: Calculating handling of company resources for establishing an effective project team. Strateg. Innov. Mark., 369–373 (n.d.)
15. Nasiopoulos, D.K.: Logistics companies reengineering. Change of leadership and human resources planning through management information systems. Bus. Intell. Model. **6**, 207–217 (n.d.)

Methodological Approaches to Assessing Regional Development

Djamily M. Maksubova, Navrat M. Umargadzhieva, and Patimat G. Aripova

1 Introduction

The methodology, understood as a set of methods used in any area of human activity, in relation to the analytical assessment of a complex object of study – the regional economy – includes various approaches, techniques, rules, and requirements that should be followed in the analysis process. Currently, an increasing number of specialists are beginning to realize that in the conditions of the information explosion that the current civilization is experiencing, considerable attention should be paid to methods of orientation in the actual material of science, methods of its application. Moreover, from the point of view of practical application, the method must correspond to such properties as efficiency, clarity and comprehensibility, and reliability. In the field of methodological analysis are the problems of everyday life of people, their communication, and behavior in the process of life. The task of the methodology is to elucidate, design, and transform schemes of social activity. To date, more (in comparison with the particular scientific methodology) is the general scientific methodology, within which the systematic approach, the modeling method, has been widely studied [5]. A necessary condition for constructing a model of socio-economic development of the region in order to predict the further course of economic processes is its scientific justification in the territorial aspect. To do this, it is necessary to reflect in the retrospective analysis and the current state of the economy the exceptional diversity of natural, geopolitical, socio-economic, national-cultural, and other conditions that ensure the effective development of the region as a single economic space open for both interregional and international economic cooperation. The methodological basis for analyzing the development of a

D. M. Maksubova · N. M. Umargadzhieva · P. G. Aripova (✉)
Federal State Budgetary Educational Institution of Higher Professional Education Dagestan State University, Republic of Dagestan, Makhachkala, Russia

D. P. Sakas et al. (eds.), *Computational and Strategic Business Modelling*, Springer Proceedings in Business and Economics,
https://doi.org/10.1007/978-3-031-41371-1_45

region is formed by a paradigm that assumes that a region is not a purely autonomous object, but a multifunctional organism that develops on the basis of vertical (center – regions) and horizontal (interregional) interactions, included in the system of world economic relations. Thus, the analysis and forecast of territorial development as interacting backbone structures should be built into the overall strategy of the country's socio-economic development. This aims at the inevitable search for the unity of regional and Russian economic policy aimed at overcoming spatial inequalities: differences in the volume of the generated gross regional product per capita, which are reflected in the standard of living and living conditions, in employment and unemployment, in the pace of development of individual regions, in conditions for entrepreneurial activity, etc. The level of economic development of the region is the result of the previous historical development of reproductive processes in the territory in the system of nationwide division and integration of social labor. To determine the achieved level of economic development of the region, first of all, it is necessary to use the method of retrospective analysis of the most important socio-economic indicators. The purpose of the retrospective analysis is to determine, on the basis of an assessment of the state of the regional economy, the ways of development and the peculiarities of the distribution of productive forces in the new market conditions, which make it possible to use the most important prerequisites and advantages of the region to increase the economic potential. It is also important to overcome and correct emerging disproportions and excessive differentiation in the levels of regional development. Regional policy should first of all determine the main vector of economic growth in its own territory and take into account the indicated disproportions and inequalities in relation to different subsystems of the region in order to eliminate them. Its goal is to minimize those inequalities that can cause social conflicts. The objectives of the analysis must be consistent with the objectives of regional economic policy. According to Larina N.I., Kiselnikova A.A., a clear statement of the tasks of regional policy and the choice of means for their solution require the identification of the causes of spatial inequalities in order to establish the objects of state intervention [6]. Among them, the authors include differences in the natural and climatic conditions of life and entrepreneurship in certain regions of the republic; the scale, quality, and directions of use of natural resources, which determine the "productivity" of the region. This factor affects not only the development of individual sectors (agriculture and forestry, fishing) but also the conditions for the location of industry and people's lives; the peripheral or deep position of the regions, as a result of which the influence of transport costs on the formation of prices and the breadth of sales markets increase – poor transport and communication links in peripheral regions hinder their economic development; outdated structure of production, delay in the introduction of innovations; advantages from proximity to agglomerations, leading to the intersection of inter-industry relations in the region, the emergence of opportunities for integrating local producers with large inter-regional production and economic structures; trends in the economic development of the country; the stage of technological development that affects certain types of production of goods (raw materials, intermediate products, end-use goods, services, etc.); political conditions, forms of all-Russian and regional policy;

institutional factors that determine the degree of regional autonomy, socio-economic development, etc.; physical location factors: the presence or absence of airports, transport systems, industrial sites, availability of telecommunications systems, etc., i.e., the level of development of the production infrastructure; greater or lesser external control over firms operating in the region (or a small share of firms founded by local entrepreneurs); socio-cultural factors: the degree of urbanization, education of the population, the presence of a scientific center, etc. These reasons can explain the different intensity of carrying out in space and time of fundamental economic transformations [3, 6]. In particular, this applies to such processes as denationalization of property; underestimation of the regional consequences of federal macropolitics; the incompleteness of the process of delimitation of rights among the federal government, the administrations of the constituent entities of the Federation and local governments in the areas of generating income for the territories, state influence on the economy and in solving social problems. The main task of the analysis of general economic indicators is to determine the economic profile of the region, its place in the national economic complex of the country, the level and trends of economic development achieved, and also to establish the correspondence between the level of development and distribution of the productive forces of the region with its economic and natural conditions and resources, territorial division labor. Ultimately, to outline the main options for the development of the region. Methodologically, there are different approaches to the analysis.

1.1 Application in the Analysis and Evaluation of Mathematical Methods

The assessment of indicators of the social and economic development of regions has been actively introduced over the past 10 years, and its formation is carried out in several areas of assessment, depending on the goals set.

The issues of building a system for assessing the indicators of the socio-economic development of the regions and the impact on them of the quality of the activities of regional executive authorities are quite relevant today. This is dictated by the introduction of a program-target approach in the distribution of budgetary funds – the so-called principle of performance-oriented budgeting.

In this regard, when distributing interbudgetary transfers by the federal center, methods of analysis and evaluation of social and economic indicators achieved by the regions began to develop. However, this process requires a systematic understanding and optimization of both the composition of the indicators used and the methods of analysis and evaluation.

A lot of scientific works of both domestic and foreign researchers are devoted to assessing the level of socio-economic development. However, in general, the developments of scientists relate to assessing the level of development of socio-economic systems based on the overall results of management in the context of solving problems of regional development.

The identification of previously unresolved problems, the active implementation of strategic planning at the regional level requires the development of an appropriate mechanism for assessing the level of socio-economic development of regions. At the same time, taking into account the multidimensional nature of the goals and, accordingly, the results of the socio-economic development of the regions, it is advisable to determine the level of development on the basis of a general indicator of the effectiveness of the functioning of the socio-economic system.

To begin with, let's make an excursion into the methodological aspects regarding the assessment of the level of socio-economic development of regions, which are the result of the developments of both domestic and foreign scientists in this area.

It should be noted that a rather serious experience in assessing the level of socio-economic development has been accumulated abroad, where, based on the results obtained, a conclusion is made about the effectiveness of the activities of power structures.

At the same time, in recent years, they are increasingly inclined toward an integral assessment – the definition of a generalized integral index for assessing the socio-economic state of the region, which, as a rule, is carried out from the standpoint of a combination of the main aspects of the development of the socio-economic system: economic, social and environmental [4].

According to this approach, information on various components of socio-economic development is reduced to a single indicator – the index of development of the region, according to which it becomes possible to compare regions in terms of their level of development.

The main aspects of the development of the socio-economic system are measured using indicators that are built into groups of the same name, economic indicators that reflect the level of development:

- Economic development of the socio-economic system at the regional level. Technical and economic results and trends in the functioning of the economic complex, the level of investment activity in the system, the level of innovation, trends in scientific and technological progress, without which progress and further functioning of the economy are impossible.
- Social indicators that characterize socio-demographic processes in socio-economic systems and the standard of living of the population, the state of labor resources, the quality of labor potential.
- Environmental indicators that indicate environmental well-being, the adequacy of measures to reduce the negative impact on the environment [2].

Depending on the components of the integral assessment, as part of the index approach to assessing the level of socio-economic development of the socio-economic system at the regional level, scientists distinguish between different methods. The most widespread are the following:

- A comprehensive assessment of the socio-economic development of the region
- Quality of life as a complex indicator (a systemic criterion for the socio-economic development of the country's regions)
- Human development index:
- World competitiveness index and others

In Russia, preference is given to a comprehensive assessment of the socio-economic development of the region, the essence of which is as follows:

- Determining the list of indicators that affect the level of socio-economic development of the region, and most fully reflects socio-economic changes
- Grouping these indicators in certain areas of socio-economic development
- Calculation of partial indices in these areas, followed by the determination of the generalizing index

The results of such a comprehensive assessment make it possible to rank regions according to the level of their socio-economic development, determine the average indicator of the integral assessment and serve as the basis for assessing the degree of depression in the state of the regions (when the value of the integral index is below the average), which in turn, according to scientists, indicates that "such regions need state assistance" [6].

The main stages of the integrated comprehensive assessment of the socio-economic development of the region are shown in Fig. 1.

As for the developments of scientists, there is no single, optimal approach to the integral assessment yet. In the scientific literature, there are quite different approaches and methods for a comprehensive assessment of the level of socio-economic development of regions, which, according to individual researchers, are

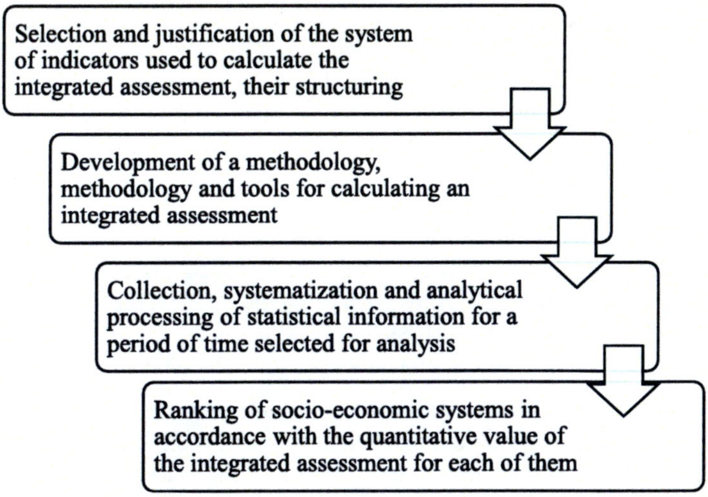

Fig. 1 The main stages of the integrated comprehensive assessment of the level of development of the region

the most appropriate in a particular case. They differ both in the areas of assessing the level of socio-economic development of the regions and in the number of indicators that make up these areas and are selected by scientists to reflect the changes taking place in the region and according to a number of other criteria.

To confirm this, as an example, we can cite the results of a study conducted by T. O. Dyukina, N. Yu. Lukyanova. So, regarding the directions of grouping indicators, scientists note the existence of different directions in the methods for assessing the socio-economic development of regions by domestic authors: there are those that offer estimates only in the context of economic and social components (Yu. A. Dmitriev., L. P. Vasilyeva); others supplement this list of the social component (Zander E. V., Smirnova T. A., Lobkova E. V.): there are also such methods that take into account indicators in the context of economic, financial and social areas (Kuzmenko N. B.).

In existing methods, the number of indicators used as input information ranges from 5 to 121 [1].

However, in our opinion, attempts to include a fairly large number of them in the integral indicator in order to take into account certain aspects of the region's activities only complicate calculations for diagnosing the level of socio-economic development of regions and ultimately also limit their practical application.

Due to the increase in the number of indicators, the amount of work on their processing on the way to obtaining the final index indicator, on the basis of which the rating of regions will be built, becomes more complicated. Therefore, it is not always appropriate to mix the entire set of indicators that measure the level of socio-economic development of the region (or a sufficiently large number of them) and reduce them into a single integral indicator.

On the one hand, the system of indicators chosen for calculating the integral assessment necessarily correlates with the goal pursued by the analysis of the socio-economic development of the region. That is, from a wide variety of indicators, it is necessary to choose exactly those that will better and more fully take into account the main changes in the socio-economic system that occurred during the analyzed period and, accordingly, act as basic information for calculating the integral assessment. On the other hand, for the possibility of practical application of the method of integral complex assessment of the level of development of the socio-economic system, the number of indicators should be reduced to a minimum.

Among the author's methods for analyzing indicators, we consider the most preferable to use the method of assessing the level of socio-economic development of the region, proposed by T. Uskova, based on the fact that the list of indicators characterizing sustainable development is wide, and they are diverse; the assessment of sustainability must be carried out on the basis of an integral indicator [3]:

1. Substantiation of indicators, including general information about the region with data on geographical location, natural conditions, population size and national composition, large settlements, main sectors of the economy, main transport routes. Special attention should be paid to the impact of competitive advantages

and socio-economic indicators in the region when choosing a development strategy.

Analysis of the socio-economic situation of the region over the previous three-year period is carried out in the following areas:

- Demographic situation (birth rate, mortality, life expectancy, health status of the population, migration movement)
- Social sphere (education, health care, science, culture, etc., employment, crime rate)
- The standard of living of the population (average per capita income, wages, subsistence minimum, and consumer basket)
- Finance (budget and taxes, off-budget funds, interbudgetary relations, financial position of enterprises)
- Macroeconomics (volume and structure of the gross regional product)
- Sectoral structure of the economy (industry, agriculture, energy, transport, communications, and informatization)
- Foreign economic activity
- Investments (fact, trends)
- Trade
- Department of Housing and Utilities
- The state of the environment, natural resources [1]

To normalize indicators and raise them to an integrated indicator, you can use formulas 1–3.

Bringing indicators for assessing the economic development of the region, which are expressed in absolute and relative terms and also have different units of measurement, to a single scale will be carried out according to the formula:

$$y_{ij}^{c} = \frac{x_{ij} - \min\left(x_{ij}\right)}{\max\left(x_{ij}\right) - \min\left(x_{ij}\right)} \tag{1}$$

where

y_{ij} is the standardized value of the *i*-th indicator of the *j*-th region.
x_{ij} is the value of the *i*-th indicator of the *j*-th region.
x_{max}, x_{min} – the maximum and minimum value of the indicator.

The standardized value of indicators – destimulators is determined by formula 2.

$$y_{ij}^{d} = 1 - y_{ij} \tag{2}$$

The total synthetic indicator for each region will be determined by the formula:

$$y_j = \frac{1}{n}\left(\sum_{i=1}^{k} y_{ij}^{c} + \sum_{i=1}^{m}\left(1 - y_{ij}^{d}\right)\right) \quad (3)$$

where

Y_j is an integral indicator of the socio-economic state of the j-th region.
y_{ij}^{c} standardized value of the i-th indicator-stimulator of the j-th region.
y_{ij}^{d} standardized value of the i-th indicator – destimulators of the j-th region.
k is the number of indicators-stimulators.
m – the number of indicators – destimulators.
n is the number of indicators for which the calculations were carried out.

2. Assessment of the level of development of the region for each indicator:

k_i – assessment of the level of development of the region/for each indicator:
$k_i = \frac{x_i}{\max\ (x_i)}$ – direct indicator
$k_i = \frac{\min\ (x_i)}{x_i}$ – reverse indicator
x_i is the value of the indicator in region i.
max (x_i), mix (x_i) is a standard indicator, which can be chosen as the optimal (or critical) values of regional development indicators.

3. Calculation of indices of economic, social, and environmental sustainability (respectively, $I_{\text{уст}} = \sqrt[3]{I_{\text{экон}} * I_{\text{соц}} * I_{\text{экол}}}$) by the method of multivariate comparative analysis.
4. Formation of an integral indicator

The integral stability index is calculated using the following formula 4:

$$I = \sqrt[3]{Iec \times Is \times Ie}; -social(Is); -economic(IEC); -ecological(Ie). \quad (4)$$

It can range from zero to one.

5. Interpretation of the integral index of sustainability of the socio-economic development of the region [2].

We also offer the following version of the method for assessing the sustainability of the development of the socio-economic system (SES) of the region:

1. Determine private indicators on the basis of which the integral indicator will be calculated, taking into account the characteristics or problems of a particular region (economic: GRP, foreign trade turnover, cost of fixed assets, etc.; social: average life expectancy, salary level, population density, etc.; environmental: MPC, MPD, MPE, etc.).
2. All indicators/indicators characterizing the level of WRM are private (measured in people, in %, in units, etc.); we propose to transform private indicators in order to achieve their comparability with each other.

The formula for transforming the indicator from private to relative:

$$y = (y_f - y_{min})/(y_{max} - y_{min}) \tag{5}$$

where

Y_f – actual indicator
Y_{max} – maximum indicator
Y_{min} – the minimum indicator

3. Aggregate transformed partial indicators (values that are already relative at the third stage of calculations) according to the formula

$$I = \frac{(y_1 + y_2 + y_3 + \dots + y_n)}{n} \tag{6}$$

where I is an integrated indicator, which is found separately for social, economic, and environmental factors.

y_1 ; y_2 ; y_3 ; y_n relative indicators included in the block of social, economic, and environmental factors;
n is the number of indicators in the block.

So, we will find integrated indicators:

- Social (I s)
- Economic (I_{EC})
- Ecological (I_e)

Knowing the values of the integrative indicators characterizing the SD of the SES of the region, we can calculate their norm or deviation from the standard indicators and then define:

(1) Integrative indicator of risks of the region (IPRK – I_{rk})
(2) Integrative indicator of the potential of the region (IPPR – I_{pr})
(3) Integrative indicator of the stability of the region (IPUR – I_{ur})

IPRK allows you to determine the initial situation in the region and identify the existing and potential losses of the subject, to minimize them.

IPPR shows the achieved level of socio-economic development of the region, the possibility of its further improvement, which requires certain resources and conditions.

The IPSD reveals the ability of the region to effectively use the available and potential resources without disturbing the balanced state of the system.

To display the ranges of changes in the integrated assessment, one should use the scale of the level of socio-economic development of regions, according to which it is proposed to consider 4 levels of development of the socio-economic system – high,

Table 1 The scale of ranking regions in accordance with the integrated comprehensive assessment of the level of their socio-economic development

Level of socio-economic	Change range
Regional development	Integrated assessment, %
Tall	[75–100]
Average	[50–74]
Short	[25–49]
Critical	[0–24]

medium, low, and critical. Each of these levels has its own range of changes in the integrated assessment (Table 1).

In Russia, until 2017, at the state level, there were two approaches to a comprehensive assessment of the socio-economic development of the region: the approach of the Ministry of Economic Development of the Russian Federation; approach of the Federal State Statistics Service of the Russian Federation [7].

The approach of the Ministry of Economic Development of the Russian Federation provides for the rating assessment (and ranking) of the regions separately for each of the 6 specific areas and the calculation of the overall rating for all areas as a whole. The main areas for a comprehensive assessment are as follows:

- Economic development: investment, scientific, and innovative activities
- Public finance
- Business support; consumer market
- Population and labor market
- Department of Housing and Utilities
- Education and health care
- Crime level
- Ecology

Based on the results of the calculation, the place of each region in the ranking is determined for all and separately taken areas of socio-economic development. The region is considered the best if the arithmetic mean value of the sum of ratings is low.

According to the second approach for calculating integral regional indices of economic development, proposed by the Federal State Statistics Service of the Russian Federation, 28 primary indicators of economic development are used to account for integral indices, which are integrated into 5 blocks, characterize the following aspects: macroeconomic efficiency: financial stability; innovation activity and development of science; market transformations and development of entrepreneurship: investment attractiveness.

The third method was the method of assessing the level of socio-economic development of the region, which is built on the basis of 13 indicators (Table 2).

Of these: 4 (32%) – labor indicators, 5 (38%) – economic, 2 (15%) – investment, 2 (15%) – financial.

The methodological part of this rating is quite interesting. The indicators achieved by the regions are compared with the average for Russia. As a result, scores are determined [2].

Table 2 Composition of indicators for determining and evaluating the level of socio-economic development of a constituent entity of the Russian Federation

No. indicator	Name of indicator	Character display.
1	Volume of gross regional product per capita	Economy
2	Investments in fixed capital (excluding budget investments) per capita	Invest
3	Profitability of sold goods, works, services	Economy
4	Unemployment rate	Work
5	The ratio of the state debt of a constituent entity of the Russian Federation to the budget revenues of a constituent entity of the Russian Federation (excluding gratuitous receipts)	Finance
6	The ratio of the number of highly productive jobs to the average annual number of employed people	Work
7	Share of organizations implementing technological innovations in the total number of surveyed organizations	Invest
8	The share of the number of highly qualified employees in the total number of qualified employees	Work
9	The number of small and medium-sized businesses operating in the territory of a subject of the Russian Federation, per 1000 people. Population	Economy
10	The share of public roads of regional or intermunicipal significance that meet the regulatory requirements for transport and operational indicators in the total length of public roads of regional or intermunicipal significance	Economy
11	Revenues of the budget of the constituent entity of the Russian Federation (excluding gratuitous receipts) per capita of the population of the constituent entity of the Russian Federation	Finance
12	The share of the employed population in the average annual population of a constituent entity of the Russian Federation	Work
13	Consumer price index	Economy

It can be concluded that the methods of analysis and assessment of the socio-economic development of regions that were in force until 2017 at the state level included, in fact, three separate methods that were used for various assessment purposes, different methodological techniques and methods of analysis were used, and often duplicated some of the indicators used.

The use of mathematical methods in the analysis of the socio-economic development of the region is due to the need to build mathematical models of economic dynamics; since when predicting the prospects for the development of the economic system, it is necessary to find reliable ways to assess the consequences of today's activities. To develop correct economic decisions, it is necessary to take into account both all past experience and the results that can be obtained from the proposed conceptual and mathematical models that adequately reflect the given economic situation. Mathematical models cannot take into account all the features of real economic processes, but they allow one to draw qualitative conclusions related to the development of the social economy over time. For the construction of mathematical models, a great difficulty is the choice of values of controlled parameters, the search for objective functions and their economic interpretation. First of all, it is necessary to clearly understand the structure of the economy as an object of mathematical modeling and an integral economic system that performs its main function through the implementation of the following actions: allocation of resources, production of products and services, distribution of consumer goods, and accumulation for subsequent investments in the public economy. Moreover, as a subsystem of human society, the economy, in turn, is a rather complex system consisting of production (commodity-producing) and non-production (commodity-distributing, financial, etc.) cells or business units located in production-technological and (or) organizational-economic relations with each other [4].

2 Conclusion

Forecasting regional development in order to improve the management system of the region is impossible without a detailed study of the economy of the region, studying the features of its development. Analysis and forecasting are carried out on the basis of scenarios that reflect changes in federal and regional economic policy. Therefore, focusing on the analysis of the development of the regional economy, which is the main link in the chain of managerial decision-making, we identify not only regional features but also advantages and threats for future development. The considered methods of analysis are not exhaustive. The regional economy is a rather complex object of study both in terms of analysis and forecasting of its development. It is necessary to take into account many different factors influencing regional development. At the same time, the main methodological approaches to the analysis of the state of the region's economy, presented in the chapter, allow using them in assessing the socio-economic processes taking place in the region, assessing the consequences of political decisions on entrepreneurial activity, the formation of the investment climate, and more. A particularly important result of the analysis is the identification of the reasons for the uneven development of Russian regions and the factors influencing the growing gap between regions in terms of GRP, investment and household income. As part of the analysis, it is also possible to assess the reaction of business to the actions of the authorities and, as a result, changes in the dynamics of investment and production efficiency in the region. As a result, using various methods of analysis, we determine the possibilities of forming a positive economic policy in order to stimulate the economic growth of the regional economy with the active influence of state regulation mechanisms. Other scientists with similar cases are studying similar problems, using modeling and simulation [9–11].

References

1. Govorova, N.V.: The modern concept of human development//modern Europe. **3**, 104–111 (2000)
2. Ivanchenko, V., Gaiduk, N.: Improvement of the social sphere – a determining factor in economic growth. The Economist. **8**, 69–76 (2011)
3. Kotilko V.V., Sanin I.I.: Regional forecasting: Proc. allowance / IREiMU. - M.: Publishing house "Saturn-S", p.191 (2011).
4. Kolemaev, V.A.: Economic and Mathematical Modeling. Modeling of Macroeconomic Processes and Systems: Textbook, p. 295. V. A. Kolemaev. – M.: UNITI-DANA (2005)
5. Larina, N.I., Kiselnikova, A.A.: Regional Policy in the Countries of Market Economy: Proc. allowance/NGAEiU, p. 172. M.OAO Publishing House Economics (2016)
6. Omarova, N.O.: The impact of digital transformation of society on the formation of value orientations and socialization of personality: a chapter in a collective monograph/Digital technologies as the basis of innovative processes in education: A collective monograph
7. Friedman, J., Ordway, N.: Analysis and valuation of income-generating real estate. Per. from English-M.: 'Delo Ltd', p. 480 (2017).
8. Rozanova, L.I.: Methodological foundations for the analysis of regional development Theor. Appl. Econ.. 4. p. 1–38 (2014).
9. Sakas, D., Vlachos, D., Nasiopoulos, D.: Modelling strategic management for the development of competitive advantage, based on technology. J. Syst. Inform. Technol. (n.d.)
10. Nasiopoulos, D.K.: The Interaction of Users' Behavior to the Development of a Distribution and Logistics Network of the Information Technology Companies. Bus. Intell. Modell. **6**, 219–227
11. Alexandros, N.K., Damianos, S.P., Dimitrios, N.K., Dimitrios, V.S.: Modelling the process of a web-based collaboration tool development. Strateg. Innov. Mark., 383–389 (n.d.)

Application of Intellectualization Methods for Educational Resources Creation

N. O. Omarova and Osmanova Amina

1 Introduction

Development of the education system in modern conditions and the identification of promising areas of research are very relevant nowadays. The introduction of digital technologies has a special impact on the educational process, and virtual network communities are an important factor that forms the value orientations of young people and the socialization of the individual.

Of particular importance are researches devoted to the study of the experience of forming the value bases for building the educational process in the context of the digital transformation of education, as well as discussing the process of forming humanitarian-oriented moral, ethical, social, and national-cultural traditional values of the multinational Russian society.

The current stage of development of society is characterized by the rapid development of the global information space, the accelerated pace of globalization, which leads to a rapid change in working conditions, public attitudes, and beliefs. The changes taking place in the world during the transition to a postindustrial society are largely associated with the emergence and development of information technologies. In turn, the development of digital technologies is becoming the driving force behind the ongoing changes. The high rates of scientific and technological progress lead to the rapid obsolescence of the knowledge of specialists, which necessitates the continuation of the educational process throughout the entire active period of life.

N. O. Omarova (✉)
Federal State Budgetary Educational Institution of Higher Professional Education Dagestan State University, Makhachkala, Russia

O. Amina
Higher School of Economics University, Moscow, Russia

D. P. Sakas et al. (eds.), *Computational and Strategic Business Modelling*, Springer Proceedings in Business and Economics,
https://doi.org/10.1007/978-3-031-41371-1_46

Further development of the education system in the context of the digital paradigm of the development of society will allow for solving many theoretical and practical problems that are at the present stage.

The following areas of application of modern technologies in education can be distinguished.

- Intelligent learning systems and chatbots: AI (artificial intelligence) helps to personalize learning work, provide students with quick feedback directly while learning.
- Automatic assessment.
- Customizable learning materials.
- Educational analytics.
- Consulting systems.
- Gamification and virtual reality. Further development of technology promises to significantly increase the visibility of educational work and make extensive use of virtual experiments.

Today, many educational organizations use information management systems (IMS) for student records, class attendance, class planning, and teacher performance. At the same time, they use learning management systems (LMS), where educational materials and assignments are located, and data on the progress and results of educational work is also presented.

Blockchain technology is becoming increasingly important for the digital transformation of education, bringing together the work of various educational organizations, and creating a good foundation for the development of education.

Digital technologies' speed of development raises the question concerning in which direction they will develop and what are the prospects for our country in this process.

Artificial intelligence (AI) technologies are rapidly penetrating all areas of human life and activity: stock markets, marketing and data mining, financial and credit risk management, industry, robotics, human resource management, recruiting, military, medicine, music, publishing, etc.

The Internet of Things has recently entered the life of every person, having carried out a merging of the physical and digital worlds, as data is collected from an increasing number of devices and then combined into the so-called big data. The number of such "Internet of Things" devices, according to experts and analysts, has already reached 50 billion by 2020. Devices include appliances, vehicles, buildings, and cameras.

The development of adaptive and intelligent robots will increase the degree of automation of production processes. Adaptive robots that currently exist in the form of prototypes are equipped with a sensory part and a set of programs, and the development of intelligent robots is associated with solving the problem of computer vision, which requires special software written in special languages focused on symbolic information processing – Lisp, Prolog, Smalltalk, Refal, etc.

In terms of practical application, the technology of "digital twins" has great prospects and is increasingly being used in industry, energy, aviation, and the automotive industry. A digital twin is a synchronized virtual model of a real object: a person, product, process, system, etc. It reflects not only the past and current state of the object but also can predict its future based on the initial data, which is achieved by a deep understanding of the processes occurring in a real object.

The use of augmented reality technologies that allow to add digital objects to objects in the real world is also promising. An example is the helmet of a military pilot, which displays various flight parameters and the state of the aircraft. Such systems are used in industry when assembling cars and other modern equipment. AI gives digital objects a greater reality and controls their behavior.

Large Russian companies are already using and improving AI technologies in their business processes. For example, using machine learning technologies, MTS predicts a subscriber's desire to use a particular service.

2 Development of Artificial Intelligence Technologies: Representation of Knowledge in Intelligent Systems

The achievements of modern society based on modern, including digital, technologies have led to radical, large-scale, and complex changes in the economy, social sphere, and public life.

In recent years, there has been a real breakthrough in artificial intelligence (AI) research. Problems like speech recognition, computer vision, and machine translation have been practically solved. To date, the quality of solving such problems by machines is comparable to that of a human and continues to improve. There are ample opportunities for practical applications – from communicating with intelligent agents in natural language to autonomous cars, drones, and robots.

The success of AI is associated with machine learning techniques called "Deep Neural Networks" or "Deep Learning" technology for short. The ability to train such deep neural networks with hundreds of millions of tuning parameters and solve complex practically important tasks is associated with the advent of sufficiently powerful highly parallel GPU processors. As a result, deep neural networks have become the main tool for updated artificial intelligence based on machine learning.

The main advantage of deep neural networks is the ability to generate hierarchies of increasingly abstract data representations. Each next layer of the neural network forms more and more abstract features of the input data, as the most significant combinations of features of the previous level. Accordingly, the more layers a neural network has, the more abstract concepts it can operate on, and the more complex tasks it can solve. Deep learning is a universal technique for automatically constructing abstractions that have advanced artificial intelligence in many directions at once. Today, in machine vision, speech recognition, and machine word processing, the same architectures of deep neural networks and algorithms for their training are used.

The deep learning revolution has dramatically changed our understanding of machine intelligence. It is no longer the intellect itself that is required to be programmed, but only the ability of machines to learn, which is much easier. Machine learning is becoming synonymous with machine intelligence.

Artificial intelligence (artificial intelligence, AI) is the science and technology of creating intelligent machines, especially intelligent computer programs, with the help of which they solve the problems of modeling human activity [1].

Since the 1950s, an important criterion for assessing machine intelligence has been the Turing test, which required a person to be unable to distinguish a machine from a live interlocutor in real-life situations. In 2014, this test was passed, and today AI is regularly used in higher education to provide round-the-clock assistance to students. For example, Deakin University uses the IBM Watson supercomputer for this purpose. The potential of AI in higher education is still not fully realized, but universities can learn from the consumer sector. For example, virtual assistants can interpret verbal cues and respond to them by imitating human speech. Siri and Cortana for smartphones are popular, and Amazon's Alexa is an example of an independent, always-on virtual assistant that uses remote microphones to search the web for information by voice command. Indeed, autonomous technologies that meet the needs of people are developing at a rapid pace.

Many fear that progress in this area is too rapid. The functions of AI are inherently complex and incomprehensible, so tools are needed to explain how AI works and increase its credibility. IBM is a leader in this field, publishing images and diagrams that explain how its AI medical systems work. In higher education, opportunities to develop virtual mentors and better adaptive learning tools are underutilized due to the widespread belief that even the most advanced technology cannot and should not replace a live teacher.

Machine learning is already driving progress in the professional world and informal learning. The Smart Flower Recognition System, a citizen science project jointly run by Microsoft Research Asia and the Chinese Academy of Sciences, aims to help Chinese botanists quickly identify plant species from smartphone photos. The use of neural networks allows you to create algorithms that automatically filter out low-quality images and determine colors from the database with a probability of over 90%. Projects like these have far-reaching implications for students and teachers, as search queries no longer must be based on text. To implement AI more widely in higher education over the next four to five years, universities should start with open-source and open-source libraries for numerical computing, available through the OpenAI and TensorFlow projects from Google.

An important place in the theory of artificial intelligence is occupied by the problem of knowledge representation, which, according to many researchers, is the key one [2].

The method (models) of knowledge representation is a set of interrelated means of formal description of knowledge and operation (manipulation) of these descriptions.

Traditional languages based on the numerical representation of data are inefficient in building knowledge bases. For this special knowledge representation, languages are used, based on the symbolic representation of data. They are divided into types according to formal models of knowledge representation. Various authors classify these models in different ways [3].

Logic models
Production models
Frame models
Network models

The main idea of building logical knowledge models is as follows – all the information necessary to solve applied problems is considered a set of facts and statements, which are presented as formulas. Knowledge is displayed by a set of such formulas, and obtaining new knowledge is reduced to the implementation of inference procedures.

Production models can be considered the most common knowledge representation models. A production model is a rule-based model that allows knowledge to be represented in sentences like:

If condition, then action

The production model disadvantage is that when a sufficiently large number (on the order of several hundred) of productions is accumulated, they begin to contradict each other.

The concept of a frame was introduced by the famous American scientist Marvin Minsky in 1975. M. Minsky considered a frame as a data structure for representing a set of stereotypical situations, events, and objects, as well as their characteristics, signs, and properties. This information (about characteristics, signs, and properties) is stored in frame slots.

The network model was proposed by the American scientist M.R. Quillian. Models of this type are based on a construction called the semantic network. The concept of a semantic network arose in 1969, when M. R. Quillian tried to describe the semantics of a verb [4, 5]. The semantic approach to the construction of artificial intelligence systems finds application in natural language understanding systems, question-answer systems, and various subject-oriented systems. This method of representing knowledge is of particular importance in connection with the development of the Internet.

The semantic web as a model is most often used to represent declarative knowledge. With the help of this model, such properties of the knowledge system as interpretability and connectivity are realized. Due to these properties, the semantic network makes it possible to reduce the amount of stored data, provides the conclusion of inferences on associative links.

In knowledge engineering, a semantic network is a graph that reflects the meaning of a holistic image. Graph nodes correspond to concepts and objects, and arcs correspond to relationships between objects. An object is a structure that contains specific information.

The use of the mathematical apparatus of graph theory for modeling the logical structure of educational material, the systematization of its concepts is the subject of research by V.P. Bespalko, S.A. Beshenkova, I.I. Logvinova, V.P. Mizintseva, and I.V. Robert.

The traditional system of education seeks to give the student as much factual material as possible at various stages of the educational process. At the same time, the assessment of the quality of knowledge is carried out by considering the number of factors (various concepts, elements of knowledge, etc.) that the student operates, and the accuracy of their reproduction. Since the concepts are interconnected and follow one from the other, connections and relationships between concepts and rules for the logical inference of specific concepts from more generalized categories of the subject area remain aside. This kind of learning leads to knowledge formalism. When solving creative problems, which include the learning process, a knowledge representation system based on a logical-semantic approach is needed, which allows you to display the conditions of the problem in the form of a structured model that considers all the links between the elements necessary to solve it.

Semantic networks use three main types of objects: concepts, events, properties.

Domain objects displayed in the semantic network can be divided into three groups: generalized, individual (concrete), and aggregate.

The introduced classification is relative. Depending on the problem being solved, the same object can be considered generalized or individual, as aggregate or non-aggregate.

One of the first well-known models based on the semantic web is the TLC model (Teachable Language Comprehender) developed in 1968. The model was used to represent the semantic relationships between concepts (words) to describe the structure of a person's long-term memory in psychology.

As a rule, extensional and intensional semantic networks are distinguished. An extensional semantic network describes the specific relationships of a given situation. The intensional semantic network describes the subject area at a generalized, conceptual level. Connections in an intensional network reflect those relationships that are always inherent in objects of a given class [3].

The semantic web is a cognitive tool that provides visual and speech tools for creating data representation maps, also known as cognitive maps. Cognitive maps are spatial representations of concepts and their interactions recorded in computer memory, that is, they are structured data (Jonassen, Beissner & Yacci, 1993). Programs such as SemNet (Fisher 1990, 1992), Learning Tool (Kozma. 1987), and TextVision (Kommers, 1989) enable students to link the concepts they are learning into multidimensional networks of representations and describe the nature of the relationships among all the concepts included in the network.

Semantic networks are similar to the representation of the human memory structure. Let us assume that human memory is organized semantically. Then it turns out that programs for organizing semantic networks are computerized tools that make it possible to make visible the semantic networks of human memory. They are made up of nodes and ordered relationships or links connecting those nodes. Nodes express concepts or assumptions, and links describe relationships between these nodes. In computer semantic networks, nodes are represented as information blocks or maps, and links as lines with symbols.

The purpose of computerized semantic networks is to represent a set of concepts or identify the basic organization of representations in a knowledge area. Therefore, semantic networks require learners to analyze the structural interactions between the contents being studied.

They can also be used by students as tools to assess changes in their thinking. If we agree that the semantic web is a complete representation of human memory, then the process of learning from this point of view can be seen as a reorganization of semantic memory. The organization of semantic networks reflects these changes in semantic memory, since semantic networks describe what the learner learns. Thus, semantic networking programs can be used to reflect knowledge acquisition.

The organization of the semantic network contributes to learning, because it forces the learners to analyze the basic structure of the concepts being studied. In the process of creating semantic networks, learners must analyze the structures of their own knowledge, which helps them to incorporate new knowledge into the structures of existing knowledge. The result is a more efficient use of acquired knowledge. Kozma (1987, 1992), one of the developers of the Learning Tool semantic networking program, believes that these tools are cognitive tools that enhance and expand human knowledge. The development of computer semantic networks requires students to:

- Reorganization of knowledge.
- An exhaustive description of concepts and relationships between them.
- Deep processing of knowledge, which contributes to better memorization and retrieval of knowledge from memory and also increases the ability to apply knowledge in new situations.
- Linking new concepts with existing concepts and ideas, which improves understanding.
- Spatial learning through the spatial representation of concepts in the field of study (Fisher, Faletn, Paterson, Lipson, Thorton, & Spring, 1990). The construction of semantic networks and cognition maps is a means to accurately represent the structure of cognition (Jonassen, 1987). This means that semantic networking helps learners map their own cognitive structures.

In geometry lessons, concept maps have been used to evaluate learning effectiveness and to monitor learner progress in a given subject (Mansfield & Happs, 1991).

The usefulness of semantic networks and concept maps is perhaps best demonstrated by their links to other higher order forms of thought. They are closely related to formal reasoning in chemistry (Schreiber & Abeg, 1991) and the ability to reason in biology (Briscoe & LeMaster, 1991; Mikulecky, 1987). Semantic networks have also been shown to be associated with research performance (Goldsmith, Johnson & Acton, 1991). After the use of semantic networks as tools for cognition, the knowledge that makes up the content of a given subject becomes more organized (Jonassen, 1993). More research is needed to test the consistent links between certain criteria for networking (listed above) and traditional methods used during study, such as exams, research papers, and targets [6].

3 Conclusion

Based on the analysis of existing knowledge engineering approaches [7–9], it is advisable to choose adaptive semantic models (ASM) as the main way of structuring knowledge in the system of teaching the disciplines of the mathematical cycle.

The semantic structuring of cognitive tasks by sections of mathematics was carried out [10]. As the study of this issue has shown, cognitive tasks can be classified according to a number of criteria. The proposed approach to the structuring and classification of educational tasks contributes to the formation of a system of knowledge for students in the disciplines of the mathematical cycle.

Based on the analysis, it was revealed that one of the main advantages of sematic networks as a model for representing knowledge and the learning process itself is the visibility of the described subject area, the possibility of changing and introducing new terms and rules, as well as high flexibility and adaptability to the goal and other factors. But the main drawback of such networks also follows from this advantage, namely, that with an increase in the size of the semantic network and the complication of connections, the knowledge base of the subject area, the visibility property is lost.

One of the important advantages of the semantic network is also the possibility of constructing a model in the form of a hierarchy of relationships, where classes of concepts are located on the first level (the topmost), generalized concepts are placed on the second, and elementary concepts are placed on the third (lowest).

This approach to the organization of knowledge in the development of training systems shows the relationship between the elements of educational material and allows you to significantly reduce training time and volume.

Regarding the disciplines of the mathematical cycle, an initial systematization is required for the current time period. Having a general idea of the current state of the academic discipline, it is possible to build a didactic learning system. To do this, it is necessary to develop a methodology for structuring and adapting existing knowledge, considering the requirements of the educational direction and social order.

The advantages of our model of the learning process are especially significant in the control of students' knowledge. The semantic network implies the semantic processing of information by a computer, which is necessary when processing trainees' answers.

Activation of the cognitive activity of trainees is possible only with an expedient sequence of presentation of cognitive tasks, as well as taking into account the characteristics of the elements of the tasks. Other scientists with similar cases are studying similar problems, using modeling and simulation [11–15].

References

1. Authors-compilers: Averkin A.N., Gaaze-Rapoport M.G., Pospelov, D.A.: Explanatory dictionary of artificial intelligence. http://www.raai.org/library/tolk/aivoc.html#
2. https://sites.google.com/site/upravlenieznaniami/nejronnye-seti
3. Bochkova E.G., Vasiliev S.S.: Models of knowledge representation. NovaInfo. **58**(6), 13–20 (2017). https://novainfo.ru/article/?nid=10897
4. Kiselev, G.M., Bochkova, R.V.: Information technologies in pedagogical education: A textbook. M: Dashkov and K. (2012).
5. ITteach.RU. http://itteach.ru/predstavlenie-znaniy/setevaya-model-predstavleniya-znaniy
6. HR-Portal. http://www.hr-portal.ru/article/primenenie-kompyuternyh-semanticheskih-setey-v-kachestve-instrumenta-poznaniya
7. Osuga, S., Saeki, Yu., Suzuki, H. et al.: Acquisition of knowledge. M.: Mir (1990).
8. Berners-Lee, T., Lassila, O.: The Semantic Web. Scientific American. http://www.sciam.com/article.cfm?articleID
9. Shuklin, D.E.: Application of a network object-oriented knowledge base in modeling an expert system based on a semantic neural network. Artificial intelligence. **4**, 403–413 (2005)
10. Shikhnabieva, T.S., Omarova, N.O.: Use of adaptive semantic models in physics and mathematics education. Scientific notes. **35**, 25–32 (2011)
11. Trivella, L., Dimitrios, N.K.: Knowledge management strategy within the higher education. The case of Greece. Procedia Soc. Behav. Sci. **175**, 488–495 (2015)
12. Nasiopoulos, D.K., Ioanna, D., Lida, G., Paraskevi, P., Giakomidou, D.S.: Modeling of B2C Communication Strategies in Electronic Commerce. Strateg. Innov. Mark., 531–537 (2017)
13. Nasiopoulos, D.K.: Implementation stages and risks of Logistics Companies reengineering. Bus. Intell. Modell. **6**, 197–206 (2020)
14. Nasiopoulos, D.K.: Web search applications in Logistics companies in the context of evolution of Management Information Systems. Bus. Intell. Modell. **6**, 241–254 (2020)
15. Dimitrios, N.K., Sakas, D.P., Vlachos, D.S.: The role of open source leadership in developing high technology companies. Key Eng. Mater. **543**, 402–405 (2013)

Big Data Technologies in the Education System

N. O. Omarova and A. A. Echilova

1 The Development of Artificial Intelligence Technologies in Modern Society

Artificial intelligence (AI) is rapidly penetrating all areas of human life and activity: stock markets, marketing and data mining, financial and credit risk management, industry, robotics, human resource management, recruiting, military, medicine, music, publishing, etc.

In 2013, the global market for AI technologies, according to the consulting company Frost and Sullivan, was 0.7 billion US dollars, but in 2017, it grew to 13.4 billion US dollars, and by 2022, it will be 52.5 billion US dollars, that is, compared to 2017, it will grow four times. It is assumed that by 2030, the use of AI technologies will increase the volume of the global market for goods and services by 15.7 trillion USD.

According to Tractica, the global AI market is expected to reach $118.6 billion by 2025.

The speed of development of AI technologies raises the question of in which direction they will develop and what are the prospects in this process.

Work on the development and application of AI technologies is being actively carried out in all developed countries, grouped around areas that can bring practical benefits.

Machine learning systems are actively developing, which include the following components: supervised learning; unsupervised learning, reinforcement learning, and others. They are models, methods, and algorithms focused on the automatic accumulation of knowledge based on the analysis and generalization of data.

N. O. Omarova (✉) · A. A. Echilova
Federal State Budgetary Educational Institution of Higher Professional Education Dagestan State University, Makhachkala, Russia

D. P. Sakas et al. (eds.), *Computational and Strategic Business Modelling*, Springer Proceedings in Business and Economics,
https://doi.org/10.1007/978-3-031-41371-1_47

Supervised learning takes place with the participation of a person, in the process of which people collect and pre-label data into certain categories. Thanks to this technology, autonomous vehicles and voice assistants have been created in the last decade. Supervised learning, on the other hand, has certain limitations, which include the laborious process of manually labeling the data. Instead of assimilation of all available data, processing algorithms are guided by their certain categories, noted in advance by the researcher.

Unsupervised learning (self-learning) is designed to eliminate this shortcoming, in which the AI system is trained to perform a task without human intervention, using the entire set of available data, and not just those that were marked in advance. This process is similar to how a person himself cognizes the world, assimilating all the data available to him without restrictions.

In self-learning, part of the input data is used by the AI as a control signal to predict the rest of the data. Many researchers believe that self-learning in the future will make it possible to create devices with intelligence close to the human level.

Another type of machine learning is reinforcement learning, in which AI interacts with some environment. The responses of the environment (and not a special control system, as happens in supervised learning) to the decisions made are reinforcement signals; hence, such learning is a special case of supervised learning, but the teacher is the environment or its model.

Much of the data processed by AI is sensitive. These, for example, include information about patients of medical institutions or clients of financial institutions. In this case, the approach when all data is collected in one central repository, from which they are then processed, is not applicable.

To preserve confidentiality, integrated learning technologies are developing, the essence of which is that the data is left in storage places, and many versions of models are sent to these places for processing. After processing in the AI field, it is not the data itself that is sent back, but the parameters of the model, which are then integrated into one overall model and, thus, the confidentiality of the original information is preserved.

A promising direction is the development of neuromorphic microcircuits and computers based on them, which differ from simple microcircuits in architecture. Conventional computing is based on the von Neumann architecture with a separate processor and memory, and the human brain is a single unit and is able to simultaneously store and process data. Neuromorphic chips are created on the basis of existing technologies, and we are talking about semiconductor systems, and not about biological devices. Chips are made from ordinary transistors, but the architecture in this case is different. In neuromorphic chips, acting as neurons, transistors can send signals to the input of other elements, changing their state. Chips with this technology are already being produced by IBM and Intel.

The next technology that has been actively developed in recent years is natural language processing. Initially, research was based on the development of human speech recognition technologies – machines were trained to understand human speech and generate it. Today, natural language generation technologies are actively developing in the processing of big data. About 80% of all existing data is available in raw form and, with the help of AI technologies, can be used to develop models for structuring and processing information.

Within the framework of this technology, speech recognition and understanding systems are being developed, which is important for services using a speech interface. It is also important for AI systems to understand the context of what was said, since it is an essential part of natural language. Research in this area is focused on the ability of AI systems to interact with a person in the form of a dialogue. Voice requests have become commonplace for smartphone users and the volume of such services will only increase. Computer agents are being created – chatbots with which you can communicate through messaging and voice call applications. They must respond to customer requests and answer questions. At this stage of development, chatbots cannot completely replace a person and, if they are unable to answer a question, they have the function of switching to an operator.

At the present stage, joint systems are being developed that allow integration of the work of various autonomous systems and people by creating algorithms that improve human efficiency through the active involvement of AI in the production process.

An example of a joint system is crowdsourcing, which makes it possible to use information systems to involve a large circle of competent specialists in solving a particular problem of innovative production activity in order to share their creative abilities, knowledge, and experience. The main disadvantage of the system, under certain conditions, may be the human factor: the deadlines for the execution of work may be violated, and the work itself is carried out poorly and poorly coordinated. To eliminate shortcomings, the system allows you to manage large amounts of data and track the main KPIs of projects.

The following prospects for the use of AI technologies can be noted:

- Management of portals, large online stores, marketplaces, and other complex web systems
- Routing of information packets during their transmission over the network
- Forecasting and optimization of loading of information transmission channels
- Management of network robots

An example is the Internet of Things, which has recently entered the life of every person and led to a merging of the physical and digital worlds, as data is collected from an increasing number of devices and then combined into the so-called big data. The number of such Internet of Things devices, according to experts and analysts, has already reached 50 billion by 2020. Devices include appliances, vehicles, buildings, and cameras.

Another actively developing area is the creation of unmanned vehicles that use computer vision technology. In Russia and abroad, there are already experimental models of vehicles that allow you to work out technologies for serial samples.

The development of adaptive and intelligent robots using AI will increase the degree of automation of production processes. Currently, existing robots are programmable manipulators and require human presence.

Adaptive robots that currently exist in the form of prototypes are equipped with a sensory part and a set of programs, and the development of intelligent robots is associated with solving the problem of computer vision, which requires special software written in special languages focused on symbolic information processing – Lisp, Prolog, Smalltalk, Refal, etc.

From the point of view of practical application, the technology of "digital twins" has great prospects and is increasingly being used in industry, energy, aviation, and the automotive industry. A digital twin is a synchronized virtual model of a real object: a person, product, process, system, etc. It reflects not only the past and current state of the object but can predict its future based on the initial data, which is achieved by a deep understanding of the processes occurring in a real object.

Recently, AI has begun to play an increasing role in cybersecurity, thanks to training to detect malware, fraud attempts, network threats, suspicious user behavior, and technology. The growing importance of accumulated data and its high vulnerability is forcing many companies to close their data with AI systems.

The use of AI in media and entertainment is expanding, where these systems can take into account the tastes of a particular user and provide recommendations, personalized advertising, commercials, and movie trailers, automate the creation of subtitles and their translation into rare languages.

It is promising to use AI in augmented reality technologies to add digital objects to objects in the real world. An example is the helmet of a military pilot, which displays various flight parameters and the state of the aircraft. Such systems are used in industry when assembling cars and other modern equipment. AI gives digital objects a greater reality and controls their behavior.

In medicine, AI is beginning to be used to visualize internal organs without traumatic procedures and analyze data from patient medical records to help make a more accurate diagnosis. AI is beginning to be used by global pharmaceutical companies to develop new drugs, allowing unwanted side effects to be detected at an early stage of research.

2 Development of the Education System Based on the Use of Big Data Technology

The rapidly gaining pace of the fourth industrial revolution requires a revision of the traditional educational process, which implies reconsideration of such concepts as "effective learning", "subject learning", as well as going beyond the educational model of transferring knowledge from teacher to student [1, 2]. According to the results of a study by the Russian Academy of Education, it should be noted that an increasing category of students is made up of "special" children, including gifted children, children with specially developed thinking, leaders, artistically gifted children, and children with motor talent. Both factors initiate the development of the educational system, actualizing the need for technologies that can draw reasonable conclusions about the need to change the approaches or actions taken in education. One of these technologies can be the technology of handling big data (BigData).

Big data (BigData) in education is a technology for analytics of the educational system, including the measurement, collection, analysis, and presentation of structured and unstructured data of huge volumes about students and the educational environment in order to understand the functioning and development of the educational system.

Historically, the education system has accumulated a significant amount of data. The question of how it is affordable to start processing a large amount of data will be removed due to the emergence and expanded use of information and communication technologies.

He identifies three major areas of BigData [2]:

(1) Related to thinking (primarily critical and creative thinking)
(2) Related to interaction with others (communication and collaboration)
(3) Related to interaction with oneself (self-regulation, reflexivity, and self-organization)

However, the results of analytics in these areas are most valuable when they reveal abnormal and borderline states of the educational system. Regulatory measures as a reaction to negative conditions are most useful for working on the development of the educational system.

By looking at how various industries, primarily finance and business, have adopted and benefited from data analytics, we may see a similar future in the education sector. Here are four key opportunities that big data provides to educational stakeholders:

First, data analysis allows you to work with individual programs of students, personalize learning. The data shows what type of learner interacts with what part of the content, how this interaction occurs, where he showed interest and where he was bored, with whom and how he interacted in the learning process, how the passage of a particular course influenced educational results, on what stage of training he needs help. Learning becomes adaptive and student-centered.

Second, educational analytics based on big data is changing the idea of the format of educational programs. Texts used in the educational process can be not only digitized but also translated into numerical data. Users move through the material with greater freedom, and then an analysis is made of how users interacted with the material: what was effective, and what was not effective. The result of such analytics is content change. Therefore, the educational program is transformed from the format of the approved text into the format of a certain set of online content, which dynamically changes through the analysis of data that appears because of interaction with students' online content. The so-called smart program and smart curriculum appear. It can be assumed that the curricula of the courses will also change: they may become meta-subjects.

Third, a change in approaches to monitoring and evaluation of both the educational process itself and educational results. Monitoring becomes permanent. The interest of students in constant monitoring is because that data analysis makes it possible to make its curriculum individual, the interest of teachers is associated with the possibility of obtaining information about productive groups, feedback from students on the content being created (interesting/not interesting, difficult/easy, etc.), for teachers – efficient allocation of resources. Evaluation of educational results can be independent and/or collective, aggregated based on all student data obtained from all interactions. The assessment is carried out in order to competently expand the educational program of the student. The dynamics of educational results is fixed constantly, based on these data patterns (repeating patterns) formed, by which one can judge the development of the student.

Fourth, new methods will become an integral part of educational analytics:

(a) Forecast, when a combination of known data will predict the desired unknown
(b) Method of revealing the structure and clustering
(c) Network analysis

The idea of "big data" (BigData) has been actively developing for the last 2–3 years, and the main factors of its development are political, economic, and social trends in the development of society. BigData allows not only to analyze the image of the consumer and the amount of planned costs, demand but also to model the future in finance, business, education, etc. They allow you to find out the status of the objects under study.

Finding new solutions and methods is also necessary for the education system especially to improve the management of educational systems and ensure the quality of education. On the one hand, it is required to use a huge amount of accumulated information that needs to be analyzed and systematized. On the other hand, BigData makes it possible for each student to build their individual educational trajectory in a new way as well as to assess the quality of education in an educational organization and choose an acceptable way of learning for themselves.

Using big data in education, it is necessary to present the material in such a way that it would be interesting to study, identify patterns, and use them.

Nowadays the essential role of the teacher remains, and no information system replaces the teacher. But for the tasks of mass and corporate education, computer systems can be very effective, additional learning tools. At the very least, they will help save time searching for information. Intelligent systems will soon be available in any educational organization. Then it will be possible to overcome the concept of educational inequality and reduce barriers to learning for people with disabilities. All this thanks to a personalized approach and a smart computer that knows everything about how the student learns the material. They will also allow to provide information about the best methods of teaching and control of knowledge, skills, and competencies acquired in various educational organizations or independently.

Big data has changed the approach to teaching a foreign language in Russia's largest online school for learning English Skyeng.ru, thanks to the active use of big data processing technology in its activities.

BigData helps to process the experience of thousands of teachers and students, based on the analysis to get an effective methodology. If traditionally a teaching methodology is created based on the personal experience of one or several teachers, then based on big data, the methodology becomes a product of mass experience. In addition to improving the quality and efficiency of the created methods, big data helps to personalize content for the needs of each student. But at the same time, one should not forget that the teacher can explain and receive feedback, while the computer cannot track the student's reaction. In addition, the teacher creates an emotional background and motivation for learning.

Big data, like any technology in education, does not save the teacher from empathy and interaction with the student, the ability of a person to empathize and motivate is always important, and such a function is not available to computers. With the help of big data, you can do, relatively speaking, three important things: create methods adapted to many students; personalize content; select the learning mode.

It should be noted that BigData will soon change the technologies of higher education, making it possible to make student learning more individual: not only to select their own course program for each, but also to give a separate homework, and also provide verification of content mastering. The method of working in groups will also be different: at Harvard, already now, in one of the courses, students are paired up with different answers to the same task so that they can come to a single solution, defending their point of view in the process of finding the right answer. Students will receive more detailed guidance on various topics and have an expanded information space. Programs can already predict how successfully a course has been completed even before the start of training. Students will have the opportunity to choose their course program, complete a separate homework assignment, and receive more detailed recommendations. With the help of big data, there will be fewer under-achievers in university groups, as technology will allow early identification of students who may be at risk, and teachers will be able to better help lagging students, as the program will indicate in which particular areas of knowledge there are problems. The system will also help teenagers in choosing a university: it is assumed that the robots themselves will select the best places to study for future students, and they do not even have to apply. The system will select the best places for future students, and by the end of the university, each student will have a digital portfolio that will help young professionals navigate the labor market, make it easier to navigate when choosing a career, and employers select specialists.

In the field of education, five main types of data are distinguished for the analysis of big data: personal data; data on the interaction of students with e-learning systems and with each other (e-textbooks, online courses, bounce rates, page view speed, page returns, number of links, distance of links, number of page views by one user, etc.); data on the effectiveness of educational materials (what type of student interacts with which part of the content, interaction results, educational results, etc.); administrative (system-wide) data (attendance, sick leave, number of lessons, etc.); predictive (estimated) data (what is the probability of the student participating in a particular activity, what is the probability of completing the task, etc.).

It is obvious that today almost all educational organizations work mainly with small data. This is due to the fact that educational organizations do not have a special electronic environment that contains a lot of online content and, as a result, a large number of content users and interactions with each other relative to it [1].

It is obvious that BigData opens up new horizons in modern education, with the development of these technologies, education reaches a higher level, but at the same time, one should not forget about the shortcomings:

(1) Lack of specialists – big data specialists spend a lot of time in the process of acquiring big data skills.
(2) Rapid change – a potential disadvantage of big data analytics is that technology changes rapidly. Educational institutions face real risks that they will invest in a particular technology, only to pay again months later for something better.
(3) Data quality – the main disadvantage of working with big data is the need to solve data quality problems. Before using big data for analysis, analysts must ensure that the information they use is accurate, up-to-date, and suitable for analysis.
(4) Confidentiality – most of the information contained in companies' big data stores is confidential or personal, which means that the processing and storage of data must comply with government standards or requirements.
(5) Difficulty in integrating legacy systems – when data is hosted across different applications and systems, integrating all these disparate data sources and moving data where it is needed also increases the time and cost of working with big data [1].

The progress we are witnessing in today's digital age provides us with unprecedented opportunities and puzzles with new challenges. It is vital for government leaders, scientists, and business leaders around the world to ensure the proper adoption of Big Data.

Let's give examples of BigData analytics in universities [3, 4].

In 2013, the English University of Nottingham Trent introduced an interactive system of descriptive analytics of student results in the form of a dashboard that showed data on student involvement in the learning process. The dashboard was designed to reduce student dropout rates, improve attendance, and increase the sense of belonging to the university community. The panel, available to students, teachers, and curators (tutors), displays the indicators of each student's involvement in comparison with his classmates: the frequency of using the library, information about the courses studied, attendance, and other learning indicators. Thus, any student can see his own activity and compare himself with fellow students in order to understand how he is included in the educational process and the life of the university in general, and also what should be paid more attention to. If a student does not show signs of activity within two weeks, the platform sends notifications to tutors so that they quickly contact the student and support him. Three years after the introduction of the system, the results of a university survey showed that 72% of freshmen used the BigData student dashboard and it inspired them to increase the amount of time they study.

Purdue University in the United States has launched a predictive analytics system that collects information about the academic history of students, their activity in the digital learning environment, and demographic data. Based on this information, the dropout risk level for each student is calculated. The value of this indicator is periodically sent to the course coordinator and the student himself. For clarity, the data is visualized: students who are most likely to successfully complete the semester are marked in green, those who are at risk in yellow, and those who require immediate assistance from the curator are marked in red. Thanks to such an interactive BigData system, it was possible to improve learning outcomes and reduce dropout rates.

Also interesting is the experience of another US university, Carnegie Mellon University, which has created an online platform with training courses that provide students with a detailed report on their progress. The results of its operation showed that detailed reviews and the opportunity to look at the analysis of one's own progress and educational activity almost doubled the time for completing online courses.

In fairness, it should be noted that today online education platforms are used in almost every domestic and foreign university. However, they are usually a digital form of the traditional learning process, where the teacher provides the material and the student studies it and completes the assignments. Feedback in the form of marks and reviews, to a greater extent, is aimed at assessing the learning outcomes of a particular student, and not the competencies of the teacher and the content of the course. Therefore, adding an element of competition with other students to the online learning platform, as well as recommendations for choosing other training courses, can significantly increase the effectiveness of such a system. This is partly implemented in independent and commercial aggregators of online courses for independent distance learning and advanced training.

But, for example, at one American university, Austin Pee, a recommender system has been introduced that helps students choose educational courses and enroll in them. The input data is the results of previous students in a particular course, the performance of each student, and information about students with similar profiles and interests. Based on the analysis of this information, the BigData system, using Machine learning algorithms, selects training courses that best suit the interests, abilities, and curriculum of an individual student. The accuracy of the recommendations is estimated at 90%.

Machine learning algorithms can be used for more than just providing relevant advice on choosing educational courses. For example, at the beginning of 2020, the University of North Carolina (USA) introduced a multi-task learning system, where Machine learning models predict the probability of a student's correct answer based on his previous behavior in the game learning process. This is useful for notifying teachers when a student may need additional instruction and facilitating adaptive learning features during the game itself. For example, changing the storyline and adding hints.

3 Conclusion

The entire data array can be conditionally divided into two groups. The first includes structured data, that is, those that we manually enter spreadsheets, traditional databases, process, and analyze. The second group consists of unstructured data, that is, those that appear in the form of various "digital traces" of a person's interaction with the electronic environment and with other people in the electronic environment. The amount of structured data is only 5% of all data. By abandoning the mess, we lose the remaining 95% of unstructured data and get only a general idea of the development processes. It is also important that the increase in the amount of data comes at the expense of an increase in the amount of unstructured data.

Education is a branch of human activity that has always produced more data than any other field. A huge number of hours of classes for five or six days a week for several years, many tasks performed by students, a variety of educational content and the number of interactions of students with it, countless interactions of participants in the educational process with each other – all this gives reason to assert that in educational organizations already have that same big data.

It is obvious that today all educational organizations work only with small data. This is because the educational institutions do not have a special electronic environment that contains a lot of online content and, as a result, a large number of user interactions with this content and with each other regarding it. In fact, the authors of online courses and e-learning work with big data.

The leader in the use of data about educational activities – Knewton USA – works with five types of data:

(1) Personal data
(2) Data on the interaction of the user with the electronic system and with each other (bounce rates, page browsing speeds, returns to pages, number of links, distance of links, number of page views by one user, etc.)
(3) Data on the effectiveness of educational materials (what type of student interacts with which part of the content, interaction results, educational results, etc.)
(4) Administrative (system-wide) data (attendance, absenteeism due to illness, number of lessons taught, etc.)
(5) Predictive (estimated) data (what is the probability of the student participating in a particular activity, what is the probability of completing the task, etc.)

These five data types are listed in order of their complexity and the database architectures they fall into.

Today, educational organizations should think about how they will work with data. Each educational organization is not required to build electronic systems to obtain five sets of data. However, each educational organization should have an answer to all questions of the five data groups. Other scientists with similar cases are studying similar problems, using modeling and simulation [5–8].

References

1. Fiofanova, O.A.: The problem of integrating digital data analytics services: the competence of a teacher in working with educational data. Bulletin of the Moscow University. Pedag. Educ. **9**(20) (2020)
2. Frumin, I.D.: Trends in the development of the content of education: key competencies and new literacy. Publishing House of the Higher School of Economics, Moscow (2020)
3. Home Page of the Scientific Electronic Library CiberLeninka, https://cyberleninka.ru/article/n/onlayn-obrazovanie-klyuchevye-trendy-i-prepyatstviya, last accessed March 22, 2022.
4. Home Page of the Electronic Scientific Journal, https://www.sciencedaily.com/releases/2020/02/200205132409.htm. Last Accessed 10 Feb 2022.
5. Nasiopoulos, D.K., Ioanna, D., Lida, G., Paraskevi, P., Giakomidou, D.S.: Modeling of B2C Communication Strategies in Electronic Commerce. Strateg. Innov. Mark., 531–537 (2017)
6. Nasiopoulos, D.K.: Implementation stages and risks of logistics companies reengineering. Bus. Intell. Modell. **6**, 197–206 (2020)
7. Nasiopoulos, D.K.: Web search applications in Logistics companies in the context of evolution of Management Information Systems. Bus. Intell. Modell. **6**, 241–254 (2020)
8. Omarova, N.O., Omarov, O.A., Osmanova, A.O. (2021). Design and Implementation of Pedagogical Innovations Based on Digital Technologies. In: Sakas, D.P., Nasiopoulos, D.K., Taratuhina, Y. (eds) Business Intelligence and Modelling. IC-BIM 2019. Springer Proceedings in Business and Economics. Springer, Cham. https://doi.org/10.1007/978-3-030-57065-1_37

Ensuring Information Security of Electronic Business

Naida O. Omarova, Elmira A. Shamkhalova, and Madina M. Osmanova

1 Introduction

Among the considered problems of the socio-economic development of Russia in the context of the formation of a global post-industrial community, the organization of sustainable use and the safety of using information systems and information and communication networks that support economic activity take the place. As complexity increases, the impact of business on the expected results of the activities of organizations, which will increase, is revealed.

The information environment is huge, but security tools track the slightest changes in system code and unauthorized access attempts, capture encrypted packets, and much more.

Solving the problem of economic security in the electronic industry is primarily associated with the emergence of issues of the protection of information technologies used in it, that is, ensuring information security.

Ensuring security is one of the main points of a security policy in business, in the implementation of which elements of protection and corporate protection, as well as software and hardware protection, must be applied.

Requirements are assigned to information security; in the presence of revealed information, they are protected from any external influences (natural and artificial). A breach of data and the confidentiality it maintains can generally harm the owner. Currently, specialists in uncovering the security of protected spaces (virtual "safes"), require security software such as antiviruses ("employ guards"), and use cryptographic methods to encrypt digital information.

When achieving the security of information, ensuring its accessibility, confidentiality, integrity and legal significance are the basic tasks. Each threat must be

N. O. Omarova (✉) · E. A. Shamkhalova · M. M. Osmanova
Dagestan State University, Makhachkala, Russia

D. P. Sakas et al. (eds.), *Computational and Strategic Business Modelling*,
Springer Proceedings in Business and Economics,
https://doi.org/10.1007/978-3-031-41371-1_48

considered in terms of how it may affect these four properties or the quality of the information. Confidentiality means that restricted information should be available only to those who own it. Under the statistics of information, it is understood that it is the property of the console in an undistorted form. Availability of the decision to submit a request to provide one's temporary unhindered access to information about persons is manifested in this appropriate security. The legal significance of information has become important recently, along with the creation of a regulatory framework for information security in our country [5].

2 Information Security in Business

The process of identifying threats to information security should be regular and systematic and should extend not only to the creation of an information security system but also to the impact of exploitation. It is necessary to establish a process for detecting and neutralizing the threat of security detection, which could prevent possible damage.

(1) Expert method. Possible security risk assessments involve the formation of an expert group that contains an analysis of vulnerabilities. Thanks to the qualitative formation of an expert group, it is possible to reduce the level of subjectivity in risk assessment.
(2) System method. A systematic method for identifying security incidents involves a continuous process aimed at identifying and identifying threats, then identifying the occurrence of occurrences and the occurrence of possible occurrences of occurrences of threats. General safety is reviewed and re-evaluated on a regular basis.
(3) Method for identifying possible sources of threats. The process of assessing the problem of information security involves the systematic identification of sources of threats, opportunities and, as a result, an assessment from this, the identification of actual problems of information security.
(4) Legal methods. Legal methods, as a rule, are aimed at eliminating the threat of an anthropogenic nature. In the case of detection of cases of infringement, legal methods allow the application of cases of discovery in relation to violations.
(5) Economic methods. Economic methods are aimed at resolving threats of an anthropogenic nature as well as at the introduction of the susceptible consequences of the implementation of threats.

The effectiveness of preventing threats to information security directly depends on an integrated approach to solving the problem under study. Ensuring the required level of security of the information resources of the enterprise requires regular modernization of the information security system, including improvements in the regulatory framework, the introduction of new software products, and training events for employees.

Many information security problems are related to the underestimation of the importance of such a threat as the confidentiality of information. As a result, this can lead to bankruptcy for the enterprise. Even a single case of negligence by the company's personnel can bring him multimillion-dollar losses and a loss of the company's reputation and customer confidence. To avoid this, enterprise security specialists use special equipment that analyzes electromagnetic radiation received while working on a computer. To ensure the cybersecurity of networks, the following methods are used:

(1) Providing physical protection of network nodes
(2) Use of the vulnerability control system
(3) Increase visibility within the network
(4) Access control
(5) Use of virtualization technologies
(6) Application of isolation from the external network
(7) Ensuring secure internetworking
(8) Implementation of platforms for creating a distributed decoy infrastructure
(9) Increasing the cybersecurity literacy of personnel, especially those maintaining the network

Other methods may be applied. Since the task of ensuring cybersecurity is complex, its solution depends on all levels. It is impossible to protect yourself from everything, from all the possible threats in the enterprise; you need to choose what is most significant and important. And the right information security system will help to keep it.

International information security depends on the information space. The information space is a field of activity that is associated with the storage of information, its creation, application, modification, transfer, and has an impact on both individual and public consciousness. The formation of modern society is associated with the processes of informatization, the improvement of information technologies and the emergence of a global information space. However, like any phenomenon, in addition to positive aspects, it also has unfavorable ones. Developed states with a high level of computerization of various spheres of society's life are constantly experiencing difficulties associated with the development of technology. The more often they massively use information technologies and the use of global networks, the more complex their information infrastructure, the lower the protection from cyberattacks and the higher the harm caused. Thus, in the United States, losses from unauthorized penetration into such systems and subsequent information leakage are estimated at tens of millions of dollars.

The rapid formation and active use of technologies led to the fact that the states turned out to be dependent on them, and this led to the likelihood of new threats. Most often, such threats are associated with the objective possibility of using information and communication technologies in order to create conflicts. First of all, the use and proliferation of information weapons and the resulting threat of information wars and information terrorism, which can violate international peace and cybersecurity, are of primary concern.

In the first three quarters of 2020, 167 attacks on public institutions were recorded (for the same period in 2019, 133 attacks were recorded). As we expected, most often attacks were carried out using phishing (49% of attacks) and malware (63% of attacks), but, in addition, cyberattacks on the websites of state-owned companies do not lose their relevance. In the first three quarters of 2020, 18% of attacks were specifically related to hacking web applications (in 2019, this figure was almost the same – 19%) [6].

Websites are attacked with the aim of stealing user personal data, deface, and infection by cryptocurrency miners. Phishing and malware are used to deliver ransomware to the infrastructure and in targeted attacks for espionage.

As before, the main ways to get to user data are social engineering and malware infection of devices. Criminals continue to use the illiteracy of people in matters of ensuring their own information security. A year earlier, guessing passwords for accounts on websites and social networks accounted for a significant share of attack methods (12% in three quarters of 2019), but in 2020, we notice that this trend is fading away: only 6% of attacks used such method. We attribute this to the fact that the vast majority of Internet services today allow the use of two-factor authentication, which makes it harder to attack – and people do use this two-factor authentication when accessing their accounts.

Data mining attacks accounted for 64% of all attacks against individuals in the third quarter of 2020. Almost half of these attacks (47%) were aimed at the theft of credentials from Internet services; 23% of the attacks were aimed at the theft of payment card data; 12% were aimed at the theft of personal data, and the rest were aimed at accessing personal correspondence [9].

Attackers are actively exploiting website vulnerabilities to attack ordinary citizens. According to statistics for 2020, 92% of web applications allow attacks on users. At the same time, 82% of the vulnerabilities that we found are due to errors in the development of the code. Security flaws in 16% of the sites studied allowed full control not only of the web application itself, but also of the server. Such control gives the attacker the ability to carry out serious attacks, such as distributing malware or injecting JavaScript sniffers 3 into the website code to steal bank card data. This is the trend of last year in the field of e-commerce, but the global problem of web application security is still relevant today. Given the rapid growth of the e-commerce market (doubling by 2023, according to Data Insight), we can expect this problem to worsen, including due to sniffer attacks [3].

Attacks using site vulnerabilities, including using JavaScript sniffers, will continue due to their high efficiency. An ordinary user is powerless to make safe the Internet resource where he pays for the purchase, the responsibility for countering attacks lies with the site owners. However, users should be more careful about those resources where they enter bank card data: if this is an untrusted site or little known, then you should weigh the pros and cons before risking your data. It is believed that larger and more well-known companies in the market are more reliable in protecting users from attacks, but brand awareness alone is not enough, there are examples of attacks when it was large brands that became victims and endangered their users.

Attacks on personal devices of users will not lose their relevance, since for most people the convenience of working with a gadget is more important than the security of personal data. Most likely, attackers will combine attacks on gadgets with classic social engineering methods (e.g., fraudulent phone calls in order to obtain banking data). Gadgets have firmly entered all areas of our lives, and now in order to find out some data or steal money from a bank account, criminals need to attack mobile devices.

These conclusions are confirmed by the data of the Prosecutor General's Office: the number of crimes in the field of IT in 2020 almost doubled compared to 2019 and by the end of the year reaches almost 270,000 just registered cases, that is, 14% of the total number of all registered crimes in Russia [11].

Considering the many different approaches and interpretations in the scientific literature of the essence of information in various fields of human activity, "information" is quite often understood as the result of data processing, addressed to a specific user and suitable for solving individual problems in achieving the goals set. Progress in the development of electronic means of storing, transmitting, and reproducing information, expanding the capabilities of the Internet have contributed to the active introduction of new information technologies in forensic activities. The most important of them was the emergence of information sources on machine media, called "electronic information resources."

In modern Russia, the problem of information security is quite acute, as the role of accumulation, processing, and dissemination of information has significantly increased. Information plays an increasingly important role in the system of public administration, and this, in turn, leads to the strengthening of the role of the media.

In modern conditions, information plays an increasingly important role in the administrative activities of internal affairs bodies. In this regard, scientific research on information security issues, as well as ensuring information security, is being updated. Federal Law No. 149-FZ of July 27, 2006 "On Information, Information Technologies and Information Protection" defines information as information (messages, data) regardless of the form of their presentation [1].

The subject of information security is an active participant in the processes in the activities of ensuring information security, influencing the object of information security, regardless of the nature of this impact: causing damage, destruction, or opposing it.

The whole set of information security subjects, similarly to security subjects, can be divided into two groups: external and internal – and each of them into two more: opposing and contributing to information security.

3 Conclusion

The problem of information security of economic objects is multifaceted and needs further study.

In the modern world, informatization is becoming a strategic national resource, one of the main wealth of an economically developed state. The rapid improvement of informatization in Russia, its penetration into all spheres of the vital interests of the individual, society, and the state, in addition to undoubted advantages, led to the emergence of a number of significant problems. One of them was the need to protect information. Considering that at present the economic potential is increasingly determined by the level of development of the information infrastructure, the potential vulnerability of the economy to information impacts is proportionally growing.

Implementation of threats to information security is to violate the confidentiality, integrity, and availability of information. From the standpoint of a systematic approach to information protection, it is necessary to use the entire arsenal of available means of protection in all structural elements of an economic object and at all stages of the technological cycle of information processing. Methods and means of protection should reliably block possible ways of illegal access to protected secrets. The effectiveness of information security means that the costs of its implementation should not exceed the possible losses from the implementation of information threats. Information security planning is carried out through the development of detailed information security plans by each service.

Other scientists with similar cases are studying similar problems, using modeling and simulation [12–17].

There is a need for clarity in exercising the powers and rights of users to access certain types of information, in ensuring control of protective equipment and immediate response to their failure.

References

1. Federal Law.: 'On Information, Information Technologies and Information Protection' dated July 27. N 149-FZ (2006)
2. Astafieva, N.S.: Development of the Information Security Management Process of an Organization. IV International Scientific-Practical Conference Management of Economy, Systems, Processes, pp. 16–20 (2020).
3. Baranova, E.K., Babash, A.V.: Information Security and Information Protection: Textbook, p. 400. Rior, Moscow (2018)
4. Blyumin, A.M., Feoktistov, N.A.: World Information Resources: Textbook Management, p. 296. Dashkov i K°, Moscow (2021)
5. Domnin, O.V.: Legal Support of Information Security in the Field of Use of Digital Technologies by Business Entities in Conditions of Remote Work, pp. 180–188. Law and Law (2021)
6. Zegzhda, P.D., Anisimov, V.G., Anisimov, E.G., Saurenko, T.N.: A model for the formation of a program for the development of an organization's information security system. Problems of information security. Comput. Syst. **2**(46), 109–117 (2021)
7. Kamsky, V.: Protection of personal information on the Internet, smartphone and computer. St. Petersburg: Sci. Technol. **272** (2019)
8. Maslakova, E.A., Zhilkin, M.G., Kachalov, V.V.: On the issue of international cooperation in the field of information security, vol. 9, pp. 75–77. Bulletin of the Moscow University of the Ministry of Internal Affairs of Russia (2015)

9. Melnikov, V.P., Kleymenov, S.A., Petrakov, A.M.: Information Security and Information Protection, p. 336. Higher School, Moscow (2020)
10. Osipova, V.A.: Information security as an element of economic security. Econ. Vector. **3**(45), 49–56 (2020)
11. Shamkhalova, E.A., Osmanova, M.M., Osmanova, A.M.: Information security in electronic business. Econ. Entrep. **12**, 112–118 (2021)
12. Nasiopoulos, D.K., Ioanna, D., Lida, G., Paraskevi, P., Giakomidou, D.S.: Modeling of B2C Communication Strategies in Electronic Commerce. Strateg. Innov. Mark., 531–537 (2017)
13. Nasiopoulos, D.K.: Implementation stages and risks of Logistics Companies reengineering. Bus. Intell. Modell. **6**, 197–206 (2020)
14. Nasiopoulos, D.K.: Web search applications in Logistics companies in the context of evolution of Management Information Systems. Bus. Intell. Modell. **6**, 241–254 (2020)
15. Dimitrios, N.K., Sakas, D.P., Vlachos, D.S.: The role of open source leadership in developing high technology companies. Key Eng. Mater. **543**, 402–405 (2013)
16. PR Damianos P. Sakas, Dimitrios K. Nasiopoulos (2019). Modeling and Simulation of the Strategic Use of Social Media Networks in Search Engines for the Business Success of High Technology Companies, Business and Economics, pp. 227-236
17. Plessias, A.A., Nasiopoulos, D.K., Giakomidou, D.S.: Modeling of market segmentation in social networks and media. Strateg. Innov. Mark., 523–529 (2017)

Developing Intercultural Communication Competences in the Conditions of Online Learning in the Modern Digital World

Grigory A. Rozhkov, Elena V. Naumova, and Alexandra P. Tumanova

1 Introduction

A number of researchers have analyzed the impact of AI on the modern approaches to educational processes in the pandemic learning environment. Some researchers highlight the fact that when AI is incorporated into the educational process, there are reoccurring technological, informational, methodological, and psychological problems. Meanwhile, there is a strikingly apparent insufficiency of methodological guidelines and recommendations on the AI incorporation in education. There is also a lack of sufficient knowledge of IT among both – the teaching staff and the students, all of which restrict the process of efficiently adapting the existing technologies to a particular educational process [1]. A.V. Kostina points out that digital technologies allowed using the self-isolation process effectively meanwhile significantly increasing the effectiveness of measures taken against the spread of the coronavirus infection in the educational process, in particular. The possibility to communicate with the students using digital platforms and to see each other – all of these factors decreased the negative effects of self-isolation [2]. The vast majority of experts agree that the pandemic situation served as a catalyst for digitizing the educational process. On the one hand, it was a difficult transition process for educational institutions, on the other hand, within just a half-year period people have made such significant breakthroughs worldwide, which, in normal life situations, would have been accomplished within a longer time-period. The challenges which the pandemic had brought with itself have penetrated all the spheres of the educational process. It has changed the interpersonal communication format. The

G. A. Rozhkov · E. V. Naumova · A. P. Tumanova (✉)
Saint-Petersburg Chemical and Pharmaceutical University, Saint-Petersburg, Russia
e-mail: grigory.rozhkov@pharminnotech.com; elena.naumova@pharminnotech.com; aleksandra.tumanova@spcpu.ru

D. P. Sakas et al. (eds.), *Computational and Strategic Business Modelling*,
Springer Proceedings in Business and Economics,
https://doi.org/10.1007/978-3-031-41371-1_49

process of attaining intercultural communicative competence (ICC) in the digital format has undergone significantly new and unexpected changes.

2 The Changing Process of Attaining ICC in Higher Educational Institutions

Before the pandemic, when nobody could predict what dismal changes the newly discovered virus would lead to, the Saint-Petersburg Chemical and Pharmaceutical University (SPCPU) had been functioning in a mode, which is typical for many other educational institutions. SPCPU has always been noted for its vast possibilities for student cooperation, practical knowledge exchange among the future experts worldwide due to the fact that development of the pharmaceutical industry is impossible without the cooperation and dialog between global partners and colleagues.

Many SPCPU students have knowledge of two, and some, even three and more foreign languages; therefore, they have always ceased the opportunity to put their language competence into practice and have taken an active part in the various international events organized by the University. In the course of many years, preceding the pandemic situation, extracurricular foreign language activities carried out among the students promoted the ICC among the students and helped them acquire soft skills as well as develop their emotional and cultural intellect (EQ).

SPCPU had long-term international relations with the University of Turku and the Abo Akademi University (Finland) and has been holding the Russian-Finnish symposium, during which students, scientists and experts in the pharmaceutical and biomedical spheres had the opportunity to discuss their scientific findings, present the results of joint research as well as to identify the current trends in the professional field. There were tours held at the SPCPU scientific research centers and the museums of Saint-Petersburg, Olympiads and foreign language contests. Members of the Student Scientific Society have annually taken part in the Young Pharmacy International Conference held in Russia. Undergraduate and postgraduate students of SPCPU have taken part in special English language workshops on the analysis of real-life situations. The main topic of these workshops was devoted to finding a solution to intercultural conflicts. These initiatives have promoted teamwork skills among the students and developed a better understanding of how to initiate a dialog effectively in a cross-cultural environment. SPCPU students also took part in the volunteering campaigns, in which they chaperoned international guests, provided translations at international events, organized the conference sessions and symposiums. The extracurricular activities promoted the development of competences, necessary for the future experts in the process of intercultural communication with international colleagues. This paper presents an overview of a successful development of such intercultural communication competences among the students through extracurricular activities in English in the pre-pandemic time-period [3]. Unfortunately, only few forms of extracurricular activities organized in a foreign language were maintained in digital format; however, it is a conclusive fact that the students have obtained valuable experience in the course of these past few years. The University has faced some critical challenges: how can the efficient methods of developing ICC be transformed in modern learning conditions and how can the ICC be developed online? A mandatory transfer to online learning served as an impulse for creating a new educational model, which was immediately put into practice.

When the global society literally had to switch to the online format of life with the use of AI due to pandemic and closed borders and when the person-to-person contacts with international partner universities and companies had to be stopped, provided with full support from the SPCPU administration, the Scientific Research and Educational Center of Foreign Languages and Intercultural Communication (SREC FLIC) began searching for new forms of organizing extracurricular activities online, which would be aimed at developing ICC among the students. The students have been actively involved in various international contests organized in an online format or in a mixed format from the very beginning of the pandemic. Examples of such activities and events include: *Public Speaking Contests of the English-Speaking Union* (ENGLISHSPEACKINGUNION), a creative contest devoted to *the International Native Language Day*, an essay contest *Parliamentary Diplomacy: Adopted Traditions and Modern Models*, international foreign language Olympiads, an annual contest *Leaders of a New World: the Skills, Vision and Mindset of Change*. SPCPU students have been awarded with prizes and have become the winners of various competitions. The students' interest in international educational projects promoted SREC FLIC to create additional educational programs designed for language-learning online. As a result, the pandemic "forced" people from across the globe to exchange positive learning outcomes in order to successfully cooperate and help each other. There have been global changes in the structure of intercultural communication which accelerated the process of global education and brought the education systems worldwide to a new level. As one of the outcomes of adapting to the pandemic, SPCPU established the LINGVAPHARM Student Scientific Society. The goal of this student society is to hold various student-oriented events, to establish partnership relations with foreign universities, to participate in international scientific conferences, to carry out volunteering campaigns at international events, to enhance the English language proficiency, international experience exchange, and scientific research in foreign languages. LINGVAPHARM is planning to improve the students' ICC in online sessions, to publish joint scientific research papers, to maintain international relations with other universities and to participate in various international conferences and seminars.

Organizing extracurricular activities in an online format allows redirecting the necessary information to the students within a short time-period. It also increases the effectiveness of the students' foreign language acquisition and gives an opportunity to organize various extracurricular activities to their full extent of student interest.

In 2021, SPCPU held the first ever series of events devoted to the World Translation Day. Translation is an integral part of international relations in the modern world and it is not surprising that the UN General Assembly has passed the Resolution NoA/RES/71/288 which "pays tribute to the role of the language professions in fostering peace, understanding and development" [4]. The Program included a series of online lectures devoted to "The Brief Introduction to Professional Translation" and "Translation in Russia: Historical Overview and Modern Trends," a professional translation Olympiad, and a creative contest devoted to "The Best Consecutive Translation."

The final event of the series included a visit by Fabian Schaller, Ambassador and Consul General, the First Secretary of the Swiss General Assembly in Saint-Petersburg, who gave a presentation for the undergraduate and post-graduate students on the role of establishing partnership relations in the modern world. He also awarded the winners of the contest with prizes. Another global achievement of the second academic year in the conditions of the pandemic was the participation of outstanding undergraduate and postgraduate students of the University in an international online project Global Case Study Chal-lenge (GCSC), which gave an opportunity to students worldwide to gain experience in intercultural communication and new knowledge through collaborative work in international teams on solving business-related cases [5]. The GCSC project is a great opportunity for the students to get acquainted with new interesting people, to learn about the lifestyle and traditions of other nations, and, certainly, to develop their language proficiency. GCSC was a two-month project. The students felt unconfident at the initial stage of the project, not only because they were worried about their language proficiency, but also because they did not have a previous experience in intercultural communication, especially in an online format. During the project, the students gave their feedback on the tasks and cooperation with the other international students and it became evident that the Project fulfilled the expectations of both – the organizing committee and the participants (students and educators) from 22 universities and 15 countries. The students did not face any problems in understanding each other; the language barrier was practically immediately overcome once the teams began working on joint projects. This was mostly achieved due to the correct teammate dispersion within the teams. In most cases, the Project participants were highly motivated and duly completed all the tasks and assignments.

3 Conclusion

The authors of this article believe it is necessary to conduct further research with the aim of evaluating the quality of education and indicating the difficulties that may occur during the education process during the pandemic, as well as of indicating effective measures for supporting and enhancing ICC in foreign language learning in the new conditions. It is also important to predict the possible intercultural learning outcomes and to carry out an in-depth study of the online learning tools.

It is important to test the extent at which the students have acquired the necessary ICC and soft skills using a practice-oriented approach. In his book, "You and I Have Come To An Understanding," G.A. Rozhkov states that "in the process of teaching the basic of intercultural communication in the new era, it is necessary to search for innovative forms and to develop new methods of promoting ICC among the students, to find new ways of constructing educational courses, to refocus the attention to significantly important aspects in the educational process." It is only possible to achieve all this by building on the existing positive outcomes. It is a widely known fact that it is only possible to step into the future by looking back at the past experience [3]. Other scientists with similar cases are studying similar problems, using modeling and simulation [6–10].

Indeed, the world is changing, innovative technologies are in the lead and the efficiency criteria are excelling together with them. For these reasons, it is essentially important to develop ICC among the students. In conclusion, it is worth noting that, on the one hand, all people are different, on the other hand, the transition to an online format has made everyone equal across the globe. Now, there is a possibility to share knowledge and experience much quicker and by doing so to improve the quality and level of education.

References

1. Donina, I., Vodneva, S., Mikhailova, M.: Iskustvenny intellekt v sovremennom obrazovanii: vozmozhnosti i ugrozy (Artificial Intelligence In Modern Education: possibilities and threats). J. Pedagog. Psychol. **1**(57), 17–29 (2021)
2. Kostina, A.: Zifrovizatsiya obrazovaniya: chto my poluchaem i chto teryaem? (Digitalization Of Education: What Do We Gain And What Do We Lose?). Knowl. Underst. Skill. J. **1**, 52–64 (2021)
3. Rozhkov, G.: Interaktivniye master-klassy na inostrannom yazike kak effektivnaya forma podgotovki studentov k vneauditornoy rabote i sredstvo formirovaniya lingvisticheskikh i mezhkulturnikh kompetentsiy u studentov neyazikovikh vuzov (Interactive Workshops In Foreign Languages As An Efficient Tool Of Preparing Students For Extracurricular Activities And A Method Of Developing Linguistic And Intercultural Competences Among The Non-linguist Students). In: Professionally-oriented language teaching: current trends and future perspectives. Proceedings of the Annual Russian Scientific Conference With International Participants, pp. 132–138. Saint-Petersburg (2021).
4. United Nations Negotiating Delegation Homepage, 'UN Declares 30 September World Translation Day', June 20, 2017., http://www.members.aiic.net. Last Accessed 22 Feb 2022.
5. Global Case Study Challenge Homepage., https://www.globalcasestudychallenge.com. Last Accessed 30 Nov 2021.
6. Trivella, L., Dimitrios, N.K.: Knowledge management strategy within the higher education. The case of Greece. Procedia-Soc. Behav. Sci. **175**, 488–495 (2015)
7. Nasiopoulos, D.K., Ioanna, D., Lida, G., Paraskevi, P., Giakomidou, D.S.: Modeling of B2C communication strategies in electronic commerce. Strateg. Innov. Mark., 531–537 (2017)
8. Nasiopoulos, D.K.: Implementation stages and risks of logistics companies reengineering. Bus. Intell. Modell. **6**, 197–206 (2020)
9. Nasiopoulos, D.K.: Web search applications in Logistics companies in the context of evolution of Management Information Systems. Bus. Intell. Modell. **6**, 241–254 (2020)
10. Dimitrios, N.K., Sakas, D.P., Vlachos, D.S.: The role of open source leadership in developing high technology companies. Key Eng. Mater. **543**, 402–405 (2013)

Study of Self-Organization Issues in Virtual Network Communities

Naida O. Omarova

1 Introduction

In the twenty-first century, there have been fundamental changes in the perception of social reality, caused by the development of digital technologies and the penetration of the Internet into all aspects of public life. The globalization of the information space has led to the emergence and subsequent development of new social practices that were previously inaccessible.

The relevance of the topic of this study is due to processes developing in several directions. First of all, the development of technology leads to the emergence of new types of network communities, which, along with meeting the needs of users in communications, lead to the emergence of problems that have not been encountered before. Another important role of Internet communities is the generation and dissemination of certain knowledge and competencies, which makes the problem of the quality of transmitted information and the assessment of the consequences of the impact on society as a whole is relevant. The third problem is a new level of self-organization in society, due to which new opportunities appear, on the one hand, for human development, on the other hand, tools for manipulating the consciousness of participants in a virtual social network appear.

In Russia and the world, there are changes in the development of information technologies and systems that change the existing reality. The main role belongs not to the information itself, but to a greater extent the quality and speed of its transmission. At the same time, the speed of change is accelerating – what used to take decades can now take place within a year. Such dynamics of transformations is largely caused by the emergence of such new forms of communication as online

N. O. Omarova (✉)
Federal State Budgetary Educational Institution of Higher Professional Education Dagestan State University, Republic of Dagestan, Russia

D. P. Sakas et al. (eds.), *Computational and Strategic Business Modelling*,
Springer Proceedings in Business and Economics,
https://doi.org/10.1007/978-3-031-41371-1_50

communities, designed to respond to the challenges and have become a new type of social interaction implemented with the help of information networks.

The emergence of network communities is the result of the transition of an industrial society to a post-industrial one [1], when bureaucratic relationships smoothly turn into communication through social networks, as an integral part of the Internet space. The Internet network combines not only social networks, but also interconnected users using information technology to interact based on common needs and interests.

The definition of the concept of a network community by various authors has both common features and some differences. Konchakovsky understands a network community as a social community that exists on the Internet and is united by common values shared by its participants [2].

Taratukhina Yu.V. and Maltseva S.V. define the network community as a certain collection of people who have common goals, values and spatial reference points, communicating through cyberspace. Common goals, interests and needs, and common resources available to all members, as well as a characteristic type of discourse and thesaurus, act as unifying factors. E-mail, mailing lists, online conferences, Internet chats, forums, etc. are used as network forms of communication [1].

In addition, Patarakin E.D. considers joint network activity as a new social institution that satisfies social needs in the joint production of knowledge and has regulatory, integrative, broadcasting and communicative functions [3]. According to the author, such an institution establishes certain rules of the game and simplifies the decision-making process.

According to Kharchenko D.A., the network community is a group of people who have common interests and are in the process of information interaction through the Internet [4]. This community is a virtual entity, but at the same time its participants, as real people, have real interests, needs, preferences and motivations, which are determined by individual personality coloring and are of a social nature.

The transformation of society and the transition to the information society, according to M. Castells, is the result of the influence of not only information technologies, but also the cultural, moral, political, and institutional component [5].

Such changes would not be possible without the global Internet, which is the core of the network society and is in continuous development [6].

As mentioned above, the network community includes a group of people who engage in joint activities and interact through the Internet using networked computer technologies.

Such associations are guided by the following main principles [7]:

- Absence of barriers for communication
- The ability to communicate (exchange messages)
- The presence of social services (network software solutions that allow group interaction)

The development of information and technical means of communication, as well as the accumulation of experience in this area, leads to changes in the scientific, professional and educational fields. There are network communities in various professional fields that exchange knowledge with the involvement of their members in joint activities.

The widespread development of Internet-related technologies undoubtedly requires the development of innovative approaches that allow the full use of the capabilities of the global network, the development of a conceptual apparatus for describing and developing the processes of information and technological progress of the modern education system.

In this context, the global network can be considered not only as a set of software and hardware, but also as a social environment that includes network communities and their members. Taking into account the social aspect allows us to understand the motives that guide a person in the telecommunications sector and take them into account when developing appropriate services.

So, for example, a group of professionals unites in professional network communities, which allows specialists located in different parts of the world or country to solve professional problems through network communication and improve their level. As a rule, such communities are based on the following goals [7]:

- Formation of a single information space available to all members of the group
- Establishing professional interaction on formal and informal topics
- Organization of network interaction, for the subsequent transfer of communication beyond the Internet
- Exchange of experience and popularization of successful practices
- Coordination and support of various initiatives

The experience of communication in professional online communities allows participants to develop critical thinking and cultivate tolerance for other community members. Participation in online communities allows you to acquire several skills that are formed when interacting with other participants in communication [8]:

- Joint thinking, which is the result of collective cognitive activity, contributing to the realization of the need to move from an egocentric position to the recognition of the significance of other people's opinions.
- Tolerance and critical thinking, as the ability to perceive someone else's point of view and reconsider one's views on the problem.
- Willingness to expand the range of communication with people from different backgrounds.
- Development of decentralized models, because there is no need to gather in one place to solve the problem.

Thus, a "professional network community" can be understood as a form of activity involving the mutual exchange of knowledge between people with the same professional interests. The motive for participation in such a community may be the desire to improve skills and abilities in their professional field [9].

The object of research is network messages and methods used to analyze the processes that occur in them.

The purpose of this work is to model the activities of network communities based on the analysis of linguistic, social and structural components.

Possible tasks are set and solved:

- To analyze the essence of network communities, operations of work and creation of goals.
- Based on the analysis of popular typologies of community networks and users, a classification is carried out.
- Describe the main characteristics and components of the subjectivity of online communities.
- General population coverage by user research methods.
- To simulate a set of connections consisting of a graph and visualization using the Gephi package.

2 Material and Research Methods

Various methods of analysis related to the use of sociology, statistics and graph theory can be used to study network communities.

It is possible to analyze social networks within the framework of the directions [10]:

- Resource
- Regulatory
- Dynamic
- Structural analysis

The resource approach differentiates communication participants by resources and allows to determine the opportunities to attract their individual and network funds to achieve their goals. As resources, in this case, knowledge, financial situation, race, gender, etc. are considered. The network resources are the volume and nature of the information owned by the participant, as well as his influence and status. Each participant is assigned an indicator: the strength of the participant's structural position.

When using this approach, it is required to analyze the content of social networks, which serves as a data source for further analysis, solving problems of clustering and classification [11].

Normative analysis studies the rules and norms that affect the process of communication between members of online communities, as well as the level of trust between them. An analysis of the social roles associated with this edge of the network is carried out. There is also a study of changes in the behavior of participants depending on various situations that accompany the communication process. When conducting this analysis, the issues of modeling the spread of influence of some participants on others are solved, based on the information available about them.

Social networks contain a significant amount of personal information about its participants (part of the biography, interests, friends, photos, videos, etc.), so there is always the possibility of its unauthorized distribution in networks. This requires the modeling of processes for maintaining confidentiality in relation to the personal data of participants.

Often, in the process of communication, in the implementation of the functional role chosen by each participant, some of them can act as experts in any field, and the other part as brokers (leaders) – people playing the role of an intermediary in a social network, establishing relationships between users and experts. This happens due to the fact that many complex tasks require collective efforts to solve them, and it is easier to achieve the set goals by cooperating with each other.

The dynamic approach considers changes over time in the structure of social networks: the emergence of new participants or the termination of interaction with others, the emergence of new connections or the obsolescence of existing ones. Changes in the structure give rise to quite logical questions: what patterns lie in the long-term changes of large communities, whether there are any permanent configurations of network communities, how communities develop over time, what changes can be predicted.

It is the prediction of the formation of connections that is an important task in this method of analyzing social networks. To solve these problems, various relational and structural models are usually built [11].

In the structural approach, all network members are considered as graph vertices that affect the configuration of edges and other network members. Structural analysis considers all network participants in the form of graph vertices, between which there are relationships that affect the configuration of edges and other network participants. The focus is on the geometric shape of the network and the intensity of interaction. The main characteristics are chosen: mutual arrangement of vertices, centrality, and transitivity of interactions. To interpret the results in this direction, structural theories and theories of network exchange are used.

In order to determine the most active communication areas of the network, much attention is paid to network communities. For this, as mentioned above, a social network is represented as a graph and the mathematical apparatus of graph theory is used [12].

The main tasks of the analysis are [13]:

- Analysis of the behavior of communication participants
- Forecasting
- Recommendation
- Clustering
- Interaction with other resources
- Visualization

Depending on their main purpose, the following groups can be distinguished [14]:

- General purpose networks (Facebook, VKontakte, Odnoklassniki)
- Niche networks (LinkedIn, Comon)
- Content networks (Twitter, YouTube, Last.fm)

- Other networks – geosocial services (FourSquare), question services (StackOverflow, Quora), social Internet bookmarks (delicious), online games (World of Warcraft), etc.

In addition to the communication function, social networking services play the role of user databases, in which each user is associated with a set of personal information that makes up his "virtual identity." The data of all users of one service forms its social graph.

From the standpoint of data analysis, a social graph is a set of heterogeneous and semi-structured user information. Despite the fact that the target audience and the functionality of a social network largely determine the structure and content of a particular social graph, the following main types of data can be distinguished [14]:

- Network data (connection relationships between users, as well as between users and objects)
- User profiles (biography, views, interests of the user)
- Text data (messages, comments)
- Multimedia data (photo, video, audio materials)
- Network level objects (communities, applications)
- Objects of the outside world (links to resources outside the network)
- User activity logs (records about the interaction of users with each other and with various objects)

In general, as already mentioned, online communities consist of users who are interconnected by a certain type of relationship: friendship, study, family, interests, values, etc.

Graph theory is used to model social networks.

Consider the graph

$$G = (V, E),$$

where V is the set of graph vertices, E is the set of graph edges, and $|V| = N$ is the number of vertices in the graph.

The vertices of the graph denote users, and the edges show the presence of connections (relationships) between them.

Links in social graphs can be of two types:

- Non-oriented (corresponding to "friendship" type bonds with edges depicted in the form of straight lines connecting two vertices)
- Oriented (corresponding to links of the "subscription" type and the edges in them are depicted as arrows directed toward the movement of information flows)

If an edge connects one vertex, then such a connection is called a loop. An undirected graph without loops is an ordinary graph [15].

Two vertices connected by one edge are adjacent, and vertices and edges adjacent to them are incident to each other.

The set of vertices adjacent to the considered vertex in the graph is its neighborhood and is denoted by $N(x)$, and the number of vertices in $N(x)$ is called the degree of the considered vertex.

For clarity, when analyzing graphs, visualization tools are used. Graph visualization programs can be divided into:

1. Editors – applications that facilitate manual drawing when building a graph.
2. Programs that calculate the position of nodes and edges based on the input data and independently build a graph.

Gephi (https://gephi.org) is an open source software that can be used to visualize and explore all kinds of graphs and analyze related data. For it to work, Java must be installed on the computer at least version 7 [16].

GraphViz, GDF, GEXF, GML, and GraphML formats can be used for input data, as well as CSV files that can be obtained by converting Excel files, which greatly simplifies the work with the program.

The open license allows programmers to make their own changes to the program encoding and create additional modules, which contributes to the rapid distribution of the package.

Gephi provides the user with a wide range of options for customizing the display of graphs. This allows the end user to independently customize the visualization of graphs and adapt it for the purposes of the analysis being carried out. In addition, there is a Russian-language version of the program, which is convenient for Russian users.

To load the necessary source data into Gephi, you need to prepare two files:

- A file of graph vertices (nodes), which requires an ID field sufficient to build an undirected graph.
- File of edges (connections -edges), which should contain source fields (source – source, starting point) and destination (target – goals, points where the flow of information on the graph is directed).

3 Results and Discussion: Typology of Online Communities and Users

Despite the short period of existence of online communities by historical standards, the growth of their influence on all aspects of society is unprecedented. The number of participants in various forums, distance learning platforms, business services is constantly increasing, and there is also a continuous increase in the activity of citizens in social networks. All this makes the problem of studying and classifying network communities one of the priority tasks for our state, the solution of which determines its further security and socio-economic development.

Network communities, being complex social and technological systems, require an integrated approach to study, since they cannot be studied with the help of any single science.

At the moment, there are several criteria for typing Internet communities, reflecting their functional features and properties. Network communities, like many relatively small social groups, have their own goals, set certain tasks for themselves, adhere to some ethical code, develop regulatory requirements for themselves, and tend to stand apart from others. The most important typification criteria are: functional value, personalization and anonymity, publicity and closeness, scale, connectivity, hierarchy, statics and dynamics [17].

The functional significance depends on the tasks that the members of network communities set for themselves. According to this criterion, they can be divided into private, commercial, cultural and entertainment, thematic and professional Internet communities [17].

Online communities can also be personalized and anonymous.

The most important characteristic of network communities is the scale of their distribution. The communities with the largest scale include national segments of the Internet in a particular country, localized in their domain zones. Another unifying principle may be the language that can be used in different countries, such as English or Russian (for all Russian speakers).

From our point of view, the most complete classification is given in the works of D.A. Kharchenko, who believes that the entire information network is structured into substructures and separate segments, including communities [4].

Based on the above and based on current practice, we can conclude the following: users of online communities, in accordance with the above characteristics, have different intensity of communication in social networks. Some users create various groups, register with each other as "friends," that is, they are in the process of continuous communication, which is repeated many times. In contrast, the most passive part of users communicates with a limited circle of users, often such communication occurs only between two participants: such users can be about half of the total number.

4 Methodological Problems of Determining the Subjectivity of Online Communities

The development of information technology leads to the transformation of social media into the main channel of communication between different groups of users. Most of the social relations that were previously formed through direct communication are moving into the virtual space, leading to the emergence of social groups and generating intragroup and intergroup interaction.

In a virtual environment, effects that were previously characteristic of face-to-face intragroup and intergroup interaction begin to appear: the emergence of a leader who broadcasts his views to other participants; acceptance of certain norms of behavior; polarization of opinions [18]. As mentioned above, the network society in a simplified form is a group of people who communicate through social media. The main problem in the study of network communities is the correct choice of the object of study – as such can be:

- Individual participants, whose properties will characterize the community as a whole
- Social media
- Participants and social networks through which communication is carried out and impose certain conditions on interaction

Another problem is to identify the nature of interaction between community members who are not connected, as before, by a common territory of residence, do not have common material interests and resources necessary to maintain life.

Currently, members of online communities communicate in their own language, using the general user content available to them, moderated by the site administration. Social networking services support the process of communication between participants, making it possible to post notes, texts, videos and photographs, likes, etc.

The definition of an online community simply as a collection of people communicating through services provided by the Internet is not exhaustive, as it does not clarify the quantitative composition of users and their location. To solve this problem, methods used in the science of networks are used, when network communities are considered as groups interconnected by a large number of connections with representatives of other communities. The participants of the society are considered as nodes of the graph, and their communications with each other as its edges. In this case, network characteristics are considered as a priority, and individual characteristics are ignored.

Network methods of analysis are used to understand the structure of a network community, the nature of the relationship between its members, determine the types of activity, boundaries and status characteristics of community members.

Another methodological problem in the study of network communities is the problem of subjectivity, which is understood as a certain set of characteristics that are inherent in them and allow them to differ from other objects of study.

In work [19], based on the study, the components of the subjectivity of online communities are given.

Linguistic and conceptual identification consists of three components (discursive markers): generality of the semantic-thematic field; "password" words and expressions and identification of oneself and other members of the community. The generality of the semantic-thematic field consists in the use of common words, phrases, expressions, metaphors, speech stamps, terms and slang that are understandable to all members of the community when communicating. In the process of communication in many communities, "password" words and expressions are used, representing the words and expressions that are most often used in this network community. They symbolize belonging to a particular community, making a distinction between “us” and “them.” Identification of oneself and other members of the community is manifested in the use of words and expressions that reflect the involvement of participants in the community and emphasize unity (we, ours, together, etc.).

The next component of online communities is their willingness to act, which is divided into: calls for cohesion and a statement of unity; calls for actions and actions; emotional impulse to action; message about readiness for action. Calls to action and actions are words and phrases that encourage community members to take action that goes beyond the Internet. In this case, there is also an attempt to include community members in joint activities.

Another marker of this component is the emotional motivation to act, which in fact are emotionally saturated calls that affect the readiness for action and change the attitude to the situation under consideration. At the same time, inspiration and confidence are broadcast, emotional infection and attention are attracted. Categorical expressions and requirements are used (obliged, must, have to). A message about the readiness for action among specific members of the network community may arise as a result of the implementation of organizational and managerial actions on the part of other members and are largely a response to calls and suggestions. In this case, the readiness for joint actions, and not for single actions, is also demonstrated.

Discussion of topics of the network community. A topic in the Internet community is a topic that is currently under discussion, receiving feedback and comments. This component consists of the following markers: expressing the significance of the topics discussed; reference to other members and publications of the online community; broadcasting by community members of their attitudes to the topics under discussion (topics).

The expression of the significance of the topics discussed lies in the fact that the participants in communication recognize the topic they are talking about as significant and important, expressing interest and satisfaction with the discussion process itself. This demonstrates a positive attitude toward the community itself and the topics discussed.

Reference to other members and publications of the online community occurs when some members of the community quote or link to other members.

In the process of discussing important topics, community members can broadcast their attitudes to the topics discussed, when their own experiences, examples from life and stories on the topic are described.

Often, communication in online communities can only be aimed at maintaining contact without a certain semantic load, then humor and irony are used as the main theme.

The “group norms and values” component consist of four markers: creating the history of the network community and maintaining its traditions; formulation of communication norms; formulation of group values; collective coping.

Each network community has its own formed traditions that need to be maintained and committed to. This goal is served by creating a history of the community: stories about the community and its members; designation of the time, reasons and motivation for creating a community. At the same time, shared experience is discussed and references are made to it.

Members of the network community formulate communicative norms that are obligatory for all who participate in communication. Desirable and undesirable topics are declared, prohibitions are established on behavior that goes beyond the given norms, rules are established for resolving contradictions between participants. Along with communicative norms, group values and attitudes of the network community, shared by all participants, are formulated. They are constantly referenced during communication, if it is necessary to return the discussion to the accepted framework.

Collective "coping" of the online community means jointly overcoming difficult situations, formulating complaints and responding to them, discussing problems and methods for overcoming them.

Such a component of the subjectivity of online communities as "planning and results of joint actions" consists of the following markers: creation and formulation of the goals of a network community; communication on joint actions and results of joint activities; discussion of community plans; distribution of roles between participants.

The creation and formulation of the goals of the network community consists in declaring its main and current goals. The main goal, as a rule, is indicated in the name of the community. Current tasks are auxiliary, are tactical in nature and serve to fulfill the main goal.

The next marker of the component under consideration is a message about joint actions and the results of joint activities, which serves to achieve the set goals. In this case, there is a discussion between the participants of the achievements of the society and the presentation of a report on the joint work done with a description of the results obtained and summing up.

The community plans discussion marker includes such components as discussing future plans and adjusting them based on the opinions, suggestions and assessments expressed by the participants. Feedback is provided between the participants on the problem that has arisen and actions are planned to solve it. During the discussion, various points of view may be expressed, including approval, criticism and skepticism. At the same time, the participants of the society are continuously informed in what state the process of solving the problem is, what circumstances and obstacles arise in this case, whether there are changes and obstacles. To solve the problem, joint events are planned at a certain time and in a specific place.

The distribution of roles between the members of the society takes place with their consent, after which a specific task is set for each and the actions necessary for its implementation are determined. Decisions made and obstacles encountered at each stage are discussed among all community members.

The next component of the subjectivity of online communities is the "removal of others," which consists of such markers as: marking "others"; discrediting "others"; generalization of estimates; insult to others.

Marking "others" means that members of the community in their comments differentiate the social, cultural, ideological component in relation to any category (e.g., officials), separating them from themselves. This emphasizes the foreignness and strangeness of the “others,” which allows you to distance yourself from them and label them. When “others” are discredited, their significance is belittled with the use of depreciating and derogatory expressions, ironic generalizations. When generalizing assessments, concepts are generalized using passive constructions (e.g., everything is the same, they say, it is considered, etc.). When insulting “others,” appropriate words and expressions are used that serve to devalue the personality of the opponent.

The “positive communication support” component consists of the following markers: support for members of the network community; involvement in communication of all participants; irony and humor; addressing everyone with a specific request; conflict resolution within the community; support for formal rules of communication. The support of the members of the online community consists in showing the consent of the members of the online community with the statements of other members, expressing their approval and support. In this case, the vocabulary of consent (yes, ok) and a positive assessment is used. Support is also provided to those community members who have asked for it.

Involving all participants in communication consists in asking or inviting all community members to express their opinion on a problem or ask a question, meaning to receive an answer from community members.

Appeal to all participants with a specific request is usually used in the form of a request for help, support and complaint.

It is often necessary to use conflict resolution within the network community, for which expressions are used to reduce the emotional stress that has arisen and direct the discussion in a rational direction.

Support for formal communication rules implies the use of polite words, greetings, congratulations, and gratitude in the communication process.

The “protecting the integrity of the community” component includes the following markers: expression of a negative attitude toward the interlocutors; prevention of unwanted activity; excitation of distrust and hostility to “foreign” interlocutors; declaring the unity of the community through the opposition of “oneself” to “others.”

The expression of a negative attitude toward interlocutors is applied to those who are not a member of the community and do not follow the established rules of communication, do not share the values of the community.

Prevention of unwanted activity of other interlocutors is used when the latter do not comply with the rules established in the community, as well as when intimidation and threats are used by others. At the same time, expressions may be used that call not to perform certain actions (it is impossible, impermissible, etc.).

To outside interlocutors (“strangers”) who do not share the values and rules of the community, excitation of distrust and hostility can be applied, which is expressed in the use of depreciating, ironic expressions. Various labels are also used to identify the foreignness of strangers.

When declaring the unity of the community through the opposition of “themselves” to “others,” the unity of the participants in the network community is emphasized as opposed to "others." For this, the vocabulary of comparing "ours" with "them" is used.

The subjectivity component of online communities is "civil identity," which consists of three markers: manifestations of civic identity; discussions of social security; declaration of an active life position.

During discussions and discussions, members of the online community often declare and discuss their civil identity, for which some attribute can be used, for example, yellow vests, red armbands, etc.

The discussion of social security occurs in the process of planning by members of the network community to participate in solving any social issues that are of interest to everyone. This may be a discussion of the procedure for filling out a certain document or a newly adopted law, etc.

The declaration of an active life position consists in the expression by community members in the process of discussion of attitudes toward change and planning actions to improve existing realities.

5 Structural Analysis of Network Communities: An Example of Graph Visualization in the Gephi Software Package

As mentioned above, the network community is a collection of users who, using the Internet, incl. with the help of social networks, carry out the process of communication with each other. Communication can be carried out in several ways, it can be messages to each other, comments, sending audio and video files, subscribing as friends, etc.

In order to simulate the process of communication in online communities in the form of a graph, a table in CSV format was compiled that includes 50 users (ID-identifier for Gephi), of which five belong to the "friends" group (group). The intensity of communication between friends significantly exceeds communication between other members of the community (by seven times). Communication is understood as the process of communication in one and the other direction, when one of the participants is a source of information (source), and the other is its recipient (target) and vice versa. This situation quite accurately models an online community in which there may be communication groups, some participants communicate with all users, and some do not show much activity, communicating with one or more users.

After loading the data into Gephi, forming nodes and edges and choosing an undirected graph, an image was obtained. In an undirected graph, the very fact of communication is taken into account, and not the direction of information flows. For greater clarity, the stacking of nodes of the Fruchterman Reingold graph was chosen. After laying, we can preliminarily say that the star-shaped form of the graph indicates that communication does not occur between all participants in

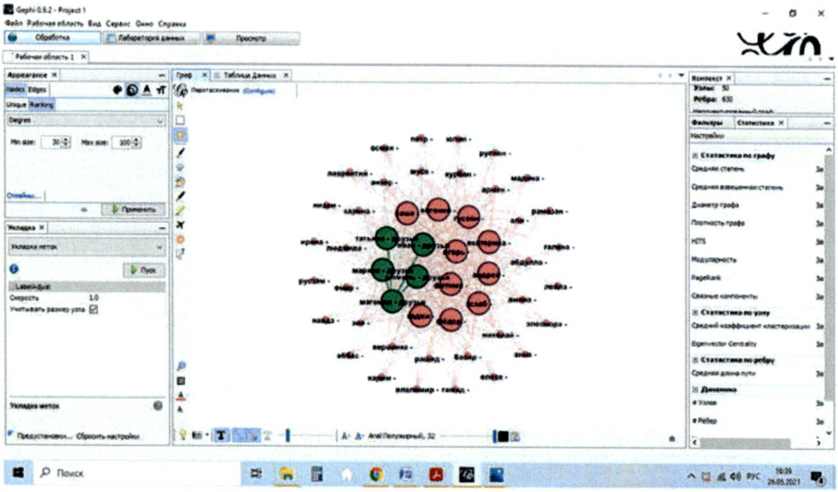

Fig. 1 Visualization of the social graph model in the Gephi software package

communication. The most intense communication occurs between nodes belonging to a group whose "friends" vertices are connected by darker edges. For even greater clarity, let's adjust the size and color of the nodes in accordance with their attributes. After adjusting the color and size of the nodes, we get the image shown in Fig. 1. As can be seen from the figure, the information content of the existing connections has increased and the structure of the graph itself has become more obvious. The "friends" group is highlighted in green, and the rest of the communication participants are highlighted in red. The size of the nodes is directly related to the intensity of communications – the more intense the communication between the nodes, the greater their diameter. Marking is carried out by pressing the black letter T, as a result of which we get the image shown in Fig. 1. After putting down the group, the image of the graph takes on its final form,

To give the graph its final form for subsequent export to PDF or PNG files in Gephi, there is a "view" tab, where the final editing and data adjustment takes place.

6 Conclusion

Modern network communities, having arisen relatively recently, have begun to actively enter into our everyday life and become an integral part of our lives. Such a situation inevitably requires the study of their essence, the basic concepts and methods of research associated with them [20].

A review of literary sources shows that the environment for the existence of network communities is the Internet and its integral part social networks. The network community includes users who turn it into a social entity, united by common values shared by its members. The number of participants in various forums, distance learning, business services is constantly increasing, and there is also a continuous increase in the activity of citizens in social networks. All this makes the problem of studying and classifying network communities one of the priority tasks for our state, the solution of which determines its further security and socio-economic development.

The study of the typology of communication participants makes it possible to identify the following users:

- Experienced users who freely navigate online communities and feel at home in them, as well as inexperienced users who are poorly oriented in the virtual space.
- Creative Internet users producing new content and passive users of the created content.
- Virtual aggressors and their victims – gullible and easily manipulated users.
- The younger generation, who feel comfortable in the world of digital communications and social networks, and the older generation, who have mastered social networks at a conscious age.

The study showed that users of online communities, in accordance with the above characteristics, have different intensity of communication in social networks. Some users create various groups, register with each other as "friends," i.e. are in the process of continuous communication, which is repeated many times. In contrast, the most passive part of users communicates with a limited circle of users, often such communication occurs only between two participants: such users can be about half of the total number.

The definition of an online community simply as a collection of people communicating through services provided by the Internet is not exhaustive, as it does not clarify the quantitative composition of users and their location. To solve this problem, methods used in the science of networks are used, when network communities are considered as groups interconnected by a large number of connections with representatives of other communities. The participants of the society are considered as nodes of the graph, and their communications with each other as its edges. In this case, network characteristics are considered as a priority, and individual characteristics are ignored.

Based on this, the main method of study on the topic of this study is the structural analysis of the network community using the social graph model and its visualization in the Gephi software package.

The conducted studies have shown that the identification of online communities is possible by analyzing the intensity of communication between a certain group of users. The intensity of communication between members of the community is always higher than with other participants in Internet communication. There are many users on social networks who communicate with only one person all the time and do not show interest in communicating with anyone else. Based on these considerations, an imitation of a network consisting of 50 people was carried out. (ID-identifier for Gephi), from which the group "friends" (group) was identified. When simulating, assumptions were made that the intensity of communication between friends significantly exceeds communication between other members of the community (seven times). The first 15 participants of communication communicate with each other and with all other participants, and the remaining participants with only one of the participants. These assumptions roughly reflect the patterns identified in the course of the study. Communication is understood as the process of communication in one and the other direction, when one of the participants is a source of information (source), and the other is its recipient (target) and vice versa. The visualization of the graph was carried out in the Gephi software package, into which data on nodes and links were loaded, generated in CSV files.

After loading the data, a graph image was obtained, which at the first stage is characterized by low information content. For the convenience of analysis and obtaining additional information, the following manipulations with the graph were carried out using the capabilities of the Gephi package:

- Fruchterman Reingold nodes were laid.
- Adjusted the size and color of nodes in accordance with their attributes: the size of nodes and edges is selected depending on the intensity of communication, and the color – on belonging to a group.
- Implemented labeling on the nodes.
- Using the "View" tab, the final adjustment of the graph image was carried out and it was prepared for export to PDF / PNG files if it is necessary to use it further.

Thus, the use of the Gephi package allows, after entering the initial numerical data on the nodes and edges of the network community, to visualize depending on the indicators of the communication process of its participants. Other scientists with similar cases are studying similar problems, using modeling and simulation [21–24].

References

1. Taratukhina, Yu.V., Maltseva, S.V.: Network communities: communication aspects. In: Automation and Modern Technologies, vol. 2, pp. 21–25 (2008)
2. Konchakovsky, R.V.: Networked Internet Community as a Sociocultural Phenomenon. Author's Abstract of Dissertation Candidate of Social Sciences (2010)
3. Patarakin, E.D.: Pedagogical Design of Joint Network Activities of Subjects of Education. Diss. Doctor of Pedagogical Sciences: 13.00.01 [FGBNU "Institute of Pedagogical Education and Adult Education of the Russian Academy of Education], St. Petersburg (2015)
4. Kharchenko, D.A.: Socio-philosophical Aspects of the Analysis of Internet Network Communities. Diss. Candidate of Philosophical Sciences: 09.00.11 [University of the Ministry of Internal Affairs of Russia], Krasnodar (2015)
5. Castells, M.: Materials for an exploratory theory of network society. Br. J. Soc. **51**(1), 5–24 (2000)
6. Klimenko, O.A.: Social networks as a means of training and interaction of participants in the educational process. In: Theory and Practice of Education in the Modern World: Materials of the International Scientific Conf. (St. Petersburg, February 2012), pp. 405–407. Renome, St. Petersburg (2012)
7. Saenko, L.A., Egorov, M.V.: Network society in the context of modern social transformations. Polythematic J. Sci. Publ. "Discussion" **7**, 88–93 (2014)

8. Chudaeva, E.V.: Professional Pedagogical Network Communities, https://nsportal.ru/blog/shkola/administrirovanie-shkoly/all/2012/01/06/professionalnye - pedagogicheskie-setevye. Last accessed 11 Oct 2020
9. Sergeev, A.N.: Education in the network communities of the Internet as a direction of informatization of education. Proc. Volgograd State Pedagogical Univ. **8**, 73–77 (2011)
10. Batura, T.V.: Models and methods of analysis of computer social networks. Softw. Prod. Syst. **3**, 130–137 (2013)
11. Aggarwal, C.C. (ed.): Social Network Data Analytics. Springer, Boston (2011)
12. Batura, T.V.: Methods of analysis of computer social networks. Bull. NSU. Ser.: Inf. Technol. **10**(4), 15–28 (2012)
13. Dyakonov, A.G.: Applied Problems of Data Analysis: Analysis of Social Networks. Moscow State University (2016)
14. Korshunov, A.V.: Studies of the Structure of User Communities in the Graphs of Online Social Networks. Dis. Candidate of Physical and Mathematical Sciences: 05.13.11 [FGBUN Institute for System Programming RAS], Moscow (2010)
15. Alekseev, V.E., Zakharova, D.V.: Graph Theory Textbook. Nizhny Novgorod. Nizhny Novgorod State University (2017)
16. Graph Methods of Analysis in Gephi. https://vc.ru/ml/216547-grafovye-methody-analiza-v-gephi. Last accessed 4 Mar 2021
17. Babushkin, A.G., Klimenko, O.A., Fedorov, R.Y.: Interdisciplinary approaches to the study of typology and connections of Internet communities. Actual Probl. Hum. Nat. Sci. **9**(56), 12–18 (2013)
18. Voronin, A.N.: Methodological problems in the study of the subjectivity of network communities. Psychol. Psychotechnics. **3**, 341–358 (2019)
19. Voronin, A.N., Kubrak, T.A.: Conceptualization of the phenomenon of subjectness of the network community using the method of grounded theory. Exp. Psychol. **13**(3), 4–14 (2020)
20. Omarova, N.O.: Formation of identity in the conditions of network development societies. In: Business Intelligence and Modelling, Springer Proceedings in Business and Economics, vol. 36, pp. 365–370 (2021). https://doi.org/10.1007/978-3-030-57065-1
21. Nasiopoulos, D.K.: Web search applications in Logistics companies in the context of evolution of Management Information Systems. Bus. Intell. Modell. **6**, 241–254 (2020)
22. Dimitrios, N.K., Sakas, D.P., Vlachos, D.S.: The role of open source leadership in developing high technology companies. Key Eng. Mater. **543**, 402–405 (2013)
23. Sakas, D.P., Nasiopoulos, D.K., Reklitis, P.: Modeling and simulation of the strategic use of social media networks in search engines for the business success of high technology companies. In: Sakas, D., Nasiopoulos, D. (eds), Strategic innovative marketing. IC-SIM 2017. Springer Proceedings in Business and Economics. Springer, Cham (2019). https://doi.org/10.1007/978-3-030-16099-9_28
24. Plessias, A.A., Nasiopoulos, D.K., Giakomidou, D.S.: Modeling of market segmentation in social networks and media. In: Kavoura, A., Sakas, D., Tomaras, P. (eds) Strategic Innovative Marketing. Springer Proceedings in Business and Economics. Springer, Cham (2017). https://doi.org/10.1007/978-3-319-33865-1_63

Part VIII
Core Workshop

Damianos P. Sakas and Dimitrios K. Nasiopoulos

Description: This workshop contains all the submissions that have not been assigned in other specific workshops. Upon the finalization of the Technical Program, submissions in the core workshop will be assigned to Presentation Slots according to their subject.

D. P. Sakas · D. K. Nasiopoulos
Agricultural University of Athens, Athens, Greece

The Impact of Stakeholder Engagement and Relationship Marketing on Aviation Market in Southeastern Europe

Christos Christopoulos and Susan Kinnear

1 Introduction

This chapter examines the challenges of the regulatory business environment in aviation and evaluates their impact on the Southeastern European market. The research approaches the subject from both a theoretical and a practical perspective. The study endeavors to fill research gaps and contribute to existing knowledge by analyzing business and management issues in a specific aviation market, providing insights, and offering practical recommendations to enhance organizational performance.

The main challenges relevant to this environment are identified and considered within the literature review, which presents a theoretical framework of systems theory, stakeholder theory and the relationship marketing concept. The research methodology provides a rationale for the research approaches deployed before the findings from data collection are thematically presented and analyzed to identify their contribution to answering the research aim and objectives. The conclusion then discusses the research results, considers the application of theoretical models in aviation, and suggests practical recommendations for organizations and potential options for future research.

C. Christopoulos (✉)
Global Operations Transformation, Swissport International Ltd, Zurich, Switzerland
e-mail: christos.christopoulos@swissport.com

S. Kinnear
Kaplan Open Learning, University of Essex, London, UK
e-mail: susan.kinnear@kaplan.com

D. P. Sakas et al. (eds.), *Computational and Strategic Business Modelling*, Springer Proceedings in Business and Economics,
https://doi.org/10.1007/978-3-031-41371-1_51

2 Background and Theory

2.1 *The Main Challenges*

The purpose of this paper is to identify the main regulatory challenges of the aviation business environment, considering the roles of regulation, deregulation, and regulated employment.

Regulations are authoritative mechanisms that control obedience and compliance and set the grounds to standardize business conduct. In aviation, regulation is primarily related to flight safety [1], certification [2], process uniformity [3], and market-entry rights [4]. *Deregulation* is broadly defined as the tendency to eliminate state-regulatory barriers in the market. Privatization in aviation has affected market access, designation, capacity, tariffs [5], and the overall financial and investment dimensions of the business [6]. *Employment* globalization allowed aviation businesses to reorganize their HRM strategies, targeting mainly labor costs and inflexible work habits [7]. EU aviation has adopted in its HR practices new labor schemes (i.e., subcontracting and outsourcing) [8], and non-standard employment types (i.e., temporary agency work) [9].

The study examines the impact of the regulatory challenges in Cyprus and considers that similar conditions broadly apply in the aviation industry in South-eastern Europe.

2.2 *The Theoretical Framework*

Responsiveness to Business Environment. Effective responsiveness and adaptation to dynamic business environments are complex and multidimensional topics, which relate to the generic concept of change management. In evolutionary change where transformation occurs through environmental adaptation, Clegg et al. [10] recommend that a systemic approach could better analyze the nature of the challenges and enable the exploration of new business models. Responsiveness and adaptation require integrative frameworks, which would be based on either an "input-process-output" model [11] or would be supported by interconnections of involved parties [12] and cultural sensitivity approaches [13] on a local market environment level. Change management, in this study, is assessed from a responsiveness and adaptation perspective and not from a leadership viewpoint. To facilitate this, the following theories are used to build a theoretical framework for analysis:

(a) *General Systems Theory*, as a driver for system thinking in managing adaptation.
(b) *Stakeholders Theory*, in defining key-players partnership relations in business.
(c) *Relationship Marketing*, as a central concept in contemporary consumer markets.

General Systems Theory (GST) and System Thinking. Organizations depend on their business environment to function; therefore, GST provides a helpful conceptual framework to analyze how organizations meet the demands, respond, and adapt to the challenges of dynamic environments.

GST developed in the early 1950s as an inter-disciplinary tendency to integrate natural and social sciences and has its origins in Aristotle's holism (the whole is more important than its parts). GST shifts the focus from the specific to the whole [14] and promotes a system thinking concept with principles applicable to all systems, as Von Bertalanffy [15] argued. Organizations are systems interacting with their environments in cyclical phases: receiving inputs, internally processing throughputs, releasing outputs back, and pursuing feedback, according to Katz and Kahn [16].

The effectiveness of these cycles is related to open organizational behavior. When there is inflow and outflow, the system is an open system (OST) and when there is zero flow the system is a closed system. Only open systems maintain homeostasis (consistency) during the adaptation and transformation phase [15].

Internal processing is central in GST because it is the process of changing component materials [15] or leading transformation in dynamic environments [14], and therefore reflects the organizational ability to adapt to changes caused by environmental effects [17]. Feedback is an essential benefit of GST because it provides a mechanism for mediating between the goal and the behaviors of the system [17] and provides information on how the organizations function [16]. Thus, feedback enables the correcting of malfunctions.

Stakeholders Theory (SH) and Value Creation. Stakeholders are groups, parties, or individuals who directly or indirectly affect or are affected by firms' market activities.

SH identifies stakeholder groups of direct relevance or of broad involuntary participation and perceives business as a set of relationships between internal and external stake-owners, aiming to understand how they interact to jointly create value and trade it [18]. Owners, investors, managers, and employees are considered "internal stakeholders". Suppliers, customers, government, society, media, and pressure groups are the "external stakeholders".

In practice, multiple stakeholders of each business environment participate in forums, committees, and working groups to improve a common understanding of issues and to develop mutual-interest understanding and community-based culture [19]. SH is considered as a dynamic model to enhance stakeholders' engagement for sustainable business [3]. This social partnership perspective considers SH management playing the role of developing the social contract, combining business with ethical behavior in market environments [10].

SH provides three distinct benefits: descriptive accuracy to observe what is done, instrumental power to review outcomes and normative validity to define changes [20]. Effective management of stakeholder relations establishes positive links between groups with mutual interests and creates value for an organization's strategy [18]. Moreover, SH offers the advantage of connecting value creation with sustainability, because it considers marketing, management, development, and financial processes under a sustainability-based mindset [21].

Relationship Marketing Concept (RM). The relationship marketing concept supports the SH approach as it considers relationships from a marketing perspective. Kotler and Keller [22] *argue that RM can be defined as how external and internal interactions influence an organization's activities and impact the effectiveness of the entire market network. These interactions are seen as value exchanges, which affect: a) the internal relationships of business units, b) the external interaction with partners, providers, suppliers, and distributors, and c) the organizational dealing with regulatory bodies, media, communities, and society at large.*

RM approach has redefined the traditional marketing-mix concept and promoted a marketing-dimension model instead [23], with three specific elements: (a) internal marketing (values, vision, and strategy), (b) external marketing (traditional 4Ps and branding), and (c) interactive marketing (relations and interactions). In a market network, there is a wide range of marketing relationships and interactions which interrelate; however, networks can only flourish when mutual benefits exist [24] because the involved parties expect to obtain positive value and maintain successful relational exchanges [25].

The relational-centric system concept of RM therefore provides performance benefits to all participants when their relational exchanges are based on mutual interests [26] and offers advantages in gaining market knowledge and increasing the understanding of capabilities, resources, objectives, and requirements [22], amplifying the potential to generate higher returns in conditions of environmental uncertainty [26].

2.3 Purpose of this Study

This study focuses on a complex business issue and seeks to address three objectives:

- To identify how contemporary regulatory challenges effect European aviation
- To assess which theoretical models can assist firms' response to market demands
- To evaluate how organizations can enhance their performance in this environment.

To this effect, the research considers the regulatory dimensions of aviation (regulation, deregulation, and regulated employment) in combination rather than individually. It examines the suitability of theoretical models to facilitate an organization's response to the challenges and regards aviation as a market-network system of interactions, open to value creation and offering growth opportunities. It further aims to recommend approaches to help organizations adapt to market demands and improve performance.

3 Methodology

Interpretivism has been selected, because the strength of this research philosophy is to enhance and extend the understanding of the social phenomena and can generate meanings from the observations [27] *to support explanations on "the how and the why"* [28]. *Additionally, interpretivism supports in-depth analysis and enables clarifying notions and behaviors* [29].

Abduction has been selected as the appropriate research approach because is a less standardized tactic and combines both empirical and rational reasoning. Abduction is sufficiently flexible to handle supplementary information, to drive different interpretations and to tolerate conclusions with alternative approaches [30].

A qualitative research design was chosen because enables drawing interpretations from social actualities and facilitates understanding of the main variables when non-numeric data is processed [27]. *Qualitative methodology is linked to the subjective approach* [31] *and focuses on participants' viewpoints and insights, highlighting the denotation of words and notions, and measuring data in natural business settings* [29].

The research has collected its own primary data. The benefits of collecting primary data are the precise fitting to a specific business problem, and the exclusivity of raw data [32]. *Data is collected through semi-structured interviews, combining the advantages of both structured and unstructured approaches* [29], *asking open-ended questions* [33], *and seeking supplementary viewpoints of informants on related topics* [34].

Convenience/purposive sampling was deployed because leverages access opportunities, enables to selection of information-rich participants, and considers restrictions in resources and time [35]. *Sampling choice targeted informants who are specialists in their field, to benefit from their expertise and their capacity to share insights. Therefore, have conducted one-to-one interviews with the Senior Air-Transport Officer of Civil Aviation in Cyprus, the Manager of the Airport Operator at Pafos, and the District Head of Unions in Cyprus, considering their ability to best provide information on regulations, deregulation and regulated employment.*

On the ethical side, confidentiality, anonymity, and data protection were safeguarded and participants' willingness and consent to contribute were preserved [36] *before any interaction with the participants was recorded.*

Thematic analysis methodology was used due to its flexibility to incorporate any research tactic and its ability to analyze any size of the dataset for all qualitative designs [37]. *The thematic analysis identifies data-driven categories (themes)* [27], *based on transcripts' codification* [28], *and facilitates sense-making and open interpretation* [38].

4 Findings: Results

The effectiveness of the results is linked to two key processes of raw data collection: analysis and interpretation. Analysis intends to extract meaningful information from raw data and is used to manipulate it, to enable thorough understanding and to clarify problems [29]. Interpretation facilitates the sense-making of raw data and supports exploring new explanations of social phenomena [27]. The interviews offered

Market Environment	Stakeholders Environment	Effects on the Business
Market (MRK) Open Market (OMK) Labor Market (LMK)	Stakeholders (STK) User (USR) Relationship (RLT) Partnership (PRT)	Product (PRD) Profit (PRF) Cost (CST) Investment (INV)

Fig. 1 Overview of themes and codes

insights into three areas: regulatory requirements, the deregulation process, and employment practices.

The thematic analysis enabled the classification of the main thematic categories from the transcripts, management and labeling of data, and codification of the interrelated items which help to answer the research questions. In qualitative research, it is crucial to structure the data units, to observe their patterns and relations [27] and to apply selective data-driven coding (with in-vivo abbreviations) [29]. The key common themes: the market environment, the stakeholders' environment, and the effects on the business, were therefore codified (see Fig. 1).

This study focuses on the analysis of those findings which relate to the *effects on the business*. The findings are presented thematically and by order of topic.

Regulatory Requirements in Aviation. The informant of the civil aviation authority referred to the effects of the regulations on the aviation business in Cyprus.

The participant highlighted that the regulatory requirements affect the product. *The role of the regulation in aviation is to ensure consistency of the activities, uniformity of passengers' processing, and to monitor compliance of the product with safety, security, environmental and fair-competitiveness requirements. Regulation also affects the* profit *because has an important role in monitoring ticket fares and tariffs, airport taxes, and servicing fees. The latter gives the opportunity to the providers and the users to predict their* costs; *however, regulatory requirements would also have negative impact on the costs of all stakeholders. Participant argued that regulations can influence the* investment *because restrictions may reduce investment potential and diminish the market. Thus, regulatory restrictions should not discourage investors; should rather accommodate investment in the industry. The participant contended that effective regulations should be designed based on mutual interests and should facilitate and promote business growth for the common benefit of all participants in the market.*

Deregulation in Aviation. The manager of the airport operator concentrated on the recent (2006) liberalization process at Cypriot airports and its impact on the business.

The participant outlined that deregulation is influential for the product *because it is driven by the satisfaction of the end-users' demands and is targeting to improve the terminal facilitation for the travelers and for the operating agencies. Such product development requires collective approach and partnership spirit by all participants to the market. The informant highlighted that airports' liberalization process has involved private entities and shifted the focus of the firms on the* profit *and on the revenue increase. Participant argued that operational workflows are affecting the costing basis of the firms and therefore improvement of the financial results also requires attention on the* costs. *Informant observed that cost efficiency has increasingly attracted low-cost carriers over the past years. The Participant underlined that when private entities invest in a business, they expect a positive return of their* investment *and noted that return of investment is the key attribute of privatization. Participant stressed that deregulation has enhanced the engagement of the stakeholders, has altered the nature of the relations between carrier and user, and has increased the attractiveness of the market.*

Regulated Employment in Aviation. The interview with the head of the trade unions was centered on the impact of employment on business conduct in Cyprus.

The participant denoted that employment relations have a direct impact on the product because labor peace allows organizations to develop their product efficiently and to be competitive. The participant indicated that satisfied staff focuses on the traveler and remains concentrated on the product. Furthermore, participants argued that businesses cannot survive without profit and labor stability in aviation workplaces can contribute to profitability improvement. The profit involves also the intensive focus of the managers on the control of the costs, which is the top priority when companies are managed rationally. The participant argued that, increasingly in the past years, unions have negotiated flexible employment conditions and have supported resource optimization practices. The participant outlined that trade unions acknowledge shareholders' investment efforts and have promoted workforce stability initiatives to allow confidence for investors. Finally, the participant contended that employment should be protected because employees are important stakeholders in business conduct and proposed that organizations should consider a partnership perception, in which employers would share part of the profit with employees to establish long-lasting and win-win relations.

5 Discussion

5.1 Discussion on the Results

Overview of the Results. The participants share a common understanding of the effects on the business and refer mainly to the impacts on product, profit, costs, and investment. Defining the key focus areas in the aviation business in Cyprus is an important contribution for the organizations, who should consider their main challenges: a) the product is regulated and at the same time seeks efficiency and development; b) the profit is impacted by regulatory and labor restrictions; c) operating costs are affected by processes when resource optimization is not possible; and d) the investment is impacted when investors cannot effectively project profitability.

The participants, regardless of their background, each identified positive and negative links to the business effects. Organizations should benefit from positive links: e.g., regulation and market accommodation; deregulation and market attractiveness; labor relations and workforce stability. Companies should refrain from the negative links: e.g., regulatory restrictions and investment; operational processes and cost base; harmful impacts on labor conditions.

The participants referred to the key stakeholders of aviation in Cyprus: regulators, airport operators, providers, travelers (end-users), investors, managers, employees, and pressure groups (unions). Organizations should consider *the different roles and the nature of their relations. The end-user role is central in defining the product demands and in specifying the scope of process development. Strong relations with the rest participants in the market are*

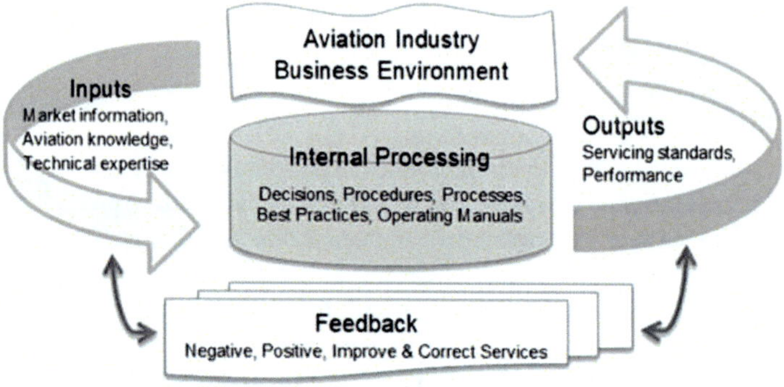

Fig. 2 Systems theory in aviation

crucial because they promote mutuality of interests, and they support a partnership perception in a framework of a joint-benefits spirit [10].

Applying Systems Theory and Systems Thinking. GST provides an appropriate conceptual approach to respond to the market environment for three reasons.

First, organizations could consider aviation business in Southeastern Europe as a cyclical system of interactions, which receives inputs from the environment, processes throughputs internally, releases back outputs, and seeks feedback on the effectiveness of its outputs (see Fig. 2*). Feedback enables businesses to identify service errors and inefficiencies and to understand how the product could be improved and developed.*

Second, companies could consider the benefits of open systems theory in the service industry to determine their relationship with the environment based on the openness in processing inflows and outflows [15]. Organizations could reflect the key assumption that closed systems remain static, while open systems allow flows and promote flexibility and dynamic responsiveness to business requirements. Organizations in Southeastern Europe could use inflows to respond to the challenges of deregulation and to adapt to the demands of the users – in line with the regulatory requirements.

Third, firms could use the GST framework to develop an organizational culture around the two key principles of holism and homeostasis. The holistic principle shifts the focus from the specific to the whole and allows organizations to prioritize system thinking in decision-making and problem-solving [17]. The homeostasis principle promotes consistency and balance. Adoption of homeostasis could assist businesses in maintaining consistency and in establishing sustainable goals related to key effects on the business: the product, the profit, the costs, and the investment.

Applying Stakeholders Theory and Value Creation. SH enables organizations to effectively interact with their stakeholders and to create value from this interaction [18].

SH provides two common concepts for the interactions of the stakeholders in Southeastern European aviation: mutuality of interests [19] *and stakeholder engagement* [3]*. Participants noted that mutuality of interests enables win-win cooperation and understanding of others' viewpoints. Engagement defines the level of commitment to interact with the remaining stakeholders. Engagement enhances a collaborative approach when stakeholders have*

Stakeholder	Role in Value Creation & Trading
Regulators	*Compliance Monitoring, Process Consistency, Activities Uniformity*
Airport Operators	*Process Development, Market Attractiveness, Market Growth*
Carriers	*Product Development, Market Attractiveness, Traffic Growth*
Providers	*Process Efficiency, Product Development, Productivity*
End-Users	*Product Definition, Process Facilitation, Cost Efficiency*
Investors	*Investment and Return of Investment, Profitability*
Managers	*Cost Efficiency, Productivity, Optimization*
Employees	*Labour Stability, Workforce Satisfaction*
Unions	*Employment Protection, Labour Peace*

Fig. 3 Aviation stakeholders and value creation

Fig. 4 How the benefits return to the market network

common interests in product development, profit increase, cost efficiency, and investment potential, and builds a spirit of win-win cooperation as the participants recommended.

Value creation is a key attribute of SH. Organizations could use the value creation concept to comprehend how engagement and mutual interests could enable creating value for all participants in the Southeastern aviation market. Stakeholders contribute individually to value creation by focusing on specific areas of interest within the sector. They may also contribute to the creation of more than one value, or multiple values [18]. For example, the airport operator mainly focuses on creating value by developing the processes; however, value in the process could also originate from end-users (for facilitation), from providers (for efficiency) and from regulators (for consistency). This example could be used to examine which of the values are interlinked and which stakeholder would better represent and trade in each value (see Fig. 3).

Applying Relationship Marketing. RM is mainly related to the business effects and suggests that the relational exchanges of participants could benefit their performance.

RM approaches could be a useful tool for aviation businesses because it plays a key role in using relational exchanges to link market activities to the end result [23] *and benefit multiple market participants. The example below (see Fig. 4) outlines the strategic advantages of applying the RM to the Southeastern European aviation market. It demonstrates how one type of positive relational exchange could positively also affect the related market partners and return the benefits to the entire market network* [22].

Aviation businesses could use the RM concept to identify their market network: customers, partners, and employees [22]. The interview participants reconfirmed RM's validity in the aviation market network, emphasizing the critical role of users, providers, investors, and the workforce. The RM perspective could also assist organizations within this relational-centric system to define the nature of the relations, as participants proposed in the cases of the relationship between carrier and user and employer and employee. Based on the quality of the relational exchanges, organizations operating in the aviation business in Southeastern Europe could develop a relational behavior approach in viewing mutual benefits and in building strong and long-lasting relations. The RM concept also enables aviation businesses to translate those benefits into a value exchange, providing performance advantages for both the product and the market.

5.2 *Recommendations*

The discussion triggers practical recommendations for firms' market performance.

First, companies should consider the challenges of the market environment combined (not isolated) and should relate them positively to organizational performance, because as per participants' views: (a) regulation can facilitate growth, (b) deregulation can increase market attractiveness, and (c) labor stability can improve profitability. Similar studies have been done in the past by other scientists [39, 40].

Second, organizations could enhance their performance through the development of a systems-thinking culture. Systems thinking provides key adaptive strengths, i.e., it: (a) translates feedback into corrective actions and best-practices development, (b) responds effectively to market demands and changes, and (c) approaches problem-solving systematically by viewing the big picture. Other scientists with similar cases are studying similar problems [41, 42].

Third, aviation businesses in Southeastern Europe should comprehend that simply dealing with stakeholders is not enough. Performance improvement strategies require partnership perception, engagement, and win-win collaboration with stakeholders of mutual interests. Additionally, value-centered interactions would enable creating and trading value for all market participants.

Fourth, companies should acknowledge the significance of relationships in contemporary market networks. The relationship marketing concept allows organizations to explore the benefits of strong and long-lasting relationships and, when it is managed strategically, could facilitate product development, market growth, enhanced performance, and financial returns. Other scientists with similar cases are studying similar problems, using modeling and simulation [43, 44].

This research focuses on participants' insights related to the effects on the business. Future research would discuss in detail the themes of market environment and market openness, stakeholders' environment and relations, and the potential to implement the profit-sharing concept in the aviation business in Southeastern Europe.

6 Conclusions

This chapter identifies the regulatory challenges of the aviation market environment in Southeastern Europe and endeavors to resolve the implications of the business and management problem with the support of theoretical frameworks. The study concludes that regulators, operators, and pressure groups are affecting the aviation business environment and can challenge the performance goals of the companies. Considering regulatory challenges combined enables firms to realize the benefits of market development, growth opportunity, and workforce stability, which are offered by regulation, deregulation, and regulated employment respectively. Regulatory challenges in the Southern European air transportation industry are affecting the market environment, the stakeholder environment, and the performance outputs.

Therefore, aviation businesses in Southeastern Europe should develop a systems thinking approach to respond effectively to the demands of their dynamic business environment. Moreover, companies should strategically focus on developing a partnership perception, based on the mutual interests of the stakeholders for value creation. Globalization trends in market networks require strategic management of relationships to establish relational exchanges and enhance their performance.

References

1. Yadav, D.K., Nikraz, H.: Implications of evolving civil aviation safety regulations on the safety outcomes of air transport industry and airports. Aviation. **18**(2), 94–103 (2014)
2. Sandi, K., Tomaž, K., Iztok, P.: Safety Management Process in Civil Aviation. No. 26 Int'l J. Econ. & L., vol. 9, p. 81 (2019)
3. Amaeshi, K.M., Crane, A.: Stakeholder engagement: a mechanism for sustainable aviation. Corp. Soc. Responsib. Environ. Manag. **13**(5), 245–260 (2006)
4. Lovelock, C.H., Yip, G.S.: Developing global strategies for service businesses. Calif. Manage Rev. **38**(2), 64–86 (1996)
5. Odoni, A.: The international institutional and regulatory environment. In: The Global Airline Industry, pp. 19–46. John Wiley & Sons Ltd., Chichester (2009)
6. Starkie, D.: The Airport Industry in a Competitive Environment: A United Kingdom Perspective. OECD Publishing, Paris (2009)
7. Gillen, D., Morrison, W.G.: Regulation, competition, and network evolution in aviation. J. Air Transp. Manag. **11**(3), 161–174 (2005)
8. Jorens, Y., Gillis, D., Valcke, L., De Coninck, J., Devolder, A., De Coninck, M.: Atypical Forms of Employment in the Aviation Sector, European Social Dialogue. European Commission (2015)
9. Helfen, M., Sydow, J., Wirth, C.: Service delivery networks and employment relations at German airports: Jeopardizing industrial peace on the ground? Br. J. Ind. Relat. **58**(1), 168–198 (2020)
10. Clegg, S.R., Kornberger, M., Pitsis, T.: Managing and Organizations: An Introduction to Theory and Practice, 4th edn. Sage, London (2019)
11. West, M.A., Hirst, G., Richter, A., Shipton, H.: Twelve steps to heaven: successfully managing change through developing innovative teams. Eur. J. Work Organ. Psychol. **13**(2), 269–299 (2004)

12. Attwood, M., Pedler, M., Pritchard, S., Wilkinson, D.: Leading Change: A Guide to Whole Systems Working. Policy Press, Bristol (2003)
13. Gladden, R.: Culturally Tuning Change Management. CRC Press, Boca Raton (2018)
14. Mele, C., Pels, J., Polese, F.: A brief review of systems theories and their managerial applications. Serv. Sci. **2**(1–2), 126–135 (2010)
15. Von Bertalanffy, L.: An outline of general system theory. Br. J. Philos. Sci. **1**, 134–165 (1950)
16. Katz, D., Kahn, R.L.: The Social Psychology of Organizations, vol. 2, p. 528. Wiley, New York (1978)
17. Chen, D., Stroup, W.: General system theory: toward a conceptual framework for science and technology education for all. J. Sci. Educ. Technol. **2**(3), 447–459 (1993)
18. Parmar, B.L., Freeman, R.E., Harrison, J.S., Wicks, A.C., Purnell, L., De Colle, S.: Stakeholder theory: the state of the art. Acad. Manag. Ann. **4**(1), 403–445 (2010)
19. Murphy, G., Efthymiou, M.: Aviation safety regulation in the multi-stakeholder environment of an airport. J. Air Transp. Stud. **8**(2), 1–26 (2017)
20. Donaldson, T., Preston, L.E.: The stakeholder theory of the corporation: concepts, evidence, and implications. Acad. Manag. Rev. **20**(1), 65–91 (1995)
21. Hörisch, J., Freeman, R.E., Schaltegger, S.: Applying stakeholder theory in sustainability management: links, similarities, dissimilarities, and a conceptual framework. Organ. Environ. **27**(4), 328–346 (2014)
22. Kotler, P., Keller, K.L.: Marketing Management, 14e, Global Edition. Pearson (2012)
23. Wilson, R.M., Gilligan, C.: Strategic Marketing Management, 3rd edn. Elsevier Butterworth-Heinemann, Oxford (2005)
24. Gummesson, E.: Making relationship marketing operational. Int. J. Serv. Ind. Manag. **5**(5), 5–20 (1994)
25. Peterson, R.A.: Relationship marketing and the consumer. J. Acad. Mark. Sci. **23**(4), 278–281 (1995)
26. Palmatier, R.W.: Relationship Marketing, pp. 1–140. Marketing Science Institute, Cambridge, MA (2008)
27. Saunders, M., Lewis, P., Thornhill, A.: Research Methods for Business Students, 8th edn. Pearson, London (2019)
28. Bell, E., Bryman, A., Harley, B.: Business Research Methods, 5th edn. Oxford University Press, Oxford (2018)
29. Ghauri, P., Grønhaug, K., Strange, R.: Research Methods in Business Studies, 5th edn. Cambridge University Press, New York (2020)
30. Awuzie, B., McDermott, P.: An abductive approach to qualitative built environment research: a viable system methodological exposé. Qual. Res. J. **17**(2), 356–372 (2017)
31. Holden, M.T., Lynch, P.: Choosing the appropriate methodology: understanding research philosophy. Mark. Rev. **4**(4), 397–409 (2004)
32. Hox, J.J., Boeije, H.R.: Data collection, primary versus secondary. In: Encyclopodia of Social Measurement, pp. 593–599. Elsevier, Amsterdam (2005)
33. Adams, W.C.: Conducting semi-structured interviews. In: Handbook of Practical Program Evaluation, vol. 4, pp. 492–505. Jossey-Bass, San Francisco (2015)
34. Zhang, Y., Wildemuth, B.M.: Unstructured Interviews. Applications of Social Research Methods to Questions in Information and Library Science, pp. 222–231. Libraries Unlimited, Exeter (2009)
35. Etikan, I., Musa, S.A., Alkassim, R.S.: Comparison of convenience sampling and purposive sampling. Am. J. Theor. Appl. Stat. **5**(1), 1–4 (2016)
36. Ellis, C.: Telling secrets, revealing lives: relational ethics in research with intimate others. Qual. Inq. **13**(1), 3–29 (2007)
37. Clarke, V., Braun, V.: Thematic analysis. In: Encyclopedia of Critical Psychology, pp. 1947–1952. Springer, New York (2014)
38. Maguire, M., Delahunt, B.: Doing a thematic analysis: a practical, step-by-step guide for learning and teaching scholars. All Ireland J. High. Educ. **9**(3) (2017)

39. Nasiopoulos, D.K., Sakas, D.P., Vlachos, D.S., Mavrogianni, A.: Modeling of market segmentation for new IT product development. AIP Conf. Proc. **1644**, 51–59 (2015)
40. Plessias, A.A., Nasiopoulos, D.K., Giakomidou, D.S.: Modeling of market segmentation in social networks and media. In: Strategic Innovative Marketing, pp. 523–529. Springer, Cham (2017)
41. Foteini, P.E., Dimitrios, N.K., Sakas, D.P., Vlachos, D.S.: The development of new ideas for IT products, through social media. In: Strategic Innovative Marketing, pp. 399–404. Springer, Cham (2017)
42. Nasiopoulos, D.K.: Build and approach of logistics companies reengineering. Modelling and simulation of the problem. In: Business Intelligence and Modelling, vol. 6, pp. 189–196. Springer, Cham (2021)
43. Sakas, D.P., Nasiopoulos, D.K., Reklitis, P.: Modeling and simulation of the strategic use of social media networks in search engines for the business success of high technology companies. In: Business and Economics, pp. 227–236. Springer, Cham (2019)
44. Plikas, J.H., Nasiopoulos, D.K., Giakomidou, D.S.: The contribution of social media in the management of social relations with customers. Modelling and simulation of the problem. In: Strategic Innovative Marketing, pp. 515–521. Springer, Cham (2017)

Exploring the Relations Between the Information System Success Model and Trust in E-Government in the Greek Public Sector During the Pandemic: An Empirical Research

Maria Athanasiadou, Anastasios Diamantidis, Thomas Fotiadis, Antonios Gasteratos, and Leonidas Hatzithomas

1 Introduction

The COVID-19 pandemic has forced many public and private organizations to undergo major transformations, rethinking the basics of their business processes and using technology to maintain their operations, while following a changing landscape of guidelines and new processes.

Through the last decade, more and more countries have gradually started providing online (digital) services to their citizens. Major financial scandals and the negative image that has been created around the concept of public administration have created an imperative need for organizations involved in public administration to become competitive and to operate in a regime governed by transparency, meritocracy, and credibility. The changes (both positive and negative) that have arisen due to the pandemic have led to the rapid advancement of e-government, with the provision of online digital governance. The transition to the online epoch for public services is essential on all levels: local, national, and international. This transformation to online services will contribute to the economic development of the public sector. It will also have a great impact on business and society at large. The decision-making process, using innovative digital solutions in relation to knowledge management practices, takes on a new importance in improving service delivery [1].

M. Athanasiadou · A. Diamantidis · T. Fotiadis (✉) · A. Gasteratos
Production & Management Engineering Department, Democritus University of Thrace, Xanthi, Greece
e-mail: tfotiadi@pme.duth.gr

L. Hatzithomas
Department of Business Administration, University of Macedonia, Thessaloniki, Greece

D. P. Sakas et al. (eds.), *Computational and Strategic Business Modelling*,
Springer Proceedings in Business and Economics,
https://doi.org/10.1007/978-3-031-41371-1_52

The aim of this work is to capture the new situation that has arisen since the beginning of the COVID-19 pandemic; to identify the positive and negative effects it has had on the public sector; to present the changes made to e-government; to examine the usefulness of the trust factor in e-government; and whether trust affects the use and success of these services, in the context of the success model of the information system and its public value [2].

2 Literature Review

2.1 E-Government

E-government can be an effective way to improve government services, provided that citizens have the ability and show willingness to use it. The trust that citizens show in government websites helps and enhances them, in order to provide better quality services. The quality of the services reflects the extent to which public sector services meet citizens' needs. Increased trust on the part of citizens in e-government websites shows that people understand and can rate the qualitative dimensions of e-government [2].

Therefore, as citizens gain more opportunities in their daily lives, it makes sense to demand more flexibility and convenience when interacting with the public sector. Using the Internet in transactions eliminates the time required to complete a task. So it should come as no surprise that citizens are looking out for similar fast, easy-to-use services, provided by the state, in order to expedite the use of these services immediately, quickly, and easily. After all, most people these days, including in Greece, are well aware of the existence of the Internet and how to use it [3]. Therefore, in this context, the state must upgrade the way it serves its citizens, using modern technology and improving the efficiency of public services.

2.2 *gov.gr*: The New Digital Portal

The digital transformation of the public administration of Greece came with the creation of the single digital portal, where businesses and citizens can search for the digital services they need/want, easily, quickly, and above all, conveniently. The single digital portal is known as *gov.gr* (which is also the URL) and has been designed in such a way that it can meet the needs of all interested parties. The basic philosophy of this portal is for public services to acquire a more friendly and united "face" for citizens. The search is done either directly from an electronic device (computer, mobile, and tablet) or with a physical presence in one of the citizens' service centers; there are 11 main categories to search from, and these include 1294 services. More specifically, there are services from 19 ministries, 71 bodies and organizations, and 9 independent authorities. Examples include the Ministry of

Justice, the Ministry of National Defense, the Ministry of Agricultural Economy and Veterinary Medicine, the Ombudsman, and the Labour Employment Office.

3 Research Hypotheses

Figure 1 presents the study's proposed research model. The dimensions of e-government include the main factors that affect the expectations a user has from the use (or not) of the services received. The model of DeLone and McLean for measuring the success of an information system has three characteristics: the quality of the information, the quality of the services, and the quality of the system.

Thus, based on the above the following hypotheses can be stated:

H1. The quality of information has a positive effect on trust in e-government.
H2. The quality of the system has a positive effect on trust in e-government.
H3. The quality of the services has a positive effect on trust in e-government.
H4. The quality of information has a positive effect on the actual use made of e-government.
H5. The quality of the services has a positive effect on the actual use made of e-government.
H6. The quality of the system has a positive effect on the actual use made of e-government.
H7. The quality of information has a positive effect on user satisfaction.

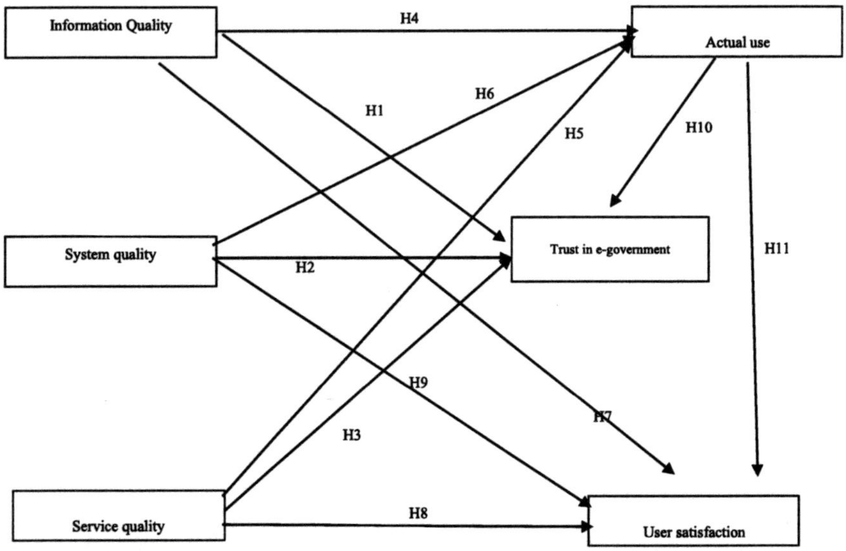

Fig. 1 Proposed research model

H8. The quality of the services has a positive effect on user satisfaction.
H9. The quality of the system has a positive effect on user satisfaction.

The actual use made of e-government includes the degree, pace, and frequency with which an individual uses electronic media for activities that might otherwise be done manually. According to the model of DeLone and McLean for measuring success, users' satisfaction is affected by their use of a system. Citizens' trust in the system increases when they can use e-government services for their day-to-day transactions with public bodies, such as paying taxes, receiving certificates, getting health referrals, enrolling children in school, and a host of other tasks related to their personal and professional activities. Previous studies have found that citizens who use an online service can determine the satisfaction it offers, and the fact that there is an interdependence between the use of the information system, user satisfaction, and the benefits gained from its use [2].

H10. The actual use made of e-government has a positive effect on trust in e-government.
H11. The actual use made of e-government has a positive effect on user satisfaction.

4 Research Methodology

4.1 Research Sample

According to the international statistical service *Internet World Stats*, on 31.12.2020, the number of Internet users in Greece was 8,115,397, a percentage of 72.90% of the total Greek population which was registered at the time (11,124,603 citizens). This population could be the target of this study, but with the additional criterion that in order to participate in this research, it is not enough just to be an Internet user: a participant must also actively use e-government services, which in this specific case is the single digital portal *gov.gr*. It is of course difficult to identify the 8,115,397 Internet users and directly ask them if they use the portal; there is also no database that has these users' details, in order to be able to ask them this question. Therefore, the population sample of the present research consists of all citizens of Greece who use the Greek government's www.gov.gr website to interact with the various bodies of the public sector and its corresponding services.

4.2 Measures

For the present work, the use of a structured questionnaire with closed-ended questions was chosen. This choice was made to avoid misinterpretations regarding participants' answers and to have a simple and comprehensible process of completion, coding, and analysis of the data. The compilation of a well-structured

Table 1 Research factors

Research factors	References	Items
Information quality	[2]	5
Service quality	[2]	5
System quality	[2]	5
Actual use	[2]	5
User satisfaction	[2]	4
Trust in e-government	[2]	3

questionnaire is very important and there should be proper planning and control measures so that the results it yields will be reliable, accurate and clear [4].

The construction of the questionnaire was initially done by dividing it into two parts. Part A contains questions that refer to the demographic characteristics of the citizens that participated in the survey, specifically: gender, age, employment, level of education, number of years of using the Internet, whether the respondents know what e-government is, and finally how often they use its services. Part B contains the construction of the model for e-government, with twenty seven (27) questions based on the six (6) different factors with six (6) subsections (Table 1).

The statements in this questionnaire should be responded to on the basis of a 5-point Likert scale, which is a closed rating system of forced choice: the respondent is asked to indicate the degree of agreement or disagreement with certain statements, where 1 indicates "I absolutely disagree," 2 "I partially disagree," 3 "I neither agree nor disagree," 4 "I partially agree," and 5 "I absolutely agree."

4.3 Sampling Method/Data Collection

The questionnaire of the present research work was prepared in printed form and then modified in digital form. The method of distribution of the questionnaire that was chosen for the present study was to send participants a link to the questionnaire by e-mail, forward it through different messaging applications from a cellphone, and at the same time to post it on various social networks, and finally ask the participants to re-send the link to their acquaintances, colleagues and friends electronically, or to repost it on their own web profile so that as many different categories of citizens as possible can be covered.

It should be noted that the sampling method chosen is the avalanche sampling technique, and the method chosen to distribute the questionnaire is electronic; after all, the use of electronic means and technology is appropriate in this case, since the study involves an investigation into e-government research. The sample size of the study is 333 Greek citizens.

4.4 Validity Check of the Questionnaire

The reliability and validity of the variables used in the present study were accepted because the design of the questionnaire was based on a previously-proven valid questionnaire already in existence. The variables and questions have been used in other surveys; they have not been converted, and no new items were added to them from the original article where the selection was made. The changes that were made to the original questionnaire concerned syntactical and grammatical corrections, as well as some rewording of a few questions, as they contained ambiguities in their Greek translation and could cause confusion to the participants.

The translation of the questionnaire was done from English into Greek, and one more person was used to translate it from Greek into English. Finally, the questions in the original English article were compared with the translated ones, and the result of the translation was considered to be excellent. The next step was to randomly distribute the questionnaire to five citizens, to determine the clarity and understanding of the questions, in order to ensure the correct outcome of the investigation. The reason that the above control method was deemed necessary is the need to adapt the questions included in the questionnaire to the particularities and idiosyncrasies of the data, according to Greek society and culture, and the inability to render all the concepts with absolute accuracy in the Greek language. After the completion of the research, the answers to the questionnaire were coded and converted into numerical variables in such a way that they could be used appropriately. The statistical analysis was performed using the IBM SPSS 27 software.

5 Results

5.1 Descriptive Statistics

The sample size of the study totaled 333 Greek citizens (189/56.80% women and 144/43.20% men) who used e-government services. Regarding the age of the respondents, the majority (153/45.90%) were 46–65 years old, 138 (41.40%) 25–45 years old, 34 (10.21%) under 25, and only 8 people (2.40%) were over 65 years old.

Concerning the participants' employment, the largest percentage of the responses (182 people/54.70%) stated that they are civil servants, while 56 (16.80%) are private employees and 46 (13.80%) are self-employed. Another 16 people (9.90%) answered that they are unemployed while 33 (4.80%) stated that their job status is "something else" – it may be inferred that there are students among the sample because 34 people under 25 also participated in the survey. The fact that the majority of the respondents were civil servants can be explained in various ways. First, the person conducting the investigation is also a civil servant; therefore, the initial

Table 2 Descriptive statistics

		Frequency	Percentage (%)
Gender	Man	144	43.20
	Woman	189	56.80
Age	≤25 years old	34	10.20
	25–45 years old	138	41.40
	46–65 years old	153	45.90
	≥65 years old	8	2.40
Occupation	Private employee	56	16.80
	Civil servant	182	54.70
	Freelancer	46	13.80
	Unemployed	16	9.90
	Other	33	4.80
Educational level	Secondary school	77	23.10
	Tertiary education	153	45.90
	Postgraduate studies	94	28.20
	PhD graduate	9	2.70
Number of years using the Internet	≤1 year	11	3.30
	1–3 years	24	7.20
	4–10 years	74	22.20
	≥10 years	224	67.30
How often do you use electronic governmental services?	Daily	107	32.10
	1–3 times a week	114	34.20
	1–3 times a month	112	33.60
Notes: Sample, $n = 333$			

sample, which, as previously mentioned, was known to her as friends and family, comprised a large percentage of her colleagues who are also civil servants (Table 2).

Concerning the education level of the participants, tertiary graduates formed the largest group (153 people/45.90%), followed by post-graduates (94/28.20%), while just 9 people (2.70%) held a doctorate degree; therefore, the majority of the respondents (256 people/76.80%) have a very high educational level, while the remaining (77/23.10%) are high school graduates. Based on the data collected from this characteristic it can be said that Internet users with a higher level of education answered the questionnaire.

Concerning Internet use, measured in number of years, the largest percentage of responses (67.30%) is more than 10 years, while only 3.30% stated less than a year. This means that a very large part of the sample has experience in the use of electronic media, utilizing it to search for information, communicate with other people, and make various transactions.

For the question concerning how often people use e-government services, the respondents' answers are almost equally divided into three categories: 107 (32.10%) citizens answered daily; 114 (34.20%) answered 1–3 times a week; and 112 (33.60%) answered 1–3 times (Table 2).

5.2 *Questionnaire Validity*

The validity of the one-dimensional (unidimensionality) structure of the elements (items – questions – identified variables) that constitute each of the six research factors was carried out by applying exploratory factor analysis (EFA) with the method of the principal components analysis (PCA). The reliability of each factor was then checked separately, with the help of Cronbach's alpha (Table 3).

The validity check was completed using this specific procedure, by examining the appropriate indicators, namely:

- The Kaiser–Meyer–Olkin statistical test was performed to check the adequacy of the factor analysis. If the KMO value is above 0.5, then the values are considered adequate, while if it is above 0.7, the values are considered satisfactory. If the values are less than 0.5, this indicates that the results of the factor analysis are inadequate [5].
- The total variance explained (TVE) index presents some data related to a number of factors. In order for the values to be acceptable, the TVE score must be over 50.
- The effect that each variable has on each factor can be identified through factor loadings. When the factor loadings are greater than 0.5, they are considered significant (sig ≤ 0.05) [5].
- The assessment of the reliability of the six research factors was done by using Cronbach's alpha statistics. In order to be considered a reliable scale criterion, its score must be above 0.6.

5.3 *Regression Analysis*

Regression is used as a prediction model which, with the help of statistical analysis, yields the possibility to predict the effect of one or more independent variables with a dependent variable, for example, the model summary table, with which we examine the R^2 coefficient of determination. This index shows the percentage (%) of the variance of the dependent variable in relation to the variance of the model of one or more independent variables. The F and sig indices from the ANOVA table (dispersion analysis table) check the statistical significance of the linear regression. For sig < 0.05, the regression model is statistically significant. The coefficients table

Table 3 Factor analysis

Factor	Items	Loadings
Information quality	1	0.841
	2	0.856
	3	0.881
	4	0.860
	5	0.839
	KMO = 0.868 Bartlett test X^2 = 1071,465 Sig = 0	Total variance explained = 73,238 Cronbach's alpha =0.909
Service quality	1	0.765
	2	0.713
	3	0.638
	4	0.831
	5	0.791
	KMO = 0.862 Bartlett test X^2 = 1181,801 Sig = 0	Total variance explained = 74.750 Cronbach's alpha = 0.915
System quality	1	0.588
	2	0.726
	3	0.839
	4	0.813
	5	0.649
	KMO = 0.828 Bartlett test X^2 = 1129,260 Sig = 0	Total variance explained = 70,278 Cronbach's alpha = 0.892
Actual use	1	0.546
	2	0.616
	3	0.731
	4	0.705
	5	0.654
	KMO = 0.859 Bartlett test X^2 = 740,391 Sig = 0	Total variance explained = 65,039 Cronbach's alpha = 0.865
User satisfaction	1	0.862
	2	0.855
	3	0.863
	4	0.845
	KMO = 0.851 Bartlett test X^2 = 1253,929 Sig = 0	Total variance explained = 85,602 Cronbach's alpha = 0.944
Trust in e-government	1	0.793
	2	0.801
	3	0.676
	KMO = 0.704 Bartlett test X^2 = 418,774 Sig = 0	Total variance explained = 75,663 Cronbach's alpha = 0.837

examines the values of the beta scores, which show the behavior of the coefficients of the model.

Hypothesis 1 Regression analysis
From the model summary, ANOVA and coefficients tables, we observe that $t = 1.150$ and sig $= 0.251 > 0.05$; thus, the quality of information is not statistically significant. The hypothesis that the quality of information positively affects trust in e-government is therefore rejected (Tables 4a, 4b, and 4c).

Hypothesis 2 Regression analysis
From the model summary, ANOVA and coefficient tables, we observe that $t = 6.206$ and sig $= 0.000 < 0.05$. According to the indices $B = 0.356$ and Beta $= 0.334$, we find that for each increase in system quality by one unit, trust in e-government increases by 0.356. The ANOVA table ($F = 140{,}653$, sig $0.000 < 0.05$) shows that linear regression is statistically significant. From the model summary table, we see that 63.2% ($R^2 = 0.632$) of the variance for trust in e-government is interpreted by the variation in the quality of information. Therefore, the hypothesis that the quality of the system positively affects the trust factor in e-government is validated (Tables 5a, 5b, and 5c).

Hypothesis 3 Regression analysis
From the model summary, ANOVA and coefficient tables, we observe that $t = 5.836$ and sig $= 0.000 < 0.05$. According to the indices $B = 0.400$ and beta $= 0.354$, we find that for every increase in the quality of services by one unit, trust in

Table 4a Model summary

Model	R	R^2	Adjusted R^2	Std. error of the estimate
1	.795	.632	.627	.50655

Table 4b ANOVA

Model	Sum of squares	*df*	Mean square	F	Sig
Regression	144.365	4	36.091	140.653	.000[b]
Residual	84.164	328	.257		
Total	228.529	332			

Table 4c Coefficients

Model		Unstandardized coefficients		Standardized coefficients		
		B	Std. error	Beta	t	Sig
1	(Constant)	−.002	.167		−.011	.991
	Quality of information	.073	.064	.065	1.150	.251

Table 5a Model summary

Model	R	R^2	Adjusted R^2	Std. error of the estimate
1	.795	.632	.627	.50655

Table 5b ANOVA

Model	Sum of squares	*df*	Mean square	*F*	Sig
Regression	144.365	4	36.091	140.653	.000[b]
Residual	84.164	328	.257		
Total	228.529	332			

Table 5c Coefficients

Model		Unstandardized Coefficients		Standardized coefficients		
		B	std. error	Beta	*t*	Sig
1	(Constant)	−.002	.167		−.011	.991
	Quality of the system	.356	.057	.334	6.206	.000

Table 6a Model summary

Model	*R*	R^2	Adjusted R^2	Std. error of the estimate
1	.795	.632	.627	.50655

Table 6b ANOVA

Model	Sum of squares	*df*	Mean square	*F*	Sig
Regression	144.365	4	36.091	140.653	.000[b]
Residual	84.164	328	.257		
Total	228.529	332			

Table 6c Coefficients

Model		Unstandardized coefficients		Standardized coefficients		
		B	Std. error	Beta	*t*	Sig
1	(Constant)	−.002	.167		−.011	.991
	Quality of services	.400	.068	.354	5.836	.000

e-government increases by 0.354. The ANOVA table (F = 140,653, sig 0.000 < 0.05) shows that linear regression is statistically significant. From the model summary table, we see that 63.2% ($R^2 = 0.632$) of the variance for trust in e-government is interpreted by the variation in the quality of services. Therefore, the hypothesis that service quality positively affects trust in e-government is validated (Tables 6a, 6b, and 6c).

Hypothesis 4 Regression Analysis

From the model summary, ANOVA and coefficient tables, we observe that $t = 2.986$ and sig = 0.000 < 0.05. According to the indices $B = 0.312$ and beta = 0.231, we find that for each increase in the quality of information by one unit, the actual use made of e-government increases by 0.312. The ANOVA table ($F = 44.618$, sig 0.000 < 0.05) shows that the linear regression is statistically significant. From the model summary table, we see that 28.9% ($R^2 = 0.289$) of the variance in the actual

Table 7a Model summary

Model	R	R square	Adjusted R square	Std. error of the estimate
1	.538	.289	.283	.83994

Table 7b ANOVA

Model	Sum of squares	df	Mean square	F	Sig
Regression	94.433	3	31.478	44.618	.000[b]
Residual	232.108	329	.705		
Total	326.541	332			

Table 7c Coefficients

Model		Unstandardized coefficients		Standardized coefficients		
		B	Std. error	Beta	t	Sig
1	(Constant)	.263	.276		.954	.341
	Quality of Information	.312	.104	.231	2.986	.003

Table 8a Model summary

Model	R	R^2	Adjusted R^2	Std. error of the estimate
1	.538	.289	.283	.83994

Table 8b ANOVA

Model	Sum of squares	df	Mean square	F	Sig
Regression	94.433	3	31.478	44.618	.000[b]
Residual	232.108	329	.705		
Total	326.541	332			

Table 8c Coefficients

Model		Unstandardized coefficients		Standardized coefficients		
		B	Std. error	Beta	t	Sig
1	(Constant)	.263	.276		.954	.341
	Quality of services	.134	.113	.099	1.180	.239

use is interpreted by the variation in the quality of the information. The hypothesis that the quality of the information positively affects the actual use is therefore validated (Tables 7a, 7b, and 7c).

Hypothesis 5 Regression Analysis

From the model summary, ANOVA and coefficient tables, we observe that $t = 1.180$ and sig $= 0.239 > 0.05$; thus, the quality of services is not statistically significant. Therefore, the hypothesis that the quality of services positively affects the actual use is rejected (Tables 8a, 8b, and 8c).

Table 9a Model summary

Model	R	R^2	Adjusted R^2	Std. error of the estimate
1	.538	.289	.283	.83994

Table 9b ANOVA

Model	Sum of squares	*df*	Mean square	F	Sig
Regression	94.433	3	31.478	44.618	.000[b]
Residual	232.108	329	.705		
Total	326.541	332			

Table 9c Coefficients

Model		Unstandardized coefficients		Standardized coefficients		
		B	Std. error	Beta	t	Sig
1	(Constant)	.263	.276		.954	.341
	Quality of the system	.331	.093	.260	3.547	.000

Table 10a Model summary

Model	R	R^2	Adjusted R^2	Std. error of the estimate
1	.875	.766	.763	.39511

Table 10b ANOVA

Model	Sum of squares	*df*	Mean square	F	Sig
Regression	167.230	4	41.807	267.799	.000[b]
Residual	51.206	328	.156		
Total	218.436	332			

Hypothesis 6 Regression Analysis

From the model summary, ANOVA and coefficient tables, we observe that $t = 3.547$ and sig $= 0.000 < 0.05$. According to the indices $B = 0.331$ and beta $= 0.260$, we find that for each increase in the quality of the system by one unit, the actual usage increases by 0.331. The ANOVA table ($F = 44.618$, sig $0.000 < 0.05$) shows that linear regression is statistically significant. From the model summary table, we see that 28.9% ($R^2 = 0.289$) of the variance in the actual use is interpreted by the variation in the quality of the system. Therefore, the hypothesis that the quality of the system positively affects the actual use is validated (Tables 9a, 9b, and 9c).

Hypothesis 7 Regression Analysis

From the model summary, ANOVA and coefficient tables, we observe that $t = 1.325$ and sig $= 0.186 > 0.05$; thus, the quality of information is not statistically significant. Therefore, the assumption that the quality of information has a positive effect on user satisfaction is rejected (Tables 10a, 10b, and 10c).

Table 10c Coefficients

Model		Unstandardized coefficients		Standardized coefficients		
		B	Std. error	Beta	*t*	Sig
1	(Constant)	−.144	.130		−1.112	.267
	Quality of information	.066	.050	.060	1.325	.186

Table 11a Model summary

Model	*R*	R^2	Adjusted R^2	Std. error of the estimate
1	.875	.766	.763	.39511

Table 11b ANOVA

Model	Sum of squares	*df*	Mean square	*F*	Sig
Regression	167.230	4	41.807	267.799	.000
Residual	51.206	328	.156		
Total	218.436	332			

Table 11c Coefficients

Model		Unstandardized coefficients		Standardized coefficients		
		B	Std. error	Beta	*t*	Sig
1	(Constant)	−.144	.130		−1.112	.267
	Quality of services	.394	.053	.357	7.371	.000

Hypothesis 8 Regression Analysis

From the model summary, ANOVA and coefficient tables, we observe that $t = 1.325$ and Sig $= 0.000 < 0.05$. According to the indices $B = 0.394$ and beta $= 0.357$, we find that for each increase in the quality of services by one unit, user satisfaction increases by 0.394. The ANOVA table ($F = 267{,}799$, sig $0.000 < 0.05$) shows that the linear regression is statistically significant. From the model summary table, we see that 76.60% ($R^2 = 0.766$) in the variance of user satisfaction is interpreted by the variation in the quality of services. Therefore, the hypothesis that the quality of services has a positive effect on user satisfaction is validated (Tables 11a, 11b, and 11c).

Hypothesis 9 Regression Analysis

From the model summary, ANOVA and coefficient tables, we observe that $t = 10.003$ and sig $= 0.000 < 0.05$. According to the indices $B = 0.448$ and beta $= 0.429$, we find that for each increase in system quality by one unit, user satisfaction increases by 0.448. The ANOVA table ($F = 267{,}799$, sig $0.000 < 0.05$) shows that linear regression is statistically significant. From the model summary table, we see that 76.60% ($R^2 = 0.766$) of the variance in user satisfaction is interpreted by the variation in system quality. Therefore, the hypothesis that the quality of the system has a positive effect on user satisfaction is validated (Tables 12a, 12b, and 12c).

Table 12a Model summary

Model	R	R^2	Adjusted R^2	Std. error of the estimate
1	.875	.766	.763	.39511

Table 12b ANOVA

Model	Sum of squares	df	Mean square	F	Sig
Regression	167.230	4	41.807	267.799	.000
Residual	51.206	328	.156		
Total	218.436	332			

Table 12c Coefficients

Model		Unstandardized coefficients		Standardized coefficients		
		B	Std. error	Beta	t	Sig
1	(Constant)	−.144	.130		−1.112	.267
	Quality of the system	.448	.045	.429	10.003	.000

Table 13a Model summary

Model	R	R^2	Adjusted R^2	Std. error of the estimate
1	.795	.632	.627	.50655

Table 13b ANOVA

Model	Sum of squares	df	Mean square	F	Sig
Regression	144.365	4	36.091	140.653	.000
Residual	84.164	328	.257		
Total	228.529	332			

Table 13c Coefficients

Model		Unstandardized coefficients		Standardized coefficients		
		B	Std. error	Beta	t	Sig
1	(Constant)	−.002	.167		−.011	.991
	Quality of services	.137	.033	.164	4.119	.000

Hypothesis 10 Regression Analysis

From the model summary, ANOVA and coefficient tables, we observe that $t = 4.449$ and sig $= 0.000 < 0.05$. According to the indices $B = 0.137$ and beta $= 0.164$, we find that for every increase in actual usage by one unit, trust in e-government increases by 0.137. The ANOVA table (F = 140,653, sig 0.000 < 0.05) shows that linear regression is statistically significant. From the model summary table, we see that 63.2% ($R^2 = 0.632$) of the variance in trust in e-government is interpreted by the variation in actual use. Therefore, the hypothesis that the actual use positively affects trust in e-government is validated (Tables 13a, 13b, and 13c).

Table 14a Model summary

Model	R	R^2	Adjusted R^2	Std. error of the estimate
1	.875	.766	.763	.39511

Table 14b ANOVA

Model	Sum of squares	*df*	Mean square	*F*	Sig
Regression	167.230	4	41.807	267.799	.000
Residual	51.206	328	.156		
Total	218.436	332			

Table 14c Coefficients

Model		Unstandardized coefficients		Standardized coefficients		
		B	Std. error	Beta	*t*	Sig
1	(Constant)	−.144	.130		−1.112	.267
	Actual Use	.126	.026	.154	4.865	.000

Hypothesis 11 Regression Analysis
From the model summary, ANOVA and coefficient tables, we observe that $t = 4.865$ and sig $= 0.000 < 0.05$. According to the indices $B = 0.126$ and beta $= 0.154$, we find that for each increase in actual use by one unit, user satisfaction increases by 0.126. The ANOVA table (F = 267,799, sig 0.000 < 0.05) shows that linear regression is statistically significant. From the model summary table, we see that 76.60% (R^2 = 0.766) of the variance in user satisfaction is interpreted by the variance in actual use. Therefore, the hypothesis that the actual use has a positive effect on user satisfaction is validated (Tables 14a, 14b, and 14c).

To summarize Table 15 presents the findings of the hypotheses testing.

5.4 *Correlations*

The correlation of the research factors is presented in Table 16. The interpretation of the numbers is as follows: values from 0.0 to 0.4 are considered to show a low correlation; values from 0.4 to 0.6 show a moderate correlation; and values from 0.6 to 1 show a high correlation.

While the actual use of e-government, as analyzed in the literature review section represents the degree, pace, frequency, and manner in which an individual uses the electronic medium for activities that could otherwise be done manually/face-to-face, this has no significant interaction with the following variables: information quality, service quality, and system quality. This may explain why, according to [6], there are difficulties in interpreting the use of this particular factor (actual use of e-government); they, therefore, suggest as already mentioned in the presentation

Table 15 Hypotheses testing results

Research hypotheses	
H1. The quality of information has a positive effect on trust in e-government.	Rejected
H2. The quality of the system has a positive effect on trust in e-government.	Verified
H3. The quality of the services has a positive effect on trust in e-government.	Verified
H4. The quality of information has a positive effect on the actual use made of e-government.	Verified
H5. The quality of the services has a positive effect on the actual use made of e-government.	Rejected
H6. The quality of the system has a positive effect on the actual use made of e-government.	Verified
H7. The quality of information has a positive effect on user satisfaction.	Rejected
H8. The quality of the services has a positive effect on user satisfaction.	Verified
H9. The quality of the system has a positive effect on user satisfaction.	Verified
H10. The actual use made of e-government has a positive effect on trust in e-government.	Verified
H11. The actual use made of e-government has a positive effect on user satisfaction.	Verified

Table 16 Correlation results

	Information quality	Service quality	System quality	Actual use	User satisfaction	Trust in e-government
Information quality	1					
Service quality	0.782 Sig: <0.001	1				
System quality	0.699 Sig: <0.001	0.752 Sig: <0.001	1			
Actual use	0.490 Sig: <0.001	0.475 Sig: <0.001	0.496 Sig: <0.001	1		
User satisfaction	0.715 Sig: <0.001	0.800 Sig: <0.001	0.816 Sig: <0.001	0.566 Sig: <0.001	1	
Trust in e-government	0.656 Sig: <0.001	0.734 Sig: <0.001	0.727 Sig: <0.001	0.529 Sig: <0.001	0.798 Sig: <0.001	1

Notes: All correlations are significant at the 0.01 level (two-tailed).

of the research variables, that it be used as an alternative measure for use in certain contexts [6]. It should be noted that the score for this factor is marginally above the limits of a low correlation.

The highest correlation lies in user satisfaction and the quality of the system (0.816). This correlation is immediately perceptible because e-government users need high-quality information systems in order to be able to interact online with the various state bodies.

The next higher correlations are for the factors of user satisfaction with service quality (0.800), followed by user satisfaction with trust in e-government (0.798). Therefore, in order for citizens to have confidence in e-government, they must be satisfied with service quality, and the government can achieve this level of satisfaction by providing quality services that will be driven by efficiency, effectiveness, and adequacy.

The variable of trust in e-government correlates with the quality of information (0.656), the quality of services (0.734), and the quality of the system (0.727). The correlation degrees are quite satisfactory, with actual use being 0.529, showing a moderate correlation, marginally further away from 0.6.

6 Conclusion

6.1 *Interpretation of Findings*

The findings indicate that all factors are very important in terms of user satisfaction and trust in e-government. The actual use factor does not have a significant interaction with the variables concerning the quality of the information, services and the system. It should be noted that the values obtained are marginally above the low correlation level. A greater correlation exists between the variables for user satisfaction and the quality of the system. What follows in the discussion is the correlation between user satisfaction with the quality of services and trust in e-government. From the above discussion, it follows that in order for citizens to have confidence in e-government, the users of the new single digital portal gov.gr must be satisfied with the provided services and the government can achieve this by providing quality services that will be driven by efficiency, effectiveness, and adequacy.

6.2 *Managerial Implications*

Every research effort has been made in order to draw various conclusions, which will be helpful in the administrative operation of both the public and private sectors for the benefit of all stakeholders.

Information Quality

In order to gain insights into the quality of the information, the participants of the study were asked to answer five questions regarding whether the information they receive from e-government is comprehensible, immediately updated, accurate, complete, and relevant to their request. The majority of respondents graded their answer with four on a Likert scale of five, that is, they "agree." It can be understood that regardless of gender, age, and educational level, the quality of information is equally important for everyone. The quality of information, according to H4, is validated, showing that the higher the quality of the information, the greater the positive impact on the actual use made of e-government. H7, which states that the quality of services has a positive effect on user satisfaction, was rejected, as was H1, thus concluding that the survey participants believe that the quality of information does not affect user satisfaction and their trust in e-government.

Service Quality

The questions concerning the quality of the services totaled five in number and were concerned with whether the services provided by e-government are reliable, "tangible" (clear, easy to understand, and easy to perform), trustworthy, valid, and whether they meet citizens' needs. The majority of respondents graded the question of whether e-government services are tangible, clear, meet citizens' needs, and are worthy of their trust with four on a Likert scale of five, that is, they "agree." At the same time, a grade of five – that is, they "completely agree" – was given by the majority to the fact that they consider the services of e-government to be reliable and valid. It should be noted that in all the above questions, the overwhelming majority of responses – over 75% – are held by these two scores (four and five on the Likert scale). This is interpreted to mean that regardless of the demographic characteristics of the participants in the study, the quality of the services is a very important factor for everyone. According to H3, the quality of services is validated and shows us that the higher the quality of the services, the greater the positive impact on trust in e-government. H8 is also validated; the quality of the services has a positive effect on user satisfaction. However, H5 was rejected; thus, it is concluded that the respondents do not believe that the quality of services has a positive effect on the use of e-government.

System Quality

Participants were called to answer five questions regarding the quality of the e-government information system. The questions were specifically concerned with whether the services provided by e-government are secure, if the website used for e-government is well organized if it is easy to navigate, and easy to use, and finally if the e-government website (gov.gr) is simple and easy to remember. The majority chose "I agree" (4), followed by "I completely agree" (5), and covering about 60% of the responses. The hypotheses tested were H2 – whether the quality of the system positively affects trust in e-government, H6 – whether the quality of the system has a positive effect on the actual use made of e-government, and H9 – whether the quality of the system has a positive effect on user satisfaction. None of the above three hypotheses were rejected; thus, the quality of the e-government information system is considered a very important factor.

Actual Use

The number of questions on the actual use made of e-government also numbered five. Participants had to answer whether they use the e-government website to search for information, make payments, and interact with the government online if they use the e-government platform to request services from the government as a whole, and if they prefer to use the e-government website to do business with the government. Two hypotheses, H10 and H11, were tested here, and both were validated. The actual use of e-government under H10 has a positive effect on trust in e-government; according to H11, it also has a positive effect on user satisfaction.

The qualitative dimensions of e-government affect the government's website (gov.gr) and its actual use. This is important because users trust the ability of technology to execute their transactions. It is worth noting that this confidence in the technological capabilities of e-government is synonymous with trust in the government itself. Trust was judged to be one of the factors influencing the public value of e-government. Trust in e-government means that citizens believe that public services are responsive and reliable, and they will understand citizen's needs when interacting with the system.

The quality of information, the quality of the system, and the quality of the services also affect the actual use made of e-government and users' satisfaction. Enhancing the features of an information system will lead to increased use and satisfaction of such a project or program. The web interface, query speed, payment speed, data security, and user information will form the basis for citizens' use of e-government [6]. It has shown that responsiveness, accuracy, reliability, and flexibility are the hallmarks of an information system which can lead to increased use of e-government. The quality of the services refers to service features such as the tangibility, responsiveness, empathy, assurance, and effectiveness of the services and their providers. The Internet interface, the speed of responding to queries, the speed of payment, the security of the data, and the information provided by the users will form the basis for citizens' use of e-government.

The results of this research show that the dimensions of quality are important factors in the actual use made of e-government and users' satisfaction. The government must continue to update and integrate digital services to simplify ease of use through the new single digital portal. Enhancing the service interface facilitates increased utilization of the system, hence enhancing user and citizen satisfaction within the digital network, as opposed to opting for traditional service delivery channels.

Confidence in e-government is an important lever for all aspects of performance in public administration. The government should build up trust in e-government to increase citizens' use of these services. When values such as public participation, good quality information, and effective communication are lacking, citizens will not be so willing to use e-government services. It is therefore prudent to reinforce these values as much as possible and in a better way.

The e-government platform should minimize, or rather eliminate the inefficiency of governmental bureaucracy in the provision of public services. It is worth emphasizing that navigation and accessibility are important in determining the quality of the system as citizens perceive it.

An important observation is that there should be more innovative procedures in the provision of services, better infrastructure capacity, more possibilities for public expense payments, as well as reliability for supporting more users, improving digital structure, and affording higher protection.

The administrators of the new single digital portal should seek to increase and enhance the understanding of the urgent need to use the platform for online services in relation to traditional face-to-face methods. Further efforts are needed to justify these technological investments and to move even further away from their dysfunctional stage in the recent past.

6.3 Research Limitations

This study used a relatively decent sample size of three hundred and thirty-three (333) citizens who use e-government services. It is worth noting that the new single digital portal, which is the central point in the provision of digital services, is in its early stages and there are still issues for citizens to get acquainted with it and understand its use and features. Finally, due to the rapid developments throughout the COVID-19 pandemic, some of the research data may have become differentiated in a very short period of time.

6.4 Suggestions for Future Research

Within the field of e-government, the Greek government must succeed by making great leaps and bounds in the implementation in the coming years of what it should have already implemented yesterday, so to speak. At the same time, it must have developed the appropriate infrastructure and information systems in order to be able to cope with the technological developments that are emerging. The present research could form the basis for a new study using the same factors and/or enriching it with additional ones. The research could be carried out for a longer period of time and with a larger sample, and it may be repeated after 2025, which is the landmark date for the digital transformation "Bible" 2020–2025 (which has been issued by the Greek Ministry of Digital Governance) [7]. Other scientists, however, with similar cases study similar problems, using modeling and simulation [8–11].

References

1. Alvarenga, A., Matos, F., Godina, R., Matias, C.O.: Digital transformation and knowledge management in the public sector. Sustainability. **12**(14), 5824 (2020)
2. Abdulkareem, A.A., Ramli, R.M.: Does trust in e-government influence the performance of e-government? An integration of information system success model and public value theory. Transform. Gov.: People Process Policy. **15**(1), 1–17 (2022)
3. Maditinos, D., Sidiropoulou, N.: Incentives for the adoption of E-government by Greek municipalities. Int. J. Econ. Bus. Admin. **8**(1), 298–326 (2020)
4. Zikmund, W.G., Babin, B.J., Carr, J.C., Griffin, M.: Business Research Methods. Cengage Learning, Andover (2013)
5. Hair Jr., J.F., Anderson, R.E., Tatham, R.L., Black, W.C.: Multivariate Data Analysis, 3rd edn. Macmillan Publishing Company, New York (1995)
6. DeLone, W.H., McLean, E.R.: Information systems success measurement. Foundations Trends Inf. Syst. **2**(1), 1–116 (2016)
7. Digital Transformation "Bible" 2020–2025. Available at digitalstrategy.gov.gr
8. Dimitrios, N.K., Sakas, D.P., Vlachos, D.S.: Analysis of strategic leadership simulation models in non-profit organizations. Proc. Soc. Behav. Sci. **73**, 276–284 (2013)
9. Dimitrios, N.K., Sakas, D.P., Vlachos, D.S.: Analysis of strategic leadership models in information technology. Proc. Soc. Behav. Sci. **73**, 268–275 (2013)
10. Dimitrios, N.K., Sakas, D.P., Vlachos, D.S.: The role of open source leadership in developing high technology companies. Key Eng. Mater. **543**, 402–405 (2013)
11. Kanellos, N., Sakas, D.P., Kamperos, I.D.G., Reklitis, D.P., Giannakopoulos, N.T., Nasiopoulos, D.K., Terzi, M.C.: The effectiveness of centralized payment network advertisements on digital branding during the COVID-19 crisis. Sustainability. **14**(6), 3616 (2022)

Trust in E-Government and Successful Information Management in the Public Sector During a Pandemic: Proposing an Extended Framework

Thomas Fotiadis, Maria Athanasiadou, Anastasios Diamantidis, Leonidas Hatzithomas, and Antonios Gasteratos

1 Introduction

The pandemic brought about changes in all aspects of everyone's lives all over the world. The economic, social, political, educational, and religious implications are highly visible, but some of them may not be so easily understood at the present time, and no doubt, many more will eventually surface in the long term [1]. Apart from the negative consequences of the health crisis caused by the COVID-19 pandemic, significant impacts were felt in a number of other sectors. Teleworking (working from home), distance learning, the digital transformation of services that used to be based solely on human contact in the past; changes in public administration with its new forms of electronic (online) communication, collaboration platforms; and most importantly, a host of innovative functions that have been developed since the outbreak of the COVID-19 pandemic are just some of those innovations that raise the standard of living in a new post-COVID-19 era.

The SARS-COV-2 (COVID-19) pandemic broke out in Greece in early 2020 and changed everyone's lives in the most irreversible way, not only in the status quo of the health sector at the time, but generally speaking, in our whole way of life as we knew it till then: our behavior, consumption patterns, the business environment, as well as the public sector and state administration. Concepts such as self-isolation, social distancing, and quarantining are just some of the terms that created a new

T. Fotiadis (✉) · M. Athanasiadou · A. Diamantidis · A. Gasteratos
Production & Management Engineering Department, Democritus University of Thrace, Xanthi, Greece
e-mail: tfotiadi@pme.duth.gr

L. Hatzithomas
Department of Business Administration, University of Macedonia, Thessaloniki, Greece

D. P. Sakas et al. (eds.), *Computational and Strategic Business Modelling*,
Springer Proceedings in Business and Economics,
https://doi.org/10.1007/978-3-031-41371-1_53

reality, which the government had to take into account, in order to be able to cope with the challenges of the pandemic. The priority of the government should be to meet the needs of its citizens through the provision of high-quality integrated services, in combination with creative and innovative actions and strategies that involve as little bureaucracy as possible [1].

Through the last decade, more and more countries have gradually started providing online (digital) services to their citizens. Major financial scandals and the negative image that has been created around the concept of public administration have created an imperative need for organizations involved in public administration to become competitive and to operate in a regime governed by transparency, meritocracy, and credibility. The changes (both positive and negative) that have arisen due to the pandemic have led to the rapid advancement of e-government, with the provision of online digital governance. The transition to the online epoch for public services is essential on all levels: local, national, and international. This transformation to online services will contribute to the economic development of the public sector. It will also have a great impact on business and society at large. The decision-making process, using innovative digital solutions in relation to knowledge management practices, takes on a new importance in improving service delivery [2].

The aim of this work is to articulate and propose a new research model based on the IS success model by adding the factor trust in e-government.

2 Literature Review

2.1 E-Government

The implementation of e-government is absolutely essential, and it utilizes the potential of modern technology. More specifically, in recent decades, the use of computers and the internet have offered unlimited prospects for growth and progress, changing the daily lives of citizens in various sectors [3]. One of the main features of these changes is the directness of communication, which has multiple dimensions since it allows people to work away from their regular workspace (teleworking), conduct remote business transactions, and execute electronic/online banking transactions, among a host of many other tasks that no longer require the physical presence of the individual [4].

It is worth pointing out that according to the official definition of the European Union, e-government is defined as the use of information and communication technologies (ICT) in public administration, in combination with organizational changes and new staff skills. The general aim is to improve public services, and furthermore to strengthen democratic processes and procedures for supporting public policies. By definition, the basis of e-government lies in the use of technology, especially the internet, to promote the connection between citizens and government agencies. According to the European Commission, e-government is a transformational government that uses modern technology to improve the country's accessibility and efficiency. In order for this transformation to be implemented successfully, organizational changes must be made to improve and strengthen the implementation of public services. Under such an implementation framework, e-government will improve the quality of the relationship between citizens and public services [5].

2.2 *The Effect of the Pandemic on the Public Sector*

Some of the problems in the field of public administration, for example, inequality in the use of information and communication technologies between the various services and delays in the management of cases, citizens, and businesses, surfaced from the moment the pandemic broke out; these issues already existed before the pandemic but were not so visible at the time. To a certain extent, these problems and the continuation of their existence are inevitable and do not appear exclusively in the public administration of Greece alone, but in the administrations of many other countries. The pandemic suddenly changed the socioeconomic conditions and the health and safety environment, as well as the operations of public administration in the modern world. Today, during extreme emergency situations, such as that of a pandemic, planning and preparing for the unpredictable and the unknown, dealing with emergencies, and responding to the needs and expectations of citizens are some of the most basic and the most difficult tasks that national authorities have to deal with [6].

As in past times during other crises, the different departments of public administration and the various economic operators (companies) and individuals (professionals) have found themselves working in an unbalanced business environment, characterized by variability, uncertainty, complexity, and ambiguity [7]. Since the COVID-19 outbreak, the public has understandably turned its attention to their elected political officials (Presidents, Prime Ministers, Ministers). The authorities were obliged to revise the situation and decide what measures needed to be taken, in order to deal with the COVID-19 pandemic, including where, when, and how to start, and to try to calculate how long this novel situation will last. At the same time, civil servants who are usually away from the limelight are now being called upon to implement these measures in a way that makes many of them invisible heroes. In the face of a pandemic crisis, civil servants find themselves at the core of the crisis strategies and plans that are being made to reduce the impact of the pandemic [8]. This is of course a difficult situation to be in. Even before the pandemic, countries with high administrative capacity were better placed to limit the spread of the pandemic to other countries. Among the various departments of the public administration, in the current crisis caused by COVID-19, those departments that had already been through other crisis situations (epidemiological, financial, etc.) present a more organized and flexible structure. They also create better performance indicators and have a more collaborative network with other public institutions and participants in civil society.

3 Conceptual Framework

The outbreak of the COVID-19 pandemic in early 2020 forced all public service sectors to change the way they operate. With the voting in of new decisions, measures, and policies, the public sector was given the incentive to meet the challenges that arose. The pandemic gave rise to a new regime that had a direct impact on the public sector, especially in terms of the performance of public services. Given that until the onset of the pandemic crisis, civil servants worked in a conventional manner (with physical presence in office spaces), their operational systems had to change radically in order to comply with the new rules, at the same time as being active and functional, all with the help of the internet.

The use of e-government is changing the way people and businesses interact with the state. The opportunities created by e-government offer a huge dynamic innovative mechanism, which allows citizens and businesses to interact with public services, easily, quickly and securely, with the support of human resources and the appropriate information systems.

The notion of e-government refers to the process by which information and services are provided electronically to the customers/users, which include citizens, businesses, and public administration. Latif and Masrek [9] point out that the key feature of e-government is the use of information systems, allowing people to access resources and services anywhere and at any time. Curtin [10] defined the purpose of e-government information systems as a tool to provide effective management of government information to all customers, providing better services to citizens through easy/fast access to information and participation in decision-making. According to [11], information systems are responsible for managing and organizing a variety of information, and their success lies in the fact that they allow the public to find the required information quickly, through an easy search system. Thus, with the help of information systems, the search function, functionality and navigation factors become important tools in the execution of e-government services.

Several studies have shown that there has been a significant increase in the presence and availability of digital e-government services in recent years. However, these digital services are often inaccessible, and they tend to be used by a small number of citizens, as there are large gaps between the expectations of information system developers and public administration. Thus, in terms of their actual use, the availability of digital e-government services provided through information systems does not necessarily translate into the effective quality of such services, the kind that citizens desire [12]. Markaki et al. [13] have highlighted the need for a comprehensive evaluation of information systems to explain their impact as a factor influencing the quality of e-government websites. It is worth noting the three specific areas they mention concerning the construction of information systems, which the competent authorities must focus on in order to be able to offer quality services are as follows: the proper allocation of resources for such implementations; and the construction and operation of an e-government website, which is directly related with the "image" that public bodies and the relevant levels of management and services want to present.

3.1 Research Factors

The aim of this work is to capture the new conditions that have been created since the outbreak of the pandemic and to identify the positive and negative effects that they have on the public sector. It is also of interest to present the changes made in Greek e-government with the implementation of the *gov.gr* platform, to examine the usefulness of the trust factor in e-government, and to what extent it affects the use and success of the services, in the context of the success model of the information system and its public value. In order to achieve the above goal, the following six factors were selected, which in essence shape the proposed research model.

Information Quality

The quality of information as a factor indicating success includes the desired characteristics of the data obtained from an information system. According to [14], standard metrics for measuring the quality of information are accuracy, adequacy, availability, completeness, content, consistency, form, relevance, reliability, the field of application, timeliness, comprehensibility, uniqueness, usability, and usefulness. To better illustrate the concept of quality, the following example is given, where employees can obtain certain information, such as up-to-date sales statistics, prices, and product offers from the information system of the company where they work; with this information, they can better focus on the sales of the company's products. If the information they receive had the characteristics to be considered as good quality, it could then become a useful tool for their work. Information quality is often seen as a key precedent for user satisfaction [15].

Service Quality

The quality of the services offered as a factor indicating success represents the quality of support that users receive from the department in charge of the information system, as well as from the staff that support these systems, such as training/education, telephone support, and information updates. This specific factor was added to DeLone & McLean's original success model. The addition of this factor is undoubtedly very important, given that the quality of the system is an important qualitative metric, but there are also some researchers, such as [16], who dispute this.

A popular model for measuring the quality of services in information systems is Servqual [17], with additional metrics added to it, such as security, empathy, flexibility, interpersonal quality, inherent quality, training in the use of the information system, reliability, and tangibility [15].

System Quality

The quality of the system as a factor indicating success embodies the desirable characteristics of an information system, which include metrics for the information system itself. These metrics usually focus on the usability aspects and the performance characteristics of the system in question. A very commonly used metric is the perceived convenience for the user who is challenged by it; this is related to the technology acceptance model. However, many additional metrics have been added and are eventually used to capture the quality structure of the system as a whole. Standard measurements of system quality are ease of access and customization, data accuracy, ease of learning the system, ease of use, efficiency, flexibility, integration, interactivity, navigation, reliability, response time, and system accuracy [15].

Actual Use

The actual use made of e-government services represents the degree, pace, frequency, and manner with which an individual uses the digital/online medium for activities that might otherwise be done manually. Measuring the use of an information system is a broad concept, which can be viewed from different perspectives. In the case of voluntary use, the actual use of an information system may be the most appropriate measure of success. In previous studies, measuring the actual use of the system was done objectively, by recording the connection time, the functions used, and/or the frequency of use. The time spent using a system is obviously not a sufficient measure of success; hence, other studies have implemented more subjective metrics, asking users about their perceived use of a system [14]. A more comprehensive approach to explaining the use of an information system is the technology acceptance model [18]. The model uses the independent variables that perceive the ease of use and the perceived usefulness, which contribute to attitudes toward the use of the system, intention to use it, and the actual use made of the system. Due to difficulties in interpreting the use of this dimension, [15] suggest that it should be used as an alternative measure in certain contexts. Some standard metrics for this success factor are actual usage, daily usage, usage frequency, intention to (re-)use, nature of usage, number of site visits, and number of transactions.

User Satisfaction

User satisfaction as a factor indicating success shows the level of satisfaction of the individual that uses the information system. User satisfaction focuses on the availability of the information required, the quality of the services, and the degree of choice available to users. It is considered to be one of the most important metrics of the success of an information system. Measuring user satisfaction becomes especially useful when the use of an information system is mandatory; thus, the amount of use is therefore not an appropriate indicator of the success of the system. The most widely used means to measure user satisfaction are those of [19, 20]. However, the metrics used by these researchers also contain elements of the system, information, and quality of the services, rather than just measuring user satisfaction. Consequently, other elements have been developed to solely measure the satisfaction a user derives from an information system; these are adequacy, efficiency, effectiveness, enjoyment derived from satisfaction with the information obtained, and overall satisfaction with the system [15] (Fig. 1).

Trust in E-Government

Trust is an important source of public value that underscores the bond between the government and the people. It is especially important in regard to services related to citizens' privacy and security when using online services. Failure to build trust between the state and the people could lead to a failure in the public value of the services. People need to trust the technological and managerial capabilities of state agencies to ensure a secure system, one in which citizens can voluntarily give their

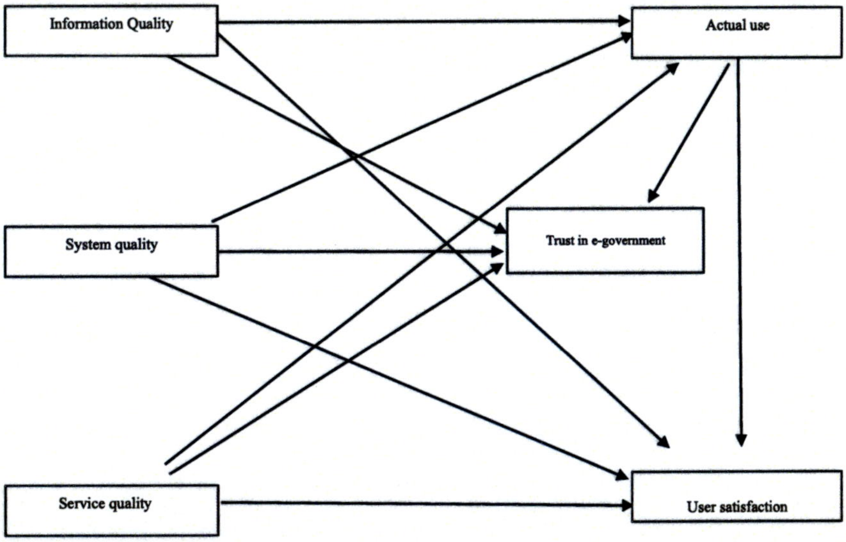

Fig. 1 Proposed research model

data to government agencies when interacting with them, which inspires confidence in e-government initiatives.

Citizens' trust, which leads to the adoption and use of e-government systems, has two dimensions: trust in the government and trust in the internet. Before trusting e-government initiatives, citizens must be convinced that the government has the administrative and technical resources needed to implement and secure them. In order to adopt e-government services, citizens must have the intention of "engaging in e-government," which includes the intention to receive and provide information through online channels. These technologies can serve a variety of different purposes: better public service delivery to citizens, improved interaction of businesses with industry, empowerment of citizens through access to information, and/or more effective government management. The resulting benefits of trust in the state may be less corruption, increased transparency, greater convenience, increased revenue, and even reduced costs [21].

4 Conclusions and Limitations

The discussed research model eventually proposes an extension of Delone and Macleans' IS Success Model by adding the dependent factor "Trust in E-government." It is stated that this factor gives a dynamic perspective of the future use of both digital platforms and its provisional services and it correlates with government effectiveness. Throughout this chapter, we have described the conceptual framework development process and presented important and operational definitions. This is an attempt to provide a useful framework for realizing the scientific literature concerning trust in e-government and additionally guides the development, implementation, and evaluation of a more comprehensive approach to public information systems.

Given the quantitative approach of this study, it can be stated that probably more research factors can be added to the model, in order its predictive perspective to be increased. Also, the overall model is not innovative since it is based on Delone and McLean's IS success model. Other scientists, however, with similar cases study similar problems, using modeling and simulation [22–25].

References

1. Fuciu, M.: Changes in the organization's marketing activity in light of the Sars-Cov-2 pandemic. Rev. Econ. **72**(2), 27–35 (2020)
2. Alvarenga, A., Matos, F., Godina, R., Matias, C.O.: Digital transformation and knowledge management in the public sector. Sustainability. **12**(14), 5824 (2020)
3. Lee, S.M., Hwang, T., Choi, D.: Open innovation in the public sector of leading countries. Manag. Decis. **50**(1), 147–162 (2012)

4. Petrakaki, D., Hayes, N., Introna, L.: Narrowing down accountability through performance monitoring technology: e-government in Greece. Qual. Res. Account. Manag. **6**(3), 160–179 (2009)
5. Nulhusna, R., Sandhyaduhita, P.I., Hidayanto, A.N., Phusavat, K.: The relation of e-government quality on public trust and its impact on public participation. Transform. Gov. People Process Policy. **11**(3), 393–418 (2017)
6. Kontiadis, X.: Pandemic, Biopolitics and Rights. Kastaniotis, Athens (2020)
7. Georgakopoulos, T.: The Ideas of the Pandemic 15 Texts on How Koronios Changes Greece and the World. Dianeosis, Athens (2021)
8. United Nations: The technological revolution: winners and losers. In: World Social Report 2021 Reconsidering Rural Development. United Nations (2021)
9. Latif, M.H.A., Masrek, M.N.: Accessibility evaluation on Malaysian e-government websites. J. E-Gov. Stud. Best Pract. **2010**, 1–11 (2010)
10. Curtin, G.G.: Issues and challenges: global E-government/E-participation, models, measurement and methodology. In: Workshop on E-Participation and E-Government: Understanding the Present and Creating the Future, Budapest, Hungary, pp. 27–28 (2006)
11. Freed, L.: American Customer Satisfaction Index E-Government Satisfaction Index. ForeSee Results, Ann Arbor (2007)
12. Jeyaraj, A.: DeLone & McLean models of information system success: critical meta-review and research directions. Int. J. Inf. Manag. **54**, 1–15 (2020)
13. Markaki, O.I., Charilas, D.E., Askounis, D.: Evaluating the quality attributes of e-government websites. In: E-Government Website Development: Future Trends and Strategic Models, pp. 65–86. IGI Global (2011)
14. DeLone, W.H., McLean, E.R.: The DeLone and McLean model of information systems success: a ten-year update. J. Manag. Inf. Syst. **19**(4), 9–30 (2003)
15. DeLone, W.H., McLean, E.R.: Information systems success measurement. Found. Trends Inf. Syst. **2**(1), 1–116 (2016)
16. Seddon, P.B.: A respecification and extension of the DeLone and McLean model of IS success. Inf. Syst. Res. **8**(3), 240–253 (1997)
17. Pitt, L.F., Watson, R.T., Kavan, C.B.: Service quality: a measure of information systems effectiveness. MIS Q. **19**(2), 173–187 (1995)
18. Davis, F.D.: Perceived usefulness, perceived ease of use, and user acceptance of information technology. MIS Q. **13**(3), 319–340 (1989)
19. Ives, B., Olson, M.H., Baroudi, J.J.: The measurement of user information satisfaction. Commun. ACM. **26**(10), 785–793 (1983)
20. Doll, W.J., Deng, X., Raghunathan, T.S., Torkzadeh, G., Xia, W.: The meaning and measurement of user satisfaction: a multigroup invariance analysis of the end-user computing satisfaction instrument. J. Manag. Inf. Syst. **21**(1), 227–262 (2004)
21. Athmay, A.A.A.A., Fantazy, K., Kumar, V.: E-government adoption and user's satisfaction: an empirical investigation. EuroMed J. Bus. **11**(1), 57–83 (2016)
22. Plikas, J.H., Nasiopoulos, D.K., Giakomidou, D.S.: The contribution of social media in the management of social relations with customers. Modelling and simulation of the problem. In: Strategic Innovative Marketing, pp. 515–521. Springer, Cham (2017)
23. Alexandros, N.K., Damianos, S.P., Dimitrios, N.K., Dimitrios, V.S.: Modelling the process of a web-based collaboration tool development. In: Strategic Innovative Marketing, pp. 383–389. Springer, Cham (2017)
24. Kanellos, N. by Sakas, D.P., Kamperos, I.D.G., Reklitis, D.P., Giannakopoulos, N.T., Nasiopoulos, D.K., Terzi, M.C.: The effectiveness of centralized payment network advertisements on digital branding during the COVID-19 crisis. Sustainability. **14**(6), 3616 (2022)
25. Foteini, P.E., Dimitrios, N.K., Sakas, D.P., Vlachos, D.S.: The development of new ideas for IT products, through social media. In: Strategic Innovative Marketing, pp. 399–404. Springer, Cham (2017)

Index

D. P. Sakas et al. (eds.), *Computational and Strategic Business Modelling*, Springer Proceedings in Business and Economics, https://doi.org/10.1007/978-3-031-41371-1

U

V

W

Z